VOLATILE ORGANIC COMPOUND REMOVAL

VOLATILE ORGANIC COMPOUND REMOVAL

TECHNOLOGIES AND FUNCTIONAL MATERIALS FOR VOC REMOVAL

Edited by

HONGQIANG REN

State Key Laboratory of Pollution Control and Resource Reuse, School of the Environment, Nanjing University, Nanjing, China

BIN GAO

Department of Civil and Environmental Engineering, Rensselaer Polytechnic University Institute, Troy, NY, United States

ELSEVIER

Elsevier
Radarweg 29, PO Box 211, 1000 AE Amsterdam, Netherlands
125 London Wall, London EC2Y 5AS, United Kingdom
50 Hampshire Street, 5th Floor, Cambridge, MA 02139, United States

ISBN: 978-0-443-15674-8

For information on all Elsevier publications visit our website at https://www.elsevier.com/books-and-journals

Publisher: Candice Janco
Acquisitions Editor: Jessica Mack
Editorial Project Manager: Mason Malloy
Production Project Manager: Bharatwaj Varatharajan
Cover Designer: Greg Harris

Typeset by TNQ Technologies

Contents

7. Traditional functional materials for VOC removal

Xirong Huang, Xuqiong Zha, Chuangye Yang, Jiawen Ding, and Zhuhong Ding

8. Biochar for VOC removal

Wei Xiang

9. Carbon nanomaterials (CNMs) for VOC removal

Zhifei Pan

10. Metal–organic frameworks (MOFs) for VOC removal

Yuxuan Zhang, Jun Luo, Xiuyan Wang, Dunhan Zhang, Hongxia Xu, and Xin Hu

11. Metal nanoparticles (MNPs) for VOC removal

Xing Liu and Jun Luo

12. Nano metal oxides (NMOs) for VOC removal

Hao Li, Shuping Wu, Guoqiang Chen, Bin Fu, Yiran Zhao, Feiyue Li, and Ping Qu

Contributors

Guoqiang Chen Institute of Agricultural Resources and Environment, Jiangsu Academy of Agricultural Sciences, Key Laboratory of Agricultural Environment in downstream of Yangtze Plain, Ministry of Agriculture and Rural Affairs, Jiangsu Collaborative Innovation Center for Solid Organic Waste Resource Utilization, Nanjing, China; College of Resources and Environment Science, Anhui Science and Technology University, Fengyang, China

Jiawen Ding School of Environmental Science & Engineering, Nanjing Tech University, Nanjing, China

Zhuhong Ding School of Environmental Science & Engineering, Nanjing Tech University, Nanjing, China

Lin Dong School of Environment, Nanjing University, Nanjing, China

Bin Fu Institute of Agricultural Resources and Environment, Jiangsu Academy of Agricultural Sciences, Key Laboratory of Agricultural Environment in downstream of Yangtze Plain, Ministry of Agriculture and Rural Affairs, Jiangsu Collaborative Innovation Center for Solid Organic Waste Resource Utilization, Nanjing, China

Bin Gao Department of Civil and Environmental Engineering, Rensselaer Polytechnic Institute, Troy, NY, United States

Xin Hu State Key Laboratory of Analytical Chemistry for Life Science, Centre of Materials Analysis and School of Chemistry & Chemical Engineering, Nanjing University, Nanjing, China

Xirong Huang School of Environmental Science & Engineering, Nanjing Tech University, Nanjing, China

Tao Jiang Key Laboratory of Karst Georesources and Environment (Guizhou University), Ministry of Education, Guiyang, China

Hao Li Institute of Agricultural Resources and Environment, Jiangsu Academy of Agricultural Sciences, Key Laboratory of Agricultural Environment in downstream of Yangtze Plain, Ministry of Agriculture and Rural Affairs, Jiangsu Collaborative Innovation Center for Solid Organic Waste Resource Utilization, Nanjing, China; Research School of Polymeric Materials, School of Materials Science & Engineering, Jiangsu University, Zhenjiang, China

Feiyue Li College of Resources and Environment Science, Anhui Science and Technology University, Fengyang, China

Xing Liu Nanjing University, Nanjing, China

Ruochen Liu School of Materials Science and Engineering, Nanchang Hangkong University, Nanchang, China

Yidan Luo School of Materials Science and Engineering, Nanchang Hangkong University, Nanchang, China

Jun Luo Nanjing University, Nanjing, China; State Key Laboratory of Pollution Control and Resource Reuse, School of the Environment, Nanjing University, Nanjing, China

Zhifei Pan School of Environmental Engineering, Xuzhou University of Technology, Xuzhou, China

Guangdou Qi School of Environmental Engineering, Xuzhou University of Technology, Xuzhou, China

Ping Qu Institute of Agricultural Resources and Environment, Jiangsu Academy of Agricultural Sciences, Key Laboratory of Agricultural Environment in downstream of Yangtze Plain, Ministry of Agriculture and Rural Affairs, Jiangsu Collaborative Innovation Center for Solid Organic Waste Resource Utilization, Nanjing, China

Hongqiang Ren State Key Laboratory of Pollution Control and Resource Reuse, School of the Environment, Nanjing University, Nanjing, China

Haiqin Wan School of Environment, Nanjing University, Nanjing, China

Bing Wang College of Resources and Environmental Engineering, Guizhou University, Guiyang, China; Key Laboratory of Karst Georesources and Environment (Guizhou University), Ministry of Education, Guiyang, China

Nana Wang Key Laboratory of Karst Georesources and Environment (Guizhou University), Ministry of Education, Guiyang, China

Xiuyan Wang State Key Laboratory of Pollution Control and Resource Reuse, School of the Environment, Nanjing University, Nanjing, China

Shuping Wu Research School of Polymeric Materials, School of Materials Science & Engineering, Jiangsu University, Zhenjiang, China

Wei Xiang School of Environmental Engineering, Jiangsu Key Laboratory of Industrial Pollution Control and Resource Reuse, Xuzhou University of Technology, Xuzhou, China

Xianchuan Xie Key Laboratory of Poyang Lake Environment and Resource Utilization, Ministry of Education, Nanchang University, Nanchang, China

Hongxia Xu Key Laboratory of Surficial Geochemistry of Ministry of Education, School of Earth Sciences and Engineering, Hydrosciences Department, Nanjing University, Nanjing, China

Mingshan Xue School of Materials Science and Engineering, Nanchang Hangkong University, Nanchang, China

Chuangye Yang School of Environmental Science & Engineering, Nanjing Tech University, Nanjing, China

Shuohan Yu School of Environmental and Chemical Engineering, Nanchang Hangkong University, Nanchang, China

Xuqiong Zha School of Environmental Science & Engineering, Nanjing Tech University, Nanjing, China

Xueyang Zhang School of Environmental Engineering, Xuzhou University of Technology, Xuzhou, China

Yuxuan Zhang State Key Laboratory of Pollution Control and Resource Reuse, School of the Environment, Nanjing University, Nanjing, China

Dunhan Zhang State Key Laboratory of Pollution Control and Resource Reuse, School of the Environment, Nanjing University, Nanjing, China

Yiran Zhao Institute of Agricultural Resources and Environment, Jiangsu Academy of Agricultural Sciences, Key Laboratory of Agricultural Environment in downstream of Yangtze Plain, Ministry of Agriculture and Rural Affairs, Jiangsu Collaborative Innovation Center for Solid Organic Waste Resource Utilization, Nanjing, China; Research School of Polymeric Materials, School of Materials Science & Engineering, Jiangsu University, Zhenjiang, China

CHAPTER

1

Introduction

Hongqiang Ren[1] *and Bin Gao*[2]

[1]State Key Laboratory of Pollution Control and Resource Reuse, School of the Environment, Nanjing University, Nanjing, China; [2]Department of Civil and Environmental Engineering, Rensselaer Polytechnic Institute, Troy, NY, United States

1.1 Volatile organic compounds

With a high vapor pressure and low water solubility, volatile organic compounds (VOCs) include a variety of organic chemicals that easily evaporate at normal temperature and atmosphere pressure. According to the World Health Organization (WHO), there are three types of VOCs of different volatility (WHO, 1989): (1) very volatile organic compounds (VVOCs) with a boiling point lower than 50–100°C, such as propane and butane; (2) VOCs with a boiling point between 50–100 and 240–260°C, such as acetone, formaldehyde, vinyl chloride, ethanol, toluene, trichloroethylene, and carbon tetrachloride; and (3) semivolatile organic compounds (SVOCs), such as pesticides, plasticizers, and fire retardants. As an air pollutant, VOCs are emitted into the atmosphere mainly from two sources (Fig. 1.1). Biogenic VOCs

FIGURE 1.1 Air pollution by BVOCs and AVOCs.

Volatile Organic Compound Removal
https://doi.org/10.1016/B978-0-443-15674-8.00005-6

(BVOCs), such as isoprene and monoterpene, are emitted from terrestrial vegetation and marine organisms under stressful conditions as a defensive mechanism. They contribute more than 80% to the total VOC emissions globally, while anthropogenic VOCs (AVOCs) only account for about 14% (Crippa et al., 2020). AVOCs are human-made chemicals that are released by anthropogenic activities, especially industrial productions. They are hydrocarbons, commonly used as industrial solvents, and are released in the production of fuels, paints, pharmaceuticals, hydraulic fluids, dry cleaning agents, and refrigerants. While this book is mainly focused on the removal of 'AVOCs', all the authors use the general term 'VOCs'.

1.2 VOC pollution

There are many types of VOCs that have different health effects depending on the level of exposure and the length of time of the exposure. In general, VOCs in the air can have an immediate effect on human health, causing symptoms, such as respiratory and eye irritation, dizziness, headaches, and internal organ (e.g., liver, kidneys, or the central nervous system) impairment. Some VOCs, such as formaldehyde and benzene, are carcinogenic and can cause cancer in people.

In addition to the adverse health effects, VOCs in the atmosphere can cause serious environmental pollution, especially with respect to the production of secondary air pollutants, such as ground level ozone (O_3) and smog (Fig. 1.2), the most widespread air pollutant globally. Under sunlight, VOC species can react with other air pollutants, such as nitrogen oxides (NOx), through a series of photochemical reactions to form ozone and ultrafine particulate matter (PM2.5), with adverse effects on human health and the environment (Sillman et al., 2003). For example, ground level ozone can be formed in an urban environment through photochemical reactions between VOCs and NOx from various emission sources. The reactions and ozone production rate are either NOx-sensitive or VOC-sensitive depending on the physical and chemical conditions. With a strong oxidation ability, ozone is not only an air pollutant that affects human health but also facilitates the formation and dispersion of

FIGURE 1.2 Production of secondary air pollutants due to VOC emission.

PM2.5 and other airborne particulate matter (aerosols), resulting in smog and reduced visibility. There are many cases of smog pollution that have caused deaths and illnesses around the world, including big cities, such as London, Los Angeles, and Beijing. In addition to health effects, such as irritations and respiratory disorders, smog can reduce air visibility and, thus, result in road accidents, flight cancellations, and traffic jams. It may also have a negative impact on tourism. In addition to ozone and smog formation, some VOCs are greenhouse gases that can absorb radiated energy from Earth to raise the surface temperature, causing global warming and climate change (David and Niculescu, 2021).

1.3 VOC treatment

Many technologies have been developed for the removal of VOCs from the air. While several methods have been applied for home uses to remove VOCs from indoor air, it is unrealistic to develop purification techniques for the outdoor atmosphere. Therefore, most of the VOC treatment technologies are designed to reduce emissions from industrial processes. In this book, VOC treatment technologies are mainly referred for industrial applications.

VOC treatment technologies can be classified into two categories that are based on VOC recovery and destruction (Fig. 1.3). Treatment technologies for VOC recoverable removal are nondestructive methods that usually rely on physical processes, such as adsorption, absorption, membrane separation, and condensation. These treatments not only remove VOCs from the inflow but also separate and condense them for potential recovery and reuse (Zhang

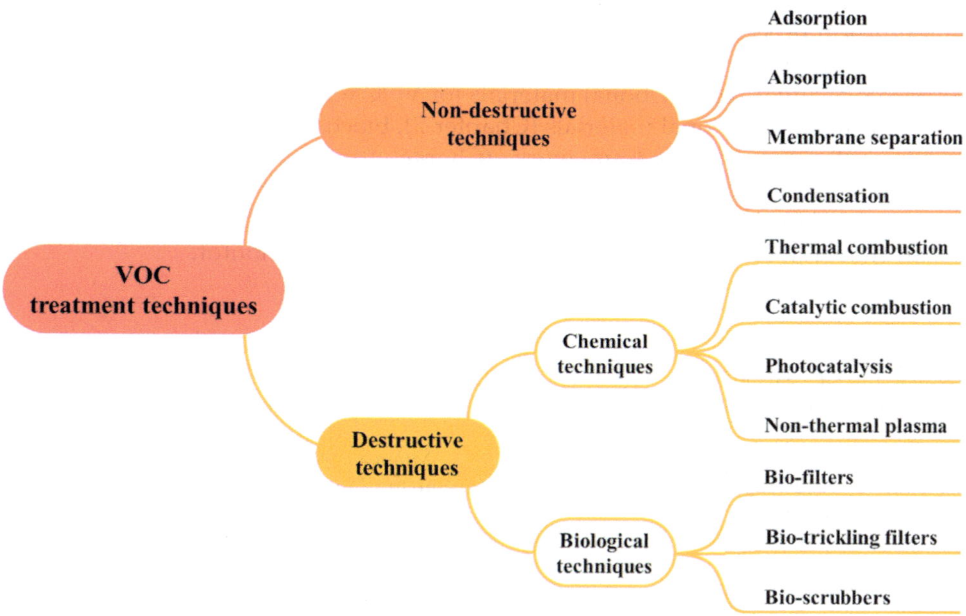

FIGURE 1.3 Classification of different VOC treatment technologies.

et al., 2017). This book covers all four nondestructive technologies, which are widely used in many industrial processes for the treatment of various types of VOCs. In addition, novel functional materials, such as nanomaterials and carbon materials, which are developed for these technologies, are also discussed in detail in this book.

Both chemical and biological degradation processes are used in the destructive VOC treatment technologies. For the chemical treatment methods, they mainly rely on the oxidation of VOCs through technologies, such as thermal combustion, catalytic combustion, photocatalysis, and nonthermal plasma (Li et al., 2023). This book only covers the first three chemical treatment technologies of VOCs and the recent development of functional materials for them. The biological treatment of VOCs is mainly carried out through biofiltration using filters, such as biofilters, biotrickling filters, and bioscrubbers (Cheng et al., 2021), which are not covered by this book.

1.4 Organization of this book

This book mainly focuses on the current technologies and functional materials for the removal of VOCs from polluted air. It not only covers the fundamental mechanisms, processes, and designs of the state-of-the-art VOC treatment technologies, but also the characteristics and applications of advanced functional materials in these technologies. The book has 12 chapters, centered on five technologies and six advanced materials. After the introduction chapter (Chapter 1), destructive chemical treatment technologies for VOCs are presented in Chapters 2 and 3. While Chapter 2 mainly focuses on thermal combustion technologies, Chapter 3 covers both catalytic combustion and photocatalysis technologies for VOC removal. Chapters 4–6 are on nondestructive VOC treatment technologies involving different physical processes, including membrane separation (Chapter 4), adsorption (Chapter 5), and absorption and condensation (Chapter 6). The next six chapters outline the recent developments of advanced functional materials for VOC removal. The functional materials in focus are traditional functional materials (Chapter 7), biochar (Chapter 8), carbon nanomaterials (Chapter 9), metal-organic frameworks (Chapter 10), metal nanoparticles (Chapter 11), and nanometal oxides (Chapter 12). This book project was initiated, developed, and finished through the collaborative work of long-time collaborators. It is the result of excellent teamwork with tremendous commitment and effort from each contributor.

References

Cheng, H.-H., Lu, I.-C., Huang, P.-W., Wu, Y.-J., Whang, L.-M., 2021. Biological treatment of volatile organic compounds (VOCs)-containing wastewaters from wet scrubbers in semiconductor industry. Chemosphere 282, 131137.

Crippa, M., Solazzo, E., Huang, G., Guizzardi, D., Koffi, E., Muntean, M., Schieberle, C., Friedrich, R., Janssens-Maenhout, G., 2020. High resolution temporal profiles in the Emissions Database for Global Atmospheric Research. Scientific Data 7, 121.

David, E., Niculescu, V.-C., 2021. Volatile organic compounds (VOCs) as environmental pollutants: occurrence and mitigation using nanomaterials. International Journal of Environmental Research and Public Health 18, 13147.

Li, S., Lin, Y., Liua, G., Shid, C., 2023. Research status of volatile organic compound (VOC) removal technology and prospect of new strategies: a review. Environmental Science: Processes & Impacts 25, 727–740.

Sillman, S., Vautard, R., Menut, L., Kley, D., 2003. O_3-NO_x-VOC sensitivity and NO_x-VOC indicators in Paris: Results from models and Atmospheric Pollution over the Paris Area (ESQUIF) measurements. Journal of Geophysical Research 108, 8563.

WHO, 1989. Indoor Air Quality: Organic Pollutants. Report on a WHO Meeting, Berlin, 23-27 August 1987. EURO Reports and Studies 111. World Health Organization Regional Office for Europe, Copenhagen.

Zhang, X.Y., Gao, B., Creamer, A.E., Cao, C.C., Li, Y.C., 2017. Adsorption of VOCs onto engineered carbon materials: a review. Journal of Hazardous Materials 338, 102–123.

CHAPTER

2

Catalytic degradation of VOCs

Haiqin Wan and Lin Dong

School of Environment, Nanjing University, Nanjing, China

2.1 Overview of catalytic degradation technologies for VOC abatement

Considering the environmental impact and toxicity of volatile organic compounds (VOCs), and the increasingly tightening regulations, more effective VOC removal technologies are accordingly required to reduce global emissions. Generally, two kinds of methods for VOC abatement, including manifold recovery technologies (e.g., adsorption, absorption, membrane separation, and condensation) and degradation approaches (catalytic oxidation, thermal incineration, biological degradation, photocatalytic decomposition, and nonthermal plasma oxidation) have been developed. Among them, catalytic oxidation typically operates at lower temperatures (200–500°C or even lower) with less formation of dioxins and noxious products and is, thus, more energy-efficient and environmentally friendly.

Catalytic VOC oxidation is the deep oxidation of VOCs into CO_2 and H_2O over a suitable catalyst in the air. It is one of the most effective and economically feasible technologies currently being applied for the removal of dilute VOC (<0.5 vol%) effluent streams. Based on the varieties of conditions of VOC oxidation, there are three kinds of catalytic oxidation technologies, including catalytic combustion, catalytic ozonation, and photocatalysis. The catalyst is the key in this technology due to the complex composition of VOCs, which can not only affect reactivity, but also affect product selectivity. Obviously, the catalysts used in the three kinds of catalytic oxidation technologies are different. Besides the catalyst, the reactor and the practical operational conditions of organic waste gas, such as halogen or sulfur-containing organic compounds, sulfur oxides, and humidity, can also affect the performance of catalytic VOC degradation (He et al., 2019).

2.2 Catalytic combustion of VOCs

Catalytic combustion is developed from thermal incineration, which can decrease the operation temperature, avoid secondary pollution, and treat dilute VOC waste gas. Compared to thermal incineration, the catalytic combustion of VOCs can be operated at lower

Volatile Organic Compound Removal
https://doi.org/10.1016/B978-0-443-15674-8.00010-X

temperatures, since the activation energy of a reaction can be reduced through different reaction routes over the catalyst. One of the major challenges of catalytic combustion is the choosing of an efficient catalyst. In general, two types of catalysts, including the supported noble metals and transition metal oxides, have been developed for the catalytic combustion of VOCs. The supported noble-metal catalysts, such as Pt, Pd, Rh, and Au, always show a higher catalytic activity than that of transition metal oxides. However, noble-metal catalysts are expensive and more sensitive to poisoning. Thus, transition metal-oxide catalysts (e.g., Cu, Mn, Ce, and Co oxides) are a cheaper alternative to a noble-metal catalyst and have received wide research attention. The performance of catalytic combustion also depends on the nature of VOCs, such as the molecular weight, molecular structure, polarity, contained heteroatoms, acidity, and redox potential (Huang et al., 2015).

2.2.1 Mechanisms of the catalytic oxidation of VOCs

It is of great importance to understand the mechanism of the catalytic oxidation of VOCs in the design of efficient catalysts. Catalytic oxidation of VOCs takes place between VOC molecules and oxygen species, which may have originated from a gas stream, catalyst surface, and catalyst lattice, depending on the free energy and diffusivity. Hence, the oxidation among different VOCs and catalysts is not identical. Based on the sources of oxygen and reaction routes, the kinetics vary and can be described by three types of models, i.e., the Mars–van Krevelen (MVK) model, the Langmuir–Hinshelwood (L–H) model, and the Eley–Rideal (E–R) model (He et al., 2019; Huang et al., 2015; Chung et al., 2019).

The MVK model considers that the reaction occurs between the adsorbed VOCs and the lattice oxygen of the catalyst rather than the oxygen in the gas phase. This model assumes that the oxidation of VOCs takes place in two steps, as shown in Fig. 2.1a. In the first step, the adsorbed VOCs react with oxygen in the catalyst, resulting in the reduction of the metal oxide. In the second step, the reduced metal oxide is reoxidized by the gas phase oxygen present in the feed. As the catalyst is first reduced and then reoxidized, this mechanism is also known as the redox mechanism. In the steady state, the rates of the reduction and oxidation steps must be equal. This model has been widely used for the kinetic modeling of the oxidation reactions of hydrocarbons over metal-oxide catalysts. According to the Mars–van Krevelen mechanism (Kamal et al., 2016), the rate of the catalytic oxidation of VOCs can be expressed by Eq. (2.1):

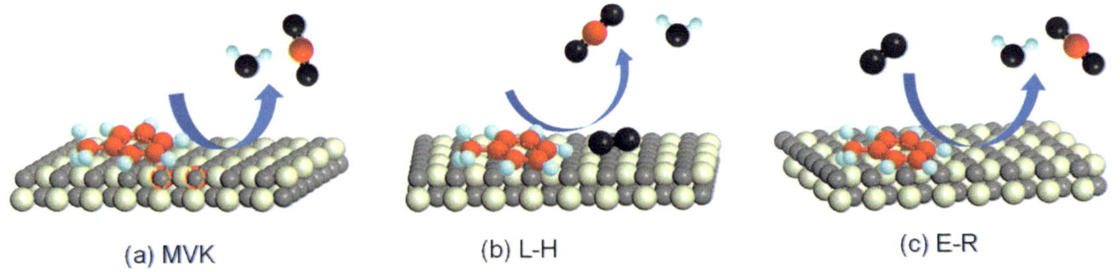

FIGURE 2.1 Mechanism of catalytic VOC oxidation: (a) Mars–van Krevelen (MVK), (b) Langmuir–Hinshelwood (L–H), and (c) Eley–Rideal (E–R).

$$-r_{VOC} = \frac{k_{O_2} k_{VOC} P_{VOC} P_{O_2}}{\gamma k_{VOC} P_{VOC} + k_{O_2} P_{O_2}} \tag{2.1}$$

where r_{VOC}, reaction rate (mol/m^3/s); P_{VOC}, partial pressure of the VOCs; P_{O2}, partial pressure of oxygen; k_{VOC}, rate constant of VOC oxidation; k_{O2}, rate constant of catalyst reoxidation; and γ, stoichiometry coefficient of O_2 in oxidation.

The L–H model, as illustrated in Fig. 2.1b, can be described as follows: two reactants, VOCs, and oxygen are first adsorbed on the catalyst surface. Next, the two adsorbed species further react with each other to form products. The last step is the desorption of the product from active sites. Therefore, it is essential for both the VOCs and the oxygen molecules (species) to be adsorbed on the surface of the catalyst. VOCs and oxygen may be adsorbed on similar types of active sites (single-site L–H model) or two different types of active sites (dual-site L–H model) (Kamal et al., 2016). Accordingly, the rate expressions for the single-site L–H model and the dual-site L–H model are different, as expressed by Eqs. (2.2) and (2.3), respectively:

$$-r_{VOC} = \frac{kK_{VOC} K_{O_2} P_{VOC} P_{O_2}}{(1 + K_{VOC} P_{VOC} + K_{O_2} P_{O_2})^2} \tag{2.2}$$

$$-r_{VOC} = \frac{kK_{VOC} K_{O_2} P_{VOC} P_{O_2}}{(1 + K_{VOC} P_{VOC})(1 + K_{O_2} P_{O_2})} \tag{2.3}$$

where K_{O2} is the equilibrium constant for the adsorption of oxygen and K_{voc} is the equilibrium constant for the adsorption of VOCs. The advantage of the L–H model is that it not only includes the reaction rate, but also takes the adsorption of VOCs and oxygen into consideration.

The E–R model (Fig. 2.1c) contains two steps; the first is the adsorption of oxygen on the catalyst surface and the second is the reaction between adsorbed oxygen and VOCs in a gas stream (or adsorbed VOCs react with oxygen gas). During the E–R mechanism reaction, the rate-limiting step is the reaction between an adsorbed molecule and a molecule from the gas phase. The following equation represents the kinetic expression based on the E–R mechanism (Eq. 2.4):

$$-r_{VOC} = \frac{kK_{VOC} P_{VOC} P_{O_2}}{(1 + K_{VOC} P_{VOC})} \tag{2.4}$$

2.2.2 Catalysts for VOC oxidation

The catalysts used for the oxidation of VOCs can be classified into two types, supported noble-metal catalysts and transition metal-oxide catalysts. Supported noble-metal (Pt, Pd, Rh, etc.) catalysts are attractive due to their high efficiency for the removal of VOCs at low temperatures. The catalytic performance of supported noble-metal catalysts is typically dependent on a number of factors, including the physicochemical properties of the active metal or support, the interaction between the active phase and support, the noble-metal precursor used in the preparation, the preparation method, and the status of metal particles and supports. However, the application of noble-metal-based catalysts is limited by the high cost, low thermal stability, and poor sulfur and chlorine resistance. Transition metal-oxide catalysts, which consist of elements from groups IIIB and IIB in the periodic table, are much cheaper with better thermal stability and higher resistance to poisoning. Therefore, great efforts have also been made to develop efficient transition metal-oxide catalysts for VOC catalytic oxidation.

2.2.2.1 *Supported noble-metal catalysts*

In spite of the expensive cost, noble-metal catalysts are recognized as the preferred ones for VOC catalytic oxidation for their high performance. Platinum (Pt), palladium (Pd), ruthenium (Ru), and gold (Au) are the most extensively studied elements out of all the noble metals. They are generally supported by Al_2O_3 and SiO_2 and also by molecular sieves and transition metal oxides (e.g., MnO_x, CeO_2, and Co_3O_4) and their mixtures to increase the dispersion of noble metals and the adsorption of reactants, as well as to reduce the loading of noble metals. The catalytic performance of the supported noble metals strongly depends on the metal's intrinsic properties, preparation method, precursors, support, the size and morphology of the particles, etc.

2.2.2.1.1 Traditional noble-metal catalysts

The metal's nature decides its catalytic performance. Platinum and palladium have been found to be highly efficient for the deep mineralization of C_2-C_8 aliphatic hydrocarbons, but gold relatively shows low activity for C_1-C_3 saturated hydrocarbons. Supported Pd catalysts have been shown to be highly active for the full destruction of xylenes. Supported Ru catalysts (Zhao et al., 2020) have been used for the deep oxidation of vinyl chloride because of their superior resistance to chlorine species and excellent activity in oxidation and reduction reactions. Au/Al_2O_3 exhibited remarkable thermal stability up to 600°C in the presence/absence of water vapor. Gluhoi et al. (2005) introduced CeO_2 into Au/Al_2O_3 for the catalytic oxidation of propane and found a high complete oxidation rate achieved at 250°C. The doped Ce over Al_2O_3 not only provides good oxygen storage capacity, but also increases the fixation and dispersion of gold particles on Al_2O_3, resulting in enhanced oxidation activity and resistance to sintering.

Moreover, the particle size of noble metals plays a crucial role in its activity. Isaifan et al. (2013) found that there was a strong relationship among the average Pt particle size, activation energy, and turnover frequency of ethylene oxidation. They proposed that activation energy decreases when the average Pt particle size decreases. This may be attributed to the stronger chemical adsorption of VOCs onto smaller Pt particles and, thus, larger adsorption heat, resulting in an increase of the turnover frequency. The industrial applications of platinum catalysts are limited due to their high cost and the likelihood of poisoning, especially when dealing with chlorinated products. Chen et al. (2015) synthesized a series of size-controllable Pt nanoparticle loading on ZSM-5 to investigate the effect of the platinum particle size on the catalytic oxidation of toluene over Pt/ZSM-5. The result showed that the Pt particle size at 1.9 nm exhibited the highest activity (T_{100} is 155°C) for toluene oxidation. This excellent performance could be attributed to the balance of Pt dispersion and Pt^0 proportion.

The properties of the support, such as specific surface area, pore structure, acidity, and crystal facet exposure, have strong influences on metal dispersion, VOC adsorption, and metal–support interaction and, thus, are critical to their catalytic performance. Okamura et al. (2003) synthesized catalysts with palladium, being supported on Al_2O_3, MgO, Nb_2O_5, SiO_2, SnO_2, WO_3, and ZrO_2, to evaluate their surface acidity and toluene oxidation efficiency. The results indicate that a higher surface acidity is unfavorable to toluene oxidation because of the lower oxygen affinity. However, for chlorinated volatile organic compounds (CVOCs), the tendency of the oxidation capability is the opposite to that of hydrocarbons. Stronger surface acidity can enhance the adsorption rate and capacity toward CVOCs, resulting in higher CVOC oxidation efficiency (Blanch-Raga et al., 2014; Assal et al., 2013). Toluene catalytic oxidation of Pt on CeO_2 with different morphologies (nanorods, nanoparticles, and nanocubes) was studied by Peng et al. (2016). Variations of the surface oxygen vacancies of catalysts was obtained by the modification of the CeO_2 morphology, and Pt/CeO_2 rods showed the best catalytic activity (T_{90} is 150°C) due to their highest oxygen vacancies.

2.2.2.1.2 Bimetallic noble-metal catalysts

Besides mono-noble metal catalysts, bimetallic noble-metal catalysts have attracted increased attention due to their unique electronic, structural, and catalytic properties that are absent in coincident monometallic nanoparticles. Compared to mono-noble metal catalysts, bimetallic noble-metal catalysts have two advantages: (1) they reduce the amount of precious metals used and (2) introduce new chemistries to improve catalytic performance. To achieve this purpose, the composition and structure of bimetallic noble-metal catalysts have been widely studied. Generally, a transition metal, such as Cu, Co, Mn, or Fe, is often introduced into the noble metal to form the bimetallic catalyst, which has a higher catalytic activity than the mono-noble metal catalyst. Bimetallic nanoparticles can be categorized into two main groups: core–shell and alloy structures. Li et al. (2018) found that silver palladium bimetallic core–shell structure catalysts supported on the TiO_2 substrate ($Ag@Pd/TiO_2$) are highly active with a low palladium loading for toluene oxidation. Additionally, a noble-based bimetal is always prepared into an alloy for the catalytic oxidation of VOCs. Its catalytic performance depends on the feature of the second metal and the ratio of the two metals. Tahsini et al. (2022) found that Pt_xCu_{100-x} alloy catalysts prepared from colloidal nanocrystals are more active than the corresponding Pt catalysts for complete propene oxidation (Fig. 2.2). The results from density functional theory (DFT) studies and structural and catalytic characterization suggest that the remarkable activity of dilute Pt_xCu_{100-x} alloys is related to the tuning of the electronic structure of Pt to reach the optimal binding energies of the C* and O* intermediates.

2.2.2.1.3 Noble-metal single-atom catalysts

Recently, single-atom catalysts (SACs) have received much attention because they have the maximum atom efficiency and minimize the usage of precious metals in catalysis. Indeed, several noble-metal single-atom catalysts have been used for VOC oxidation. For example, $Pt_1/OMS\text{-}2$, $Pt_1/meso\text{-}Fe_2O_3$, and Pt_1/TiO_2 SACs are more efficient than their nanocatalyst counterparts in benzene/C_3H_6 oxidation. Different from the traditional noble nanoparticle catalysts showing a strong capacity for activating O_2 due to the donation of electrons from the metal to O_2, noble-metal SACs promote the activation of O_2 by constructing surface oxygen vacancies (Zhang et al., 2022). Furthermore, these SACs show better sulfur and chlorine

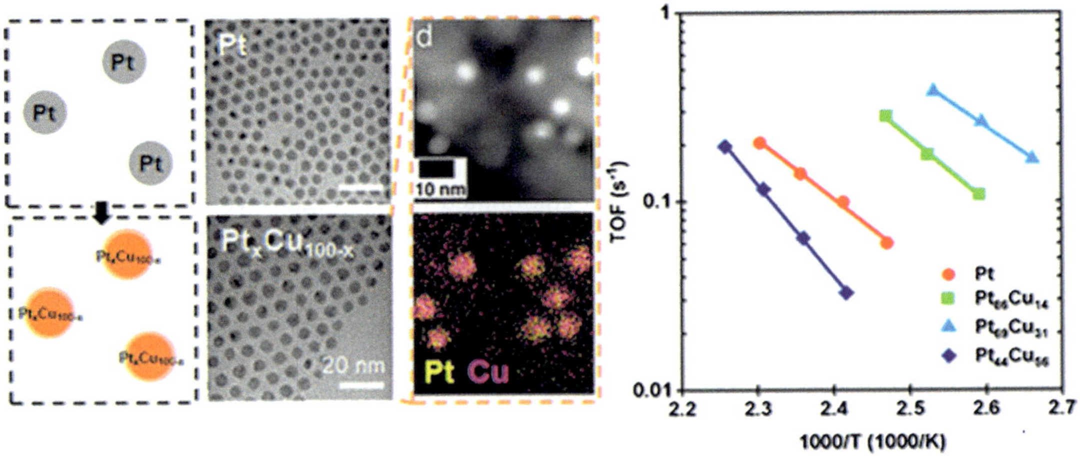

FIGURE 2.2 TEM (transmission electron microscopy) images, HAADF–STEM (high-angle annular dark-field scanning transmission electron microscopy) images, and turnover frequencies (TOFs) of Pt_xCu_{100-x}/Al_2O_3 catalysts for propene combustion. *Reprinted from Tahsini, N., Yang, A. C., Streibel, V., Werghi, B., Goodman, E. D., Aitbekova, A., Bare, S. R., Li, Y.J., Abild-Pedersen, F., Cargnello, M., 2022. Colloidal platinum–copper nanocrystal alloy catalysts surpass platinum in low-temperature propene combustion. Journal of the American Chemical Society 144, 1612–1621. Copyright 2022 American Chemical Society.*

resistance than nanocatalysts due to the easier decomposition of sulfides on single metal atoms and the more efficient desorption of the formed HCl. As a result, noble-metal SACs have shown a promising application potential in the catalytic oxidation of VOCs.

2.2.2.2 *Transition metal-oxide catalysts*

Transition metal-oxide catalysts are widely studied as alternates to noble-metal catalysts because of limited natural resources and the high price of noble metals. They are usually composed of oxides of elements in the fourth period of the periodic table, such as Cr, Cu, Mn, Co, and Fe. Due to the good dispersity of the active components, transition metal-oxide catalysts show excellent activity in the catalytic degradation of VOCs. In addition, because of their good electronic fluid oxidation and low-temperature resistance, they have become a promising catalyst for VOC removal. Generally, Mn-based, Co-based, and Ce-based single metal-oxide catalysts not only show excellent catalytic performance for hydrocarbons, such as toluene, but also have the advantages of being low cost, strong reducibility, excellent thermal stability, and toxicity resistance. However, single metal oxides are prone to phase transformation; their thermal stability is worse than that of precious metals, and the activity of catalysts with different valence states is very different. Multiple composite catalysts are, thus, needed to be developed vigorously. It's well acknowledged that mixed oxides usually exhibit higher activity than single oxides in most catalytic reactions. This is considered as the result of the higher surface mobility of oxygen and/or activated species, as well as electron transport through the lattice for the multiple energy levels of metals and the abundant number of associated oxygen anions.

Besides the advantage of price, transition metal-oxide catalysts show better resistance to chlorine and sulfur poisoning than noble-metal catalysts. Wang et al. (2019) found that manganese oxides (i.e., γ-MnO_2, Mn_2O_3, and Mn_3O_4) showed good stability for 1,2-dichloroethane (DCE) oxidation. In detail, it was found that the activity of manganese oxide catalysts decreased continuously during 1,2-DCE oxidation under 350°C. Only when the reaction temperature exceeded 350°C, could the activity of the catalyst be stable. It suggests that a high reaction temperature may facilitate the removal of the chlorine species. Zhao et al. (2020) found that $Ru/Ti_xSn_{1-x}O_2$ catalysts present high stable activity on the stream feed containing 1000 ppm dichloromethane at 300°C for 50 h. The authors proposed that the Lewis acidity of Ru/Ti_xSn_{1-x} catalysts are promoted by the formation of Ti—O—Sn, which promotes catalytic activity by transferring surface oxygen to the active sites, while Ru—O—Ti and Ru—O—Sn enhance the resistance of Cl. It is found that the reasonable adjustment of the catalyst metal-composition ratio, preparation method, and load mode has a positive impact on the catalytic performance.

2.3 Catalytic ozonation of VOCs

Ozone, as a strong oxidant, has been applied to VOC oxidation. Compared to conventional catalytic oxidation techniques, the catalytic ozonation of VOCs greatly reduces the reaction temperature, significantly saves energy, and, thus, has gradually attracted much attention in recent years. Ozone spontaneously decomposes to oxygen and leaves no residues, which is an advantage for VOC oxidation. However, it needs to be generated on-site from a flow of either air or pure oxygen for the catalytic ozonation of VOCs.

Over the past years, ozone has been successfully used to degrade organic compounds in polluted water. Inspired by this, it has also been applied in removing gaseous pollutants, such as VOCs. Ozone displays low activity toward alkanes, aldehydes, and aromatics in the gas phase. However, catalysts can boost its degradation of VOCs. Therefore, catalytic ozonation oxidation is a promising technology for VOC elimination.

2.3.1 Mechanisms of catalytic VOC ozonation

The mechanism of the catalytic ozonation of VOCs is similar to that of the catalytic oxidation of VOCs. Typically, catalytic ozonation follows the L—H and MVK mechanisms (Liu et al., 2022), and active oxygen species, including atomic oxygen and surface oxygen, are the key factors in deciding the type of the mechanism used. For the L—H mechanism, ozone decomposes into atomic oxygen and reacts with the adsorbed VOCs to form carbon dioxide and water, as shown in Eqs. (2.5) and (2.6):

$$O_3 \rightarrow O_2 + O \tag{2.5}$$

$$VOC + O \rightarrow \ \rightarrow \ \rightarrow CO_2 + H_2O \tag{2.6}$$

Therefore, this mechanism is based on the information that atomic oxygen is the major active species and ozone decomposition should be the determining step in the catalytic ozonation of VOCs.

For the MVK mechanism, a surface oxygen species, like lattice oxygen (O_{lat}), partially oxides the adsorbed VOCs into some intermediates, such as alcohols and/or carboxylates (Eq. 2.7), and the atomic oxygen species oxides the intermediate into dicarbon oxide and water (Eq. 2.8). Finally, ozone replenishes the consumed lattice oxygen (Eq. 2.9):

$$VOC + O_{lat} \rightarrow V_o + RCOOH \text{ or } ROH \tag{2.7}$$

$$RCOOH \text{ or } ROH + O \rightarrow \rightarrow \rightarrow CO_2 + H_2O \tag{2.8}$$

$$O_3 + V_O \rightarrow O_2 + O_{lat} \tag{2.9}$$

In addition, the active oxygen species should be changed when water vapor is present in the catalytic ozonation of VOCs. In this case, reactive hydroxyl groups are formed due to the interaction of adsorbed water and atomic oxygen species (Eq. 2.10). The reactive hydroxyl group oxides the intermediates into carbon dioxide and water (Eq. 2.11):

$$H_2O + O \rightarrow 2OH \tag{2.10}$$

$$RCOOH \text{ or } ROH + OH \rightarrow \rightarrow \rightarrow CO_2 + H_2O \tag{2.11}$$

Moreover, researchers have proposed an autoxidation reaction in VOC ozonation, suggesting that oxygen molecules also participate and oxide radical intermediates (R•) into CO_2 and CO, as shown in Eq. (2.12) (Liu et al., 2022; Jin et al., 2011):

$$R\bullet + O_2 \rightarrow \rightarrow \rightarrow CO_2, CO \tag{2.12}$$

2.3.2 Catalysts for VOC ozonation

Based on the mechanisms discussed above, atomic oxygen and surface oxygen species (lattice oxygen and hydroxyl groups) are identified as the reactive oxygen species for the catalytic ozonation of gaseous VOCs. Therein, catalysts with strong Lewis acid sites and rich surface oxygen vacancies can boost the catalytic ozonation of VOCs.

2.3.2.1 *Zeolite-based catalysts*

Due to its high specific surface area, rich pore structure, large pore volume, and strong Lewis acid sites, zeolite is a favorable catalyst for VOC ozonation. Among all the zeolites, 13X, MCM-41, and HX have a higher removal efficiency in the catalytic ozonation of toluene because of their advantages in enhancing the adsorption of reactants, prolonging the residence time, and providing more active sites for ozone decomposition to form more active oxygen species. However, zeolite materials suffer from relatively rapid deactivation due to the accumulation of intermediates, including acetaldehyde, benzaldehyde, formaldehyde, and organic acids, which result from the incomplete oxidation of toluene (Chao et al., 2017).

Moreover, water molecules can occupy the catalyst active sites when they are formed during VOC oxidation. Therefore, some metal oxides are added into zeolites to improve their catalytic performance. For example, Sugasawa and Ogata (2011) introduced Fe, Ni, Ag, Co, and Mn into ZSM-5 for the catalytic ozonation of toluene. They found that Mn/ZSM-5 had the best performance for toluene removal and Ag/ZSM-5 exhibited the highest CO_2 selectivity, promoting the oxidation of CO to CO_2 by O_3. It can be concluded that supported manganese oxide is one of the most active catalysts for the catalytic ozonation of toluene at low temperatures.

2.3.2.2 Manganese-based catalysts

Like in other catalytic oxidations, most transition metal oxides are also effective in the catalytic ozonation of VOCs. Especially, manganese oxides have shown outstanding performance in the ozonation of many kinds of VOCs. Formaldehyde (HCHO) is one of the representative indoor VOC air pollutants and human carcinogens. Zhao et al. (2012) found that amorphous manganese oxide could completely oxidize HCHO into carbon dioxide and water under ambient conditions with the assistance of ozone, comparable to noble-metal catalysts. In addition, manganese oxides have been used for the catalytic ozonation of benzene. Jin et al. (2011) reported that mesoporous α-Mn_2O_3 showed excellent performance in the ozonation of benzene, and benzene conversion and the selectivity of CO_2 reached 70% and 80% at 80°C, respectively. However, its catalytic performance was negligible without ozone. Generally, a manganese oxide's outstanding performance depends on its abundant oxygen vacancies, high specific surface area, and mesoporous structure.

To combine the advantages of zeolite and manganese oxide, manganese oxides have been loaded on zeolites to create composite catalysts for VOC ozonation, such as Mn/13X for toluene ozonation and MnO_x/MCM-41, MnO_x/SBA-15, and MnO_x/USY for benzene ozonation. Huang et al. (2015) reported that MnO_2/ZSM-5 could promote the ozonation of benzene at room temperature with a benzene conversion of 100% and CO_2 selectivity of 84.7%. They proposed that the presence of Lewis acid sites and highly dispersed MnO_2 nanoparticles, which generated a large number of active oxygen species, play a key role in ozone decomposition. Furthermore, the high surface area and abundant pores of ZSM-5 can retain intermediates and residual ozone in the adsorption phase, thereby extending the contact and reaction time.

Additionally, manganese oxides have been supported on other carriers, such as Al_2O_3, SiO_2, active carbons, and carbon nanotubes. Shao et al. (2020) found that, compared to MnO_x/SiO_2 and MnO_x/TiO_2, MnO_x/γ-Al_2O_3 possessed the highest total acid content and most amount of Mn^{3+} and, thus, realized the highest conversion (nearly 100%) of toluene and ozone. In their study, Mn^{3+} was a key species for VOC ozonation in supported manganese oxide catalysts. Reed et al. (2006) also found that a lower oxidation state (3+) of Mn in MnO_x/SiO_2 was favorable for delivering oxygen, accepting electrons, and boosting acetone ozonation. Furthermore, the dispersion of manganese oxide also affects its performance. Rezaei and Soltan (2012) proposed that only MnO_x, with the Mn^{3+} oxidation state and optimal dispersion, was preferable for the catalytic ozonation of toluene, while a high oxidation state and high metal dispersion were unfavorable for promoting this reaction. Furthermore, Jin et al. (2015) found that MnO_x/CNTs exhibited good catalytic performance for chlorobenzene ozonation.

2.3.2.3 *Noble-metal catalysts*

Noble-metal catalysts have also been used for the catalytic ozonation of VOCs. Differing from thermal catalytic oxidations, Pt and Pd have poor catalytic performance for the ozonation of VOCs, while Ag shows good catalytic performance. Pt and Pd were introduced into MnO_x/Al_2O_3, in which Pt enhanced the activity for toluene ozonation but Pd was inefficient. This is because Pt interacted with Mn to form the Pt–O–Mn bond while Pd did not (Rezaei et al., 2013). Ag was introduced into Mn/HZSM-5 for toluene ozonation at room temperature, which showed outstanding activity (Li et al., 2014). Toluene conversion was about 90%, ozone almost completely decomposed, and the selectivity of CO_x was above 80%. The authors thought that the addition of Ag contributed to the activity of O_3 decomposition to generate reactive oxygen species. Moreover, Ag also improved oxygen mobility followed by improving the reducibility of manganese oxide, thus Mn–Ag/HZSM-5 had a lower manganese oxidation state, which is conducive to the catalytic ozonation of toluene. MnAg/HZSM-5 also exhibited good performance for benzene ozonation.

Chlorinated or sulfur-containing VOCs are difficult to catalytically oxidize due to the high amount of poisoning as compared to the catalyst. Noble-metal catalysts have been applied to the ozonation of chlorinated or sulfur-containing VOCs. Pt/Al_2O_3 was used for the ozonation of dichloroethane; the results indicated that T_{50} dropped from 309 to 157°C with the introduction of ozone and CO_2 selectivity was enhanced. It is attributed to the active oxygen species formed by ozone decomposition and the autoxidation process over Pt particles on the catalyst surface, facilitating the oxidation of chloride-containing and carbon-containing intermediate species and, thus, enhancing mineralization (Yuan et al., 2015). Ag supported on three-dimensional MnO_2 porous hollow microspheres (Ag/MnO_2) was applied in the ozonation of methyl mercaptan and showed good performance (Fig. 2.3). Its great performance can be attributed to better reducibility and higher oxygen mobility, which caused more oxygen to be adsorbed and further excited to reactive oxygen species, resulting from the low oxidation state of Mn and a large amount of lattice oxygen due to the incorporated Ag (Xia et al.,

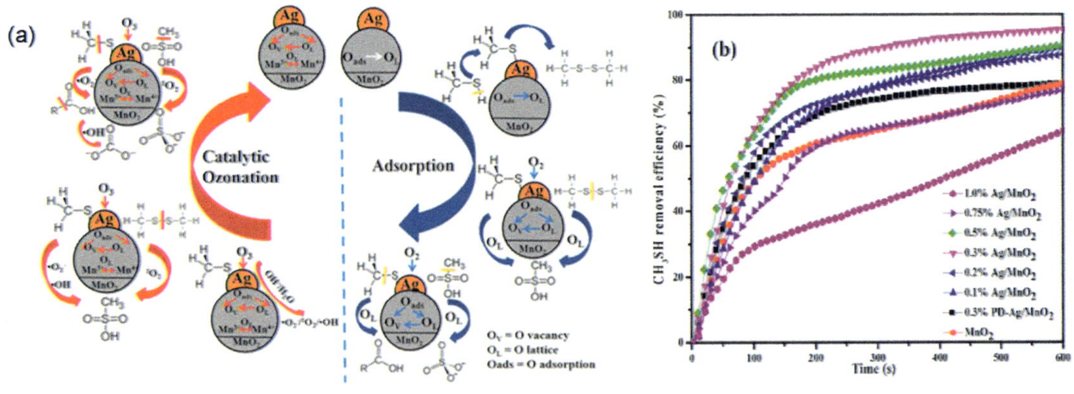

FIGURE 2.3 Mechanism (a) and activity (b) of the catalytic ozonation of CH_3SH over Ag/MnO_2. *Reprinted from Xia, D., Xu, W., Wang, Y., Yang, J., Huang, Y., Hu, L., He, C., Shu, D., Leung, D. Y. C., Pang, Z., Enhanced performance and conversion pathway for catalytic ozonation of methyl mercaptan on single-atom Ag deposited three-dimensional ordered mesoporous MnO_2. Environmental Science & Technology 52, 13399–13409. Copyright 2018 American Chemical Society.*

2018). Besides, the Ag species on the catalyst surface also facilitated the chemisorption of CH_3SH and transformed CH_3SH into CH_3SAg/CH_3S-SCH_3, which was further oxidized into SO_4^{2-} and CO_2 by hydroxyl radicals and $^1O_2$.

2.4 Photocatalytic oxidation of VOCs

In recent years, the indoor air quality (IAQ) of residential units and workplaces has attracted increasing attention due to human beings spending more than 80% of their lifetimes indoors. Thus, VOCs, as a classic indoor pollutant, become a serious concern. Indoor VOCs are produced from a variety of sources, including the utilization of consumer household products, emissions from adhesives and building materials, and combustion processes. VOCs are easily absorbed by the skin and mucous membranes, causing damaging consequences to human organs and metabolic systems. Therefore, the removal of indoor VOCs is significant to protect human health. Many VOCs are ubiquitous in the indoor environment because of the presence of typical indoor emission sources, and the concentration of indoor VOCs is very low. Typically, adsorption is a conventional method used to remove indoor VOCs. However, adsorbents not only adsorb VOCs but also water and need to be replaced after their saturation in a short time. Therefore, the photocatalytic oxidation (PCO) of indoor VOCs has attracted widespread attention due to the advantage of the utilization of solar energy, which can reduce energy cost during VOC abatement and deeply oxidizing VOCs into inorganic molecular compounds, including CO_2 and H_2O.

2.4.1 Mechanisms of the photocatalytic oxidation of VOCs

Photocatalysis, as a promising technology developed since 1972, is defined as an ideal technology for low-concentration VOC elimination under ambient conditions, especially in indoor areas. The mechanism of the photocatalytic oxidation of VOCs is one of the advanced oxidation processes. Oxidation occurs on the surface of the photocatalyst (a semiconductor). The catalyst is excited by the absorption of a photon with an energy greater than or equivalent to the bandgap energy of the semiconductor, resulting in the electron transition from the valence band to the conduction band, as shown in Eq. (2.13). Radiation can consequently produce electrons and holes (e^-/h^+) in the conduction band (CB) and valence band (VB), respectively. Following irradiation, the electrons and holes can undergo redox reactions with the adsorbed reactants on the photocatalyst's surface that lead to the formation of intermediates and products, as shown in Eqs. (2.14–2.16) (Zhang et al., 2022; Zhao et al., 2022):

$$\text{photocatalyst} + h\nu \rightarrow e^- + h^+ \tag{2.13}$$

$$h^+ + H_2O \rightarrow \bullet OH + H^+ \tag{2.14}$$

$$e^- + O_2 \rightarrow O_2^- \tag{2.15}$$

$$VOCs + \cdot OH(O_2^-) \rightarrow \rightarrow \rightarrow CO_2 + H_2O \tag{2.16}$$

Generally, the light source and the property of the photocatalyst are the key factors controlling the efficiency of the photocatalytic oxidation of VOCs. Ultraviolet light is more effective than visible light due to its higher energy, but visible light occupies a major content of sunlight. Enhancing the utilization of sunlight is, thus, a key topic in the photocatalytic oxidation of VOCs. Designing a heterojunction photocatalyst in response to the full spectrum and to enhance the separation of photogenerated electron–hole pairs is one of the efficient strategies. When the two selected semiconductors have a staggered band structure configuration, the heterojunction can be classified into type-II and Z-scheme heterojunctions. In a type-II heterojunction (Fig. 2.4a), photogenerated electrons and holes are preferred to be transferred to their respective low-energy bands; therefore, electrons accumulate in semiconductor B. Meanwhile, the holes gathering in semiconductor A lead to an increased electron–hole pair separation. Different from the type-II heterojunction, a direct Z-scheme is developed once two semiconductors (SCs) with staggered band alignments are intimately contacted. Due to the direction of the internal electric field, electrons in the CB of semiconductor B can move to the VB of semiconductor A and recombine with the holes there (Fig. 2.4b). Thus, photogenerated holes are left at the VB of semiconductor B with electrons at the CB of semiconductor A, resulting in the efficient migration and separation of photogenerated charge carriers.

2.4.2 Catalysts for the photocatalytic oxidation of VOCs

TiO_2 has been widely used in the photocatalytic oxidation of VOCs due to its good chemical stability, excellent photocatalytic performance, and low cost. However, the band gap of TiO_2 ($E_g = 3.2$ eV) limits its application in visible-light irradiation. In addition, the fast recombination of photogenerated charge carriers seriously restricts its photocatalytic activity toward VOC oxidation. Therefore, the modification of TiO_2 with other photocatalysts has been often used to enhance the photocatalytic oxidation of VOCs.

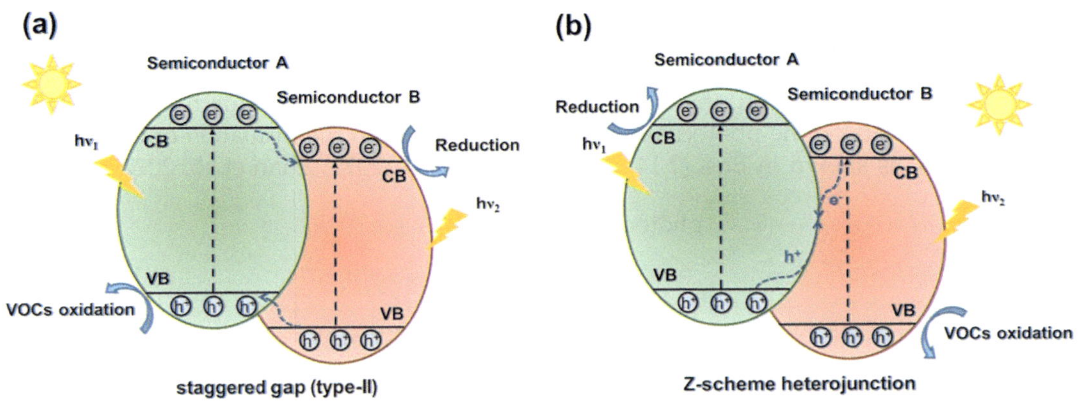

FIGURE 2.4 Photocatalytic mechanism of heterojunction photocatalysts: (a) type-II heterojunction and (b) Z-scheme heterojunction.

2.4.2.1 TiO_2-based catalysts

Both anatase and rutile TiO_2 have been widely used in the photocatalytic oxidation of indoor VOCs. Anatase's bandgap energy is higher than that of rutile (3.23 vs. 3.02 eV). It has shown better performance than rutile in PCO processes because of its more favorable conduction band configuration and stable surface peroxide groups. The photocatalytic oxidation of VOCs is a surface catalytic process; the specific surface area, morphology, and crystallinity of TiO_2 also affect its activity. Compared with the {110} and {101} planes, the exposed {001} crystal plane of TiO_2 has more unsaturated coordinated Ti and O atoms, which can easily trap VOCs, such as acetaldehyde, acetone, and methyl acetate (Tasbihi et al., 2011). DFT results indicate that photogenerated electrons and holes migrate to the {101} and {001} crystal planes, respectively (Yao et al., 2017).

To enhance the adsorption of reactants on the catalyst, TiO_2 has been modified to create porous structures and increase the surface area and functionality. TiO_2 with porous structures has better performance because of its more effective catalytic sites, superior pollutant adsorption capacity, and better mass transfer performance. Phosphate groups are often used to modify the photocatalyst. Li et al. (2015) found that the phosphate groups anchored to anatase TiO_2 can increase the adsorption of O_2, resulting in the better separation of photogenerated carriers. The surface modification strategy can also prevent the adsorption of intermediates, thus preventing catalyst deactivation.

Noble-metal deposition on TiO_2 is another strategy to enhance the separation of photogenerated carriers, by forming a Schottky barrier to facilitate the transfer of photogenerated charges. As the semiconductor is activated by light irradiation, the photogenerated electrons are transferred to the surface of noble-metal nanoparticles and participate in the reduction reaction with O_2 to form reactive oxygen species. On the other hand, the holes move to the surface and participate in the oxidation reaction with VOCs and H_2O. In this way, noble metals coupled with a TiO_2 semiconductor can effectively enhance the efficiency of charge separation and, thus, improve the photocatalytic properties. For example, Pt on TiO_2 acts as a charge receptor and promotes the formation of reactive oxygen species. Moreover, it changes the photocatalytic oxidation pathway of the surface reaction and inhibits the formation of CO by the direct oxidation of VOCs to CO_2 (Selishchev et al., 2017).

Metal- or nonmetal-ion doping can tune the electronic and band structure of TiO_2, thus changing its light adsorption ability. For example, Ce-doped TiO_2 shows the enhanced photocatalytic degradation of benzene and ethylbenzene because the transition of Ce^{3+}/Ce^{4+} promotes the separation of photogenerated electrons and holes (Korologos et al., 2012; Lee et al., 2016). The photocatalytic degradation of VOCs by TiO_2 doped with N is obviously better than that by undoped TiO_2 under visible light irradiation. This can be attributed to bandgap narrowing by mixing the p states of N with O's 2p states, forming a nitrogen-induced midgap, slightly above the O 2p VB (Nakamura et al., 2004).

Constructing a TiO_2-based composite photocatalyst can also enhance its catalytic performance for PCO, mainly through three kinds of methods: (1) coupling with a photosensitizer to expand the light absorption spectrum of TiO_2; (2) combining with visible-light-driven (VLD) photocatalysts; and (3) forming a heterojunction structure to efficiently separate e^-/h^+ at the interface. Accordingly, $BiVO_4$, with a narrower bandgap, has been coupled with TiO_2 for removing VOCs under visible-light irradiation. Metal chalcogenide semiconductors, such as SnS_2, CdS, CdSe, PbS, and PbSe, are good photosensitizers for coupling them with TiO_2 to broaden the light-absorption range from ultraviolet (UV) to visible light and the near-infrared region. The $g\text{-}C_3N_4-TiO_2$ Z-scheme photocatalyst has also been developed and used for the decomposition of formaldehyde in the air under the irradiation of UV lamps.

2.4.2.2 *Other catalysts*

Besides TiO_2-based photocatalysts, many new photocatalysts have also been explored for the PCO of VOCs, especially those materials with advantages in visible-light absorption, organic adsorption, and the pore structure and surface area. For example, hydroxide photocatalysts, such as $In(OH)_3$, $ZnSn(OH)_6$, GaOOH, and $CaSn(OH)_6$, show excellent photocatalytic performances for benzene degradation under UV irradiation. Polymorphs of Ga_2O_3 (α-, β-, and $\gamma\text{-}Ga_2O_3$) photocatalysts have a stronger redox ability than TiO_2, which is beneficial for the abatement of the highly adsorbed intermediates of gaseous benzene (Xu et al., 2022). Bismuth-based photocatalysts (Li and Li, 2020), such as $BiPO_4$, Bi_2WO_6, and $BiVO_4$ and bismuth oxyhalides, like BiOBr, BiOCl, and BiOI, have been used to photocatalytically degrade VOCs under Ultraviolet–visible (UV-vis) and visible light irradiation. Structure control and surface engineering are performed on bismuth-based photocatalysts to improve their photocatalytic activity. Owing to the visible light-induced photocatalytic activity and suitable VB edge, WO_3 can decompose organic pollutants by oxidative OH radicals and holes. Metal–organic frameworks (MOFs) have also been used for PCO.

2.5 Case study of reactors for catalytic VOC degradation

Different types of reactors (e.g., fixed-bed reactors and fluidized-bed reactors) have been applied in the catalytic destruction of VOCs. Fixed-bed reactors can be subdivided into continuous-flow fixed-bed reactors and membrane reactors. A series of structured catalysts, such as monolithic honeycomb catalysts and foam catalysts, have been designed to replace conventional granular catalysts with high diffusion resistance in continuous-flow fixed-bed reactors. This can improve the gas–solid contact, enhance the attrition resistance of a given catalyst, and reduce pressure drops across the system.

Depending on the detailed catalytic oxidation technology, the reactor used could be different. Here, a photocatalytic oxidation reactor is used as a case. In general, four main components matter the most in a typical photocatalytic reactor, including the photocatalyst, light source, reactor frame, and air-flow channel. The photoreactors' designs play an essential role in the efficiency of the photocatalytic process, particularly regarding the intensification of the lighting efficiency and the contact between the pollutant molecules and the photocatalyst. Thus, the greatest challenge of a photocatalytic reactor is to increase mass transfer while still maintaining a good photon transfer. In general, photoreactors can be divided into four main types: flat-plate, annular, fixed-bed, and monolith reactors. Some other types of reactors have also been reported, such as microfluidic reactors, fluidized reactors, and Trombe walls and so on (Zhang et al., 2022).

A flat-plate reactor is one of the most widely used reactors for the PCO of VOCs, which is usually assembled with UV lamps at the top and a catalyst film at the bottom of the reactor. For example, Qiu et al. (2018) designed a flat interlaid reactor for photocatalytic toluene degradation (Fig. 2.5a). They found that the catalysts coated on both the top and bottom showed

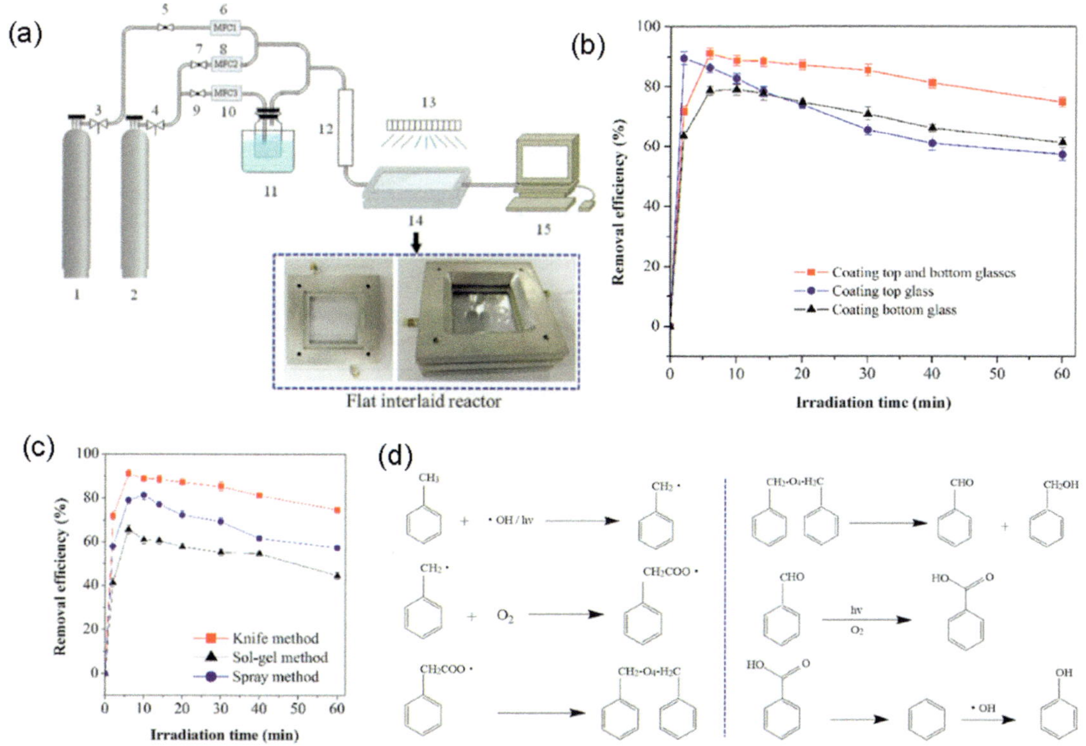

FIGURE 2.5 (a) Schematic diagram of the flat-plate reactor; the photocatalytic oxidation of toluene with different (b) catalyst-coating locations and (c) catalyst-coating methods, and (d) possible generation pathways of intermediates in the PCO of toluene. *Reprinted from Qiu, L., Wang, Y., Li, H., Cao, G., Ouyang, F., Zhu, R., 2018. Photocatalytic oxidation of toluene on fluorine doped TiO_2/SiO_2 catalyst under simulant sunlight in a flat reactor. Catalysts, 8 (12), 596. Copyright 2018 MDPI.*

higher activity than those just coated on one side, which was due to a larger surface area and more active sites for the air pollutants to react with the catalysts. In addition, the removal efficiency was related with the height of the reactor, the irradiation time, and the catalyst coating methods. They proposed the possible generation pathways of intermediates (i.e., benzaldehyde and benzoic acid) in the process of the photocatalytic oxidation of toluene in the reactor.

Additionally, other PCO conditions were studied by da Costa Filho et al. (2019). They designed a mili-photoreactor based on the NETmix technology and used it for the PCO of n-decane (Fig. 2.6). They found that back- or front-side illumination had a minimal effect on the reaction rate (Fig. 2.6d) due to the highly porous structure of the catalyst film. However, the film thickness and feed flow rate had significant effects on the performance of the reactor (Fig. 2.6e and f). When the film thickness increased from 1.4 to 4.2 μm, reaction rates and the rate of mineralization both increased, indicating that the reaction by-products were oxidized into CO_2 and H_2O. Moreover, the light source also affected the removal efficiency. In this case, light emitting diodes (LEDs) were a more efficient light source than sunlight and lamps, being able to illuminate a larger catalyst-coated surface.

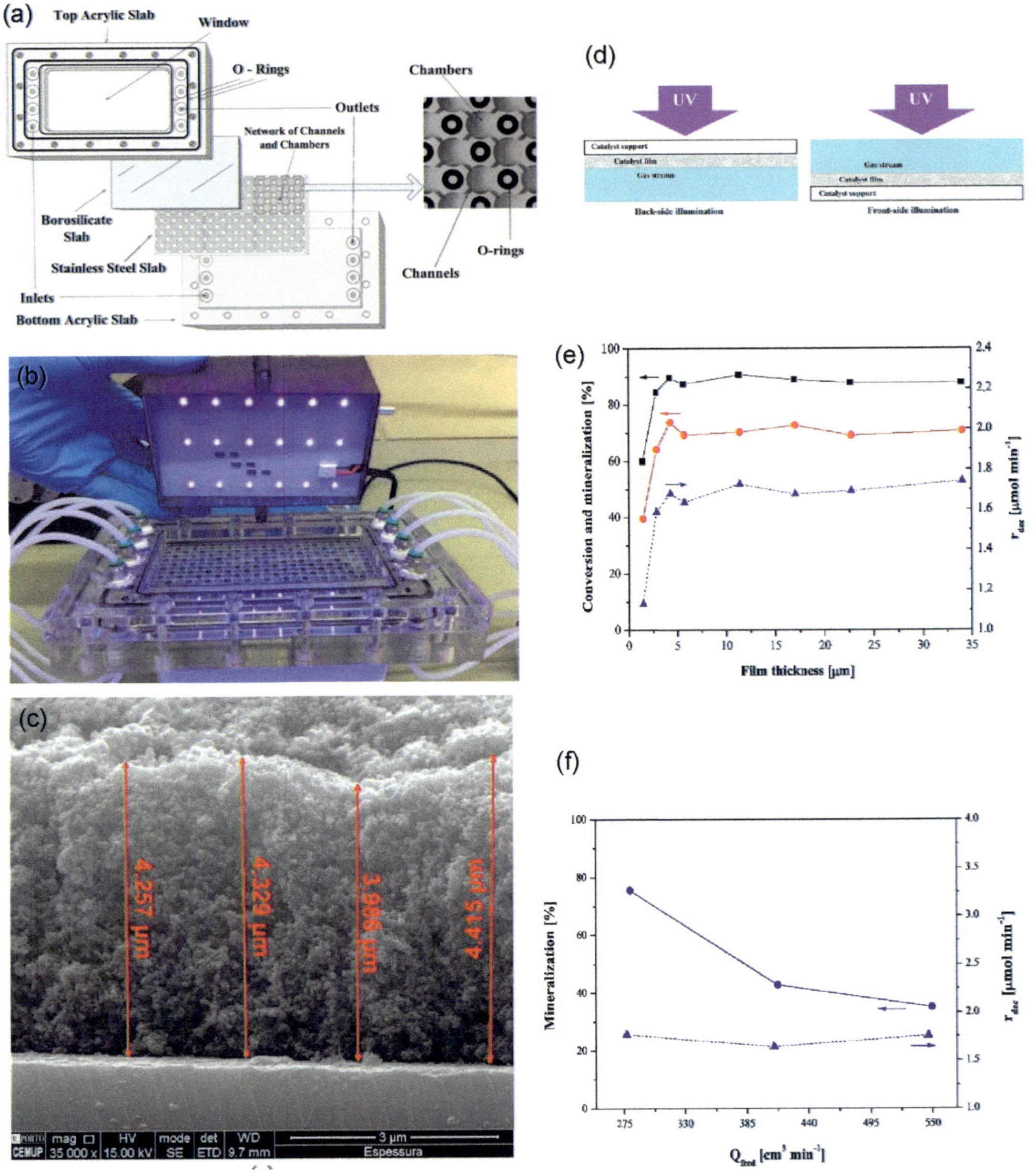

FIGURE 2.6 (a) Schematic diagram of the NETmix components, (b) stainless steel slab NETmix flat reactor, (c) SEM micrographs of the catalyst film of TiO_2 on the borosilicate slab surface (side view), (d) schematic diagram of back- and front-side illumination, and the effects of (e) catalyst film thickness and (f) feed-flow rate on the PCO of n-decane performance. *Reproduced from da Costa Filho, B. M., Araujo, A. L., Padrão, S. P., Boaventura, R. A., Dias, M. M., Lopes, J. C. B., Vilar, V. J., 2019. Effect of catalyst coated surface, illumination mechanism and light source in heterogeneous* TiO_2 *photocatalysis using a mili-photoreactor for n-decane oxidation at gas phase. Chemical Engineering Journal 366, 560–568. Copyright 2019 Elsevier.*

2.6 Summary and perspectives

Catalytic oxidation is one of the most promising technologies for VOC removal. It is more effective and economical than conventional thermal incineration techniques, especially for application in low-concentration pollutant streams (<0.5 vol%). Several catalytic oxidation technologies have been developed for VOCs from different emission sources. For example, thermal catalytic oxidation is always used for industry organic waste gas, which exhibits good removal efficiency. Sometimes, catalytic ozonation is used when the waste gas has relatively low VOC concentrations and includes some odorants. It can operate at low temperatures, even room temperature, and displays an outstanding removal performance for low-concentration VOCs. Photocatalytic oxidation is a promising technology for the purification of indoor VOCs. It can be designed into a small and portable reactor.

The mechanisms of VOC oxidation are classified as L−H, MVK, and E−R. Adsorptive and active VOCs and oxygen are the key functions of the catalyst. Generally, the supported noble metals, such as Pt- and Pd-based catalysts, are the most widely applied ones for thermal catalytic oxidation due to their high efficiency for VOC destruction at relatively low temperatures. Manganese oxide-based catalysts exhibit outstanding performance for the catalytic ozonation of VOCs because of their abundant oxygen vacancies, high specific surface area, and mesoporous structure. Titanium dioxide-based catalysts are the most extensively used ones for the photocatalytic oxidation of VOCs owing to their low cost, anticorrosion property, chemical stability, and high photocatalytic activity.

In the future, efforts should focus on the following aspects: (1) developing better catalysts to enhance the efficiency of VOC oxidation technologies and exploring these mechanisms; (2) designing and optimizing VOC oxidation reactors that can maximize the performance of catalysts and, thus, the efficiency of VOC removal in practical applications; and (3) developing coupling technologies, such as photothermal synergistic catalytic oxidation and plasma-assisted catalytic oxidation, to enhance VOC removal in a more efficient and cost-effective way.

References

Assal, Z.E., Pitkäaho, S., Ojala, S., Maache, R., Bensitel, M., Pirault-Roy, L., Brahmi, R., Keiski, R.L., 2013. Total oxidation of dichloromethane over metal oxide catalysts. Topics in Catalysis 56, 679−687.

Blanch-Raga, N., Palomares, A.E., Martínez-Triguero, J., Puche, M., Fetter, G., Bosch, P., 2014. The oxidation of trichloroethylene over different mixed oxides derived from hydrotalcites. Applied Catalysis B: Environmental 160−161, 129−134.

Chao, C.Y.H., Kwong, C.W., Hui, K.S., 2017. Potential use of a combined ozone and zeolite system for gaseous toluene elimination. Journal of Hazardous Materials 143, 118−127.

Chen, C.Y., Chen, F., Zhang, L., Pan, S.X., Bian, C.Q., Zheng, X.M., Meng, X.J., Xiao, F.S., 2015. Importance of platinum particle size for complete oxidation of toluene over Pt/ZSM-5 catalysts. Chemical Communications 15, 5936−5938.

Chung, W., Mei, D., Tu, X., Chang, M., 2019. Removal of VOCs from gas streams via plasma and catalysis. Catalysis Reviews 61 (2), 270−331.

da Costa Filho, B.M., Araujo, A.L., Padrão, S.P., Boaventura, R.A., Dias, M.M., Lopes, J.C.B., Vilar, V.J., 2019. Effect of catalyst coated surface, illumination mechanism and light source in heterogeneous TiO_2 photocatalysis using a mili-photoreactor for n-decane oxidation at gas phase. Chemical Engineering Journal 366, 560−568.

Gluhoi, A.C., Bogdanchikova, N., Nieuwenhuys, B.E., 2005. Alkali (Earth)-Doped Au/Al_2O_3 catalysts for the total oxidation of propene. Journal of Catalysis 232, 96–101.
He, C., Cheng, J., Zhang, X., Douthwaite, M., Pattisson, S., Hao, Z., 2019. Recent advances in the catalytic oxidation of volatile organic compounds: a review based on pollutant sorts and sources. Chemical Reviews 119, 4471–4568.
Huang, H., Xu, Y., Feng, Q., Leung, D.Y.C., 2015. Low temperature catalytic oxidation of volatile organic compounds: a review. Catalysis Science & Technology 5, 2649–2669.
Huang, H.B., Huang, W.J., Xu, Y., Ye, X.G., Wu, M.Y., Shao, Q.M., Ou, G.C., Peng, Z.R., Shi, J.X., Chen, J.D., Feng, Q.Y., Zan, Y.J., Huang, H.L., Hu, P., 2015. Catalytic oxidation of gaseous benzene with ozone over zeolite-supported metal oxide nanoparticles at room temperature. Catalysis Today 258, 627–633.
Isaifan, R.J., Ntais, S., Baranova, E.A., 2013. Particle size effect on catalytic activity of carbon-supported Pt nanoparticles for complete ethylene oxidation. Applied Catalysis A: General 464–465, 87–94.
Jin, M.S., Kim, J.W., Kim, J.M., Jurng, J., Bae, G.N., Jeon, J.K., Park, Y.K., 2011. Effect of calcination temperature on the oxidation of benzene with ozone at low temperature over mesoporous α-Mn_2O_3. Powder Technology 214, 458–462.
Jin, D., Ren, Z., Ma, Z., liu, F., Yang, H., 2015. Low temperature chlorobenzene catalytic oxidation over MnO_x/CNTs with the assistance of ozone. RSC Advances 5, 15103–15109.
Kamal, M., Razzak, S., Hossain, M., 2016. Catalytic oxidation of volatile organic compounds (VOCs) - a review. Atmospheric Environment 140, 117–134.
Korologos, C.A., Nikolaki, M.D., Zerva, C.N., Philippopoulos, C.J., Poulopoulos, S.G., 2012. Photocatalytic oxidation of benzene, toluene, ethylbenzene and m-xylene in the gas-phase over TiO_2-based catalysts. Journal of Photochemistry and Photobiology A 244, 24–31.
Lee, C.H., Shie, J.L., Yang, Y.T., Chang, C.Y., 2016. Photoelectrochemical characteristics, photodegradation and kinetics of metal and non-metal elements co-doped photocatalyst for pollution removal. Chemical Engineering Journal 303, 477–488.
Li., Q., Li., F.T., 2020. Recent advances in surface and interface design of photocatalysts for the degradation of volatile organic compounds. Advances in Colloid and Interface Science 284, 102275.
Li, J., Na, H., Zeng, X., Zhu, T., Liu, Z., 2014. In situ DRIFTS investigation for the oxidation of toluene by ozone over Mn/HZSM-5, Ag/HZSM-5 and Mn-Ag/HZSM-5 catalysts. Applied Surface Science 311, 690–696.
Li, Z., Luan, Y., Qu, Y., Jing, L., 2015. Modification strategies with inorganic acids for efficient photocatalysts by promoting the adsorption of O_2. ACS Applied Materials & Interfaces 7, 22727–22740.
Li, Y.F., Liu, F.F., Fan, Y., Cheng, G., Song, W., Zhou, J.L., 2018. Silver palladium bimetallic core-shell structure catalyst supported on TiO_2. Applied Surface Science 462, 207–212.
Liu, B.Y., Ji, J., Zhang, B.G., Huang, W.J., Gan, Y.L., Leung, D.Y.C., Huang, H.B., 2022. Catalytic ozonation of VOCs at low temperature: a comprehensive review. Journal of Hazardous Materials 422, 126847.
Nakamura, R., Tanaka, T., Nakato, Y., 2004. Mechanism for visible light responses in anodic photocurrents at N-doped TiO_2 film electrodes. The Journal of Physical Chemistry B 108, 10617–10620.
Okamura, K., Kobayashi, T., Tanaka, H., Niwa, M., 2003. Toluene combustion over palladium supported on various metal oxide supports. Applied Catalysis B: Environmental 44, 325–331.
Peng, R.S., Sun, X.B., Li, S.J., Chen, L.M., Fu, M.L., Wu, J.L., Ye, D.Q., 2016. Shape effect of Pt/CeO_2 catalysts on the catalytic oxidation of toluene. Chemical Engineering Journal 306, 1234–1246.
Qiu, L., Wang, Y., Li, H., Cao, G., Ouyang, F., Zhu, R., 2018. Photocatalytic oxidation of toluene on fluorine doped TiO_2/SiO_2 catalyst under simulant sunlight in a flat reactor. Catalysts 8 (12), 596.
Reed, C., Lee, Y.K., Oyama, S.T., 2006. Structure and oxidation state of silica-supported manganese oxide catalysts and reactivity for acetone oxidation with ozone. The Journal of Physical Chemistry B 110, 4207–4216.
Rezaei, E., Soltan, J., 2012. Low temperature oxidation of toluene by ozone over MnO_x/γ-alumina and MnO_x/MCM-41 catalysts. Chemical Engineering Journal 198–199, 482–490.
Rezaei, E., Soltan, J., Chen, N., Lin, J., 2013. Effect of noble metals on activity of MnO_x/γ-alumina catalyst in catalytic ozonation of toluene. Chemical Engineering Journal 214, 219–228.
Selishchev, D.S., Kolobov, N.S., Pershin, A.A., Kozlov, D.V., 2017. TiO_2 mediated photocatalytic oxidation of volatile organic compounds: formation of CO as a harmful by-product. Applied Catalysis B 200, 503–513.
Shao, J., Lin, F., Wang, Z., Liu, P., Tang, H., He, Y., Cen, K., 2020. Low temperature catalytic ozonation of toluene in flue gas over Mn-based catalysts: effect of support property and SO_2/water vapor addition. Applied Catalysis B: Environmental 266, 118662.

Sugasawa, M., Ogata, A., 2011. Effect of different combinations of metal and zeolite on ozone-assisted catalysis for toluene removal. Ozone: Science & Engineering 33, 158–163.

Tahsini, N., Yang, A.C., Streibel, V., Werghi, B., Goodman, E.D., Aitbekova, A., Bare, S.R., Li, Y.J., Abild-Pedersen, F., Cargnello, M., 2022. Colloidal platinum–copper nanocrystal alloy catalysts surpass platinum in low-temperature propene combustion. Journal of the American Chemical Society 144, 1612–1621.

Tasbihi, M., Štangar, U.L., Černigoj, U., Jirkovsky, J., Bakardjieva, S., Tušar, N.N., 2011. Photocatalytic oxidation of gaseous toluene on titania/mesoporous silica powders in a fluidized-bed reactor. Catalysis Today 161, 181–188.

Wang, J., Zhao, H.N., Song, J.F., Zhu, T.Y., Xu, W.Q., 2019. Structure-activity relationship of manganese oxide catalysts for the catalytic oxidation of (chloro)-VOCs. Catalysts 9, 726.

Xia, D., Xu, W., Wang, Y., Yang, J., Huang, Y., Hu, L., He, C., Shu, D., Leung, D.Y.C., Pang, Z., 2018. Enhanced performance and conversion pathway for catalytic ozonation of methyl mercaptan on single-atom Ag deposited three-dimensional ordered mesoporous MnO_2. Environmental Science & Technology 52, 13399–13409.

Xu, Z.M., Ren, Y.Q., Deng, X.M., Xu, M.L., Chai, W., Qian, X.F., Bian, Z.F., 2022. Recent developments on gas-phase volatile organic compounds abatement based on photocatalysis. Advanced Energy and Sustainability Research 3, 2200105.

Yao, M., Ji, Y., Wang, H., Ao, Z., Li, G., An, T., 2017. Adsorption mechanisms of typical carbonyl-containing volatile organic compounds on anatase TiO_2 (001) surface: a DFT investigation. Journal of Physical Chemistry C 121, 13717–13722.

Yuan, M., Chang, C., Chang, C., Liao, W., Tu, W., Tseng, J., Ji, D., Shie, J., Chen, Y., 2015. Ozone-catalytic oxidation for gaseous 1, 2-dichloroethane in air over Pt/Al_2O_3 catalyst. Journal of the Taiwan Institute of Chemical Engineers 53, 52–57.

Zhang, L.N., Xue, L.L., Lin, B.Y., Zhao, Q.G., Wan, S.L., Wang, Y., Jia, H.P., Xiong, H.F., 2022. Noble metal single-atom catalysts for the catalytic oxidation of volatile organic compounds. ChemSusChem 15, e202102494.

Zhang, Y.G., Wang, Y.F., Xie, R.J., Huang, H.B., Leung, M.K.H., Li, J.T., Leung, D.Y.C., 2022. Photocatalytic oxidation for volatile organic compounds elimination: from fundamental research to practical applications. Environmental Science & Technology 56, 16582–16601.

Zhao, D.Z., Ding, T.Y., Li, X.S., Liu, J.L., Shi, C., Zhu, A.M., 2012. Ozone catalytic oxidation of HCHO in air over MnO_x at room temperature. Chinese Journal of Catalysis 33, 396–401.

Zhao, J., Xi, W., Tu, C., Dai, Q., Wang, X., 2020. Catalytic oxidation of chlorinated VOCs over Ru/TixSn1-x catalysts. Applied Catalysis B: Environmental 263, 118237.

Zhao, W.C., Adeel, M., Zhang, P., Zhou, P.F., Huang, L.L., Zhao, Y.W., Ahmad, M.A., Shakoor, N., Lou, B.Z., Jiang, Y.Q., Lynch, I., Rui, Y.K., 2022. A critical review on surface-modified nanocatalyst application for the photocatalytic degradation of volatile organic compounds. Environmental Science: Nano 9, 61–80.

CHAPTER

3

Thermal oxidation of VOCs

Yidan Luo[1], Ruochen Liu[1], Shuohan Yu[2], Mingshan Xue[1] and Xianchuan Xie[3]

[1]School of Materials Science and Engineering, Nanchang Hangkong University, Nanchang, China; [2]School of Environmental and Chemical Engineering, Nanchang Hangkong University, Nanchang, China; [3]Key Laboratory of Poyang Lake Environment and Resource Utilization, Ministry of Education, School of Resources and Environment, Nanchang University, Nanchang, China

3.1 Overview of VOC thermal oxidation technologies

The thermal oxidation technology for volatile organic compounds (VOCs) is also known as VOC combustion technology. It is a treatment technology that refers to the use of direct combustion or catalytic combustion (using catalysts to accelerate chemical reactions) to reduce VOCs. The elementary composition of VOCs is carbon, hydrogen, and oxygen. Eventually, they are decomposed into H_2O and CO_2 (van der Vaart et al., 1994).

VOCs include many organic compounds, such as alkanes, alcohols, amines, ketones, halogenated hydrocarbons, aromatic hydrocarbons, esters, aldehydes, ethers, carboxylic acids, and sulfur-containing organic compounds (Zhou et al., 2020). Most of them are toxic substances in the air, causing serious air pollution. Since most of the VOCs are combustible, they can be converted into nontoxic substances, such as water and carbon dioxide, after combustion. Therefore, the treatment of VOCs through combustion is an attractive method. It is also the most economical method among all VOC abatement techniques, including biological degradation, photocatalytic oxidation, adsorption, membrane separation, and condensation, and has the widest range of applications (Zhang et al., 2017).

3.1.1 Classification of VOC thermal oxidation techniques

There are two types of VOC thermal oxidation technologies. One is thermal combustion, which can achieve 99% VOC removal efficiency (van der Vaart et al., 1994). Thermal combustion includes both direct combustion and regenerative thermal combustion. Another type is catalytic combustion, using catalysts to accelerate the VOC reaction. Catalytic combustion

Volatile Organic Compound Removal
https://doi.org/10.1016/B978-0-443-15674-8.00008-1

TABLE 3.1 Comparison of direct combustion, regenerative combustion, and catalytic combustion methods for VOC removal.

	Direct combustion	Regenerative combustion	Catalytic combustion	Ref.
Temperature (°C)	600–900	750–850	250–500	Zhang et al. (2022), Song et al. (2020)
Reaction method	High-temperature flame burning	High-temperature flame burning	Catalysis without flame	Song et al. (2020)
NO_x generation	Small	Small	Little	Song et al. (2020), Zhang et al. (2022)
Removal efficiency (%)	≥99	≥99	≥99	Liu (2017, 2022), Xie (2021)
Heat recovery rate (%)	Low	High	Medium to high	Zhang et al. (2022), Xie (2021), Song et al. (2020)
Catalyst	No	No	Yes	Wang (2021)
Concentration of VOCs	High	Medium to high	Low	Liu (2017), Xie (2021), Song et al. (2020)
Fuel cost	High	Low	Low	Liu (2017)
Device cost	Low	High	Low	Xie (2021)

requires lower temperatures and discharges fewer pollutants than that in thermal combustion (Kamal et al., 2016) and is usually used for low-concentration VOC treatment. A comparison of these methods is shown in Table 3.1.

3.1.2 Research status of VOC thermal oxidation technology

The results of a recent search on the "Web of Science" using "VOCs" and "combustion" as keywords are shown in Fig. 3.1. As one of the most widely used VOC elimination technologies, VOC combustion technology has been widely used for the elimination and treatment of VOCs. The number of published articles, thus, shows an increasing trend.

The distribution of published articles related to the thermal oxidation technology of VOCs in countries around the world is shown in Fig. 3.2, in which one point in the graph represents 10 published articles. While research on the thermal oxidation of VOCs is widely distributed around the world, China, the United States of America, and France are more active than the other countries. The wide distribution of VOC combustion technology research in the world further confirms its popularity as the mainstream VOC abatement technology.

3.1.3 Overview of advantages and disadvantages

The advantages of thermal oxidation technology for treating VOCs are obvious. It has a maximum treatment efficiency of 99% and its thermal oxidation units are inexpensive (Kamal et al., 2016). Besides, it can be applied to almost all types of VOCs. However, it is only suitable

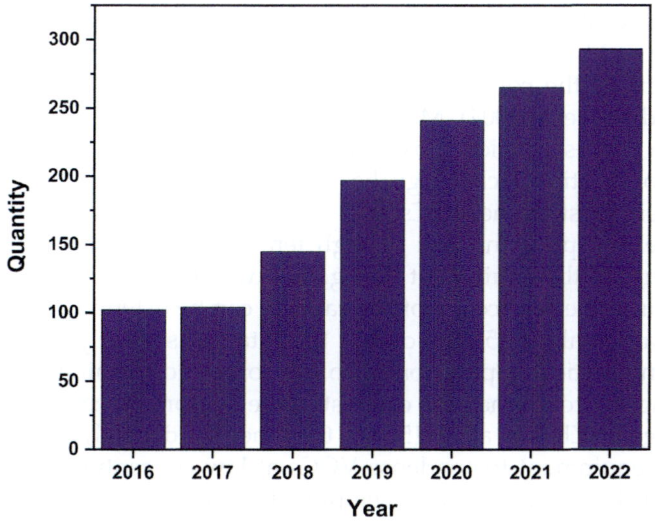

FIGURE 3.1 Number of annual publications using "VOCs" and "combustion" as keywords (2016–2022).

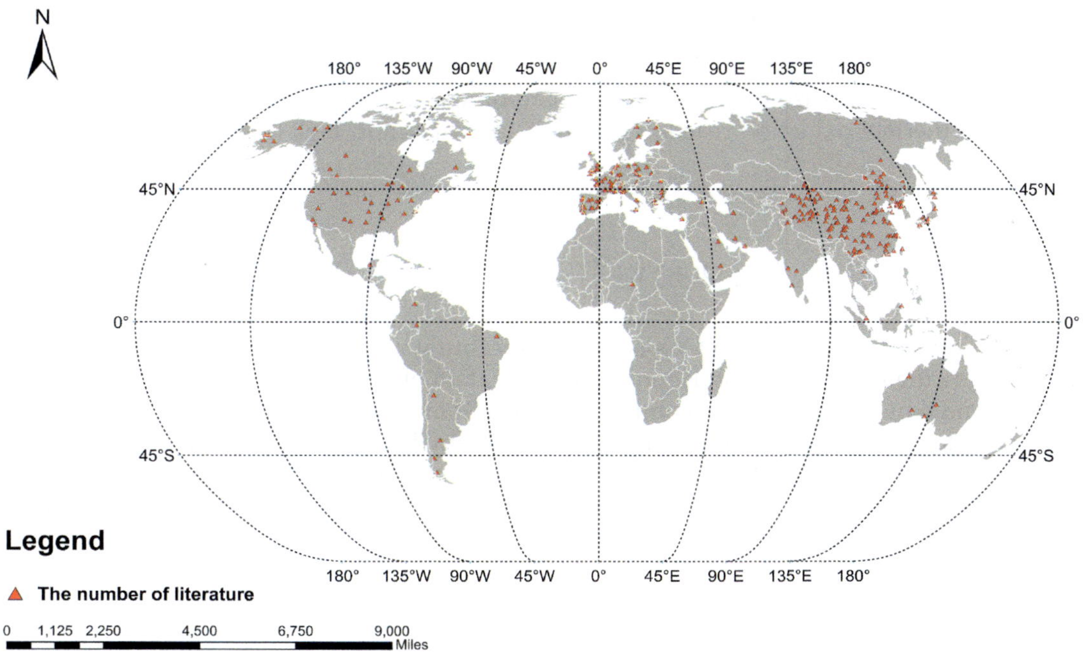

FIGURE 3.2 Distribution of the number of published articles related to the combustion technology of VOCs in the world.

for treating high concentrations of VOCs, while incomplete combustion tends to occur at low concentrations. In addition, the temperature of thermal combustion needs to exceed 800−1200°C to achieve the complete degradation of VOCs, which leads to high operating costs (Everaert and Baeyens, 2004). Moreover, carbon monoxide and NO_x may be produced in the combustion process, which causes secondary environmental pollution (Zhang et al., 2017). Catalytic combustion technology has the highest efficiency for VOC removal. This method can convert most of the VOCs into harmless substances. The catalytic combustion unit is highly safe and poses no risk of high temperature or dust explosion. In addition, the internal pressure of the equipment is negative, which can effectively avoid the problem of odor leakage. Since the service life of the catalyst can be extended by physical or chemical regeneration (Huang et al., 2015), the cost of the catalyst is greatly reduced. However, catalytic combustion in practical applications also has some shortcomings. On the one hand, it is likely for poisoning to occur when the catalyst in the equipment comes into contact with water, sulfur, or halogens (Zhao et al., 2015). On the other hand, catalytic combustion units often require additional equipment to condense VOCs at low concentrations, which increases the cost and overhead. In addition, the development of catalysts for the efficient combustion of VOCs is still in the research stage.

3.2 Thermal oxidation reactions and mechanisms

3.2.1 Combustion reaction mechanisms

During combustion removal, VOCs are heated to a high enough temperature in the presence of oxygen to oxidize them to harmless carbon dioxide and water (Van der Vaart et al., 1991). The thermal oxidation reaction equations are shown in Eqs. (3.1) and (3.2):

$$\mathrm{VOC} + \mathrm{O_2} \rightarrow \mathrm{CO_2} + \mathrm{H_2O} \tag{3.1}$$

$$\mathrm{C}_x\mathrm{H}_y\mathrm{O}_z + \left(x + \frac{y}{4} - \frac{z}{2}\right)\mathrm{O_2} \rightarrow x\mathrm{CO_2} + \frac{y}{2}\mathrm{H_2O} \tag{3.2}$$

The combustion reaction of VOCs is exothermic, but it requires a certain amount of energy to stimulate the reaction. Combustion is complete if all of the VOCs are converted to carbon dioxide and water. Incomplete combustion can result in some VOCs being left unchanged or converted to other organic compounds, such as aldehydes and acids (Stone et al., 1992). As shown in Eqs. (3.3) and (3.4), both CO and CO_2 can be combustion products (Choi and Yi, 2000). As the temperature increases, the ratio of CO to CO_2 will continue to increase. At the same time, CO will be further oxidized to CO_2:

$$\mathrm{C}_n\mathrm{H}_m + \left(\frac{n}{2} + \frac{m}{4}\right)\mathrm{O_2} \rightarrow n\mathrm{CO} + \frac{m}{2}\mathrm{H_2O} \tag{3.3}$$

$$\mathrm{CO} + \frac{1}{2}\mathrm{O_2} \rightleftharpoons \mathrm{CO_2} \tag{3.4}$$

On the other hand, the combustion mechanism of VOCs is related to their speciation (methane, toluene, acetone, ethyl acetate, etc.). For example, many studies have investigated methane combustion and have developed several mechanisms, including GRI-Mech 3.0 (Wang, 2016) and DRM19 (Graça et al., 2013). Among them, the GRI-Mech 3.0 mechanism is recognized internationally for methane combustion reactions. Because the combustion kinetics of each type of VOC is also quite complex, it is necessary to understand the combustion kinetics of every individual to understand their specific thermal oxidation mechanism. Four typical VOCs (methane, toluene, acetone, and ethyl acetate) are, thus, selected to discuss their combustion mechanisms below. Methane (CH_4) is the simplest hydrocarbon compound and a major component of gas fuel. The GRI-Mech 3.0 mechanism, developed by Smith et al., is the most famous mechanism for methane and other high-carbon fuels (such as propane and nitric oxide) (Wang, 2016). The detailed combustion mechanism contains 53 components and 325 elementary reactions. This mechanism can be applied in a wide range of applications, such as different burners and combustion environments, due to its elementary reactions being detailed and accurate. Its scope of application is as follows (Wu, 2012): (1) temperature range, 726.85–2226.85°C; (2) pressure range, 10 Torr–10 atm; and (3) equivalent ratios for premixed reaction systems, 0.1–5. Although the mechanism of GRI-Mech 3.0 is quite detailed for CH_4 combustion, it cannot be directly used to simulate the combustion of other fuels yet.

As a member of the aromatic hydrocarbon group, toluene is one of the most abundant substances in VOCs, and many of its properties are similar to those of benzene. Although the mechanism of toluene oxidation is very complex, many mechanisms have still not been investigated. The oxidation of toluene in a jet-stirred reactor (JSR) in the temperature range of 727–1102°C was studied by Dagaut et al. (2002). They developed a detailed kinetic reaction mechanism to simulate the high temperature oxidation of toluene. During combustion, toluene is decomposed to benzyl and H atom or reacts with H to form benzene and methyl. Benzyl can also be oxidized to benzaldehyde and further react to form phenyl. Benzyl thermal decomposition yields acetylene and cyclopentadienyl, while cyclopentadienyl can further produce vinylacetylene. Colket et al. performed experiments on the decomposition of toluene in a single-pulse shock tube (Colket and Seery, 1994). The experiment was conducted with a toluene concentration of 1%, a temperature range of 927–1577°C, a total pressure of about 10 atm, and a residence time of nearly 600 μs. Toluene was pyrolyzed to hydrogen and hydrocarbons, such as methane to pyrene. Cyclopentadiene and benzyl cyclopentadienyl were formed at low temperature during benzyl radical decomposition. Methylphenyl radicals played an important role in the formation of isomers of dimethyldiphenyl and anthracene. The C–C cleavage decomposition of toluene to benzene and methyl rarely happened. The results indicate the sequences of ring growth and rapid anthracene/phenanthrene isomerization.

Acetone is a representative compound of aliphatic ketones. The combustion mechanism of ketones has been heavily investigated. Tsuboi et al. observed the thermal oxidation and thermal decomposition of acetone using ultraviolet absorption and infrared emission after the reflected shock wave (Tsuboi et al., 1999). They obtained the oxidation pattern of acetone by comparing the experimental curve with the computer simulation curve. Its simple decomposition path is shown in Eq. (3.5):

$$CH_3COCH_3 \rightarrow CH_3COCH_2(CH_3CHO, CH_3) \rightarrow CH_2CO \rightarrow CH_2O \rightarrow CO \rightarrow CO_2 \quad (3.5)$$

The products of acetone combustion at low and high temperatures are different. Hoare investigated the combustion mechanism of ketones at low temperatures (Hoare and Li, 1968a). They found that the main products of acetone combustion at 330°C were CO and formaldehyde, accompanied by small amounts of CO_2, methanol, and hydrogen peroxide. In addition, similar results were obtained from the combustion of acetone at 400°C (Hoare and Li, 1968b). However, the only difference between burning acetone at high and low temperatures is that the combustion products of acetone at high temperatures can produce methane and ethylene oxide. Later, an increasing number of acetone oxidation mechanisms were developed. Sato and Hidaka experimentally investigated the oxidation and pyrolysis of acetone behind reflected shock waves and developed a mechanism consisting of 164 elementary reactions and 51 products (Sato and Hidaka, 2000). The submechanisms include the oxidation of methane, acetylene, ethylene, ethane, formaldehyde, and ketene. Capelin et al. studied the decomposition of acetone at low temperatures (Capelin et al., 1974). They found that the oxidation of acetone was a unimolecular homogeneous reaction, and the main decomposition products of acetone at 850°C were hydrogen, carbon monoxide, and methane. Ernst et al. researched the decomposition of acetone at 1076.85–1376.85°C using reflected shock waves (Ernst et al., 1976) and found that acetone combustion decomposed into CH_3COCH_2-, CH_3-, and CH_3CO-, followed by the reaction to produce carbon monoxide, methane, ethane, and formaldehyde.

As a fast-drying solvent with excellent solubility, ethyl acetate (EA) is one of the most widely used VOCs. The oxidation mechanism of EA at low temperatures has been explored. For example, Fish and Waris researched the slow combustion of EA between 250 and 450°C, and listed the detailed decomposition mechanism of EA (Fish and Waris, 1962). They found that the mechanism of EA can be divided into two categories: the first mainly occurs below 320°C, where formaldehyde and acetic acid are produced, and the second occurs above 350°C, where acetaldehyde is produced. Salooja explored the combustion reactions mechanism of EA at 200–600°C (Salooja, 1966). They described two different oxidation modes of EA (Gasnot et al., 2005). The specific steps are shown in Eqs. (3.6)–(3.9):

$$CH_3COOC_2H_5 + O_2 \rightarrow HO_2\cdot + \cdot CH_2COOC_2H_5 \tag{3.6}$$

$$\cdot CH_3COOC_2H_5 + O_2 \rightarrow CH_2O + CH_3CHO + CO + \cdot OH \tag{3.7}$$

$$CH_3COOC_2H_5 + O_2 \rightarrow HO_2\cdot + CH_3COO\cdot CHCH_3 \tag{3.8}$$

$$CH_3COO\cdot CHCH_3 + O_2 \rightarrow CH_2CO + CH_3COOH + \cdot OH \tag{3.9}$$

Recently, the high temperature oxidation mechanism of EA has also been investigated. The combustion kinetics of EA at high temperatures was systematically studied by Morsch et al. (2022). They performed productivity and sensitivity analysis for constant volume simulation at 977°C and 40 bar and carried out the conversion analyses at different fuel concentrations (1%, 20%, and 50%). It was demonstrated that most of EA was consumed by the six-membered elimination process, producing ethylene and acetic acid. Acetic acid is decomposed into ketones and water at the beginning of ignition and is consumed by the abstraction reaction of the H atom at the primary carbon atom. This subsection introduces the current research status and mechanisms of the four representative VOC molecules and summarizes their important reactions and parameters. Understanding the mechanisms of these representative molecules can improve the development and applications of VOC thermal oxidation technology. It is worth noting that the combustion mechanism of industrial VOCs is much more complex because they contain various organic molecules. Until now, the reaction mechanisms of VOC mixtures have rarely been studied.

3.2.2 Catalytic thermal oxidation reaction mechanisms

The equation of the catalytic thermal oxidation reaction for VOCs is shown in Eq. (3.10). The thermal oxidation reaction of VOCs is the oxidation reaction of VOCs with active oxygen. The addition of a catalyst can significantly reduce the activation energy required for the thermal oxidation reaction. VOCs are combusted at a lower temperature in the catalytic oxidation reaction and eventually decomposed into carbon dioxide and water (Everaert and Baeyens, 2004):

$$C_xH_y + \left(x + \frac{1}{4}y\right) O_2 \rightarrow xCO_2 + \frac{y}{2}H_2O \tag{3.10}$$

Similar to direct combustion methods, the catalytic combustion of VOCs has been widely investigated. Catalytic combustion is controlled by both the reaction kinetics and reaction equilibrium (Everaert and Baeyens, 2004). There are some catalytic combustion mechanisms that were reported, such as the Mars–van Krevelen (MVK) model, the Langmuir–Hinshelwood (L–H) model, and the Eley–Rideal (E–R) model (Behar et al., 2015).

The MVK model, also known as the redox mechanism, is one of the most accepted mechanisms for VOC oxidation (Kamal et al., 2016). It was first reported by Mars and van Krevelen in 1954. The schematic diagram of the MVK model is shown in Fig. 3.3. In this mechanism, the lattice oxygen of the catalyst is an active part. The VOCs are absorbed on the surface of the catalyst and activated by combining the lattice oxygen to produce CO_2 and H_2O and oxygen vacancies. In the second step, the oxygen molecules fill the oxygen vacancies and the active oxygen in the catalysts is replenished. According to this mechanism, the oxidation rate of VOCs can be expressed as Eq. (3.11) (Kamal et al., 2016):

$$-r_{\mathrm{voc}} = \frac{k_{O_2} \cdot k_{\mathrm{VOC}} \cdot P_{O_2} \cdot P_{\mathrm{VOC}}}{\gamma k_{\mathrm{VOC}} \cdot P_{\mathrm{VOC}} + k_{O_2} \cdot P_{O_2}} \tag{3.11}$$

where $-r_{\mathrm{VOC}}$ is the reaction rate (mol/m^3/s), k_{O_2} is the rate constant of catalyst reoxidation, k_{VOC} is the rate constant of VOC oxidation, P_{O_2} is the partial pressure of oxygen (mol/m^3), P_{VOC} is the partial pressure of the VOCs (mol/m^3), and γ is the stoichiometry coefficient of O_2 in oxidation.

The L–H mechanism is a multiphase catalytic mechanism with a surface reaction of two adsorbed molecules. The reaction is shown in Fig. 3.4. Reactants A and B are adsorbed on the surface of the solid catalyst and react there to produce an adsorbed state of A* and B* (Becker, 2018). Then, A* reacts with B* to form the final product. Finally, desorption is performed. A and B may be adsorbed on a similar type of active site (single-site L–H model) and two different types of active sites (duel-site L–H model). The kinetic expressions of L–H mechanisms are shown in Eq. (3.12) (single-site L–H model) and Eq. (3.13) (duel-site L–H model):

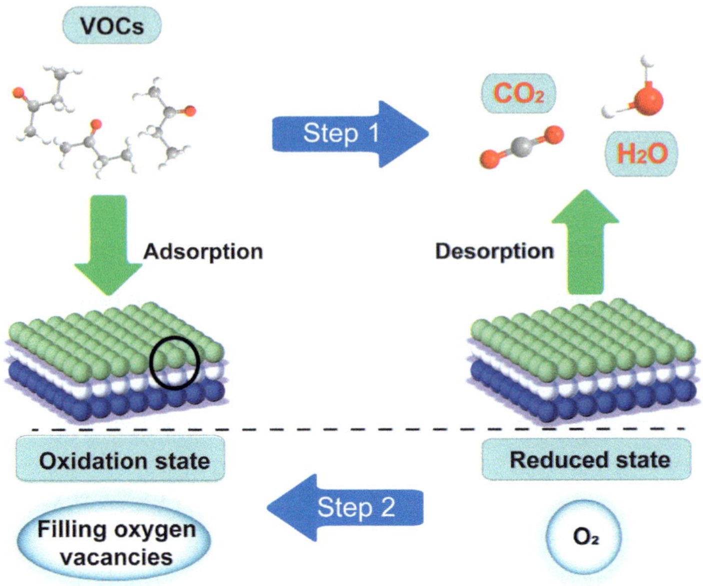

FIGURE 3.3 Schematic diagram of the MVK model.

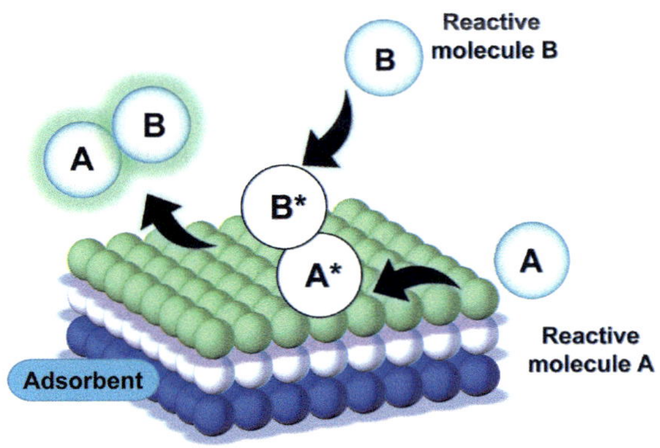

FIGURE 3.4 Schematic diagram of the L–H mechanism.

$$-r_{\text{voc}} = \frac{k \cdot K_{\text{VOC}} \cdot K_{\text{O}_2} \cdot P_{\text{VOC}} \cdot P_{\text{O}_2}}{(K_{\text{VOC}} \cdot P_{\text{O}_2} + K_{\text{VOC}} \cdot P_{\text{VOC}} + 1)^2} \tag{3.12}$$

$$-r_{\text{voc}} = \frac{k \cdot K_{\text{VOC}} \cdot K_{\text{O}_2} \cdot P_{\text{VOC}} \cdot P_{\text{O}_2}}{(K_{\text{VOC}} \cdot P_{\text{VOC}} + 1)(K_{\text{O}_2} \cdot P_{\text{O}_2} + 1)} \tag{3.13}$$

where $-r_{\text{VOC}}$ is the reaction rate (mol/m^3/s), k is the rate constant (s^{-1}) (Behar et al., 2015), K_{VOC} is the equilibrium constant for the adsorption of VOCs, K_{O_2} is the equilibrium constant

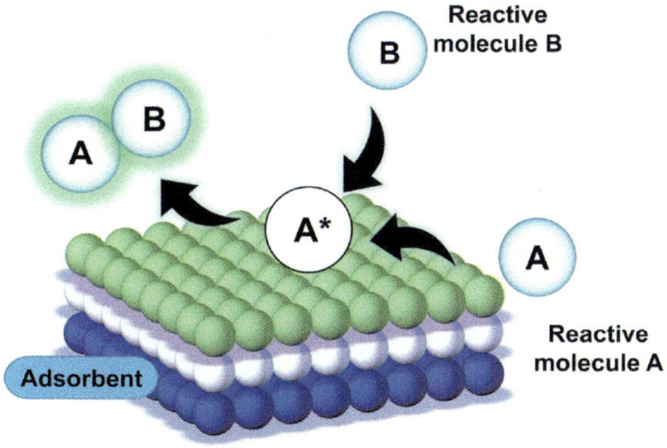

FIGURE 3.5 Schematic diagram of the E–R mechanism.

for the adsorption of oxygen, P_{VOC} is the partial pressure of the VOCs (mol/m^3), and P_{O_2} is the partial pressure of oxygen (mol/m^3) (Kamal et al., 2016).

The E–R mechanism is shown in Fig. 3.5. In this mechanism, only a single molecule is involved in the reaction process. Molecule A is first adsorbed, then the intermediate A* is formed. Substance B reacts with this intermediate A* to form the product. The following equation represents the kinetic expression based on the E–R mechanism, as shown in Eq. (3.14) (surface reaction between the adsorbed VOCs and gas-phase oxygen) and Eq. (3.15) (surface reaction between the adsorbed oxygen and gas-phase VOCs).

$$-r_{voc} = \frac{k_s \bullet K_{VOC} \bullet P_{O_2} \bullet P_{VOC}}{(K_{VOC} \bullet P_{VOC} + 1)} \tag{3.14}$$

$$-r_{voc} = \frac{k_s \bullet K_{O_2} \bullet P_{O_2} \bullet P_{VOC}}{(K_{O_2} \bullet P_{O_2} + 1)} \tag{3.15}$$

where $-r_{voc}$ is the reaction rate (mol/m^3/s); k_s is the surface reaction rate constant; K_{O_2} is the adsorption equilibrium constant of oxygen; K_{VOC} is the adsorption equilibrium constant of the VOCs; P_{O_2} and P_{VOC} are the partial pressures of the O_2 and VOCs of the reaction system, respectively (Aghazadeh et al., n.d.).

3.3 Design, characteristics, and key factors

3.3.1 VOC thermal oxidation systems

As shown in Fig. 3.6, thermal oxidizer (TO) is one of the simplest thermal oxidation devices. In the TO, VOC gas enters through inlet 1 and mixes with the air in mixing room 3. Then, the mixture is directly fed into combustion chamber 4 for oxidation to produce harmless H_2O and CO_2. Ultimately, exhaust gases can be emitted directly into the air. Additionally, the temperature of this type of combustion furnace is quite high and the temperature

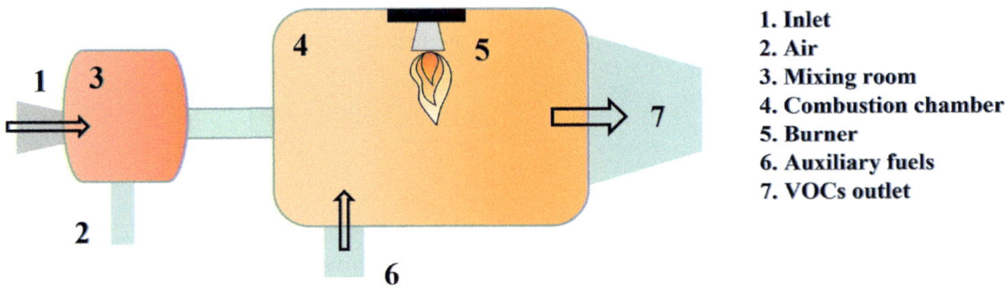

FIGURE 3.6 Schematic diagram of the TO system.

of the exhaust gas coming out of the combustion chamber can reach up to 600–900°C (Song et al., 2020). The TO equipment has a low cost and small footprint, which is suitable for high-concentration and small-volume exhaust gas treatment. In addition, the device also has high treatment efficiency, and the removal rate of high concentrations of VOCs exceeds 99%. However, when dealing with low-concentration VOCs, TO must use more fuel to maintain combustion, significantly increasing the operating costs. Moreover, TO produces substances, such as nitrogen oxides and dioxins, which cause secondary pollution to the environment. The heat generated by combustion is directly emitted into the air, leading to the loss of energy. Therefore, it is indispensable to add a heat recovery system to TO. In recent years, the application of TO in VOC combustion has become less popular.

Regenerative thermal oxidizer (RTO) uses regenerative heat transfer technology to increase the temperature of VOCs after treatment, which can effectively oxidize VOCs into H_2O and CO_2 and achieve high efficiency and low-consumption treatment. The RTO mainly consists of regenerative chambers, a combustion chamber, and a switching valve (Wang et al., 2017). Its working principle is to directly heat the VOC gas to a high temperature above 800°C and decompose it. The heat generated by the combustion is stored in the regenerative chamber. The regenerative chamber is mostly made up of ceramic material, which can keep the furnace at a consistently high operating temperature to save fuel consumption for heating.

The RTO may have two or more regenerative chambers, which discharge gas through on–off valves. The two-chamber RTO is shown in Fig. 3.7. The working principle of this system is as follows: first, regenerative chamber A is heated to 800°C. Then, VOCs are discharged into regenerative chamber A through inlet 1 for preheating. Finally, VOCs are burned in combustion chamber 3. The heat released by combustion flows into regenerative chamber B for storage. Then, the valve is changed. The VOCs are heated by passing them through chamber B, being burned in chamber 3, and then being discharged through chamber A. Each regenerative chamber experiences the process of heat storage and heat release. In this cycle, regenerative chamber A and regenerative chamber B alternately flow into the VOCs, enabling the two-chamber RTO to make full use of the oxidation heat to treat VOCs (Wang et al., 2017). The removal efficiency of the two-chamber RTO is above 95%, and the temperature difference between the inlet and outlet can reach 45°C. In addition, the pressure fluctuation range in the exhaust pipe is ±500 pa when switching its valve. The two-chamber RTO can remove VOCs stably and efficiently, which has been widely used in many industries.

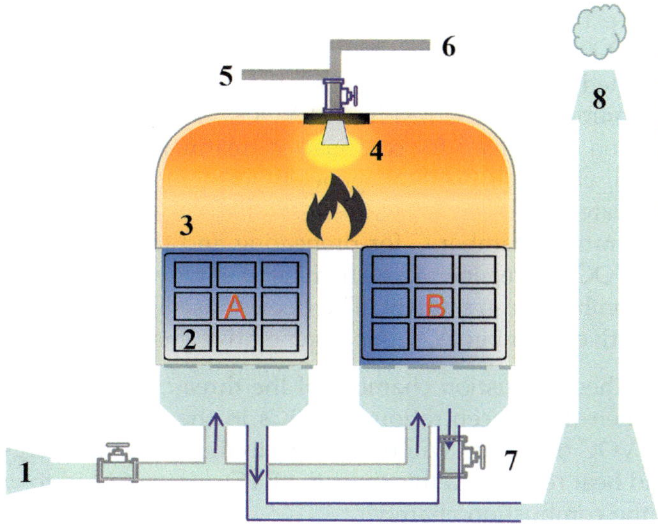

FIGURE 3.7 Schematic diagram of the two-chamber RTO system.

Three-chamber RTO is shown in Fig. 3.8. The main difference between a three-chamber RTO and a two-chamber RTO is the addition of a regenerative chamber and a blowing system (Guo et al., 2019). This device works in the following three phases:

Phase I: VOCs are preheated in regenerative chamber A and they then enter combustion chamber 3 for combustion. The residual of VOCs in regenerative chamber C is blown back to combustion chamber 3 for combustion treatment (the blowing system).

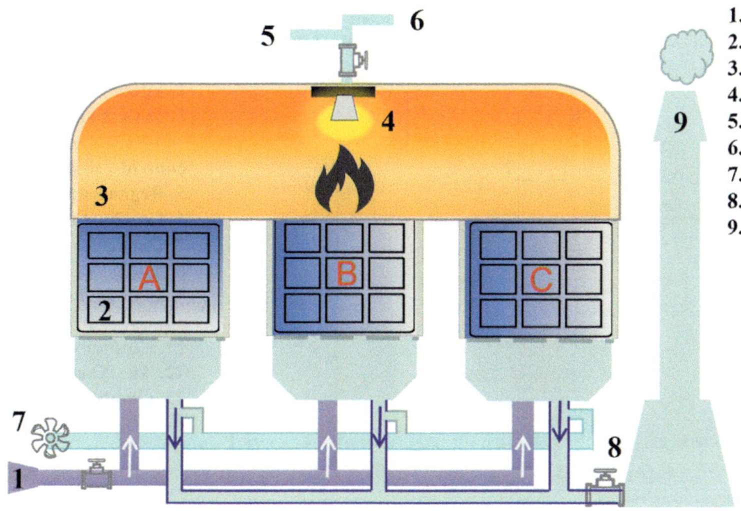

FIGURE 3.8 Schematic diagram of the three-chamber RTO system.

Eventually, the decomposed exhaust gas is discharged through regenerative chamber B, while regenerative chamber B is heated.
Phase II: VOCs enter the heated regenerative chamber B for preheating and then enter combustion chamber 3 for combustion. The residual of VOCs in regenerative chamber A is blown back to combustion chamber 3 for combustion treatment. Finally, the decomposed exhaust gas is discharged through regenerative chamber C, while regenerative chamber C is heated.
Phase III: VOCs enter regenerative chamber C for preheating and then enter combustion chamber 3. The residual of VOCs in regenerative chamber B is blown back to combustion chamber 3 for combustion treatment. The decomposed exhaust gas is discharged through regenerative chamber A, while regenerative chamber A is heated.

The internal temperature of the combustion chamber of the three-chamber RTO is maintained at a constant value. When the concentration of VOCs in the inlet reaches a certain value, the energy released by VOC oxidation will be enough to maintain the energy reserve of the RTO for heat storage and heat release. At this time, the RTO does not need to use fuel to maintain the temperature of the combustion chamber.

The rotary RTO is an advanced RTO equipment (Fig. 3.9). It has evolved from two-room to three-room, multi-room, and rotating formats. This device typically has only one combustion chamber and multiple regenerative chambers (4–12 or more). These regenerative chambers are arranged in a circle (Shuai et al., 2021). The device changes the direction of the airflow in the different zones of regenerative chamber 5 by the rotation of the rotary directional control valve 4 (Xiao, 2021). Finally, VOCs are fully oxidized in combustion chamber 6. Compared with other types of RTOs, the rotary RTO has only one moving part (rotary directional control valve). Therefore, the operation of the equipment is more reliable, and the removal rate of VOCs can reach up to more than 99%. The RTO has the characteristics of

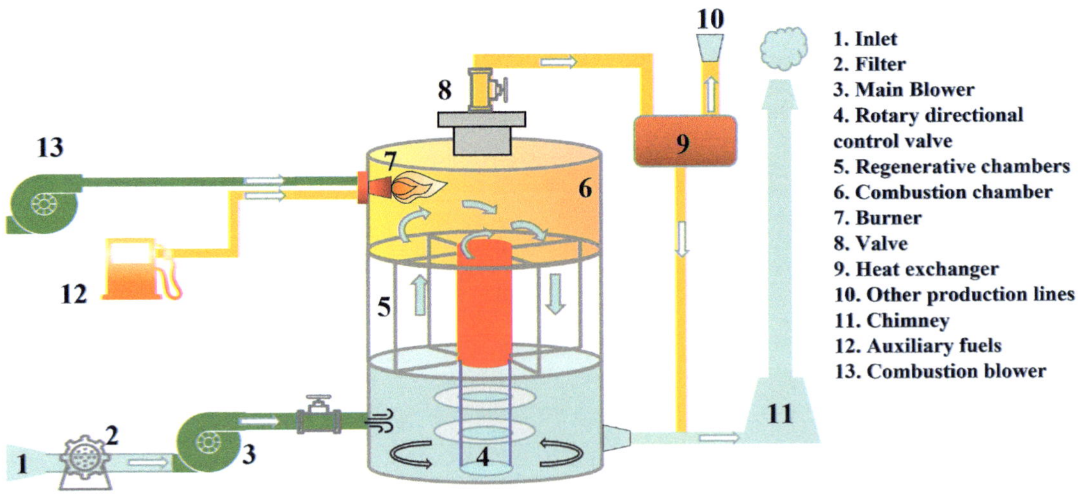

FIGURE 3.9 Schematic diagram of the rotary RTO system.

strong adaptability, high removal efficiency, low energy consumption, and high automation (Fu et al., 2017). At present, it is widely used for the treatment of VOCs in various industries.

3.3.1.1 *Regenerative catalytic oxidation*

Regenerative catalytic oxidation (Zhao et al., 2008) is carried out in the RTO system equipped with a catalytic device (such as a catalytic chamber or a catalytic bed). It mainly uses the combustion chamber to burn VOCs in the presence of a catalyst to achieve VOC treatment (Strots et al., 2000). This device is shown in Fig. 3.10, which mainly consists of regenerative chamber 3 and catalytic chamber 4. The purification principle is as follows: the unpurified gas is first preheated and then fed into catalytic chamber 4. The gas then undergoes a flameless oxidation reaction in the presence of a heater and a catalyst. After purification, the gas is finally discharged from chimney 9. The excess energy is discharged into regenerative chamber 3 for storage. Compared to the RTO, regenerative catalytic oxidization (RCO) has the following advantages:

I The adsorption of VOC molecules by the catalyst increases the concentration of the reactants.

II The catalytic oxidation of VOCs reduces the activation energy of the reaction and increases the VOC reaction rate.

III The equipment has a low temperature and low energy consumption. In some cases, it does not require external heating after reaching the reaction temperature.

RCO is an ideal device for the treatment of organic pollutants by catalytic combustion reactions (Strots et al., 2000). With the lower combustion temperature, it is more cost-effective than traditional combustion devices. RCO has the advantages of a wide application range, simple structure, high purification efficiency, energy saving, and no secondary pollution.

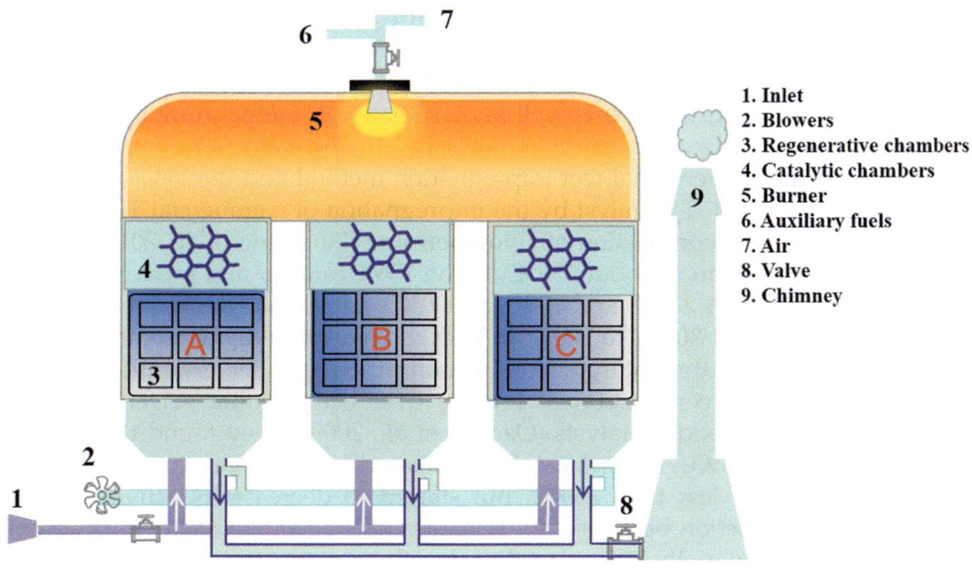

FIGURE 3.10 Schematic diagram of the RCO system.

3.3.2 Key factors affecting thermal oxidation

3.3.2.1 *Temperature*

When the combustion temperature of VOCs is low, the removal effect of VOCs will be reduced and a large amount of incomplete combustion products will be produced. Thus, the operating temperature of the thermal oxidizer is the most important factor in VOC destruction. In general, the thermal oxidation temperature is kept in the range of 760–1204°C, except for the total reduced sulfur (TRS) compounds, such as hydrogen sulfide, methyl mercaptan, and dimethyl sulfide, which can be effectively destroyed at temperatures as low as 649°C (Donley and Lewandowski, 2000). In the thermal oxidation process, the reaction temperature can affect the type of intermediate products produced from organic gases. Wang et al. conducted the low-temperature oxidation experiment of acetylene at an atmospheric pressure of over 326.85–726.85°C with an equivalence ratio $\phi = 0.5$ to 3.0 using Jet-stirred reactors (Wang et al., 2017). The reaction produced 17 intermediates, including aromatic compounds, such as toluene, styrene, and ethylbenzene. A detailed kinetic mechanism containing 295 species and 1830 reactions has been established. Benzene is formed through the $C_2 + C_4$ channel when the reaction is below 726.85°C: $C_2H_2 + iC_4H_5 \rightarrow$ fulvene $\rightarrow C_5H_5CH_2$ radicals $\rightarrow C_6H_6$. When the reaction temperature is above 726.85°C, $C_1 + C_5$ is the main pathway: $CH_3 + C_5H_5 \rightarrow C_5H_5CH_3 \rightarrow$ (fulvene and $C_5H_5CH_2$ radicals) $\rightarrow C_6H_6$. The effects of the temperature of combustion-supporting air on combustion were verified by fuel oil combustion under different operating conditions (20 m/s inlet air, 5 m/s fuel velocity, and standard atmospheric pressure at both the inlet and outlet) by Yin and Zhang (2007). It was found that when the temperature of combustion-supporting air was higher, the flame thickness was thicker and the temperature distribution was more uniform. Moreover, the air temperature of the inlet should not be too high because the NO_x concentration produced from combustion increases exponentially with the increase of the inlet temperature. The temperature of combustion-supporting air should be below 166.9°C.

Temperature is also important for catalytic combustion. Low temperatures can easily lead to the deactivation of the catalyst. It is well-known that high temperature increases the conversion rate of VOCs. For example, Finocchio et al. studied the 5% wt V_2O_5 on $WO_3–TiO_2$ (VWT) catalyst by the conventional wet impregnation method and the V_2O_5 on $MoO_3–TiO_2$ (3.5%–5.5%–91% wt) (VMoT) catalyst by the impregnation of commercial TiO_2 anatase and used them in the catalytic combustion of chlorobenzene (Finocchio et al., 2011). The conversion of chlorobenzene began at about 150°C on the VWT catalyst and increased with the reaction temperature (95% at 270°C and 100% at 300°C). Moreover, the conversion on VMoT also increased from 65% at 300°C to 100% at 350°C. However, a higher combustion temperature cannot guarantee better catalytic combustion efficiency. For example, Özçelik et al. investigated the catalytic combustion of toluene over clinoptilolite (CLT)-loaded Fe, Co, and Mn transition metal-oxide catalysts (Özçelik et al., 2009). They found that the catalytic performance of the 9.5 MnO_2/NaCLT catalyst increased with the increasing temperature at reaction temperatures of less than 350°C, but started to decrease its activity after 350°C. This is due to the interaction between the exchange ions and the zeolite. Furthermore, too high of a temperature can easily lead to catalyst deactivation due to sintering. Chen et al. prepared Ru–Ag alloy catalysts supported on composite TiO_2 (P25) for propane catalytic combustion (Chen et al., 2021). They found that the $Ru_{0.9}Ag_{0.1}/P25$ catalyst lost its activity at 800°C with an air stream containing 10% H_2O due to the collapse of its support structure and the sintering of the metal. The removal efficiency of VOCs is highly dependent on each compound. The ignition temperatures of some of the VOCs in reference are summarized in Table 3.2 (van der Vaart et al., 1994).

3.3.2.2 VOC type and concentration

The VOC type is a key factor that affects the thermal oxidation efficiency. VOCs contain numerous types of pollutants, including alkanes, lipids, aldehydes, ketones, benzene, organic chlorides, Freon series, organic ketones, acids, esters, and amines. Their physical and chemical properties and toxicity are different, leading to complex combustion mechanisms and influencing factors. Hermia and Vigneron demonstrated that the higher the molecular weight, the more difficult it is to be oxidized (Hermia and Vigneron, 1993). This is consistent with the relative destructibility of VOC molecules reported by Palazzolo and Tichenor (alcohol < aldehyde < aromatic < ketone < acetate < alkane) (Tichenor and Palazzolo, 1987).

Generally, the efficiency of VOC removal decreases with the increase of the VOC concentration in catalytic combustion, which may be due to the competition between VOCs and oxygen molecules for adsorption sites. The increased concentration of VOCs inhibits the adsorption of oxygen molecules, and the removal efficiency of VOCs is reduced. Wu et al. prepared a series of Mn_xCe_y/Al_2O_3 catalysts using the wet impregnation method for the

TABLE 3.2 Combustion points of common VOC compounds.

No.	Type	Chemical formula	Ignition temperature (°C)
1	Benzene	C_6H_6	497.8
2	Toluene	C_7H_8	535
3	Xylene	C_8H_{10}	525
4	Acetone	C_3H_6O	465
5	Methyl ethyl ketone	C_4H_8O	404
6	Dichloromethane	CH_2Cl_2	556
7	Ethyl acetate	$C_4H_8O_2$	426.7
8	Trichloroethylene	C_2HCl_3	420
9	Butadiene	C_4H_6	415
10	Propylene	C_3H_6	460
11	Cyclohexanone	$C_6H_{10}O$	420
12	Bromobenzene	C_6H_5Br	565
13	N-Hexane	C_6H_{14}	225
14	Anisole	C_7H_8O	475

catalytic combustion of chlorobenzene (Wu et al., 2010). They found that with the increase of the chlorobenzene concentration from 500 to 2000 ppm, the conversion of chlorobenzene decreased to varying degrees on different catalysts at 300°C. Especially for Mn_8Ce_2/Al_2O_3, the conversion of toluene decreased from 77% at 500 ppm to 37% at 2000 ppm.

3.3.2.3 *Residence time*

Residence time is a factor that affects the emission of VOCs. Zheng et al. carried out the thermal decomposition experiments of N,N-dimethylformamide (DMF) (Zheng et al., 2022). In these experiments, the jet-stirred reactor (JSR) was used. They found that when the residence time increased from 1.5 to 7 s at 700–900°C, the pyrolysis products N_2, C_2H_4, CH_4, H_2, and HCN of DMF increased by 8.04%, 11.31%, 6.15%, 11.17%, and 42.13%, respectively. This proves that the increase in residence time promotes the combustion of DMF. If the residence time is appropriate, the removal rate of VOCs can reach up to more than 98% (Cheng et al., 2002). The residence time determines whether the reaction is complete combustion or not. If the residence time is too short, VOCs will not be fully combusted. This is because combustion products must have enough time to react. When 99.99% or a higher VOC removal efficiency is required, a residence time of 1.0 s or more is typically required (Donley and Lewandowski, 2000). However, O'Reilly found that longer residence times may require larger-sized burners for the thermal incineration of CO and VOCs, which then increase the cost of fuel (O'Reilly, 1998). Therefore, choosing a residence time of 0.5 s or less requires the least amount of incinerator size and fuel requirements. Considering the purification effect combined with the actual operation, the residence time of VOCs is generally not less than 0.75 s in industrial applications. The specific VOC residence time in the combustion chamber can be calculated according to Eq. (3.16):

$$t_{\mathrm{VOC}} = \frac{V_{\mathrm{R}} T_{\mathrm{N}}}{V_{\mathrm{N}} T_{\mathrm{R}}} \tag{3.16}$$

where t_{VOC} is the residence time of VOCs in the combustion chamber (s), V_{R} is the volume of the combustion chamber (m^3), V_{N} is the volumetric flow rate of the exhaust gas in the standard state (Nm^3/s), T_{N} is the temperature in the standard state (°C), and T_{R} is the temperature of the combustion chamber (°C).

For catalytic combustion, the effect of the residence time is also important for the effect of the removal efficiency of VOCs. Ma et al. prepared a series of AMn_2O_x (A = Co, Cu, Ni, Zn, and Ce) spinel catalysts by the sol–gel method and used them in the catalytic combustion experiments of light alkane VOCs (Ma et al., 2023). They found that the conversion of alkanes over $NiMn_2O_4$ catalysts at the same temperature decreased continuously as the gas hourly space velocity (GHSV) increased from 30,000 $mL \cdot g^{-1}\,h^{-1}$ to 90,000 $mL \cdot g^{-1}\,h^{-1}$. This was due to the fact that a high GHSV caused the residence time of alkanes on the catalyst surface to be shorter, thus leading to a decrease in its catalytic combustion efficiency.

Moreover, the effects of the residence time on the VOC removal efficiency are also influenced by the temperature. The effect of the residence time on the VOC removal efficiency is obvious when the combustion temperature is low, while the effect of the residence time on the VOC removal efficiency is negligible when the combustion temperature is high. Therefore, if the residence time needs to be reduced, the combustion temperature needs to be increased to compensate for it. In short, the determination of the residence time requires a balance among the VOC removal efficiency, cost, and temperature.

3.3.2.4 *Turbulence*

Turbulence is an irregular flow phenomenon in a fluid and gas, which has irregularity, randomness, and diffusivity. Turbulence and combustion interact with each other during VOC combustion. On the one hand, the heat released by combustion affects the flow. On the other hand, turbulence also affects the flame by stretching, rotating, and changing the chemical transport during the process.

Turbulence is used to ensure the complete mixing of oxygen and VOC molecules (Donley and Lewandowski, 2000). Therefore, it is essential to maintain a high degree of turbulence within the combustion equipment. Turbulence is usually defined by the gas Reynolds number, which is calculated as shown in Eq. (3.17):

$$\mathrm{Re} = \frac{DV\rho}{\mu} \tag{3.17}$$

where D is the inner diameter of the oxidizer, V is the gas velocity, ρ is the gas density, and μ is the gas viscosity. The Reynolds number should be greater than 10,000 to ensure complete turbulence.

In the thermal combustion of VOCs, increasing the turbulence level during the combustion process contributes to the combustion effect of VOCs, and the location of turbulence is also important. For example, Salvador et al. designed a turbulator device to study the effect of turbulence on VOC combustion (Salvador et al., 2005). The results showed that CO emissions decreased from 1990 to 780 ppmv when the turbulator was used, which proves that increasing the turbulence intensity can accelerate the reaction rate of CO oxidation to CO_2. In addition, CO emission increased with the distance between the turbulator and the combustion chamber. When the turbulator was placed 500 mm downstream of the burner, the CO emission was the lowest. Similar findings were reported in a study about the effect of the turbulence intensity on the efficiency of ethylene combustion using a plug-flow reactor (PFR) at three different turbulent flow states (Procaccini et al., 1998). It indicates that the rate of ethane combustion is accelerated by the increase in the turbulence intensity.

3.3.2.5 *Oxygen concentration*

The amount of air required for exhaust gas combustion is composed of two components: theoretical air volume and excess air volume. Their sum determines the oxygen concentration during incineration, while the amount of excess air determines the oxygen content of the final flue gas.

An appropriate increase in the oxygen concentration has a positive effect on the combustion of VOCs and, thus, improves the combustion efficiency. Mao et al. used a small laboratory incinerator to analyze the combustion of chlorobenzene (Mao et al., 1992). They discovered that chlorobenzene could be completely oxidized at 900°C when the inlet oxygen concentration was 32.1%. In addition, as the inlet oxygen concentration increased from 19.8% to 32.1%, the destruction and removal efficiency (DRE) of chlorobenzene increased from 88.97% to 99.95%. On the other hand, the concentration of the combustion products of chlorobenzene also changed. The emission concentration of benzene and CO also decreased under the excessive oxygen ratio. However, increasing the oxygen concentration does not always promote the combustion of VOCs. Schöbel-Ostertag et al. investigated the effect of the oxygen concentration on the combustion of benzene with a heated laminar-flow reactor at an atmospheric pressure range of 576.85−686.85°C (Schöbel-Ostertag et al., 2005). When the oxygen concentration increased from 0.6% to 2.3%, the conversion of benzene decreased from 95% to 82% at high temperature (686.85°C). At medium temperature, the conversion of benzene was the highest (80%), when the oxygen concentration was 0.5%, and then decreased to 55% with the increase of the oxygen concentration (639.85°C). However, the effect of the oxygen concentration is small at low temperature (579.85°C). The toluene conversion efficiency was 38% at the oxygen concentration of 0.5% and was 30% at the oxygen concentration of 4.5%. This proves that the effect of the oxygen concentration on the combustion efficiency of VOCs is related to the temperature.

Additionally, the oxygen concentration is also important for catalytic combustion. Krishnamoorthy et al. studied the catalytic combustion of 1,2-dichlorobenzene (o-DCB) over V_2O_5/TiO_2 catalysts (Krishnamoorthy et al., 1998). They found that the conversion of o-DCB increased (12%−53%) with the increasing oxygen concentration at low oxygen concentrations below 20,000 ppmv. However, when the oxygen concentration was higher than 20,000 ppmv, the conversion of o-DCB did not increase significantly. Maldonado-Hódar also reported the same results by conducting the catalytic combustion experiments of toluene, xylene, and acetone using platinum−carbon aerogels to explore the effect of the oxygen content on combustion (Maldonado-Hódar, 2011). The results showed that the catalytic efficiency was significantly improved when the oxygen concentration increased from 5% to 10%. However, when the concentration of O_2 in the air increased from 10% to 30%, the catalytic performance was not improved.

3.3.2.6 *Water vapor*

Most industrial oxidation processes produce flue gas containing water vapor, thus the effect of water vapor on the VOC combustion process needs to be paid more attention. Water is often considered to be a toxicant for VOC catalytic combustion. Shan et al. researched the effects of water vapor on the adiabatic flame temperature, premixed flame temperature, and the burning rate of the premixed flames of methane combustion using Chemkin 17.0 (Shan et al., 2019). The premixed flame temperature, adiabatic flame temperature, and laminar premixed flame velocity of methane combustion decreased significantly after the water vapor content was increased. The burning rate of premixed flames decreased from 36.999 cm/s to 10.247 cm/s, and the adiabatic flame temperature decreased from 1916.55°C to 1547.95°C. Additionally, the chain reaction rate and the methane combustion reaction rate were also reduced to varying degrees. Mazas et al. studied the effect of water vapor on the combustion rate of the $CH_4/O_2/N_2/H_2O$ mixture (Mazas et al., 2011). The authors found that the burning velocity of methane decreased when the water vapor concentration increased, especially for highly oxygen-enriched flames.

For catalytic combustion, humidity may interfere with the adsorption of VOCs and reduce the adsorption capacity of catalysts. Kullavanijaya et al. found that the adsorption capacity of all zeolite catalysts for VOCs was significantly reduced in the presence of 0.9% water (Kullavanijaya et al., 2000). Even if they used a high silica zeolite, it did not greatly improve the adsorption capacity of VOCs. This is because water molecules compete with VOCs for adsorption sites and inhibit the catalytic activity. Lahousse et al. conducted a catalytic combustion experiment of VOCs on the γ-MnO_2 catalyst and observed the effect of water vapor on the combustion effect (Lahousse et al., 1998). The research discovered that the complete conversion temperature of EA increased by 10°C and the complete conversion temperature of hexane increased by 40°C in the presence of water vapor. This proves that the presence of water leads to a decrease in the conversion rate of VOCs. Abdullah et al. investigated the catalytic combustion of dichloromethane (DCM), trichloromethane (TCM), and trichloroethylene (TCE) over bimetallic chromium and the copper-supported H-ZSM-5 catalyst (Abdullah et al., 2006). They found that a water vapor concentration of 9000 ppm in the VOC stream decreased the conversion of DCM from 98.3% to 85.2%, TCM from 99.3% to 86.6%, and TCE from 94.2% to 88.5% and increased the yield of CO_2 from 47.5% to 68.4% at 400°C. The increase in CO_2 production was due to the inhibition of chlorine transfer by water vapor.

However, the presence of water vapor may be beneficial in some cases. Bertinchamps et al. investigated the effect of H_2O on the oxidation of chlorobenzenes on VO_x/TiO_2 and checked the effect of H_2O on the total oxidation of chlorobenzenes on VO_x/TiO_2 (Bertinchamps et al., 2006). They found that the conversion of chlorobenzene with VO_x as a catalyst increased from 29% to 44% and 42% in the presence of 1% and 2% water, respectively. This may be because water vapor can remove chlorine species from the catalyst surface, thereby improving the catalyst activity and HCl selectivity (Wang et al., 2008). However, in the case of an increase of the water concentration to 5%, the conversion of chlorobenzene decreased to 26%. The effect of water in VOC oxidation varies with the concentration of water vapor and the catalyst. The effect of water should be considered when designing the catalyst for VOC combustion.

3.3.2.7 *Catalysts' performance*

The catalyst is an indispensable key factor for catalytic combustion. The effectiveness of VOC removal by catalytic combustion is closely related to the performance of the catalyst, which determines the efficiency of the removal technology. The catalysts used for the oxidation of VOCs include noble-metal catalysts and nonnoble metal catalysts. Table 3.3 lists some of the commonly used catalysts for VOC removal and their performances.

The first group of noble-metal catalysts includes oxides or composites of Pd, Pt, Au, etc (Brummer et al., 2022). They are currently the most used catalysts in the industry due to their high catalytic performance at low temperatures, resistance to deactivation, and water resistance. The VOC removal performance of noble-metal-based catalysts varies. For example, the catalytic combustion activity of the Pd catalyst for polychlorinated benzenes ($PhCl_x$) is lower than that of the Pt catalyst (Van den Brink et al., 2000). However, for trichloroethylene and dichloromethane, the Pd catalyst has a higher activity than the Pt catalyst (González-Velasco et al., 1998). Noble-metal-based catalysts may have certain disadvantages; they are too expensive and tend to be poisoned when dealing with chlorinated VOCs (Cl—VOCs) or S-containing VOCs (S—VOCs).

The second category is nonnoble catalysts, including transition metals and their oxides, perovskite oxide catalysts, spinel-type oxide catalysts, and zeolite-loaded catalysts (Li and Luo, 2023). Transition metal elements, including Mn, Fe, Cu, Co, Ni, and Ce, have been widely studied for their low price and high stability (Li et al., 2009). Perovskite oxides are

TABLE 3.3 Examples of the catalysts used in the catalytic combustion of VOCs.

Catalyst	Support	Type of VOC	Reaction temperature (°C)	Conversion rate (%)	Ref.
Au	CeO_2	Propane	230	100	Solsona et al. (2011)
Pd	Nb_2O_5	Toluene	215	90	Rooke et al. (2015)
Pt	Carbon nanotube (CNT)	O-xylene	104	100	Joung et al. (2014)
Pt	Mesoporous Cr_2O_3	Toluene	144	90	Chen et al. (2018)
Pd–Pt	γ-Al_2O_3	Toluene	238	90	Yang et al. (2020)
Cu	CeO_x	Chlorobenzene	328	99	He et al. (2015)
Fe–Mn oxides	–	O-xylene	236	90	Mei et al. (2022)
Co_3O_4	Carbon xerogel	Ethyl acetate	245	100	Chen et al. (2013)
Mn–Co oxides	Cordierite monolith	Ethyl acetate	180	100	Zhao et al. (2022)
δ-MnO_2	–	Toluene	280	100	Wu et al. (2022)
SiO_2	$Ni_2Co_1O_x$	Toluene	298	90	Dong et al. (2022)

composite oxides with natural perovskite types formed by rare earth and transition metal oxides under certain conditions, such as $LaMnO_3$ and $LaCoO_3$. A spinel-type oxide catalyst is an important combustion catalyst with the general formula AB_2O_4, such as $ZnCo_2O_4$ and $CuCo_2O_4$. Zeolite-loaded catalysts usually consist of zeolite supports and catalytic active components (such as Zn/ZSM-5 and Fe-MFI), which have been widely used in the catalytic combustion of various VOC molecules due to their adsorption and catalytic properties. In short, nonprecious metal-based catalysts have been observed by numerous scholars because of their low price, good reducibility, good thermal stability, and strong antipoisoning ability. However, their catalytic performances may be lower than that of noble-metal catalysts.

3.3.3 Key factors of the VOC abatement device design

Device or system improvements are effective ways to improve the efficiency of VOC removal. Stone et al. researched the commercially available flare systems that can control VOC-containing waste streams (Stone et al., 1992). The flare systems can be used to control the fluctuations in the concentration, flow rate, heating value, and inert content of almost all types of VOC streams. This makes it possible to control the cost of the treatment of waste gas-containing VOCs. The flare types are categorized by the mixing method at the flare tips, including the steam-assisted, air-assisted, pressure-assisted, and nonassisted types. The flare is suitable for a batch, variable, and continuous flow vents, which can be applied in most chemical plants and refineries. Gas streams containing a high concentration of sulfur or halogenated compounds are not normally combusted due to the secondary contamination caused by the combustion products. The factors affecting flare combustion are gas flammability, autoignition temperature, heating value, density, and flame zone mixing. The minimum diameter of the flare tip is 1 inch, and the maximum size commercially available is 90 inches. The flare height is based on the thermal radiation intensity, noise, luminosity, height of the surrounding structures, and the dispersion of the exhaust gases.

On the other hand, the cost of VOC removal is also a factor that must be considered when designing the device. Reducing heat loss from the equipment is one of the simple ways to reduce cost. Dvorak et al. invented an RCO that integrates a combustion chamber, catalytic reactor, and heat exchanger into a single unit, which is shown in Fig. 3.11 (Dvorak et al., 2007). A normal RCO requires pipeline connections in the combustion chamber (catalytic chamber) and the heat exchanger. Thus, this unit minimizes energy loss in the transfer and reduces the fuel cost of the RCO. The device treated the exhaust gas volumes of up to 1600 Nm^3/h and exhaust gas concentrations of up to 5000 ppm at 250–400°C. Chen et al. designed a VOC treatment unit for the RCO in a fluidized bed (Chen et al., 2022). The RCO's prefilter component, blower, and adsorption tower are connected via a pipeline. Three activated carbon transport units are set between the adsorption and desorption towers. The desorption tower is connected to a heater, and the heater is equipped with a heat exchanger

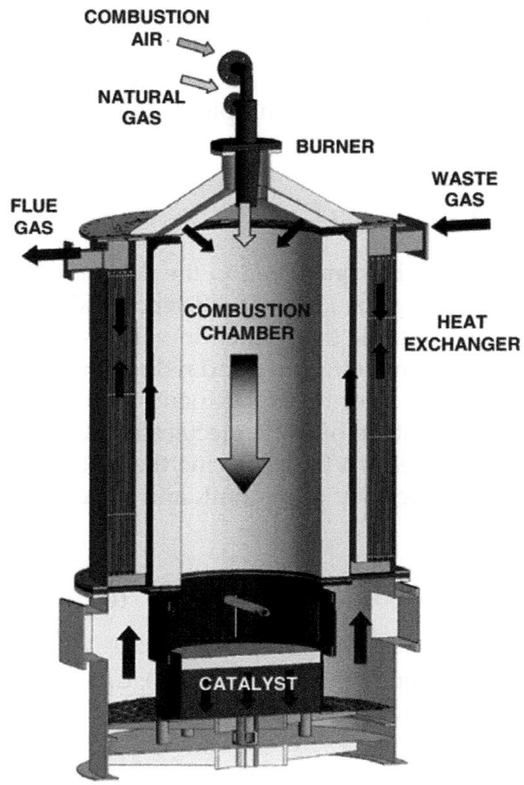

FIGURE 3.11 Schematic diagram of a new RCO reported by Dvorak et al. (Dvorak et al., applied thermal engineering, 27, 1150–1157, 2007).

with a blower between them. The common RCO can only be used to treat VOCs of small volumes because of its high cost, while this unit can be used to treat VOCs of high volumes. The equipment has a small footprint and low cost. Furthermore, it can recycle heat energy, thus reducing energy consumption and saving incineration costs.

Furthermore, for the design of the equipment, the local policy on VOC management is also one of the factors to be considered. In China, the "Technical specifications for industrial organic waste gas treatment by regenerative thermal oxidation" (HJ 1093–2020) are the national requirements for the RTO equipment. At present, the deep emission reduction of VOCs is being carried out all over the world. The design requirements for RTOs will also be further increased. This will increase the difficulty of the RTO design. According to the "Regenerative catalytic oxidizer for the treatment of industrial organic waste gas" (JB/T 13733–2019), the design requirements for the RCO are as follows:

I The operating temperature of the RCO is 250–500°C, depending on the exhaust gas composition and catalyst type.
II The air velocity of the catalytic bed of the RCO should be 10,000–40,000/h, depending on the exhaust gas composition and catalyst type.
III The temperature rise rate of the catalytic bed should be between 5 and 10°C/min.

In conclusion, industries need to think about many parameters of VOC abatement devices when designing thermal oxidation equipment. At the same time, they need to be optimized according to the policy requirements. In this way, the dual purpose of being both efficient and economical is achieved. In addition, multiple factors, such as safety and practicality, must also be considered.

3.3.4 Models of VOC abatement devices

There are numerous aspects to be considered in the design of VOC devices. Most of these design principles aim at improving the removal efficiency and reducing costs. Among them, analysis models have been a hot topic of research in various countries, and various models can be used by companies to enhance efficiency and reduce costs according to the parameters. The design of the model needs to refer to the DRE and the heat recovery efficiency (TRE).

The DRE represents the percentage of the molecules of VOCs removed in the incinerator relative to the number of molecules of VOCs entering the system. The parameter can be used to evaluate the VOC removal efficiency of the unit, and the calculation formula is shown in Eq. (3.18):

$$\mathrm{DRE} = \frac{C_{\mathrm{in}}Q_{\mathrm{in}} - C_{\mathrm{out}}Q_{\mathrm{out}}}{C_{\mathrm{in}}Q_{\mathrm{in}}} \times 100\% \tag{3.18}$$

where C_{in} is the VOC concentration in the inlet (mg/m^3); C_{out} is the VOC concentration in the outlet (mg/m^3); Q_{in} is the gas flow rate in the inlet (m^3/h); Q_{out} is the gas flow rate in the outlet (m^3/h).

The TRE is a significant parameter of the cost. The TRE equation (Eq. 3.19) can be used to evaluate the purification cost of the equipment for VOC removal:

$$\mathrm{TRE} = \frac{T_{\mathrm{com}} - T_{\mathrm{out}}}{T_{\mathrm{com}} - T_{\mathrm{in}}} \times 100\% \tag{3.19}$$

where T_{com} is the combustion chamber temperature of the RTO (°C), T_{out} is the outlet temperature (°C), and T_{in} is the inlet temperature (°C).

New technologies and models can be useful for the system optimization of the RTO. Choi et al. invented a computational fluid dynamics (CFD) technique that can be used to predict the temperature, flow field, and composition distribution of VOCs in a three-bed RTO (Choi and Yi, 2000). The simulation results of this technology can inform the design and operate the RTO. According to the simulation, a concentration of 1.0% of VOCs is sufficient to provide energy for oxidation through heat exchange by the ceramic bed. The height of the ceramic bed designed at 0.2 m is sufficient for proper operation under these conditions. A steam switching time of 5 s is optimal for the heat exchange.

Other models are also beneficial to reduce the cost of VOC removal. Warahena and Chuah introduced two models: equipment cost factor (ECF) and heat recovery factor (HRF) (Warahena and Chuah, 2009). These two models can separately evaluate the cost-effectiveness and energy efficiency of VOC systems or devices. The HRF and ECF model are shown in Eqs. (3.20) and (3.21):

$$\mathrm{ECF} = \frac{C}{\dot{V}t\beta} \tag{3.20}$$

$$\mathrm{HRF} = \dot{Q}_{\mathrm{R}}\frac{-}{\dot{V}} = \eta\dot{Q}_{\mathrm{T}}\frac{-}{\dot{V}} = \eta\dot{q} \tag{3.21}$$

where C is the cost of the equipment; $\dot{V}$ processes the airflow rate (m^3/h); t is the equipment nominal age (h); β is the usage time factor (ratio of the actual operated hours to the equipment age); η is the heat recovery efficiency; $\dot{q}$ is the specific heat rate (Watt/m^3/h). Furthermore, they also developed a model of the equipment cost relative to heat recovery (ECHR). It is shown in Eq. (3.21). The model calculates the ratio of the ECF to the HRF, which can comprehensively evaluate and compare the equipment cost of the heat recovery system:

$$\mathrm{ECHR} = \frac{\mathrm{ECF}}{\mathrm{HRF}} = \frac{C}{\eta\dot{q}\dot{V}t\beta} \tag{3.21}$$

Ultimately, the application examples show that the TO has a lower ECF while the RTO has a higher HRF and ECHR. This demonstrates that the TO has lower equipment costs while the RTO has a higher heat recovery efficiency. Consequently, it is extremely important to improve the heat recovery efficiency of the TO and reduce the equipment cost of the RTO. In conclusion, the development of these models can be applied to the analysis of equipment cost and energy consumption, which is an outstanding idea to reduce the cost of VOC thermal oxidation.

3.4 Case studies

Thermal oxidation is widely used to treat VOCs from waste gas because of its highly practical application capability. The various types of combustion devices are also widely used in various plants, including chemicals, dye, and leather. In particular, the RTO and RCO have become the main methods in industrial production today due to their unique advantages (Kamal et al., 2016). In this section, the application cases of these two devices are presented.

3.4.1 RTO

The first case is on a mica tape production line in China (Cao et al., 2016). There are mainly three processes generating exhaust gases in this line: preparation of glue, dipping in glue, and drying. Among them, the drying process generates the highest waste gas. Toluene, xylene, and acetone are the main components of exhaust gas, which have serious effects on the human body. Before the installation of the exhaust gas treatment system, the whole production line's toluene emissions reached more than 200 mg/Nm^3 and nonmethane hydrocarbons reached more than 300 mg/Nm^3. This company chose to introduce an RTO to treat the VOCs generated by the production line. Otherwise, the direct emission of exhaust gas would greatly exceed the limits according to the "Integrated emission standard of air pollutants" of China (the maximum allowable emission concentration of toluene is 40 mg/Nm^3 and the maximum allowable emission concentration of nonmethane hydrocarbons is 120 mg/Nm^3). In this production line, the VOCs are collected and sent to the RTO by the induced draft fan and then discharged at a high altitude after treatment. The inlet and outlet of the RTO are monitored. The monitoring items are toluene and nonmethane hydrocarbons. The measurement results are shown in Table 3.4 (case 1). The removal efficiency of toluene was 99.49% and 99.46%, respectively, and the removal efficiency of nonmethane hydrocarbons was 99.02% and 99.01%, respectively. The results show that the VOCs in the outlet of the RTO can fully meet the emission standards, indicating that the RTO is a very efficient VOC treatment method.

Moreover, Xiang et al. analyzed the treatment of VOCs in the pharmaceutical and chemical industry (Xiang et al., 2014). The VOCs are first alkali sprayed and then introduced into the

TABLE 3.4 VOC removal by the RTO in two industrial cases.

Test items		Toluene			Nonmethane hydrocarbons		
		RTO inlet	RTO outlet	Removal rate (%)	RTO inlet	RTO outlet	Removal rate (%)
Case 1 (mg/Nm^3)	First time	205.3	1.056	99.49	376.6	3.7	99.02
	Second time	212.8	1.153	99.46	345.2	3.4	99.01
Case 2 (mg/m^3)	First time	255	27	89.4	2267	79.6	96.5
	Second time	234	28	88.0	2028	82	96.0
	Third time	253	36.7	85.5	2269	84	96.3

Source: Cao et al., Mechanical engineer, 12, 153–155, 2016; Wang et al., Engineering, 52, 01, 36–43, 2022.

RTO for incineration. The exhaust gas after combustion is then cooled in an emergency cooling tower. Finally, it is released into the air after passing through a secondary spray of alkali and water. The inlet and outlet are equipped with exhaust gas detection devices, and the test results are shown in Table 3.4 (case 2). After RTO treatment and three tests, the maximum concentration of toluene was reduced from 253 mg/m^3 to a minimum of 36.7 mg/m^3, with a maximum removal rate of 85.5%. The maximum concentration of nonmethane hydrocarbons was reduced from 2269 mg/m^3 to a minimum of 84 mg/m^3, with a maximum removal rate of 96.3%. The emission is also far less than that required by China's "Integrated emission standard of air pollutants" (GB 16297–1996) issued in 1996.

3.4.2 RCO

A container manufacturer produces a large number of VOCs during the curing process of the paint of the container's surface, and the total exhaust gas concentration reaches about 3000 mg/m^3 (Wang, 2015). Therefore, the RCO was applied for exhaust gas treatment. Before the treatment, the catalytic bed was preheated to reach the required temperature for the catalytic reaction. The exhaust gases were passed in and decomposed by catalysis. Eventually, four measurements were taken to count the RCO removal effect. The exhaust gas concentrations at the RCO inlet and outlet are shown in Table 3.5 (case 1). The concentration of nonmethane hydrocarbons was reduced from a maximum of 2660 mg/m^3 to a minimum of 15.0 mg/m^3, and the removal rate can reach a maximum of 99.1%, indicating that RCO technology has excellent removal efficiency for VOCs.

A styrene–butadiene rubber plant (SBR) (Table 3.5, case 2) produces a total of 200,000 Nm^3/h of waste gas emissions (Zhang et al., 2022). The exhaust gas was passed through the alkali washing tower and then discharged into the air. However, the alkali often leaks out from the alkali washing tower, which pollutes the surrounding environment. Accordingly, the company introduced an RCO unit to treat the exhaust gas. After RCO treatment and several tests, the concentration of nonmethane hydrocarbons in the exhaust gas

TABLE 3.5 Removal of nonmethane hydrocarbons by RCO in two industrial cases.

	Case 1			Case 2	
Number of tests	Inlet volume (mg/m^{-3})	Outlet volume (mg/m^{-3})	Removal rate (%)	Inlet volume (Nm^3/h)	Outlet volume (mg/m^{-3})
First time	2240	23.9	98.9	176,590	6.1
Second time	1670	15.0	99.1	180,010	6.1
Third time	2660	45.3	98.3	167,600	7.2
Fourth time	2530	38.7	98.5	176,000	7.4
Fifth time	–	–		187,450	6.6
Sixth time	–	–		173,780	6.6

Source: Wang, Technology & Engineering Application, 08, 16–18, 2015; Zhang et al., Contemporary Chemical Industry, 51, 04, 980–984, 2022.

was reduced to less than 50 mg/m^3. It eliminates odor in the styrene–butadiene rubber plant and achieves the purpose of clean production and energy saving.

3.4.3 Others

The above examples demonstrate that the thermal oxidation technologies represented by the RTO and catalytic thermal oxidation are widely applied in the industry (Zhao et al., 2008). The RTO is widely used in various industries for the treatment of various VOCs in exhaust gases due to its adaptability, high treatment efficiency, low energy consumption, and low operating costs. According to the application case above, the RTO can achieve more than 99% removal of VOCs from exhaust gas. The RCO is designed to lower the oxidation temperature by modifying the RTO with a catalytic unit. As a result, the fuel consumption required for VOC combustion and the production of NO_x is reduced (Kim, 2011). The RCO has obtained the VOC removal efficiency of 96%. In conclusion, VOC thermal oxidation technology has been widely used in many industries through continuous improvement and development. Industries should consider the characteristics of their installations and their influencing factors when selecting the thermal oxidation equipment for VOCs. Additionally, we have listed more case applications of the thermal oxidation of VOCs in Table 3.6.

TABLE 3.6 Cases of VOC thermal oxidation and VOC combustion devices.

Enterprise	VOC types	Equipment	Handling air volume	Removal efficiency (%)	Ref.
Chemical enterprise	Toluene and ethanol.	RTO	10,000 m^3/h	≥98	Dai (2020)
Pharmaceutical company	Toluene; ethyl acetate; acetone; N,N-dimethylformamide; dichloromethane; tetrahydrofuran; acetonitrile; acetic acid; and nonmethane hydrocarbons.	RTO	7950 m^3/h	≥97	Fu et al. (2017)
Coating workshop	Toluene; xylene; N,N-dimethylformamide; tert-butanol; methyl acetate; ethanol; acetone; chlorobenzene; etc.	RTO	20,000 Nm^3/h	≥99	Bi (2022)
Electric vehicle coating line	Benzene, toluene, xylene, and nonmethane hydrocarbons.	RTO	31,500 m^3/h	≥99	Guo and Chen (2019)
Synthetic rubber factory	Styrene and nonmethane hydrocarbons.	RCO	239,000 m^3/h	≥97	Liu (2020)
Iron printing plant	Benzene, toluene, butyl acetate, ethylbenzene, paraxylene, styrene, O-xylene, and n-undecane.	RTO + RCO	–	≥95	Zhu et al. (2020)

3.5 Summary and perspectives

Thermal oxidation is an effective method to treat VOC pollution. At present, thermal oxidation methods with TO, RTO, and RCO systems are the mainstream VOC treatment methods in the industry. A TO has the simplest device structure and the lowest equipment cost (Xie, 2021). However, if the concentration of VOCs is low, the removal efficiency of the TO is not obvious. Furthermore, the inability to recover heat is an obvious disadvantage of the TO. Consequently, it is gradually replaced by other thermal combustion methods.

The advantages of an RTO system include high removal effectiveness, low maintenance requirements, and safe operation (Chen, 2022). It is an essential device for the thermal oxidation removal of VOCs. The RTO uses ceramic materials to store heat from hot exhaust gases, which produces better heat storage efficiency. As a result, the RTO is frequently used and selected by various industries for the treatment of waste gas. Nevertheless, the RTO equipment has high combustion temperatures and is not safe to operate. Excessive combustion temperatures also consume a large amount of fuel and increase costs (Deng et al., 2014). In addition, the RTO must use high-temperature resistant materials, which undoubtedly increases the difficulty and cost of production (Brummer et al., 2022).

RCO can effectively reduce or even eliminate NO_x generation by reducing the combustion temperature of VOCs with the use of catalysts (Song et al., 2020). As a widely used thermal oxidation equipment, RCO has advantages, such as the high treatment effect, low fuel consumption cost, and high automation. In recent years, it has been used in industries such as printing and chemical plant production, and it is suitable for treating medium and low concentrations of VOCs (Han, 2022). However, the choice of catalyst in RCO is a factor that needs careful consideration. In addition, catalyst poisoning during the reaction needs to be noted. There are many ways to improve the thermal oxidation of VOCs, including device upgrade, technology combination, and catalyst improvement (Fig. 3.12), which are outlined below.

For thermal oxidation technology, improving device performance and developing new devices are effective ways to upgrade the device. Many researchers from various countries have developed a series of ways for device improvement. For example, the efficiency of VOC combustion is improved by introducing a turbulator from the perspective of increasing turbulence levels (Salvador et al., 2005). It is also a good choice to develop a new burner, such as a torch system, to control the combustion temperature of VOCs (Stone et al., 1992). When the combustion temperature is low, the reaction temperature can be reached without a lot of fuel consumption and the operation is safer. Improving the heat recovery capacity of the device to increase fuel consumption is an important way. The best way to reduce heat loss is to design a new type of combustion furnace (Chen et al., 2022) or to change the distribution of components in traditional combustion furnaces (Dvorak et al., 2007) which can reduce the fuel cost. Therefore, it is necessary to upgrade equipment to improve efficiency, control temperature, and save fuel costs.

Thermal oxidation technology has some limitations on some types of VOCs. An effective way to improve the efficiency of all VOC treatment technologies is to combine thermal oxidation technology with other technologies. Photothermal catalysis is a new type of degradation method combining thermal catalysis and photocatalysis (Wang et al., 2017). This method solves the high energy consumption problem of thermal catalysis and the inefficiency problem

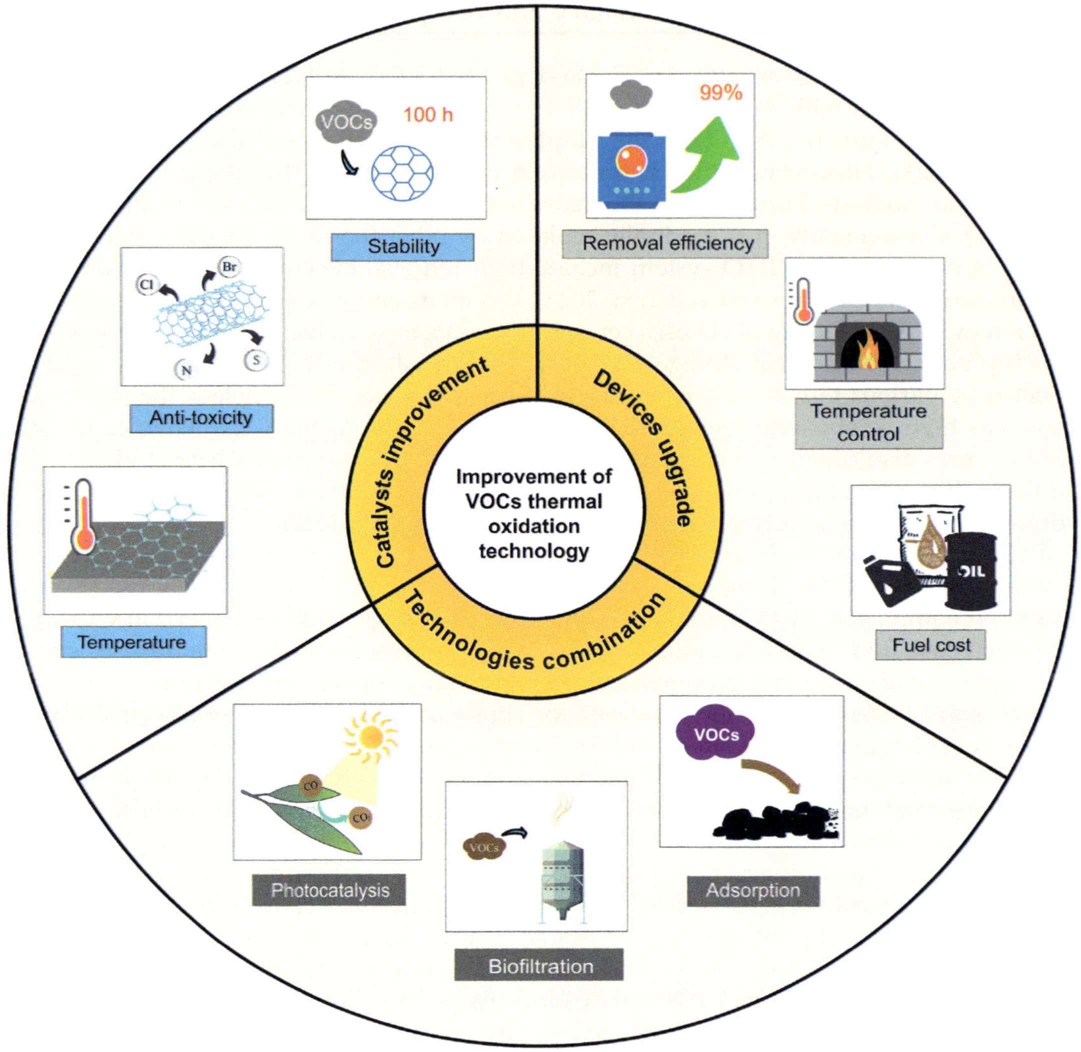

FIGURE 3.12 Improvement strategies of VOC thermal oxidation technology.

of photocatalysis and, thus, is a research hotspot in the field of catalysis (Li et al., 2020). The infrared photothermal catalysis of toluene by noble metal-free $CeO_2/LaMnO_3$ hybrid materials was carried out by Li et al. (2019). They found that toluene conversion reached 89% and the CO_2 yield reached 87% under an infrared (IR) irradiation intensity of 280 mW/cm^2. Furthermore, the catalysts had an excellent stability of about 30 h. All this proves that photothermal catalysis is a promising way for the treatment of VOCs.

As a new type of technology, biological methods have the advantages of convenient operation and low cost of investment and operation, which are cost-effective and environmentally friendly VOC removal technologies. Leethochawalit et al. studied the possibilities of linking catalytic and biofiltration technologies together by the oxidation of 2-propanol (IPA) with catalytic combustion and biofiltration, separately (Leethochawalit et al., 2001). They prepared Zn, Cu, and Cr oxide catalysts by the sol–gel method and carried out the catalytic combustion of IPA in the temperature range of 25–165°C. In addition, the distribution and concentration of compounds at the catalytic combustion outlet are consistent with the inlet conditions of the biofilter. This makes it possible to remove VOCs by biofiltration–catalytic combustion systems. Furthermore, the by-products of catalytic combustion are often toxic, which also provides opportunities for posttreatment using biofilters. Accordingly, the combination of the two technologies is possible and creates synergy.

Another effective method has been developed by combining activated carbon adsorption and catalytic combustion (Wang et al., 2009). This method first uses activated carbon to adsorb VOCs and then introduces the exhaust gas into the combustion equipment for combustion treatment. The desorption concentration of VOCs in the adsorbents can be condensed up to 10 times, which reduces the consumption of fuels in the small concentration of VOCs. Currently, this method has been widely used in RCO units. Chen used an activated carbon adsorption–RCO catalytic oxidation unit to purify VOCs on a spray-painting production line (Chen, 2019). The results showed that the final discharged concentration of VOCs using the device was reduced significantly from 265 mg/m^3 to 18 mg/m^3. All the above examples confirm that combining various technologies with the thermal oxidation of VOCs is one of the future directions of VOC treatment research.

The catalytic combustion of VOCs is currently the most studied method, in which the performance of the catalyst is crucial. Catalytic combustion requires a lower temperature than that in thermal combustion, but its reaction temperature is usually higher than 200°C. At this temperature, the reaction has a possibility of producing NO_x and the catalyst tends to sinter (Huang et al., 2015). Consequently, the catalysts for reducing the reaction temperature have been investigated. Álvarez-Galván et al. prepared an alumina-supported manganese–palladium oxide catalyst for VOC catalytic combustion (Álvarez-Galván et al., 2004). It was shown that the complete combustion of formaldehyde/methanol on the Mn/Al_2O_3 catalyst with manganese loading of 18.2 wt% was observed at 220°C. However, when 0.1% Pd was added to the catalyst, this process achieved the same result at only 90°C. Luo et al. prepared Cu–Mn oxide catalysts with tunnel, layer, and transition structures by the one-step hydrothermal method (Luo et al., 2019). The results showed that the catalyst had effective toluene removal at low temperatures. In particular, the $Cu–Mn_4$ catalyst achieved 90% of toluene conversion at 169°C. The low temperature catalyst should, thus, be studied more widely in the future.

Since VOCs may contain S, Cl, and Br elements, the antitoxicity of the catalyst is also one of the research directions. Li et al. used a homogeneous precipitation method to add the Sn element to $MnO_x–TiO_2$ to form a $SnO_x–MnO_x–TiO_2$ catalyst (Li et al., 2014). They discovered that both $MnO_x–TiO_2$ and $SnO_x–MnO_x–TiO_2$ catalysts exhibited outstanding activity in the catalytic combustion of chlorobenzene with 90% of chlorobenzene conversion below 180°C. The Sn–Mn–Ti catalyst had higher stability because the introduction of Sn can inhibit the formation of MnO_xCl_y and reduce the energy required to desorb chlorine species. S-containing VOCs are usually generated in industrial activities, which are harmful to human health (Guo et al., 2021). Therefore, the S resistance of the catalyst also needs further improvement. Cellier et al. reported the catalytic combustion experiment of methanethiol on $MnSO_4 \cdot H_2O$ and compared it with the activity of the traditional MnO_2 catalyst (Cellier et al., 2004). They found that $MnSO_4$ can achieve 100% conversion of methanethiol at 150°C, while MnO_2 needs to be at over 250°C. This indicates that the $MnSO_4$ catalyst has higher activity and S resistance in the catalytic combustion of methanethiol than the traditional MnO_2 catalyst.

If the catalyst has low stability, the catalyst will be deactivated in a short time, greatly reducing the removal efficiency of VOCs. Therefore, improving the stability of the catalyst is an important research area. Pt nanoparticles were embedded in the ZMS-5 monolayer through the electrostatic adsorption of ammonium ions in a diquaternary ammonium surfactant and calcinated to form sandwich-structured Pt@ZSM-5 (Pt@PZN-2) nanosheets (Liu et al., 2019). Eventually, a sandwich-structured Pt@ZSM-5 catalyst was prepared and used in VOC catalytic combustion experiments. It was found that Pt@PZN-2 still had 100% toluene conversion at 230°C for more than 360 h, proving its high metal stability. Zhang et al. prepared silicotungstic acid-modified CeO_2 catalysts using four cerium salts (Zhang et al., 2021) Among them, the CeO_2 catalyst prepared by $Ce(CH_3COO)_3$ (Cat-A) was certificated to have the best catalytic performance in the catalytic combustion of chlorobenzene. Additionally, Cat-A can continuously catalyze the combustion of chlorobenzene for 100 h at 235 and 295°C, demonstrating high stability (Becker, 2018).

References

Abdullah, A.Z., Bakar, M.Z., Bhatia, S., 2006. Combustion of chlorinated volatile organic compounds (VOCs) using bimetallic chromium-copper supported on modified H-ZSM-5 catalyst. Journal of Hazardous Materials 129 (1–3), 39–49.

Álvarez-Galván, M.C., Pawelec, B., de la Peña O'Shea, V.A., Fierro, J.L.G., Arias, P.L., 2004. Formaldehyde/methanol combustion on alumina-supported manganese-palladium oxide catalyst. Applied Catalysis B: Environmental 51 (2), 83–91.

Aghazadeh, F., Omidifar, D., & Niaei, A., n.d. Oxidation of 2-Propanol at Low Temperature over Pt/γ-Al_2O_3 Catalyst, pp. 375–378.

Becker, C., 2018. From Langmuir to Ertl: the "nobel" history of the surface science approach to heterogeneous catalysis. In: Encyclopedia of Interfacial Chemistry, pp. 99–106.

Behar, S., Gómez-Mendoza, N.-A., Gómez-García, M.-Á., Świerczyński, D., Quignard, F., Tanchoux, N., 2015. Study and modelling of kinetics of the oxidation of VOC catalyzed by nanosized Cu–Mn spinels prepared via an alginate route. Applied Catalysis A: General 504, 203–210.

Bertinchamps, F., Attianese, A., Mestdagh, M.M., Gaigneaux, E.M., 2006. Catalysts for chlorinated VOC abatement: multiple effects of water on the activity of VO_x based catalysts for the combustion of chlorobenzene. Catalysis Today 112 (1), 165–168.

Bi, D.-W., 2022. Application of three compartment RTO in fine waste gas treatment. Chemical Engineering Management Business School for Chemical Entrepreneur (10), 58–61 (in China).

Brummer, V., Teng, S.Y., Jecha, D., Skryja, P., Vavrcikova, V., Stehlik, P., 2022. Contribution to cleaner production from the point of view of VOC emissions abatement: a review. Journal of Cleaner Production 361, 132112–132133.

Cao, J.-P., Hou, J.-Z., Peng, Z.-W., ShangGuan, J.-H., 2016. Application study of regenerative thermal oxidizer in VOC treatment. Mechanical Engineer 12, 153–155 (in China).

Capelin, B.C., Ingram, G., Kokolis, J., 1974. The pyrolytic identification of organic molecules: II. A quantitative evaluation. Microchemical Journal 19 (3), 229–252.

Cellier, C., Gaigneaux, E.M., Grange, P., 2004. Sulfur resistance and high activity of hydrated manganese sulfate in the catalytic oxidation of methanethiol. Journal of Catalysis 222 (1), 255–259.

Chen, C.-F., 2022. Distribution and composition characteristics of volatile organic compounds in petrochemical enterprises based on navigation observation. Shanxi Chemical Industry 42 (03), 260–262 (in China).

Chen, J., Lv, X.-L., Xu, W.-J., Li, X.-L., Chen, J., Jia, H.-P., 2021. Utilizing Cl coordination to facilitate Ru–Ag self-assembling into alloy and recover thermally-inactivated catalyst for propane combustion. Applied Catalysis B: Environmental 290, 119989–120012.

Chen, L., 2019. Application of activated carbon adsorption and concentration-RCO catalytic oxidation device in a painting production line exhaust gas purification system. Modern Mining 35 (06), 255–257 (in China).

Chen, T., Tang, Y., Huang, J., 2022. Volatile organic compounds (VOC) processing device of regenerative catalytic oxidization (RCO) for fluidized bed, has heat exchanger that is fixed with chimney, where pipeline of chimney and heat exchanger is provided with fan. China, 202210262451.7 [P].

Chen, X., Carabineiro, S.A.C., Bastos, S.S.T., Tavares, P.B., Órfão, J.J.M., Pereira, M.F.R., Figueiredo, J.L., 2013. Exotemplated copper, cobalt, iron, lanthanum and nickel oxides for catalytic oxidation of ethyl acetate. Journal of Environmental Chemical Engineering 1 (4), 795–804.

Chen, X., Chen, X., Cai, S.-C., Chen, J., Xu, W.-J., Jia, H.-P., Chen, J., 2018. Catalytic combustion of toluene over mesoporous Cr_2O_3-supported platinum catalysts prepared by in situ pyrolysis of MOFs. Chemical Engineering Journal 334, 768–779.

Cheng, W.-H., Chou, M.-S., Lee, W.-S., Huang, B.-J., 2002. Applications of low-temperature regenerative thermal oxidizers to treat volatile organic compounds. Journal of Environmental Engineering 128, 313–319.

Choi, B.S., Yi, J., 2000. Simulation and optimization on the regenerative thermal oxidation of volatile organic compounds. Chemical Engineering Journal 76 (2), 103–114.

Colket, M.B., Seery, D.J., 1994. Reaction mechanisms for toluene pyrolysis. Symposium (International) on Combustion 25 (1), 883–891.

Dagaut, P., Pengloan, G., Ristori, A., 2002. Oxidation, ignition and combustion of toluene: experimental and detailed chemical kinetic modeling. Physical Chemistry Chemical Physics 4, 1846–1854.

Dai, Q., 2020. Problems analysis and design optimization of two-chamber regenerative oxidation furnace in VOC treatment. Shanghai Chemical Industry 45 (04), 25–28 (in China).

Deng, R., Zhou, F., Tang, J., Hu, Y.-X., Chu, Y.-H., 2014. Emission and control measures and suggestions of VOC. Materials Reports 28 (S2), 378–384 (in China).

Dong, F., Han, W., Han, W., Tang, Z., 2022. Assembling core-shell SiO_2@NiaCobO_x nanotube decorated by hierarchical NiCo-Phyllisilicate ultrathin nanosheets for highly efficient catalytic combustion of VOCs. Applied Catalysis B: Environmental 315, 121524–121537.

Donley, E., Lewandowski, D., 2000. Optimized design and operating parameters for minimizing emissions during VOC thermal oxidation. Metal Finishing 98 (6), 446–458.

Dvorak, R., Stulir, R., Cagas, P., 2007. Efficient fully controlled up-to-date equipment for catalytic treatment of waste gases. Applied Thermal Engineering 27 (7), 1150–1157.

Ernst, J., Spindler, K., Wagner, H.G., 1976. Untersuchungen zum thermischen Zerfall von Acetaldehyd und Aceton. Berichte der Bunsen-Gesellschaft für Physikalische Chemie 80 (7), 645–650.

Everaert, K., Baeyens, J., 2004. Catalytic combustion of volatile organic compounds. Journal of Hazardous Materials 109 (1), 113–139.

Finocchio, E., Ramis, G., Busca, G., 2011. A study on catalytic combustion of chlorobenzenes. Catalysis Today 169 (1), 3–9.

Fish, A., Waris, A., 1962. Gaseous oxidation of aliphatic esters. Part I. Slow combustion of ethyl acetate. Journal of the Chemical Society (Resumed) 4513–4517.

Fu, S.-Q., Fang, X.-B., Zhu, J.-Q., 2017. Analysis on application of RTO. Environmental Science and Management 42 (09), 132–136.

Gasnot, L., Decottignies, V., Pauwels, J.F., 2005. Kinetics modelling of ethyl acetate oxidation in flame conditions. Fuel 84 (5), 505–518.

González-Velasco, J.R., Aranzabal, A., Gutiérrez-Ortiz, J.I., López-Fonseca, R., Gutiérrez-Ortiz, M.A., 1998. Activity and product distribution of alumina supported platinum and palladium catalysts in the gas-phase oxidative decomposition of chlorinated hydrocarbons. Applied Catalysis B: Environmental 19 (3), 189–197.

Graça, M., Duarte, A., Coelho, P.J., Costa, M., 2013. Numerical simulation of a reversed flow small-scale combustor. Fuel Processing Technology 107, 126–137.

Guo, L.-H., Chen, C.-Y., 2019. Brief discussion on VOC treatment in electric vehicle coating line. Modern Finishing 22 (04), 57–58+67 (in China).

Guo, Y.-L., Wen, M.-C., Li, G.-Y., An, T.-C., 2021. Recent advances in VOC elimination by catalytic oxidation technology onto various nanoparticles catalysts: a critical review. Applied Catalysis B: Environmental 281, 119447–119485.

Guo, Y., Wang, W., Li, M., 2019. Innovative RTO for VOC treatment. Coating and Protection 40 (07), 40–45 (in China).

Han, D.-D., 2022. Hazard and treatment technology analysis of industrial VOC. Leather Manufacture and Environmental Technology 3 (07), 6–8 (in China).

He, C., Xu, B.-T., Shi, J.-W., Qiao, N.-L., Hao, Z.-P., Zhao, J.-L., 2015. Catalytic destruction of chlorobenzene over mesoporous $ACeO_x$ (A = Co, Cu, Fe, Mn, or Zr) composites prepared by inorganic metal precursor spontaneous precipitation. Fuel Processing Technology 130, 179–187.

Hermia, J., Vigneron, S., 1993. Catalytic incineration for odour abatement and VOC destruction. Catalysis Today 17 (1), 349–358.

Hoare, D.E., Li, T.-M., 1968a. The combustion of simple ketones I—mechanism at 'low' temperatures. Combustion and Flame 12 (2), 136–144.

Hoare, D.E., Li, T.-M., 1968b. The combustion of simple ketones II—mechanism at 'high' temperatures. Combustion and Flame 12 (2), 145–154.

Huang, H.-B., Xu, Y., Feng, Q.-Y., Leung, D.Y., 2015. Low temperature catalytic oxidation of volatile organic compounds: a review. Catalysis Science and Technology 5 (5), 2649–2669.

Joung, H.-J., Kim, J.-H., Oh, J.-S., You, D.-W., Park, H.-O., Jung, K.-W., 2014. Catalytic oxidation of VOCs over CNT-supported platinum nanoparticles. Applied Surface Science 290, 267–273.

Kamal, M.S., Razzak, S.A., Hossain, M.M., 2016. Catalytic oxidation of volatile organic compounds (VOCs) – a review. Atmospheric Environment 140, 117–134.

Kim, B.R., 2011. VOC emissions from automotive painting and their control: a review. Environmental Engineering Research 16, 1–9.

Krishnamoorthy, S., Baker, J.P., Amiridis, M.D., 1998. Catalytic oxidation of 1,2-dichlorobenzene over V_2O_5/TiO_2-based catalysts. Catalysis Today 40 (1), 39–46.

Kullavanijaya, E., Trimm, D.L., Cant, N.W., 2000. Adsocat: adsorption/catalytic combustion for VOC and odour control. In: Corma, A., Melo, F.V., Mendioroz, S., Fierro, J.L.G. (Eds.), Studies in Surface Science and Catalysis. Elsevier, pp. 569–574.

Lahousse, C., Bernier, A., Grange, P., Delmon, B., Papaefthimiou, P., Ioannides, T., Verykios, X.E., 1998. Evaluation of γ-MnO_2 as a VOC removal catalyst: comparison with a noble metal catalyst. Journal of Catalysis 178, 214–225.

Leethochawalit, M., Bustard, M., Wright, P., Meeyoo, V., 2001. Novel vapor-phase biofiltration and catalytic combustion of volatile organic compounds. Industrial & Engineering Chemistry Research 40, 5334–5341.

Li, J.-J., Yu, E.-Q., Cai, S.-C., Chen, X., Chen, J., Jia, H.-P., Xu, Y.-J., 2019. Noble metal free, $CeO_2/LaMnO_3$ hybrid achieving efficient photo-thermal catalytic decomposition of volatile organic compounds under IR light. Applied Catalysis B: Environmental 240, 141–152.

Li, J.-W., Zhao, P., Liu, S.-T., 2014. SnO_x–MnO_x–TiO_2 catalysts with high resistance to chlorine poisoning for low-temperature chlorobenzene oxidation. Applied Catalysis A: General 482, 363–369.

Li, K., Luo, X.-Q., 2023. Research progress on catalytic combustion of volatile organic compounds in industrial waste gas. Catalysts 13 (2), 268–288.

Li, W.-B., Wang, J.-X., Gong, H., 2009. Catalytic combustion of VOCs on non-noble metal catalysts. Catalysis Today 148 (1), 81–87.

Li, Y.-Z., Wu, S.-W., Wu, J.-C., Hu, Q.-Q., Zhou, C.-Y., 2020. Photothermocatalysis for efficient abatement of CO and VOCs. Journal of Materials Chemistry A 8 (17), 8171–8194.

Liu, C.-Y., 2017. Technology and equipment for thermal oxidation of industrial VOCs. Sichuan Chemical Industry 20 (02), 42–44 (in China).

Liu, G.-Z., Tian, Y.-J., Zhang, B.-F., Wang, L., Zhang, X.-W., 2019. Catalytic combustion of VOC on sandwich-structured Pt@ZSM-5 nanosheets prepared by controllable intercalation. Journal of Hazardous Materials 367, 568–576.

Liu, S.-B., 2020. Introduction and hazard analysis of the three bed type waste gas treatment process in butadiene styrene rubber unit. Qilu Petrochemical Technology 48 (04), 289–293 (in China).

Liu, X.-Y., 2022. Application research on RCO in waste gas treatment of high pressure polyethylene. China Coal 48 (08), 82–86.

Luo, M.-M., Cheng, Y., Peng, X.-Z., Pan, W., 2019. Copper modified manganese oxide with tunnel structure as efficient catalyst for low-temperature catalytic combustion of toluene. Chemical Engineering Journal 369, 758–765.

Ma, S., Hou, Y.-Q., Li, Y.-F., Ding, X.-X., Yang, Y.-T., Tian, J., Cui, Y., Huang, Z.-G., 2023. Regulation of A-site cations in AMn_2O_x spinel catalysts on the deep oxidation of light alkanes VOCs. Fuel 334, 126785–126797.

Maldonado-Hódar, F.J., 2011. Removing aromatic and oxygenated VOCs from polluted air stream using Pt–carbon aerogels: assessment of their performance as adsorbents and combustion catalysts. Journal of Hazardous Materials 194, 216–222.

Mao, Z.-X., McIntosh, M. J., Demirgian, J. C., 1992. On-line emissions monitoring of chlorobenzene incineration using Fourier transform infrared spectroscopy.

Mazas, A.N., Fiorina, B., Lacoste, D.A., Schuller, T., 2011. Effects of water vapor addition on the laminar burning velocity of oxygen-enriched methane flames. Combustion and Flame 158 (12), 2428–2440.

Mei, J., Shen, Y., Li, Y.-M., Zhang, S.-H., Shen, Y., Li, W., Cheng, Z.-W., Zhao, J.-K., Chen, J.-R., 2022. Combined experimental and theoretical study of o-xylene elimination on Fe–Mn oxides catalysts. Chemosphere 292, 133442–133450.

Morsch, P., Döntgen, M., Heufer, K.A., 2022. Kinetic investigations on the high- and low-temperature chemistry of ethyl acetate. Combustion and Flame 243, 111995–112004.

O'Reilly, A., 1998. Estimation of residence time in Co incinceration. Process Safety and Environmental Protection 76 (2), 166–176.

Özçelik, Z., Soylu, G.S.P., Boz, İ., 2009. Catalytic combustion of toluene over Mn, Fe and Co-exchanged clinoptilolite support. Chemical Engineering Journal 155 (1), 94–100.

Procaccini, C., Kraft, M., Fey, H., Bockhorn, H., Longwell, J.P., Sarofim, A.F., Smith, K.A., 1998. PIC formation during the combustion of simple hydrocarbons in inhomogeneous incineration systems. Symposium (International) on Combustion 27 (1), 1275–1281.

Rooke, J.C., Barakat, T., Brunet, J., Li, Y., Finol, M.F., Lamonier, J.-F., Giraudon, J.-M., Cousin, R., Siffert, S., Su, B.L., 2015. Hierarchically nanostructured porous group Vb metal oxides from alkoxide precursors and their role in the catalytic remediation of VOCs. Applied Catalysis B: Environmental 162, 300–309.

Salooja, K.C., 1966. Studies of combustion processes leading to ignition of some oxygen derivatives of hydrocarbons. Combustion and Flame 10 (1), 11–21.

Salvador, S., Kara, Y., Crussol, J.M., 2005. Improving VOC recuperative incinerators performances by increasing turbulence levels inside the combustion chamber—experimental results. Applied Thermal Engineering 25 (13), 1871–1881.

Sato, K., Hidaka, Y., 2000. Shock-tube and modeling study of acetone pyrolysis and oxidation. Combustion and Flame 122 (3), 291–311.

Schöbel-Ostertag, A., Braun-Unkhoff, M., Wahl, C., Krebs, L., 2005. The oxidation of benzene under conditions encountered in waste incinerators. Combustion and Flame 140 (4), 359–370.

Shan, T.-X., Cui, G., Li, Z.-L., Wang, S., 2019. Numerical simulation of water vapor's effect on methane combustion. Journal of Petrochemical University 32 (05), 62–68 (in China).

Shuai, Q.-F., Dong, X.-P., Lu, J.-G., Wei, T., Gong, X.-H., Zhou, J.-H., Ma, H.-L., Zhang, Y., 2021. Research progress in regenerative oxidation technology of industrial VOCs exhaust treatment. Environmental Science & Technology 44 (01), 134–140 (in China).

Solsona, B., Garcia, T., Aylón, E., Dejoz, A.M., Vázquez, I., Agouram, S., Davies, T.E., Taylor, S.H., 2011. Promoting the activity and selectivity of high surface area Ni–Ce–O mixed oxides by gold deposition for VOC catalytic combustion. Chemical Engineering Journal 175, 271–278.

Song, Y.-C., Wang, Y., Li, J.-H., Yue, W., Guo, X.-L., 2020. Overview of VOC combustion treatment method. Contemporary Chemical Industry 49 (06), 1180–1183 (in China).

Stone, D.K., Lynch, S.K., Pandullo, R.F., Evans, L.B., Vatavuk, W.M., 1992. Flares. Part I: Flaring technologies for controlling VOC-containing waste streams. Journal of the Air & Waste Management Association 42 (3), 333–340.

Strots, V.O., Bunimovich, G.A., Roach, C.R., Matros, Y.S., 2000. Chapter 10 – Regenerative catalytic oxidizer technology for VOC control. In: Abraham, M.A., Hesketh, R.P. (Eds.), Reaction Engineering for Pollution Prevention. Elsevier Science, Amsterdam, Noord-Holland, Netherlands, pp. 113–126.

Tichenor, B.A., Palazzolo, M.A., 1987. Destruction of volatile organic compounds via catalytic incineration. Environmental Progress 6 (3), 172–176.

Tsuboi, T., Ishii, K., Tamura, S., 1999. Thermal oxidation of acetone behind reflected shock wave.

Van den Brink, R.W., Krzan, M., Feijen-Jeurissen, M.M.R., Louw, R., Mulder, P., 2000. The role of the support and dispersion in the catalytic combustion of chlorobenzene on noble metal based catalysts. Applied Catalysis B: Environmental 24 (3), 255–264.

Van der Vaart, D.R., Marchand, E.G., Bagely-Pride, A., 1994. Thermal and catalytic incineration of volatile organic compounds. Critical Reviews in Environmental Science and Technology 24 (3), 203–236.

Van der Vaart, D.R., Vatvuk, W.M., Wehe, A.H., 1991. Thermal and catalytic incinerators for the control of VOCs. Journal of the Air & Waste Management Association 41 (1), 92–98.

Wang, B.-Y., Liu, Y.-X., Weng, J.-J., Glarborg, P., Tian, Z.-Y., 2017. New insights in the low-temperature oxidation of acetylene. Proceedings of the Combustion Institute 36 (1), 355–363.

Wang, B., Ma, R., Xue, G.-C., Zhang, L., 2017. Research progress on thermal oxidation technology for industrial organic waste gas. Chemical Industry and Engineering Progress 36 (11), 4232–4242.

Wang, G.-G., 2015. Application of regenerative catalytic oxidation in exhaust gas treatment of besmearing-baking house. Technology & Engineering Application (08), 16–18 (in China).

Wang, G.-Z., 2021. Analysis of VOC exhaust gas treatment measures in the chemical industry. Leather Manufacture and Environmental Technology 2 (23), 23–24+27 (in China).

Wang, H., Guo, G.-Y., Zhou, Y.-Y., Liang, Z.-Y., 2009. Status and progress of treating volatile organic compounds. Chemical Industry and Engineering Progress 28 (10), 1833–1841 (in china).

Wang, L.-F., Sakurai, M., Kameyama, H., 2008. Catalytic oxidation of dichloromethane and toluene over platinum alumite catalyst. Journal of Hazardous Materials 154 (1), 390–395.

Wang, L.-M., Wang, L.-Q., Zhang, Y.-C., Su, H.-Q., 2017. Photothermal synergistic catalytic technology in energy field. Chemical Industry and Engineering Progress 36 (07), 2457–2463 (in China).

Wang, W., 2016. Studies on the Efficient Reduction Methods for the Combustion Chemical Kinetic Mechanism of Fuel. Chongqing University, China.

Warahena, A.S.K., Chuah, Y.K., 2009. Energy recovery efficiency and cost analysis of VOC thermal oxidation pollution control technology. Environmental Science & Technology 43 (15), 6101–6105.

Wu, M., Wang, X.-Y., Dai, Q.-G., Gu, Y.-X., Li, D., 2010. Low temperature catalytic combustion of chlorobenzene over Mn–Ce–O/γ-Al_2O_3 mixed oxides catalyst. Catalysis Today 158 (3), 336–342.

Wu, P., Zhao, S.-Q., Jin, X.-J., Chong, Y., Li, Y.-F., Li, A.-A., Lin, J.-J., Qiu, Y.-C., Ye, D.-Q., 2022. Acid-activated layered δ-MnO_2 promotes VOC combustion. Applied Surface Science 574, 151707–151713.

Wu, Y.-F., 2012. NO_x Production Mechanism Simplified Method Research and Analysis with Pulse Combustion Based on GRI-MECH 3.0. Zhejiang University of Technology, China.

Xiang, Z.-B., Xia, G.-H., Ye, J., 2014. A research on the application of regenerative thermal oxidizer technology in dealing with volatile organic exhaust gas project. Journal of Green Science and Technology 174–177 (in China).

Xiao, X.-L., 2021. The Study on Key Factors of Regenerative Rotary Thermal Oxidizer Furnace. Harbin Institute of Technology, China.

Xie, Z.-Y., 2021. Hazard of industrial VOC and treatment technology. Chemical Management 35, 31–32 (in China).

Yang, Y., Wang, G., Fang, D., Han, J.-N., Dang, F.-L., Yang, M., 2020. Study of the use of a Pd–Pt-based catalyst for the catalytic combustion of storage tank VOCs. International Journal of Hydrogen Energy 45 (43), 22732–22743.

Yin, H.-C., Zhang, W., 2007. The numerical simulation of preheated air temperature on combustion process. Energy Conservation (09), 4–6+2 (in China).

Zhang, S.-W., Guan, H., Duan, Y., Wang, H., Hui, C., 2022. Application of RCO in odor control of styrene-butadiene rubber plant. Contemporary Chemical Industry 51 (04), 980–984 (in China).

Zhang, X.-J., Wei, Y.-H., Song, Z.-X., Liu, W., Gao, C.-X., Luo, J.-W., 2021. Silicotungstic acid modified CeO_2 catalyst with high stability for the catalytic combustion of chlorobenzene. Chemosphere 263, 128129–128141.

Zhang, X.-Y., Gao, B., Creamer, A.E., Cao, C.-C., Li, Y.-C., 2017. Adsorption of VOCs onto engineered carbon materials: a review. Journal of Hazardous Materials 338, 102–123.

Zhao, H.-Y., Wang, H., Qu, Z.-P., 2022. Synergistic effects in Mn–Co mixed oxide supported on cordierite honeycomb for catalytic deep oxidation of VOCs. Journal of Environmental Sciences 112, 231–243.

Zhao, L., Zhang, Y.-F., Li, R.-H., Ma, Z.-C., 2015. Harms, recycling and treatment technology of VOC. Chemical Education 36 (16), 1–6 (in China).

Zhao, Z.-W., Chaos, M., Kazakov, A.F., Dryer, F.L., 2008. Thermal decomposition reaction and a comprehensive kinetic model of dimethyl ether. International Journal of Chemical Kinetics 40, 1–18.

Zheng, S.-J., Qian, Y., Wang, X.-B., Vujanović, M., Zhang, Y.-J., Ur Rahman, Z., Yang, P.-H., Duan, F., Tan, H.-Z., De Toni, A., Li, Y., Mikulčić, H., 2022. Experimental investigation of the NO_x formation and control during the self-sustaining incineration process of N-containing VOCs (dimethylformamide). Fuel 315, 123149–123158.

Zhou, L.-L., Ma, C., Horlyck, J., Liu, R.-J., Yun, J., 2020. Development of pharmaceutical VOC elimination by catalytic processes in China. Catalysts 10 (6), 668–717.

Zhu, R., Dong, L.-B., Liu, Z.-Z., 2020. A new method for combustion treatment of VOC organic waste gas in Iron Printing factory. Energy Conservation 39 (08), 120–122 (in China).

CHAPTER

4

Membrane separation of VOCs

Guangdou Qi

School of Environmental Engineering, Xuzhou University of Technology, Xuzhou, China

4.1 Overview of membrane technology for VOC separation

Volatile organic compounds (VOCs) released into the atmosphere are regarded as toxic pollutants, as they can cause serious harm to both the ecological environment and human health. It has been reported that the emission amount of VOCs from industrial processes in China increased to 25.9 Tg from 2005 to 2020, with an average growth rate of 5.9% per year (Gan et al., 2023). Recently, a large number of VOC recovery technologies, including adsorption, membrane separation, absorption, and condensation, have been developed and have received widespread attention due to the advantages of recycling valuable VOCs and reducing carbon emissions. Among them, membrane technology realizes the separation of VOCs based on their different transmission rates through the membrane caused by the different selective permeability of each component under the pressure difference. Compared with other VOC recovery technologies, the membrane separation process is regarded as one of the most environmentally friendly approaches due to the following merits (Chen et al., 2020; Hong et al., 2022; Petrusová et al., 2019):

(1) High efficiency and selectivity: The process can work continuously with a low flow rate or variable flow rate with a more than 80% removal of VOCs in the feed.
(2) Low energy demand: The VOC gas can be directly separated, benefiting from the no need of any reaction and no change in the molecular structure; thus, it does not consume organic solvents.
(3) Simple process and operation: Furthermore, it can be coupled with other technologies for a higher VOC separation efficiency.

In the membrane separation process, membrane performance is a crucial factor which is typically evaluated in the terms of two parameters: permeability and selectivity. The productivity of the membrane can be determined through its permeability, whereas selectivity defines its gas separation capability. Gas separation in the membranes is dependent on the membrane structure, its chemical and physical properties, and the characteristics of the penetrant gas, such as the size, shape, polarity, and their interactions with the membrane

Volatile Organic Compound Removal
https://doi.org/10.1016/B978-0-443-15674-8.00012-3

material. General membrane materials can be primarily divided into two types, rubbery or glassy polymers, which are used to synthesize nonporous (dense) and porous membranes, respectively. Polydimethylsiloxane (PDMS, silicone rubber) and polyether block amide (PEBA) are the representative rubbery materials. Polyethersulfone (PES); polyimide (PI), such as Matrimid 5218; and polyamide (PA) are the common examples of glassy polymers. The separation membrane can be produced from a single material or a combination of two materials to obtain the preferable performance. Different membrane structures also determine the different separation mechanisms for gas transport through the membrane: microporous diffusion for porous membranes and solution–diffusion for nonporous (dense) membranes.

For the separation performances of membranes for VOCs, the primary factors are the permeate flux (J_i, $L/m^2 \cdot h$), gas permeability (P_i), selectivity (α), and/or VOC rejection (R), which are calculated by the following formulas:

$$J_i = \frac{V_i}{At} \tag{4.1}$$

$$P_i = l \times \frac{J_i}{\Delta p} \times \frac{273.15}{T} \times \frac{P_0}{76} \tag{4.2}$$

where V_i is the volume (L), A is the effective membrane area (m^2), t is the time, P_i is the gas permeability in Barrer (1 Barrer $= 1 \times 10^{-10}$ cm^3 (STP)$\cdot$cm/($cm^2 \cdot s \cdot$cmHg)), l is the thickness of the selective layers of the membrane, Δp is the transmembrane pressure difference of the component i, T is the temperature, and P_0 is the upper stream pressure (usually is defined as the atmospheric pressure, kPa):

$$\alpha = \frac{P_{VOC}}{Pi} \tag{4.3}$$

where P_{VOC} is the permeability of VOCs, and P_i is the permeability of the component i:

$$R = \left(\frac{C_{feed} - C_{perm}}{C_{feed}}\right) \times 100\% \tag{4.4}$$

where C_{feed} is the feed VOC concentration, and C_{perm} is the permeate VOC concentration, which is usually measured by gas chromatography.

The permeability–selectivity trade-off has been the biggest challenge in the membrane gas separation field, which was quantified by Robeson in 1991 (Robeson, 1991) and revised in 2008 (Robeson, 2008) by plotting log P_i (permeability of the more permeable gas) versus log α_{ij} (separation factor $= P_i/P_j$) for a large number of polymers. The "upper bound" is used to assess the potential of a new polymer as a membrane material for the separation of the two gases. On the other hand, there are still some challenges, such as membrane swelling and regeneration after long-term use, as well as the indistinctness of the comprehensive influence of various factors (i.e., temperature, transmembrane pressure, flow rates, and the composition of the feed stream) on the performance.

4.2 VOC separation mechanisms

During VOC treatment, membranes act as permselective barriers to realize separation based on the different transmission rates of each penetrant through the membrane. The separation mechanisms can be distinguished according to the structure and chemical constitution of the membranes. At present, there are two mechanisms of membrane separation: microporous diffusion and dissolution–diffusion (Mulder, 1991; Richard, 2004). Typically, microporous diffusion occurs in porous membranes, while solution–diffusion predominates in the dense (nonporous) membranes. The VOC transport rate in porous membranes depends on the pore size and pore-size distribution. Dense membranes are usually composed of organic polymers, and the transmembrane pressure and operating conditions are the main factors affecting the VOC permeation rate (Farnam et al., 2021).

4.2.1 Microporous diffusion

Gas transport through porous membranes mainly occurs via two mechanisms: Knudsen diffusion and molecular sieving, as shown in Fig. 4.1. The interaction between the gas molecules and the porous membrane is diverse due to the difference in the pore size and the surface properties, thus showing various transfer characteristics. Knudsen diffusion occurs when the mean free path of gas molecules is greater than the pore size. In this case, the interactions between the gas molecule and the pore wall are more frequent than the interaction between the gas molecules; the diffusion of a small molecule is faster than that of a large molecular (Li et al., 2018). Separation selectivity depends on the inverse square root of the molecular weight

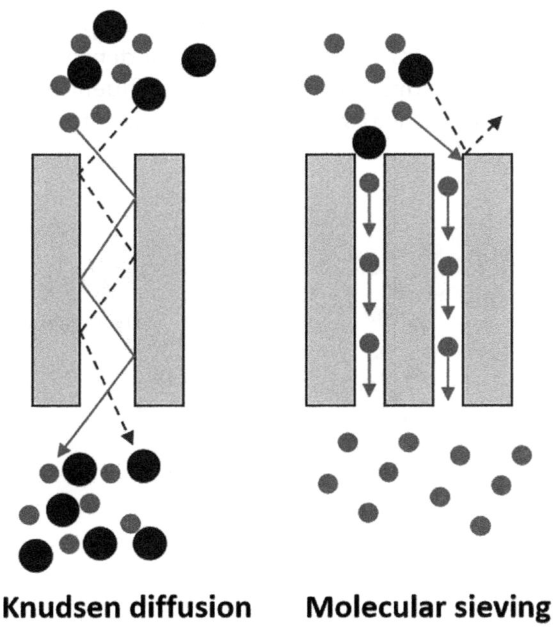

FIGURE 4.1 Schematic diagrams of the microporous diffusion mechanism.

of the permeating gas components. If the pore size decreases to the order of a subnanometer, such as micropores in zeolites, a sharp molecular discrimination known as "molecular sieving" is observed.

4.2.2 Dissolution–diffusion

Gas transport through nonporous (dense) membranes is generally explained by the dissolution–diffusion mechanism. The permeability of a gas is a function of its solubility and diffusivity. Solubility is a thermodynamic parameter which quantifies the amount of permeating gas molecules sorbed by the membrane under equilibrium conditions, whereas diffusivity is a kinetic parameter which measures how fast these molecules are transported through the membrane. Differences in solubility arise from the disparities in the molecular-level interactions between the gas molecules and membrane materials, while diffusivity is controlled by the kinetic diameters and shapes of the gas molecules. Gas separation via a solution–diffusion mechanism, therefore, leverages these differences to induce dissimilar gas permeability and achieve separation selectivity (Chuah et al., 2018).

The diffusion process can be divided into three steps: (1) adsorption process, gas molecules are adsorbed and dissolved on the surface of the upstream side of the membrane; (2) diffusion process, the adsorbed and dissolved gas molecules diffuse through the membrane driven by the difference in the concentration; and (3) desorption process, the gas molecules are desorbed from the downstream surface of the membrane. Generally, the diffusion process is slow, while the adsorption and desorption process can reach equilibrium quickly. Therefore, the diffusion process is the rate-control step for separation.

The dissolution–diffusion model assumes that molecular transport through the membrane is driven by a concentration gradient across the membrane, rather than a pressure gradient. Fig. 4.2 shows the schematic diagram of penetrant transport across a dense polymeric membrane. The upstream gas-phase partial pressure of each gas, p_i(u), influences the concentrations of the gases sorbed into the membrane. The resulting concentration gradient

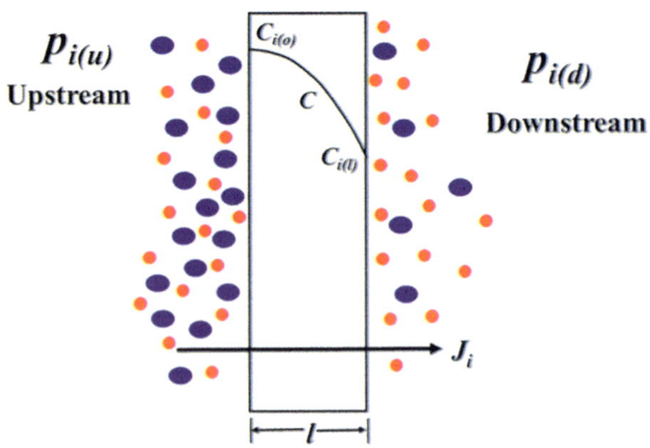

FIGURE 4.2 Schematic diagram of penetrant transport across a dense polymeric membrane (Qian et al., 2020).

within the membrane material drives transport through the membrane. The downstream gas-phase partial pressure of each gas, p_i(d), influences penetrant desorption from the membrane.

4.3 Design, characteristics, and key factors of VOC membrane separation systems

For the membrane separation of VOCs, the controlling factors mainly come from the properties of the membrane and separation conditions. Among the membrane properties, membrane thickness can affect separation performance, while the separation conditions, including temperature, feed pressure, flow rate, and concentration, are crucial factors that have been widely studied (Liu et al., 2006a).

4.3.1 Membrane thickness

PDMS/PA composite membranes with three different PDMS thicknesses (1.0, 1.2, and 1.9 μm) were tested for cyclohexane/nitrogen separation (Shen et al., 2022). The results indicated that a higher membrane thickness led to a slightly higher cyclohexane and nitrogen permeability, which was attributed to the alteration of the free volume with the increase of the thickness. Thinner membranes usually exhibit a higher density and, thus, a lower free volume. As a result, fewer channels and adsorption sites would be available for the transport of cyclohexane and nitrogen through the membrane, lowering their permeability. When PDMS thickness was larger, cyclohexane and nitrogen permeability increased simultaneously. Selectivity seems to be independent of the membrane thickness due to the homogeneous morphology of the membrane (Zong et al., 2021; Zhou et al., 2010).

Another study also confirmed the effect of thickness on membrane separation performance (Zhang et al., 2021). A microporous PI VOC-rejective membrane showed a constant rejection value with a different membrane thickness. If the membranes are defect-free, selectivity should be independent of the selective layer thickness according to Fick's law (Koops et al., 1994). However, different results were reported for the separation of VOCs/N_2 by the PEBA/polysulfone (PSF) composite membrane (Liu et al., 2009b). The permeance of both VOCs (methanol and ethanol) and N_2 increased with a decrease in the PEBA selective layer thickness. The increasing trend of VOC permeance became less significant with a decreasing layer thickness and gradually leveled off when the PEBA layer was sufficiently thin. In contrast, the permeance of N_2 had a good linear correlation with the reciprocal PEBA layer thickness. Therefore, membrane selectivity decreased sharply at a lower layer thickness, indicating that a dense polymeric layer ensured the selective separation of the composite membrane, and the substrate offered little selectivity because of its microporous structure.

4.3.2 Feed temperature

The effects of the feed temperature on the gas separation performance of membranes are controlled by the membrane structure and properties as well as the gas properties. Liang et al.

(2023) studied the vapor permeation processes of $VOCs/N_2$ through the PDMS composite membrane at temperatures ranging from 25 to 40°C. Their results showed that both the flux and permeability of VOCs decreased with the increasing temperature. During the VOC vapor permeation process on a dense membrane, dissolution is typically the rate-limiting step (Zong et al., 2021). As the temperature increases, the adsorption of VOCs on the membrane surface is weakened as the adsorption process is exothermic (Dai et al., 2020; Yu et al., 2017), thereby resulting in reduced fluxes and the permeability of VOCs. Different from VOCs, the flux and permeability of nitrogen increased slightly with the increase in the temperature. As previously reported, the rate-limiting step for nitrogen in vapor permeation is the diffusion step (Dumont et al., 2016). The increasing temperature led to the enhanced chain mobility of the polymer segment in the rubbery PDMS membrane, resulting in an enlarged free volume, higher flux, and higher permeability of the nitrogen mixture. Finally, the membrane selectivity of $VOCs/N_2$ decreased with an increase in the temperature. Similar results were reported for the separation of cyclohexane/nitrogen by the PDMS composite membrane (Shen et al., 2022) and the permeation of N_2, C_2H_6, and C_3H_8 through asymmetric polyetherimide hollow-fiber membranes (Wang et al., 2002).

The effects of feed temperature cyclohexane/nitrogen separation through triptycene-based microporous PI membranes have been studied (Chen et al., 2020; Dai et al., 2020; Zhang et al., 2021). These studies have found that feed temperatures had a little impact on membrane performance, constant cyclohexane and nitrogen permeability, and VOC rejection rates (Fig. 4.3). This may be ascribed to the rigid 3D structure of triptycene in the polymer chain, which might interlock with itself, restricting the thermal movement of polymer chains as the feed temperature increases (Cho and Park, 2011). Furthermore, the network structure endowed by tetramethyl moieties might also prevent chain movement. Therefore, the membrane performance remained constant across the increasing temperatures.

The operation temperature can affect the performance of liquid membranes as it modifies the viscosity of ionic liquids, which in turn affects the diffusion coefficient. According to Zhang et al. (2014), the viscosity of N-butylpyridinium tetrafluoroborate and 1-butyl-3-methylimidazolium tetrafluoroborate was lowered by increasing the temperature, resulting in the improved diffusion of toluene into the ionic liquid phase of a membrane. As a result, the toluene removal efficiency of cyclohexane was higher at 323 K than that at 298 K.

4.3.3 Feed pressure

Feed pressure plays an important role in gas separation through a membrane, as a high feed pressure can cause membrane plasticization and, thus, weaken dense membrane performance. Jiang and Kumar (2005) found that the permeances of alkene (C2–3) and alkyne (C2–3) through the PDMS/PSF composite membrane increased with the increasing pressure, whereas the permeances of noncondensable gases (oxygen and nitrogen) were pressure independent (Fig. 4.4a). With the aforementioned feed pressure of 114.6 kPa, the permeances of propylene and propane increased more significantly than ethane and ethylene, indicating that propylene and propane sorbed into PDMS and continuously caused plasticization above this pressure (Jiang and Kumar, 2005). Similar results were also reported for the PDMS-dense membrane by Liu et al. (2009b). They found the permeability of N_2 and C_2H_4 were almost

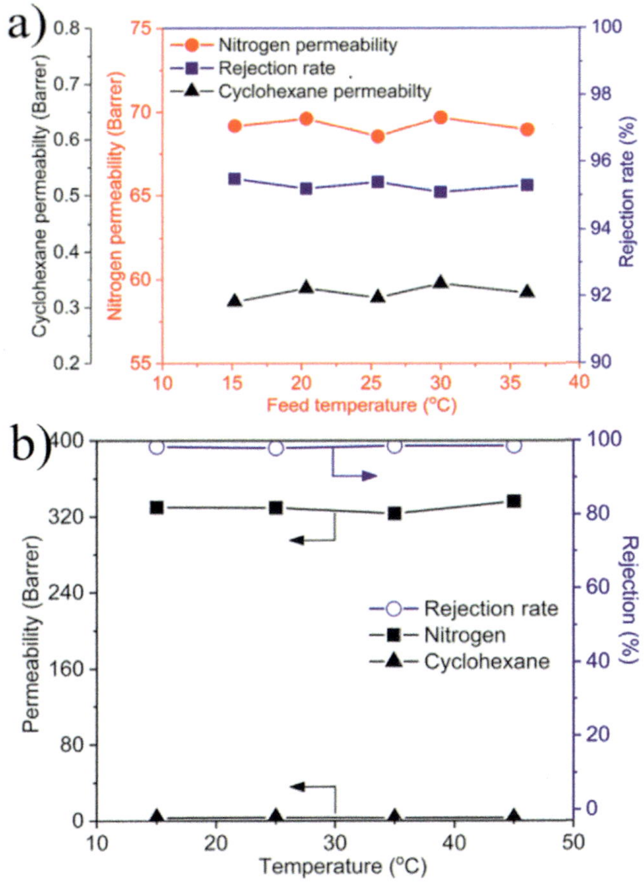

FIGURE 4.3 Effect of the feed temperature on the cyclohexane/nitrogen separation performance of (a) 2,6,14-triaminotriptycene (Trip)-tetramethylbenzidine (TMB) composite membranes (Dai et al., 2020) and (b) HTPA6 (80) membranes (Chen et al., 2020).

independent with the feed pressure, whereas the permeability of C_3H_6 increased with an increase in the feed pressure (Fig. 4.4b). Another study indicated that the permeability of ethane and propane increased at higher pressures, which was attributed to capillary condensation that occurred at higher pressures (Wang et al., 2002).

Liu et al. (2009b) also reported that the permeation behaviors of nitrogen, oxygen, and ethylene in the PDMS/porous polyetherimide (PEI) hollow-fiber composite membrane were also similar to that in PDMS dense membranes (Fig. 4.4c). However, the permeance of propylene through the composite membrane was different in dense PDMS membranes, and it decreased substantially as the feed pressure increased. This was attributed to the different resistance from the membrane substrate (PEI). Propylene was more permeable than ethylene, thus the resistance of the membrane substrate accounted for over 90% of the total resistance, which was higher than 77% for ethylene permeation. The contribution

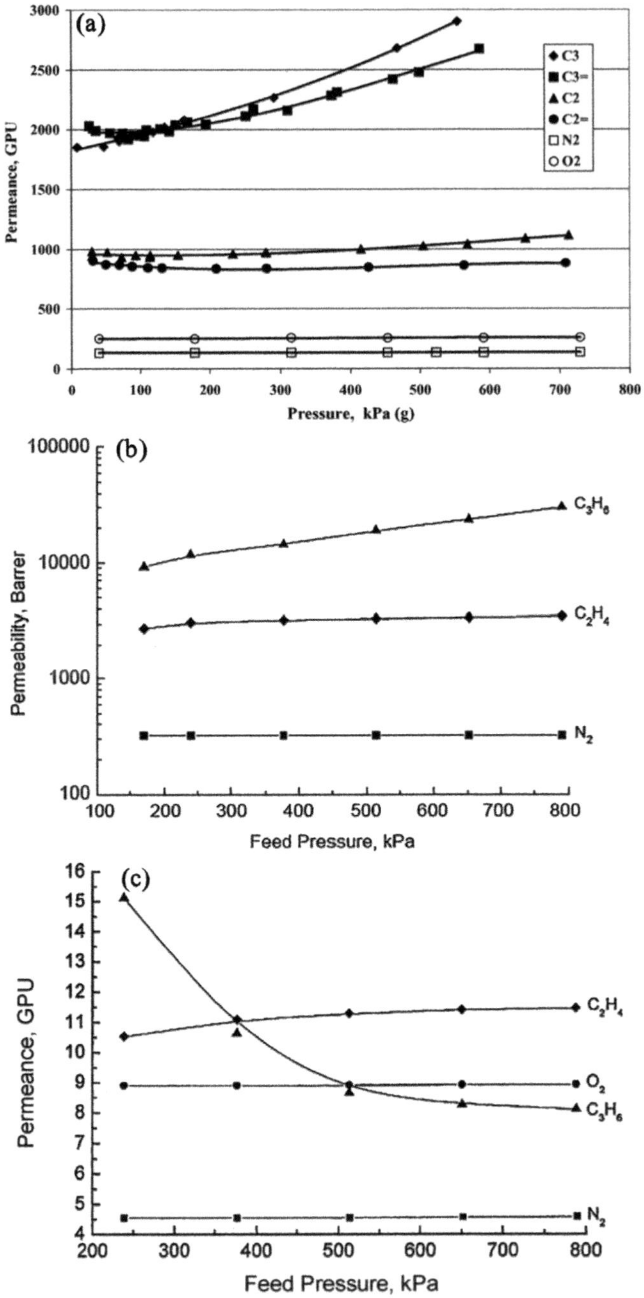

FIGURE 4.4 Effects of the feed pressure on the permeance/permeability of (a) PDMS/PSF composite membranes (Jiang and Kumar, 2005), (b) PDMS dense membranes, and (c) PDMS/PEI hollow-fiber composite membranes (Liu et al., 2009b).

of substrate resistance increased with the increasing feed pressure. Thus, the permeance of propylene decreased with an increase in the feed pressure, leading to decreased propylene/nitrogen selectivity. Several studies have confirmed decreased membrane selectivity with the increasing feed pressure. Shen et al. (2022) found that high feed pressure weakened the cyclohexane flux through the PDMS/PA composite membrane, while it had little influence on the nitrogen flux due to their different saturated vapor pressures. Meanwhile, PDMS selectivity for C_{3+} hydrocarbon declined significantly with the feed pressure (Vaughn et al., 2017). On the contrary, the increased selectivity of PDMS thin-film composite membranes with an increase in the feed pressure for propylene/nitrogen separation was reported by Shi et al. (2006). These results have demonstrated the strong dependence of membrane selectivity on the feed pressure.

Although constant nitrogen permeability with an increased feed pressure has been widely reported, Dai et al. (2020) showed decreased nitrogen permeability with the feed pressure for a porous PI membrane. As the feed pressure enhanced their permeability, many more cyclohexane molecules were pushed into the membrane pores with the help of the feed pressure. They occupied transport channels in the membrane and blocked the transport of nitrogen, leading to increased cyclohexane permeability and decreased nitrogen permeability.

4.3.4 Feed flow rate

Generally, VOC permeability increased with the increasing feed flow rate. The permeability of VOCs (ethyl, ethanol, and acetone) and N_2 through a PDMS/ceramic composite membrane were studied at the feed flow rate of 3–7 L min^{-1} (Yang et al., 2018). The increase in VOC permeability along with the decrease in nitrogen permeability with the increasing feed flow rate was detected for all VOC/N_2 mixtures (Fig. 4.5a). When the feed flow rate increased, the VOC concentration gradient layer became thinner and more VOC molecules came in contact with the membrane surface, resulting in a higher permeability. As a result, more VOC molecules were adsorbed on the membrane surface, which prevented the adsorption of N_2 on the membrane surface. Hence, less N_2 diffused through the membrane. On the other hand, when the VOC concentration in the gradient layer increased, the N_2 concentration was diluted, also leading to reduced N_2 permeability (Fig. 4.5b). Finally, VOC/N_2 selectivity increased with the enhanced feed flow rate. Similar trends of VOC and N_2 permeability were detected for binary heptane/N_2, cyclohexane/N_2, and hexane/N_2 mixtures. Similarly, Yeom et al. (2002a) found that, with the increasing feed flow rate, the membrane performances in terms of the permeation rate and selectivity were enhanced in the permeation of VOCs/N_2 by increasing VOC solubility in the PDMS membrane.

Zhang et al. (2023) investigated the separation performances of the four microporous PA membranes for the molecular sieving of nitrogen over cyclohexane at three different permeate fluxes. All four membranes exhibited over 99% rejection of cyclohexane when the permeate flux was $1\ L \cdot m^{-2} \cdot min^{-1}$. Increasing the permeate flux to 2 and $3\ L \cdot m^{-2} \cdot min^{-1}$, a significant increase in cyclohexane permeability was observed, while nitrogen permeability decreased slightly (Fig. 4.5c). When more cyclohexane molecules were squeezed into the upstream side of the membrane at a higher permeate flux (which also

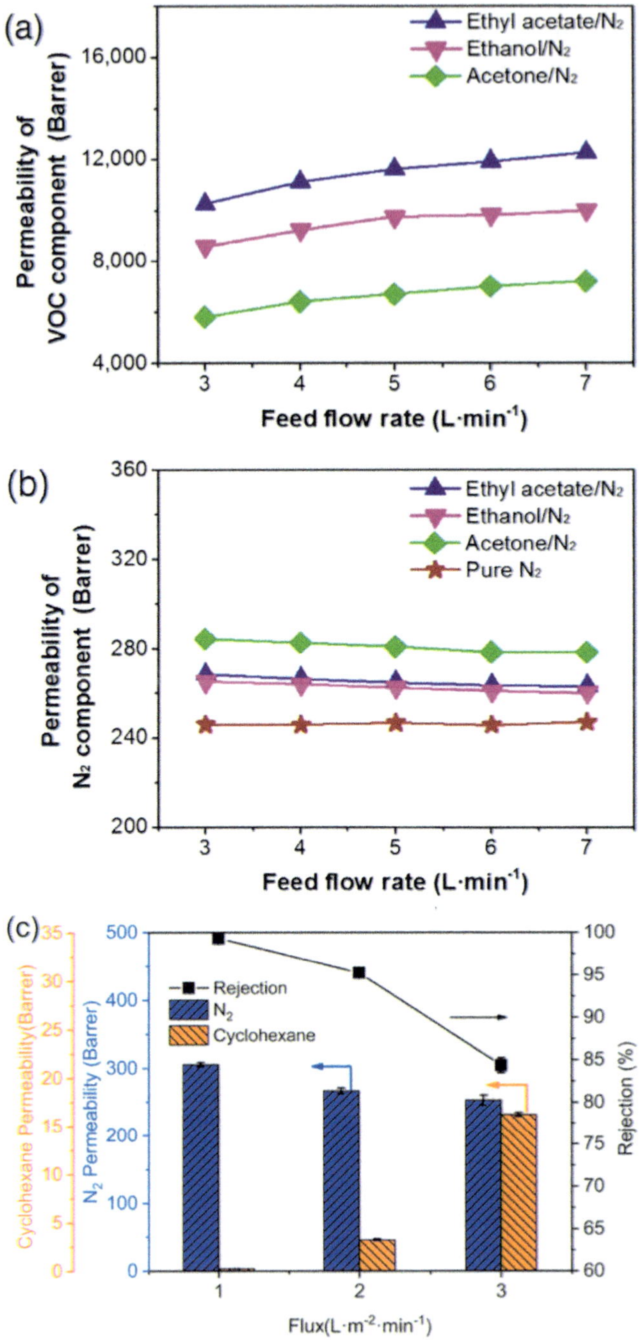

FIGURE 4.5 Effects of the feed flow rate on the permeability of (a) VOCs and (b) N_2 through the PDMS composite membrane (Yang et al., 2018) and (c) cyclohexane and N_2 permeability through the microporous PA membrane (Zhang et al., 2023).

meant a higher operating pressure), more transport channels were occupied by cyclohexane molecules with a lower diffusion rate because of their larger kinetic diameters than that of nitrogen molecules.

4.3.5 Feed concentration

Jiang et al. (2008) studied the effects of the feed concentration on permeance and selectivity for ethylene/nitrogen at 1.7%–37.7% and propylene/nitrogen at 1.5%–29.1% through the polyurethane–poly(vinylidene fluoride) (PU-PVDF) thin-film composite membrane. Their results showed that ethylene and propylene permeances increased considerably and nitrogen permeances remained constant (Fig. 4.6a and b), which resulted in 45.3% and 105% increases in the selectivity of nitrogen, respectively. Another study also confirmed the increase in VOC permeability and decrease in N_2 permeability with the increasing feed VOC concentration (Liang et al., 2023). It appears that nitrogen has a low affinity for VOC molecules as its inert property, thus a small coupling effect exists between nitrogen and VOCs. Therefore, a higher VOC feed concentration could not enhance nitrogen permeability. On the other hand, a high feed concentration increases the possibilities for the VOC molecules to be solubilized in and to permeate through the membrane; more transport channels are occupied by larger VOC molecules, and nitrogen transport through the membrane is blocked (Shen et al., 2022). As a result, VOC/N_2 selectivity increases with an increase in the feed VOC concentration (Liu et al., 2009a).

Nevertheless, the opposite results were obtained from the molecular sieving of N_2 from cyclohexane by Dai et al. (2020). They found that increases in the feed cyclohexane concentration improved nitrogen permeability but decreased cyclohexane permeability (Fig. 4.6c). The authors explained this phenomenon from two aspects: one is that a higher feed concentration increased VOC adsorption in the membrane, leading to membrane swelling. Nitrogen possesses a lower molecular kinetic diameter than cyclohexane, thus a nitrogen molecule can be transported more easily in a swollen membrane with a higher permeability. The other reason is that although a high feed cyclohexane concentration can cause more cyclohexane to be transported through the membrane, the adsorption of cyclohexane onto the membrane also increases. Thus, cyclohexane permeability decreased.

4.4 Different types of membranes for VOC separation

The membrane properties and separation performance are highly dependent on the synthesis materials; therefore, the key issues in membrane separation research are related to the development of advanced membrane materials. An appropriate gas separation membrane should exhibit excellent permeability and selectivity, good chemical and thermal stability, and high mechanical strength. When choosing the membrane materials, the following parameters need to be considered (Farnam et al., 2021):

(i) Inherently highly permeable and selective materials are desirable for gas separation membranes since they can decrease the required material and capital cost.

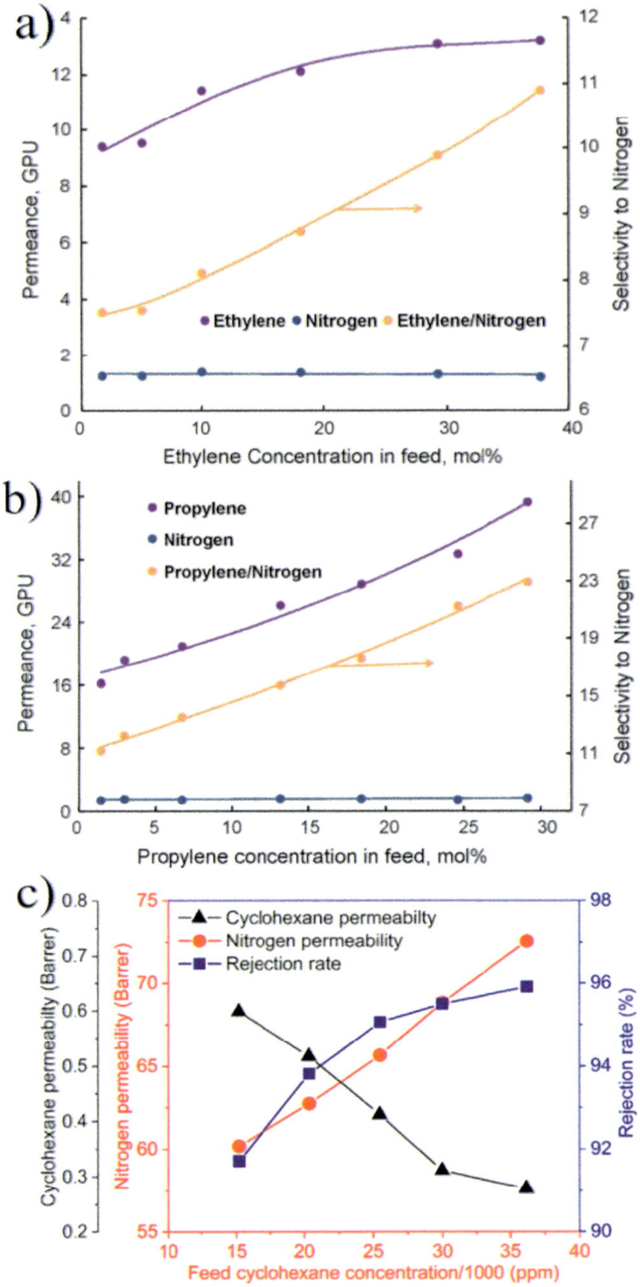

FIGURE 4.6 Effects of the feed concentration on the membrane performance of (a) ethylene and nitrogen, (b) propylene and nitrogen (Jiang et al., 2008), and (c) cyclohexane and nitrogen (Dai et al., 2020).

(ii) Good chemical and thermal stability: Membrane materials may be degraded easily by aromatic VOCs, such as benzene, toluene, ethylbenzene, and xylenes, which lead to the reduced lifespan of membranes. In addition, membrane materials need to resist high operating temperatures in some applications. Good chemical and thermal stability is important to prolong membrane usage as well as to extend its application in industry processes.

(iii) High mechanical stability is another important selection parameter for membrane materials, since gas separation processes are usually operated under high pressure to let the gas pass through it, which may break membranes with a low mechanical strength.

(iv) Besides the parameters mentioned above, the membrane materials need to be readily available in the market, and they should have the capability to be processed easily with low cost, especially for an industrial-scale membrane separation process.

The permeability—selectivity trade-off has been the biggest challenge in the membrane gas separation process. The permeability and selectivity of a membrane are directly related to the interactions between the membrane materials and the gas molecules. Membrane materials can be primarily divided into two types: organic (polymeric) or inorganic materials. Organic (polymeric) materials are either rubbery or glassy based on their glass transition temperature (T_g) and operating temperature. The glass transition temperature is defined as a temperature at which the first polymer chain movements start to take place and cause a transition in the polymer from the glassy, solid, and hard state to the soft, flexible, and rubbery state.

Rubbery polymers need to be operated above their T_g to keep them soft and flexible. Some examples of rubbery polymers widely used for membrane fabrications are polyvinyl acetate (PVAc), polydimethylsiloxane (silicone rubber), polyether block amide, poly(vinylidene fluoride), and polyurethane. While polyethersulfone; polyimide, such as Matrimid 5218; polycarbonate (PC); polysulfone; and polyamide are the common examples of glassy polymers, which need to be operated at temperatures lower than their T_g to maintain their hard and rigid structures.

Due to their easy and low-cost processing as well as the capability to be scaled up for industrial applications, rubbery and glassy membranes have been commonly used to synthesize nonporous (dense) and porous membranes, respectively. However, both rubbery and glassy polymer membranes have the limitation of the trade-off between selectivity and permeability. Rubbery membranes possess high permeability and low selectivity as their polymer chains are soft and can produce more fractional free volume (FFV) for gas molecules to pass. In rubbery polymers, the loose polymer chain packing and the random movements of polymer chain segments allow gas molecules to diffuse through the matrix more freely and cause the gas permeability to increase. While glassy membranes always show high selectivity and low permeability because of the densely packed polymer chains in their structures providing less room for macromolecules to pass through.

For inorganic membranes, silica; zeolites; metal-organic frameworks (MOFs), like zeolitic imidazolate frameworks (ZIFs); and carbon molecular sieves (CMSs) are usually used. Organic (polymeric) materials may also be combined with inorganic materials as fillers to produce mixed matrix membranes (MMMs). A composite membrane can be formed by loading a layer of dense organic polymer on a porous support. A suitable liquid can also be loaded on a porous support to form a liquid membrane with an improved ability to capture gas molecules (Friess et al., 2012). Some advanced membrane materials in the recent literature are elaborated in detail in the following sections.

4.4.1 Polydimethylsiloxane membrane

PDMS is known as a rubber polymer with good flexibility, which has been widely studied as membrane materials for the permeation and separation of VOCs, including C5–C7 alkanes (Lin et al., 2012), C2–C4 olefins (Choi et al., 2007), methanol, toluene, m-xylene, isopentane (Lue et al., 2014), chlorinated hydrocarbons (Dotremont et al., 1995; Yeom et al., 2002a,b), and acetone, butanol, ethanol, and ethyl acetate (Kujawska et al., 2016).

PDMS dense membranes are mainly prepared by mixing the polymer and crosslinking agent in a certain ratio (usually 10:1), and membranes with different permeabilities can be obtained by adjusting this ratio. VOC separation by the dense membrane is controlled by the dissolution–diffusion mechanism, thus solubility and diffusion coefficients affect the permeability. Choi et al. (2007) prepared PDMS membranes from the crosslinking siloxane oligomer with vinyl groups and the siloxane oligomer with silane groups at a mixing ratio of 10:1, and we then examined their sorption and permeation performance for olefins (ethylene, propylene, and butene) and N_2 over a pressure range of 1–20 atm and a temperature range of −20–50°C. Their results showed that the solubility of olefins was mainly influenced by the critical temperatures of olefins and operation conditions, such as the operation feed pressure and temperature. With the increasing feed pressure and/or decreasing temperature, the solubility of olefins increased, thus the permeability of more condensable olefin gases increased, but nitrogen permeability changed only slightly. As a result, olefin/nitrogen selectivity increased greatly.

In another study (Yeom et al., 2002b), the permeation performances of chlorinated hydrocarbons, including methylene chloride, chloroform, 1,2-dichloroethane, and 1,1,2-trichloroethane/nitrogen, through the PDMS membrane were investigated at various operation conditions. The authors found that the permeability coefficient of VOCs increased with the increasing VOC content in the feed but decreased with the increasing operating temperature. Analyses of the diffusion and solubility coefficients could help understand the change in the permeation coefficient. The diffusion coefficient decreased slightly with the increasing VOC content, as a higher content in the feed led more VOC molecules to be absorbed into the membrane, and the diffusing permeant size could increase, resulting in the slowed diffusion of the permeants. The solubility coefficient increased with the increasing VOC content in the feed, and this trend was more noticeable for the larger VOCs, such as $C_2H_4Cl_2$ and $C_2H_3Cl_3$. The results also showed that VOC/N_2 selectivity increased with the increase in the critical temperature or molecular size of VOCs, as nitrogen permeability was depressed due to the preferential sorption of VOCs into PDMS membranes.

Moreover, PDMS was selected as a commercially used membrane material in VOC/N_2 separation with a high permeability, above 20,500 Barrer for hexane, 13,500 Barrer for cyclohexane, 12,900 Barrer for 2,2,4-trimethylpentane, and 11,100 Barrer for ethanol vapors (Jirsáková et al., 2021). At the same time, PDMS was applied on the shell side of the hollow fiber on an industrial scale. The study found that the PDMS-based hollow-fiber membrane could remove 98.8%–99.9% of acetone, toluene, and methanol from the polluted air (Obuskovic et al., 2003). Furthermore, the long-term stability of PDMS-based membranes was confirmed to be observed at least 6 months after. Similarly, PDMS hollow fibers can successfully separate acetone, ethyl acetate, and ethanol from the air (Gales et al., 2002).

Due to the sorption of VOCs in the membrane during the separation process, plasticization and swelling may occur and lead to the destruction of polymer chains, which results in an increase in permeability and a decrease in selectivity, thus reducing the membrane separation performance (Suleman et al., 2016). Park and Lee (2002) found that the toluene/N_2 selectivity of PDMS was as high as 140. However, toluene vapors would cause the plasticization of the membrane, resulting in the gradual decrease of selectivity. Similar results of the plasticization of the PDMS coating in a composite membrane, caused by propylene and/or propane, were reported in another study (Jiang and Kumar, 2005).

4.4.2 Poly(ether block amide) membrane

PEBA is a commercial block copolymer formed by the copolymerization of polyether segments and polyamide segments; thus, it combines the hardness and mechanical strength of a glassy polymer and the softness of a rubbery polymer, which makes PEBA a promising membrane material for VOC separation (Liu et al., 2004). Furthermore, PEBA possesses high VOC permeability due to its amorphous and flexible polyether segments.

PEBA 2533 contains 80 wt% of rubbery polyethers in PEBA; thus, it mainly behaves like rubbery polymers. Nevertheless, its separation properties are specific due to the glassy PA presence. The sorption, diffusion, and permeation of three olefins (i.e., C_2H_4, C_3H_6, and C_4H_8) in PEBA 2533 membranes were investigated (Liu et al., 2006a). The results showed that the favorable permselectivity of olefin/N_2 was primarily attributed to solubility and selectivity. Olefin permeability was in the order of C_2H_4, C_3H_6, and C_4H_8, the same order as their solubility in the membrane, whereas the effect of diffusivity–selectivity on permselectivity depended on the operating temperature and pressure. Similar results were also reported for various VOCs, including n-pentane, n-hexane, cyclohexane, n-heptane, methanol, ethanol, n-propanol, n-butanol, acetone, dimethyl carbonate, and methyl tert-butyl ether, by Liu et al. (2009a). The separation performance of PEBA membranes for both binary VOCs/N_2 and multicomponent VOCs/N_2 gas mixtures were investigated. It was found that the permeability of these VOCs was mainly dominated by their solubility in the membrane. The membranes exhibited good separation performance, and a VOC concentration of more than 90 mol% was achieved at a feed VOC concentration of 5 mol%.

As mentioned above, PEBA contains rubbery polyethers; thus, it undergoes plasticization or swelling under high pressure. As a result, the permeability of N_2 in gas mixtures is higher than that of pure N_2 (Liu et al., 2009a), which may lead to a decrease in selectivity. Nevertheless, different results were reported for propylene/N_2 separation by a PEBA/PSF composite membrane. The interactions between propylene and N_2 improved the selectivity of propylene over N_2. Moreover, N_2 permeability was lower in the binary mixture than that measured for pure gas; thus, the propylene/N_2 mixture can be more selectively separated than what is expected from the permeability of pure VOCs and N_2 (Liu et al., 2006b). The difference in N_2 permeability may be attributed to the difference in the membrane structures.

4.4.3 Glassy polymer membrane

Glassy polymer membranes (GPMs) from polymers of intrinsic microporosities (PIMs) have been widely researched for gas separation because of their unique micropore structure and excellent separation performance (Dai et al., 2020; Zhang et al., 2021). As mentioned above, glassy polymers always show high selectivity and low permeability properties because of the densely packed polymer chains in their structures which give less room for macromolecules to pass through. Therefore, a small molecule (such as N_2) easily passes through the GPM, and VOCs can be selectively separated from the gas mixture (Iyer et al., 2020; Mc Keown and Budd, 2006). Poly(1-trimethylsilyl-1-propyne) [PTMSP] first developed glassy polymers for VOC separation, which showed n-butane/methane selectivity as being 30 (Merkel et al., 2000; Pinnau and Toy, 1996). Wang et al. (2002) applied the porous polyetherimide hollow-fiber membrane to separate H_2/(CH_4, C_2H_6, and C_3H_8) and obtained a selectivity as high as 300 for H_2/C_3H_8.

PI is one of the most widely researched polymers as it exhibits good gas selectivity and permeability due to its good thermal stability, chemical stability, excellent mechanical properties, and high free volume (Vanherck et al., 2013). The separation performance of olefins and paraffins of C_2 and C_3 for several PI membranes and other polymer membranes were investigated at pressures up to 8 atm and temperatures from 308 to 423 K (Tanaka et al., 1996). Olefins were more permeable to the corresponding paraffins due to their preferential diffusion based on the difference in their molecular size. Generally, the diffusion coefficients of VOCs in the GPM depend on their molecular structures, where the diffusion coefficients of small VOC molecules are high (Vergadou and Theodorou, 2019). Moreover, elongated or flattened alkane molecules find it easier to pass through the micropores of the GPM than spherical molecules due to their smaller diameters (Berens and Hopfenberg, 1982). It was found that PI prepared from 4,4′-(hexafluoroisopropylidene)diphthalic anhydride (6FDA) and 2,4,6-trimethyl-1,3-phenylenediamine (TrMPD) displayed relatively high performance, with the permeability coefficient of propylene as $P_{C3H6} = 20-40$ Barrer and the ideal selectivity of $\alpha(C_3H_6/C_3H_8)$ was 11 at 323 K and 2 atm. PI from 6FDA and dimethyl-3,7-diaminodiphenylthiophene-5,5-dioxide (DDBT) displayed low permeability ($P_{C3H6} = 0.8$ Barrer) and high permselectivity ($\alpha(C_3H_6/C_3H_8) = 27$) at 323 K and 2 atm. However, their performance was much better than that of other polymers, such as poly(2,6-dimethyl-1,4-phenylene oxide) (Tanaka et al., 1996).

In recent years, conjugated microporous polymers (CMPs) have also been researched as a porous organic polymer with the advantages of a large specific surface area and easy-to-control micro–nano pore structure. A self-supporting conjugated microporous polymer membrane (SS-CMP-M) was synthesized by the Sonogashira–Hagihara (S–H) crosscoupling reaction using KBr tables as a template (Wang et al., 2022). The resulted SS-CMP-M was used to treat organic vapors with a formaldehyde (HCHO) concentration higher than 1.999 mg/m^3 and VOC concentration higher than 9.999 mg/m^3. After SS-CMP-M filtration, the HCHO concentration was reduced to 0.05–0.07 mg/m^3 with a removal efficiency above 97%, and the VOC concentration after filtration was reduced to 0.4–0.5 mg/m^3 with a removal efficiency above 95%.

Although the GPM has excellent separation performance, its membrane aging has to be considered (Bakhtin et al., 2018). The free volume of PI is always reduced after the first exposure of a gas or any VOCs, which is probably caused by the swelling of glassy polymers. In the work of Žák et al. (2015), the selective removal of butanol from an aqueous solution by pervaporation with a PIM-1 membrane was studied. The freshly cast and aged (370 days) PIM-1 membranes were used for pervaporation experiments at 50°C. The aged PIM-1 membrane exhibited a decrease in the permeability coefficient compared to the freshly cast membrane, from 9.0×10^{-5} to 1.1×10^{-5} $mol\ m^{-1}\ kPa^{-1}\ h^{-1}$ for 1-butanol and from 3.2×10^{-5} to 2.2×10^{-6} $mol\ m^{-1}\ kPa^{-1}\ h^{-1}$ for water, and showed a simultaneous increase in selectivity, from 2.8 to 5.1, for the separation process at 2 wt% of butanol. If the problem of aging is resolved, glassy polymers can be used in membrane technology for industrial applications. Besides PI, some examples of glassy polymers, widely used recently for membrane fabrications, are polyethersulfone, polycarbonate, polysulfone, and polyamide. These glassy polymers are usually used to fabricate composite membranes for VOC separation, which are expatiated in section "4.4.5".

4.4.4 Mixed matrix membrane

MMMs are typically prepared by dispersing inorganic micro- or nanosized particles in a continuous organic polymer phase. Recently, attention has been given to various MMMs due to their potential to combine the advantages of polymer matrices and inorganic fillers to simultaneously enhance gas permeability, selectivity, and mechanical strength (Chung et al., 2007). The comparison between the characteristics of the polymeric membrane, inorganic membrane, and MMMs are shown in Table 4.1. In MMMs, the low cost and easy processing of polymeric membranes and the high separation performance of inorganic membranes are combined.

The selection of inorganic fillers must be very precise as their properties can greatly affect membrane structures and separation performances. Previous studies have shown encouraging results by dispersing inorganic fillers, such as MOFs (Basu et al., 2011; Wang et al., 2017), zeolites and their derivatives (Adams et al., 2011; Mashhadikhan et al., 2020), carbon nanotubes (CNTs) (Ismail et al., 2011), carbon molecular sieves (Vu et al., 2003), covalent organic frameworks (COFs), and graphene oxide (GO) (He et al., 2019), in MMMs.

Sedláková et al. (2014) studied the separation characteristics of aliphatic hydrocarbons (hexane), aromatic compounds (toluene), and alcohol (ethanol) on elastomeric ethylene–octene copolymers with CNT and carbon fiber (CF) membranes. The results showed that the permeability of VOCs was much higher than that of a permanent gas, and the permeability of toluene and hexane was about two orders of magnitude higher than that of ethanol, which was due to the lower solubility of ethanol in the intrinsic polymer matrix.

Among fillers, zeolites have been widely researched as appropriate fillers for VOC separation. Two types of zeolites (the ZSM-5 type) with SiO_2/Al_2O_3 ratios of 34.5 (T-15) and 71 (T-1000) were used to synthesis the PEBA/zeolite membrane for VOC (benzene, toluene,

TABLE 4.1 Comparison between the characteristics of polymeric, inorganic, and mixed matrix membranes (Imtiaz et al., 2022).

Characteristics	Polymeric membrane	Inorganic membrane	MMMs
Thermal and chemical stability	Moderate	Superior	Superior
Mechanical strength	Good	Low	Superb
Cost of fabrication	Low	High	Moderate
Process ability and synthesis	Easy and simple	Tough	Easy and simple
Separation performance	Below the Robeson upper bound	Above the Robeson upper bound	Above the Robeson upper bound
Resistant toward high pressure	Moderate	Higher	Higher
Plasticization	Susceptible	Unsusceptible	Unsusceptible

and methanol) separation from CO_2 or N_2 (Friess et al., 2011). In comparison with the pure PEBA membrane, PEBA/T-15 showed higher values of permeability and VOC/(CO_2/N_2) selectivity. The maximum selectivity reached 3003.4 for toluene/N_2. However, the permeability coefficients and selectivity of PEBA/T-1000 were far away from the other two membrane samples, suggesting that it is necessary to select a suitable filler for MMMs with enhanced separation performance.

ZIFs belong to the family of MOFs which are formed by the self-assembly of molecules, in which divalent cations, i.e., Zn or Co, are coordinated tetrahedrally and are connected with imidazolate linkers to form topologies that are similar to those of aluminosilicate zeolites (Kontos et al., 2014). ZIF-based MMMs with polyphenylene oxide (PPO) were synthesized and tested for the permeability of a gas with different kinetic diameters (He, O_2, N_2, CH_4, C_2H_4, C_2H_6, C_3H_6, and C_3H_8) (Kunjattu and Kharul, 2023). The authors found that the permeability of various gases increased with the lowering of the permeant size. Compared to the pristine PPO, there is a 3.4-3.7 fold enhancement of the permeability of C_3H_6, observed for PPO/ZIF membranes. However, C_3H_8 permeability did not exhibit any substantial difference between PPO/ZIF and PPO. Therefore, PPO/ZIF showed excellent C_3H_6/C_3H_8 separation performance, even exceeding the Robeson's upper bound. The enhanced selectivity was attributed to the absence of defects due to metal–polymer interactions. Propylene/propane separation was also conducted by a membrane fabricated using 6FDA–DAM polyimide and ZIF-8 (Zhang et al., 2012). The ZIF-8/6FDA–DAM membrane showed significantly enhanced separation performance over the pure 6FDA–DAM membrane. The C_3H_6 permeability and C_3H_6/C_3H_8 selectivity of the ZIF-8/6FDA–DAM membrane were 258% and 150% higher than that of the pure 6FDA–DAM membrane, respectively. In another study, ZIF-8/PDMS MMMs were prepared for the separation of the propane/nitrogen mixture (Fang et al., 2015). High-resolution scanning electric microscopy (SEM) and energy dispersive X-ray spectroscopy (EDS) confirmed the homogeneous particle dispersion and intimate interface interaction, which led to improved separation selectivity. ZIF-8/PDMS MMMs demonstrated enhanced separation performance (38% higher than that of pure PDMS membranes) and propane permeance over of 100 GPU. In addition, propylene/propane was efficiently separated by polycrystalline ZIF-8 using ceramic support (e.g., alumina disk) (Sun et al., 2018). The tubular ZIF-8 membranes showed excellent propylene/propane separation factors of ~80. MMMs were formed by incorporating ZIF-71 into (PEBA), which showed enhanced separation performance for aromatic compounds (phenol and aniline) from aqueous solutions by pervaporation (Cao et al., 2022).

Mordenite framework inverted (MFI) zeolite membranes have also been widely applied for the separation of xylene isomers. Banihashemi and Lin (2022) synthesized high quality b-oriented MFI zeolite membranes with an organic template on the modified silica substrate. For vapor permeation with a binary p-/o-xylene feed, both the permeance (for p-xylene and o-xylene) and p-/o-xylene separation factor increased with the separation factor reaching a maximum of 515 at 150°C. Sakai et al. (2001) fabricated self-supporting MFI-type zeolites on a Teflon plate, and the separation factors of *p*- to *m*-xylenes and *p*- to *o*-xylenes showed the maximum value of 250 at 473 K.

In MMMs, inorganic fillers offer excellent separation performance of the membrane; thus, the contents of fillers in MMMs play an important role in MMM fabrication and, therefore, need to be carefully designed according to the data of the experiments. The performance of zeolite/PEBA MMMs for VOCs (benzene, toluene, and methanol) and CO_2/N_2 separation was investigated at two zeolite-filled ratios of 10 wt% and 30 wt%, respectively (Friess et al., 2011). At the filled ratio of 10 wt%, MMMs showed much lower selectivity than that of the unfilled PEBA membrane, while VOC permeability remained almost constant. However, increasing the zeolite content to 30 wt%, MMMs changed the VOC permeability considerably (+113% for benzene, +82% for toluene, and +73% for methanol) and, thus, led to an increase of VOC/N_2 selectivity. Another study investigated the performance of PA@GO MMMs in the separation of cyclohexane/nitrogen at four GO contents of 0.15, 0.3, 0.45, and 0.6 wt% (Mao et al., 2022). A higher GO content enhanced nitrogen permeation because of the increasing amorphous domain in the GO-added mixed polymers. Meanwhile, with the increasing GO content, the surface area increased and membrane density decreased, indicating that more transporting channels were available for small molecule nitrogen permeation through the membrane. As for cyclohexane, the highest permeability was observed at GO content of 0.3 wt%, and thus the highest rejection of 99.4% was obtained. Ordoñez et al. (2010) fabricated ZIF-8/Matrimid MMMs with loadings up to 80% (w/w), which are much higher than the typical loadings achieved with select zeolite materials. The ZIF-8/Matrimid MMMs permeability values increased as the ZIF-8 loading increased to 40% (w/w). However, at higher loadings of 50% and 60% (w/w), the permeability decreased, and the selectivity increased consistent with the influence of the ZIF-8 additive. In addition, with 50% (w/w) ZIF-8 loading, the ideal selectivity of the gas pairs of CO_2/CH_4, CO_2/C_3H_8, and H_2/C_3H_8 showed improvement.

The fabrication of MMMs usually faces the issues of incompatibility among fillers and polymers, which is generally considered as the key factor that results in interfacial defects. Generally, MMMs are fabricated by mechanically mixing polymers and fillers in a solvent to form a membrane solution, which often causes poor filler dispersion and the formation of voids in the polymer matrix (Tien-Binh et al., 2018; Zhang et al., 2019). Therefore, studies have been conducted focusing on the synthesis method to enhance filler dispersion in the polymer matrix and interface compatibility. Mao et al. (2022) develop an in situ polymerization method, in which 2,6,14-triaminotriptycene (Trip), octanedioyl chloride, and graphene oxide reacted in the polymerization process to form a PA@GO MMMs. The MMMs exhibited a higher separation performance than the mechanically mixed PA + GO membrane. Furthermore, the VOC rejection of the mechanically mixed PA + GO membrane was even lower than that of the pure PA membrane, which was attributed to the aggregation of GO sheets that led to the formation of nonselective interfacial voids. These voids permit the escape of VOC molecules and result in the reduction of the separation performance (Goh et al., 2011; Mahajan and Koros, 2000). Another research attempted to tune the molecular sieving properties of ZIF-8 fillers by doping them with 2-ethylimidazole (eIm) linkers (Park and Jeong, 2020). The eIm-doped ZIF-8 was filled in PI using a novel polymer modification-enabled in situ MOF formation (PMMOF) process. The resultant PI/eIm-doped ZIF-8 MMMs enhanced the propylene/propane separation factor owing to eIm linkers. The examples of VOC separation by MMMs are summarized in Table 4.2.

4.4.5 Composite membrane

A composite membrane is commonly formed by coating a dense organic polymer as the selective skin layer on a porous support (Fig. 4.7). Therefore, a composite membrane combines the advantages of both a dense polymer and porous support, showing high permeability and selectivity. In the composite membrane, a dense polymeric layer ensures selective separation, and the porous support allows for the transport of separated VOCs. During synthesis, the selective layer should be as thin as possible to reduce transmission resistance, while the porous support is usually 10–100 times more permeable. The transmission resistance of the support layer is usually negligible compared with that of the skin layer. As a result, the membrane flux is determined by the dense polymer. PDMS and PEBA are the most widely used dense polymers for composite membranes, while PSF, PEI, PA, and PVDF are the common porous support materials (Gan et al., 2023; Petrusová et al., 2019).

PDMS/ceramic composite membranes have a good potential for the separation of condensable VOCs from N_2. The separation performance of the representative VOCs, such as acetone, p-xylene, ethanol, and ethyl acetate, from N_2 was investigated by a modified constant volume/variable pressure method to avoid the effects of VOC condensation in the permeate (Yang et al., 2018). The N_2 permeability depended on the properties of the VOCs. The higher condensability of VOCs led to a higher VOC flux and a lower N_2 flux. A higher feed concentration and feed flow rate increased VOC permeabilities and weakened those of N_2 through the PDMS/ceramic membrane, leading to higher VOC/ N_2 selectivity.

The PDMS thin-film composite membrane was adapted for propylene/nitrogen separation by Shi et al. (2006). They found that propylene permeability through the composite membrane was independent of the nitrogen presence, while nitrogen permeability increased with the propylene presence due to polymer plasticization. Consequently, the actual permeability ratio of propylene and nitrogen in a gas mixture was lower than that obtained from

TABLE 4.2 VOC separation performance of MMMs.

Polymer matrix	Filler	Permeability (barrer)	Selectivity		Refs.
Polyurethane (PU)	Silica nanoparticles	C_3H_8: 118	C_3H_8/CH_4: 7.01		Khosravi et al. (2014)
Polyurethane (PU)	Zeolite 4 Å and ZSM-5	C_2H_6: 23.2–71.3 C_3H_8: 34.7–117.2	C_2H_6/CH_4: 1.2–2.2 C_3H_8/CH_4: 1.4–3.6		Tirouni et al. (2015)
PEBA	ZSM-5	CO_2: 131.8 N_2: 7.4	C_6H_6/CO_2: 100.6 C_7H_8/CO_2: 146.5 CH_3OH/CO_2: 70.4	C_6H_6/N_2: 2075.5 C_7H_8/N_2: 3003.4 CH_3OH/N_2: 1227.0	Friess et al. (2011)
PPO	ZIF-8 ZIF-67	C_3H_6: 12 C_3H_6: 13	C_3H_6/C_3H_8: 27.5 C_3H_6/C_3H_8: 25		Kunjattu and Kharul (2023)
6FDA–DAM	ZIF-8	C_3H_6: 56.2	C_3H_6/C_3H_8: 31		Zhang et al. (2012)
6FDA–DAM	ZIF-67	/	C_3H_6/C_3H_8: 29.9		An et al. (2017)
Poly(4-methyl-2-pentyne)	Silica	/	C_4H_{10}/CH_4: 26		He et al. (2002)

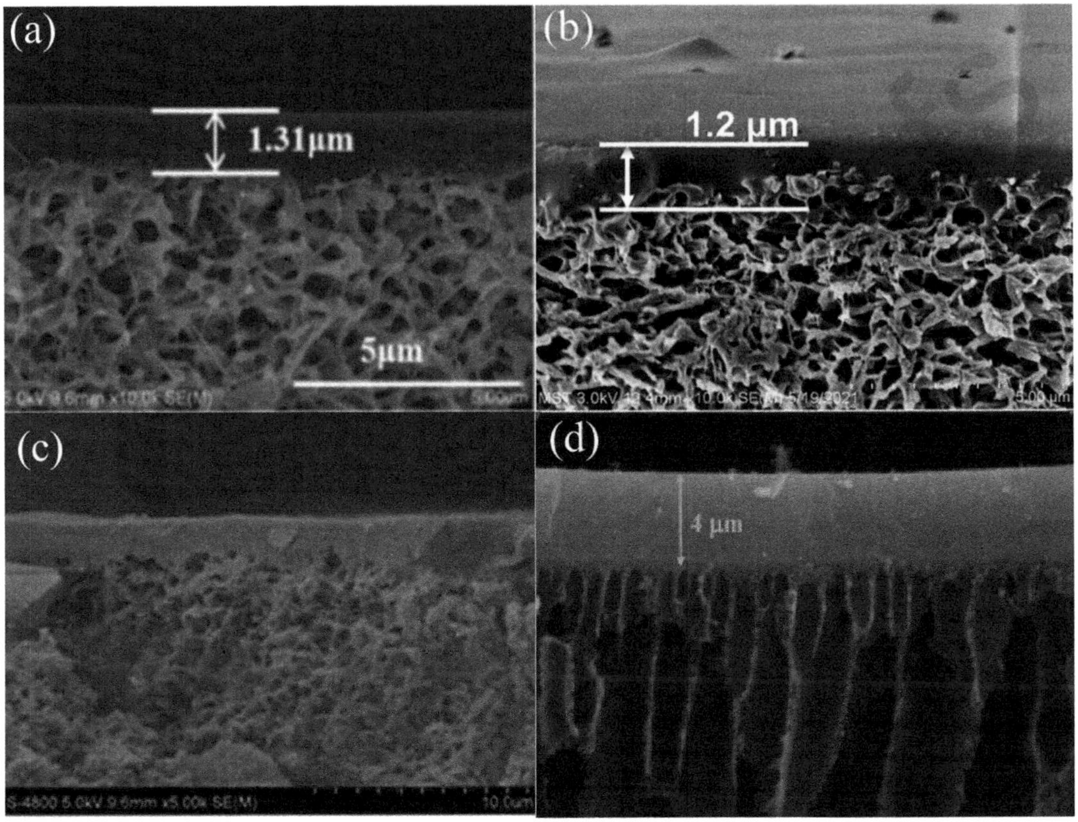

FIGURE 4.7 (a) PA composite membrane (Zhang et al., 2023), (b) PDMS/PA composite membrane (Liang et al., 2023), (c) PA composite membrane (Chen et al., 2020), and (d) PDMS/PES composite membrane (Sadrzadeh et al., 2009).

pure gas. However, it was still highly efficient for practical propylene/nitrogen separation. The membrane exhibited a propylene/nitrogen permeability ratio as high as 125, and membrane selectivity increased with the increasing feed pressure and decreasing temperature.

Jiang and Kumar (2005) studied the separation of VOCs (C_2H_4, C_2H_6, C_3H_6, and C_3H_8) and nitrogen by a PDMS/PSF composite membrane. They found that C_3H_8 and/or C_3H_6 significantly plasticized the PDMS coating in the pure as well as gas mixture permeation experiments. At the applied pressures that were higher than the plasticization pressures, the observed permeability order was $C_3H_8 > C_3H_6 > C_2H_6 > C_2H_4 > N_2$, which was in order of their solubilities. However, C_3H_6 was more permeable than C_3H_8 below the plasticization pressures, while the order of other gases did not change.

During VOC/nitrogen or air separation, a third impurity, such as water vapor, may coexist in the gas mixture and affect membrane performance. Liang et al. (2023) studied the effect of water vapor on the separation performance of the PDMS/PA composite membrane for binary VOC/nitrogen and ternary VOC/water/nitrogen mixtures. The VOCs tested include hydrophilic ethanol, hydrophobic cyclohexane, and butanol. The results indicated that water inhibited the permeation of hydrophobic VOCs but promoted the permeation of the hydrophilic ones due to the different interactions between the VOCs and water. As a result, nitrogen permeability increased in the ternary cyclohexane/water/nitrogen mixture due to the decrease in hydrophobic cyclohexane. On the contrary, it decreased in the ternary hydrophilic ethanol/water/nitrogen mixture. Therefore, a higher selectivity of a hydrophilic VOC over nitrogen can be observed in the ternary mixture rather than that in the binary mixture.

PDMS/PEI hollow-fiber composite membranes were successfully tested for light olefin (C_2H_4 and C_3H_6) separation from nitrogen (Liu et al., 2009b) and acetone and ethyl acetate and ethanol separation from the air (Gales et al., 2002). Furthermore, PDMS/PA composite membranes were used for cyclohexane/nitrogen separation (Shen et al., 2022). These studies have shown that the permeability of N_2 was almost constant, while the permeability of VOCs was affected by many factors, such as porous substrate resistance and permeate pressure. PDMS as a selective skin layer with different porous supports for VOC separation is shown in Table 4.3.

PEBA is also an attractive polymer for VOC separation. Liu et al. (2006b) studied the separation of propylene from nitrogen by a PEBA 2533 skin layer on a microporous PSF support. The solubility of VOCs significantly influenced the diffusion–solution transport through the PEBA/PSF composite membrane, which was similar with a rubbery membrane. The permeability of propylene was influenced by the presence of nitrogen, due to the interaction between two permeating components and the competition in the sorption in the PEBA skin layer. As a result, both propylene permeability and propylene/nitrogen selectivity were lower than that of the pure gas permeation process.

TABLE 4.3 VOC separation performance of composite membranes.

Organic polymer	Porous support	Permeability (barrer)	Selectivity	Refs.
PDMS (4 μm)	PES	C_3H_8: 24,550	C_3H_8/CO_2: 5.1 C_3H_8/CH_4: 14.7 C_3H_8/H_2: 22.1	Sadrzadeh et al. (2009)
PDMS (0.20 μm)	PSF	C_3H_8: 1970 GPU[a] C_3H_6: 1700 GPU	C_3H_8/N_2: 16.8 C_3H_6/N_2: 16.2	Jiang and Kumar (2005)
PDMS (0.45 μm)	PSF	C_3H_8: 143 GPU C_3H_6: 218 GPU	C_3H_8/N_2: 5.9 C_3H_6/N_2: 8.9	Jiang and Kumar (2005)
PDMS (0.20 μm)	PSF	C_2H_4: 1546 GPU C_3H_6: 1468 and 2443 GPU	C_2H_4/N_2: 9.1 C_3H_6/N_2: 9.1 and 12.2	Jiang and Kumar (2006)
PDMS based PU (12.5 μm)	PVDF	C_3H_6: 37.5 GPU	C_3H_6/N_2: 19.0	Jiang et al. (2008)
PDMS based PU (25 μm)	PVDF	C_3H_6: 40.3 GPU	C_3H_6/N_2: 17.0	Jiang et al. (2008)

[a] *1 GPU = 1×10^{-6} cm^3 (STP)/(cm^2·s·cmHg).*

In a composite membrane, the dense polymeric layer ensures selective separation; thus, polymeric layer thickness is a critical factor for membrane separation performance. For example, the selectivity of the PEBA/PSF membrane for the separation of the methanol/N_2 and ethanol/N_2 mixtures decreased as the PEBA skin-layer thickness reduced (Liu et al., 2009b). VOC properties also affect membrane performance; the permeability order of three VOC representatives through a PEBA/PVDF composite membrane was heptane >cyclohexane >hexane, corresponding to their condensability (Liu et al., 2006). Similarly, the permeability of light hydrocarbons (ethylene, ethane, propylene, propane, n-butane, and methane) through a multilayer PEI/PDMS/PEBA 1657/PDMS composite membrane increased with their condensability; olefins were more permeable than the corresponding paraffins. The multilayer composite membrane exhibited VOC/N_2 selectivity that was as high as 67.1 (Ren et al., 2013).

A VOC-rejective membrane refers to a membrane that lets smaller molecules, such as nitrogen, to permselectively travel through the membrane over larger VOC molecules, such as the cyclohexane molecule for the separation of a nitrogen/cyclohexane mixture (Zhang et al., 2021). Recently, a research group at Nanjing Tech University have synthesized VOC-rejective membranes for the separation of VOCs and the nitrogen mixture.

A microporous VOC-rejective composite membrane was synthesized by coating PI on the surface of the nylon 6 substrate for the molecular sieving of nitrogen over VOCs (cyclohexane, toluene, hexane, and acetone) (Zhang et al., 2021). The membrane exhibited high nitrogen permeability ranging from 2000 to 2140 Barrer. The highest VOC permeability was observed for acetone (about 5 Barrer), followed by hexane, toluene, and cyclohexane, which oppositely correspond to the order of their molecular dimensions. Therefore, the highest rejection was observed for the nitrogen/cyclohexane mixture, which was higher than 99%. Furthermore, the stability of the composited membrane for nitrogen/cyclohexane separation was investigated for 100 days. During the first 40 days, the membrane maintained a constant VOC rejection (over 99%) and nitrogen permeability, and then, the VOC rejection started to decline and the permeability for nitrogen and cyclohexane increased. This indicates that the composite membrane had good stability for at least 40 days.

A triptycene-based linear PA composite membrane also exhibited high rejection rates of VOCs (cyclohexane, hexane, and heptane) from nitrogen (Dai et al., 2020). Its lowest VOC permeability was obtained for cyclohexane (>96%), as it has relatively large molecular dimensions. The rejection rates of hexane and heptane were both about 95%, as the two have similar molecular dimensions. Chen et al. (2020) investigated the effect of the crosslinking temperature on the microporous PA composite membrane structure and cyclohexane/nitrogen separation performance. Their results showed that a higher temperature led to higher crosslinking and a smaller pore size, so that 99% rejection was obtained when the crosslinking temperature was over 80°C, while the permeability of nitrogen slightly decreased with the increasing crosslinking temperature from 50 to 120°C. To improve the stability membrane, an ultrastable fully aromatic microporous PA composite membrane was synthesized by the polymerization of triptycene with terephthaloyl chloride, which was fabricated into a membrane using a solution process to evaluate its separation performance and stability for cyclohexane/nitrogen separation (Zhang et al., 2023). The results showed constant rejection (>99%) and permeability (~50 Barrer) during the 100-d experiments, suggesting good membrane separation performance and stability.

4.4.6 Liquid membrane

Ionic liquids are organic salts whose melting points are typically lower than 100°C (Ortiz-Martínez and Salar-García, 2019). Recently, a combination of membrane separation and ionic liquids has been applied in VOC removal due to its enhanced separation efficiency (Wang et al., 2016; Yan et al., 2019). For ionic liquid membrane fabrication, an ionic liquid is either spread on a porous support as a thin selective layer or is distributed evenly inside the porous support due to capillary forces. When the ionic liquid is distributed inside the pores of the support membrane, it is called the supported ionic liquid membrane (SILM). According to their configurations, SILMs can be classified into flat sheet (FSILM) or hollow-fiber module (HFSILM) ones.

SILMs are attractive for VOC separation due to their adjustable properties, negligible volatility, and good sorption ability of ionic liquids. The physical properties of an ionic liquid, such as density, viscosity, and polarity, as well as the solubility and diffusivity of VOC molecules control the liquid membrane performance. Generally, VOC molecules are transported slower in a more viscous liquid, thus having lower permeability. On the other hand, the properties of the porous support influence the transport process by its porosity and pore-size distributions. Some commonly used ionic liquids and porous supports for VOC separation are summarized in Table 4.4.

Because of its nonvolatility, triethylene glycol (TEG) is a suitable candidate for ionic liquid membranes for VOC separation (Shimoyama and Ito, 2011). Ito et al. (2008) separated acetaldehyde, acetone, benzene, and toluene from the air using SILMs containing four kinds of TEGs, i.e., TEG, TEG−DBE (dibutyl ether), TEG−DME (dimethyl ether), and TEG−MBE (monobutyl ether), on PVDF support. The resulting permeability of the VOCs and selectivity over air depended on the hydrophobic or hydrophilic property of the liquid membrane material. The hydrophobic TEG−DBE membrane was effective for removing hydrophobic VOCs, such as benzene and toluene. While the hydrophilic TEG liquid membrane had selectivity for hydrophilic methanol and acetone. Thus, the VOC selectivity of a TEG liquid membrane could be explained by the solubility or absorption in an ionic liquid.

Sui et al. (2009) synthesized a novel surface-soaked liquid membrane by soaking a hydrophilic surface Durapel membrane in a TEG ionic liquid. The membrane was tested for the separation of VOC vapors and showed comparable separation performance to a conventional adsorption process with active carbon. The TEG membrane process was effective for the separation of benzene, toluene, acetone, or methanol from a room's air.

Vegetable and lubricant oils can be used as cheaper alternatives to ionic liquids in membranes. Ozturk et al. (2015) immobilized them in flat-sheet hydrophilic microporous cellulose acetate for VOC/N_2 separation. They tested C_6H_6, CCl_4, and CH_3OH with different properties. C_6H_6 and CCl_4 are nonpolar and hydrophobic and have higher solubility in oils; the membrane permeabilities of C_6H_6 and CCl_4 were much higher than that of CH_3OH, which is polar and hydrophilic. Between C_6H_6 and CCl_4, CCl_4 has higher solubility in oils than that of C_6H_6, but its branched chemical structure and interaction with cellulose acetate reduced its diffusivity through oil, resulting in lower permeability. The results also showed that the permeability of N_2 obtained by feeding binary VOCs/N_2 was much higher than that of pure N_2, which was contributed to the swelling effect of VOCs (Gales et al., 2002). The

TABLE 4.4 Ionic liquids and support materials for VOC separation.

Ionic liquids	Support materials	VOCs	Refs.
[BMIM][PF_6]: 1-n-Butyl-hexafluorophosphate [HMIM][PF_6]: 1-n-Hexyl-hexafluorophosphate [OMIM][PF_6]: 1-n-Octyl-3-methylimidazolium hexafluorophosphate	PVDF	Benzene, toluene, and *p*-xylene	Matsumoto et al. (2005)
[C_4mim][PF_6]: 1-Butyl-3-methylimidazolium hexafluorophosphate [C_6mim][PF_6]: 1-n-Hexyl-3-methylimidazolium hexafluorophosphate [C_8mim][PF_6]: 1-n-Octyl-3-methylimidazolium hexafluorophosphate	PVDF	Benzene and cyclohexane	Matsumoto et al. (2009)
[BMIM] [BF_4]: 1-Butyl-3-methylimidazolium tetrafluoroborate (containing Ag^+)	Iolitec	Propane and propylene	Fallanza et al. (2012)
[C_4C_1im][BF_4]: 1-Butyl-3-methylimidazolium tetrafluoroborate [C_4C_1im][PF_6]: 1-Butyl-3-methylimidazolium hexafluorophosphate [C_6C_1im][PF_6]: 1-Hexyl-3-methylimidazolium hexafluorophosphate	Matrimid	Ethanol and cyclohexane	Dahi et al. (2014)
[C_2min][Tf_2N]: 1-Ethyl-3-methylimidazolium bis(trifluoromethylsulfonyl)imide [C_2min][TfO]: 1-Ethyl-3 methylimidazolium trifluoromethanesulfonate [C_4min][Tf_2N]: 1-Butyl-3-methylimidazolium bis(trifluoromethylsulfonyl)imide [C_4min][TfO]: 1-Butyl-3-methylimidazolium trifluoromethanesulfonate [C_6min][Tf_2N]: 1-Hexyl-3-methylimidazolium bis(trifluoromethylsulfonyl)imide	Polypropylene (PP)	Toluene	Cichowska-Kopczyńska et al. (2018)

stability of the membrane was also confirmed by no deterioration in the selectivity after 150 h of operation.

Ionic liquid membranes can be combined with other technologies, such as photocatalysis, for VOC removal. Li et al. (2021) embedded Nd (neodymium)−TiO_2 nanoparticles in a SILM to functionalize the membrane with a photocatalytic capability. The membrane was installed in a photocatalytic membrane-based reactor under UV−Vis light illumination for removing VOCs. The results showed that at a loading of 50 wt%, Nd−TiO_2 nanoparticles (1 wt% Nd content) on 1-hexyl-3-methylimidazolium hexafluorophosphate ([HMIm]PF6) SILMs ([HMIm]PF6/Nd−TiO_2 SILMs) exhibited a higher removal efficiency of VOCs from air (60%−80% removal after 10 h) than that of pure SILMs or photocatalytic systems. In another study, composite polymeric membranes were synthesized using thermoplastic polyurethane (TPU) and several silver salts/silver salts with ionic liquids (Wang et al., 2020). The study showed that silver bis(trifluoromethanesulfonyl)imide (Ag[Tf2N])-containing membranes outperformed other silver salt-containing membranes in the terms of permeability and propane/propylene selectivity. C_3H_6 permeability was 590.4–8512.3 Barrer with a C_3H_6/C_3H_8 selectivity of 1.8–38.0.

The challenge faced by liquid membranes is their stability because the liquid is easily leaked from the membrane at elevated transmembrane pressures (Wang et al., 2016). In order to prevent the leakage of liquid, a silicone coating is usually applied. Obuskovic et al. (2003) coated PDMS on the surface of the microporous polypropylene membrane and then introduced silicone oil into the pores of the PDMS skin layer. The resultant ionic liquid membrane showed high selectivity for VOCs (toluene, methanol, and acetone) through N_2 and a long stability of 6 months and even to 2 years without sacrificing the separation performance. Another method is based on bonding the liquid to a polymer matrix to synthesize a poly(ionic liquid) membrane, which can overcome the stability limitations (Fallanza et al., 2013).

4.5 Summary and perspectives

VOCs are among the most common air pollutants discharged into the atmosphere. Recently, the use of membrane separation in VOC removal has received widespread attention globally. In this chapter, the separation mechanisms, key factors affecting membrane performance, and the various types of membranes are summarized. The membrane VOC separation mechanisms mainly include microporous diffusion and solution–diffusion depending on the materials used for fabricating membranes. Membrane performance is influenced by the membrane properties, such as thickness, and operation conditions, including feed temperature, pressure, flow rate, and VOC concentration. The effects of these factors on the membrane performance vary based on the membrane structure and VOC properties. The commonly used membrane materials mainly include PDMS, PEBA, glassy polymers, and their combinations for the fabrication of MMMs and composite membranes. Ionic liquid membranes also show excellent VOC separation performance.

One of the notable challenges in the membrane VOC separation field is the permeability–selectivity trade-off, which is simply defined as an imbalance between permeability and selectivity. Other challenges include the following: (1) plasticization causes declined selectivity and increased permeability in polymeric membranes; (2) incompatibility between a filler and a polymer is generally considered as a notorious issue for MMMs, which results in interfacial defects and nonselective interfacial voids; (3) a similar problem of incompatibility between the selective skin layer and the porous support limits the development of a composite membrane; (4) the liquid is easily leaks from the ionic liquid membrane at an elevated transmembrane pressure; and (5) the long-term stability of all the membranes is still uncertain and more pilot plan tests are required. For efficient VOC removal by membrane separation, further investigations are encouraged in the following fields: (1) the exploration of new membrane materials with high separation and permeability; (2) a combination of different membrane materials for a synergetic effect in the separation process; (3) the optimization of the separation process by adjusting the operation conditions; and (4) the improvement of membrane stability for large-scale industrial applications.

References

Adams, R.T., Lee, J.S., Bae, T.H., Ward, J.K., Johnson, J.R., Jones, C.W., Nair, S., Koros, W.J., 2011. CO2-CH4 permeation in high zeolite 4A loading mixed matrix membranes. Journal of Membrane Science. https://doi.org/10.1016/j.memsci.2010.10.059.

An, H., Park, S., Kwon, H.T., Jeong, H.K., Lee, J.S., 2017. A new superior competitor for exceptional propylene/propane separations: ZIF-67 containing mixed matrix membranes. In: Sep. Div. 2017 - Core Program. Area 2017 AIChE Annu. Meet.

Bakhtin, D.S., Kulikov, L.A., Legkov, S.A., Khotimskiy, V.S., Levin, I.S., Borisov, I.L., Maksimov, A.L., Volkov, V.V., Karakhanov, E.A., Volkov, A.V., 2018. Aging of thin-film composite membranes based on PTMSP loaded with porous aromatic frameworks. Journal of Membrane Science. https://doi.org/10.1016/j.memsci.2018.03.001.

Banihashemi, F., Lin, J.Y.S., 2022. B-oriented MFI zeolite membranes for xylene isomer separation - effect of xylene activity on separation performance. Journal of Membrane Science. https://doi.org/10.1016/j.memsci.2022.120492.

Basu, S., Cano-Odena, A., Vankelecom, I.F.J., 2011. MOF-containing mixed-matrix membranes for CO2/CH4 and CO2/N2 binary gas mixture separations. Separation and Purification Technology. https://doi.org/10.1016/j.seppur.2011.06.037.

Berens, A.R., Hopfenberg, H.B., 1982. Diffusion of organic vapors at low concentrations in glassy PVC, polystyrene, and PMMA. Journal of Membrane Science. https://doi.org/10.1016/S0376-7388(00)81415-5.

Cao, X., Wang, K., Feng, X., 2022. Incorporating ZIF-71 into poly(ether-block-amide) (PEBA) to form mixed matrix membranes for enhanced separation of aromatic compounds from aqueous solutions by pervaporation. Separation and Purification Technology. https://doi.org/10.1016/j.seppur.2022.121924.

Chen, Y., Qin, J., Tong, T., Zhou, H., Cao, X., Jin, W., 2020. Study on the effect of crosslinking temperature on microporous polyamide membrane structure and its nitrogen/cyclohexane separation performance. Separation and Purification Technology 252, 117401. https://doi.org/10.1016/j.seppur.2020.117401.

Cho, Y.J., Park, H.B., 2011. Gas separation properties of triptycene-based polyimide membranes. American Chemical Society.

Choi, S.H., Kim, J.H., Lee, S.B., 2007. Sorption and permeation behaviors of a series of olefins and nitrogen through PDMS membranes. Journal of Membrane Science 299, 54–62. https://doi.org/10.1016/j.memsci.2007.04.022.

Chuah, C.Y., Goh, K., Yang, Y., Gong, H., Li, W., Karahan, H.E., Guiver, M.D., Wang, R., Bae, T.H., 2018. Harnessing filler materials for enhancing biogas separation membranes. Chemical Reviews 118, 8655–8769. https://doi.org/10.1021/acs.chemrev.8b00091.

Chung, T.S., Jiang, L.Y., Li, Y., Kulprathipanja, S., 2007. Mixed matrix membranes (MMMs) comprising organic polymers with dispersed inorganic fillers for gas separation. Progress in Polymer Science. https://doi.org/10.1016/j.progpolymsci.2007.01.008.

Cichowska-Kopczyńska, I., Joskowska, M., Debski, B., Aranowski, R., Hupka, J., 2018. Separation of toluene from gas phase using supported imidazolium ionic liquid membrane. Journal of Membrane Science. https://doi.org/10.1016/j.memsci.2018.08.058.

Dahi, A., Fatyeyeva, K., Langevin, D., Chappey, C., Rogalsky, S.P., Tarasyuk, O.P., Benamor, A., Marais, S., 2014. Supported ionic liquid membranes for water and volatile organic compounds separation: sorption and permeation properties. Journal of Membrane Science. https://doi.org/10.1016/j.memsci.2014.01.031.

Dai, S., Liao, R., Zhou, H., Jin, W., 2020. Synthesis of triptycene-based linear polyamide membrane for molecular sieving of N2 from the VOC mixture. Separation and Purification Technology 252, 117355. https://doi.org/10.1016/j.seppur.2020.117355.

Dotremont, C., Brabants, B., Geeroms, K., Mewis, J., Vandecasteele, C., 1995. Sorption and diffusion of chlorinated hydrocarbons in silicalite-filled PDMS membranes. Journal of Membrane Science. https://doi.org/10.1016/0376-7388(95)00019-9.

Dumont, E., Couvert, A., Amrane, A., Couriol, C., Darracq, G., Le Cloirec, P., 2016. Equivalent Absorption Capacity (EAC) concept applied to the absorption of hydrophobic VOCs in a water/PDMS mixture. Chemical Engineering Journal 287, 205–216. https://doi.org/10.1016/j.cej.2015.11.020.

Fallanza, M., Ortiz, A., Gorri, D., Ortiz, I., 2013. Polymer-ionic liquid composite membranes for propane/propylene separation by facilitated transport. Journal of Membrane Science. https://doi.org/10.1016/j.memsci.2013.05.015.

Fallanza, M., Ortiz, A., Gorri, D., Ortiz, I., 2012. Experimental study of the separation of propane/propylene mixtures by supported ionic liquid membranes containing Ag +-RTILs as carrier. Separation and Purification Technology. https://doi.org/10.1016/j.seppur.2012.01.044.

Fang, M., Wu, C., Yang, Z., Wang, T., Xia, Y., Li, J., 2015. ZIF-8/PDMS mixed matrix membranes for propane/nitrogen mixture separation: experimental result and permeation model validation. Journal of Membrane Science. https://doi.org/10.1016/j.memsci.2014.09.040.

Farnam, M., bin Mukhtar, H., bin Mohd Shariff, A., 2021. A review on glassy and rubbery polymeric membranes for natural gas purification. ChemBioEng Reviews 8, 90–109. https://doi.org/10.1002/cben.202100002.

Friess, K., Hynek, V., Šípek, M., Kujawski, W.M., Vopička, O., Zgažar, M., Kujawski, M.W., 2011. Permeation and sorption properties of poly(ether-block-amide) membranes filled by two types of zeolites. Separation and Purification Technology 80, 418–427. https://doi.org/10.1016/j.seppur.2011.04.012.

Friess, K., Jansen, J.C., Bazzarelli, F., Izák, P., Jarmarová, V., Kačírková, M., Schauer, J., Clarizia, G., Bernardo, P., 2012. High ionic liquid content polymeric gel membranes: correlation of membrane structure with gas and vapour transport properties. Journal of Membrane Science. https://doi.org/10.1016/j.memsci.2012.05.072.

Gales, L., Mendes, A., Costa, C., 2002. Removal of acetone, ethyl acetate and ethanol vapors from air using a hollow fiber PDMS membrane module. Journal of Membrane Science. https://doi.org/10.1016/S0376-7388(01)00628-7.

Gan, G., Fan, S., Li, X., Zhang, Z., Hao, Z., 2023. Adsorption and membrane separation for removal and recovery of volatile organic compounds. Journal of Environmental Sciences 123, 96–115. https://doi.org/10.1016/j.jes.2022.02.006.

Goh, P.S., Ismail, A.F., Sanip, S.M., Ng, B.C., Aziz, M., 2011. Recent advances of inorganic fillers in mixed matrix membrane for gas separation. Separation and Purification Technology. https://doi.org/10.1016/j.seppur.2011.07.042.

He, R., Cong, S., Wang, J., Liu, J., Zhang, Y., 2019. Porous graphene oxide/porous organic polymer hybrid nanosheets functionalized mixed matrix membrane for efficient CO2 capture. ACS Applied Materials & Interfaces. https://doi.org/10.1021/acsami.8b17599.

He, Z., Pinnau, I., Morisato, A., 2002. Nanostructured poly(4-methyl-2-pentyne)/silica hybrid membranes for gas separation. Desalination. https://doi.org/10.1016/S0011-9164(02)00463-0.

Hong, T., Li, Y., Wang, S., Li, Y., Jing, X., 2022. Polyurethane-based gas separation membranes: a review and perspectives. Separation and Purification Technology 301, 1–12. https://doi.org/10.1016/j.seppur.2022.122067.

Imtiaz, A., Othman, M.H.D., Jilani, A., Khan, I.U., Kamaludin, R., Samuel, O., 2022. ZIF-filler incorporated mixed matrix membranes (MMMs) for efficient gas separation: a review. Journal of Environmental Chemical Engineering 10, 108541. https://doi.org/10.1016/j.jece.2022.108541.

Ismail, A.F., Rahim, N.H., Mustafa, A., Matsuura, T., Ng, B.C., Abdullah, S., Hashemifard, S.A., 2011. Gas separation performance of polyethersulfone/multi-walled carbon nanotubes mixed matrix membranes. Separation and Purification Technology. https://doi.org/10.1016/j.seppur.2011.03.031.

Ito, A., Sui, G., Yamanouchi, N., 2008. VOC vapor permeation through a liquid membrane using triethylene glycols. Desalination. https://doi.org/10.1016/j.desal.2007.09.094.

Iyer, G.M., Liu, L., Zhang, C., 2020. Hydrocarbon separations by glassy polymer membranes. Journal of Polymer Science. https://doi.org/10.1002/pol.20200128.

Jiang, X., Ding, J., Kumar, A., 2008. Polyurethane-poly(vinylidene fluoride) (PU-PVDF) thin film composite membranes for gas separation. Journal of Membrane Science 323, 371–378. https://doi.org/10.1016/j.memsci.2008.06.048.

Jiang, X., Kumar, A., 2006. Silicone-coated polymeric membrane for separation of hydrocarbons and nitrogen at subambient temperatures. Journal of Membrane Science. https://doi.org/10.1016/j.memsci.2006.10.007.

Jiang, X., Kumar, A., 2005. Performance of silicone-coated polymeric membrane in separation of hydrocarbons and nitrogen mixtures. Journal of Membrane Science 254, 179–188. https://doi.org/10.1016/j.memsci.2004.12.041.

Jirsáková, K., Stanovský, P., Dytrych, P., Morávková, L., Přibylová, K., Petrusová, Z., Jansen, J.C., Izák, P., 2021. Organic vapour permeation in amorphous and semi-crystalline rubbery membranes: experimental data versus prediction by solubility parameters. Journal of Membrane Science. https://doi.org/10.1016/j.memsci.2021.119211.

Khosravi, A., Sadeghi, M., Banadkohi, H.Z., Talakesh, M.M., 2014. Polyurethane-silica nanocomposite membranes for separation of propane/methane and ethane/methane. Industrial & Engineering Chemistry Research. https://doi.org/10.1021/ie403322w.

Kontos, A.G., Likodimos, V., Veziri, C.M., Kouvelos, E., Moustakas, N., Karanikolos, G.N., Romanos, G.E., Falaras, P., 2014. CO2 captured in zeolitic imidazolate frameworks: Raman spectroscopic analysis of uptake and host-guest interactions. Chem Sus Chem. https://doi.org/10.1002/cssc.201301323.

Koops, G.H., Nolten, J.A.M., Mulder, M.H.V., Smolders, C.A., 1994. Selectivity as a function of membrane thickness: gas separation and pervaporation. Journal of Applied Polymer Science 53, 1639–1651. https://doi.org/10.1002/app.1994.070531210.

Kujawska, A., Knozowska, K., Kujawa, J., Kujawski, W., 2016. Influence of downstream pressure on pervaporation properties of PDMS and POMS based membranes. Separation and Purification Technology. https://doi.org/10.1016/j.seppur.2015.12.057.

Kunjattu, H.,S., Kharul, U.K., 2023. PPO-ZIF MMMs possessing metal-polymer interactions for propane/propylene separation. Journal of Membrane Science 668, 121208. https://doi.org/10.1016/j.memsci.2022.121208.

Li, J., Li, B., Sui, G., Du, L., Zhuang, Y., Zhang, Y., Zou, Y., 2021. Removal of volatile organic compounds from air using supported ionic liquid membrane containing ultraviolet-visible light-driven Nd-TiO_2 nanoparticles. Journal of Molecular Structure 1231, 2–10. https://doi.org/10.1016/j.molstruc.2021.130023.

Li, L., Xu, R., Song, C., Zhang, B., Liu, Q., Wang, T., 2018. A review on the progress in nanoparticle/C hybrid CMS membranes for gas separation. Membranes. https://doi.org/10.3390/membranes8040134.

Liang, Y., Lei, H., He, X., Zhou, H., Jin, W., 2023. Influence of water vapor on the separation of volatile organic compound/nitrogen mixture by polydimethylsiloxane membrane. Chinese Journal of Chemical Engineering. https://doi.org/10.1016/j.cjche.2023.06.007.

Lin, D., Ding, Z., Liu, L., Ma, R., 2012. Experimental study of vapor permeation of C5C7 alkane through PDMS membrane. Chemical Engineering Research and Design.

Liu, L., Chakma, A., Feng, X., 2006a. Sorption, diffusion, and permeation of light olefins in poly(ether block amide) membranes. Chemical Engineering Science. https://doi.org/10.1016/j.ces.2006.05.045.

Liu, L., Chakma, A., Feng, X., 2006b. Propylene separation from nitrogen by poly(ether block amide) composite membranes. Journal of Membrane Science. https://doi.org/10.1016/j.memsci.2005.12.058.

Liu, L., Chakma, A., Feng, X., 2004. A novel method of preparing ultrathin poly(ether block amide) membranes. Journal of Membrane Science. https://doi.org/10.1016/j.memsci.2003.12.025.

Liu, L., Chakma, A., Feng, X., Lawless, D., 2009a. Separation of VOCs from N2 using poly(ether block amide) membranes. Canadian Journal of Chemical Engineering. https://doi.org/10.1002/cjce.20181.

Liu, L., Jiang, N., Burns, C.M., Chakma, A., Feng, X., 2009b. Substrate resistance in composite membranes for organic vapour/gas separations. Journal of Membrane Science 338, 153–160. https://doi.org/10.1016/j.memsci.2009.04.019.

Liu, Y., Feng, X., Lawless, D., 2006. Separation of gasoline vapor from nitrogen by hollow fiber composite membranes for VOC emission control. Journal of Membrane Science. https://doi.org/10.1016/j.memsci.2005.07.012.

Lue, S.J., Chen, W.W., Wang, S.F., 2014. Vapor permeation of toluene, m-Xylene, and methanol vapors on poly(dimethylsiloxane) membranes. Separation Science and Technology. https://doi.org/10.1080/01496390903212615.

Mahajan, R., Koros, W.J., 2000. Factors controlling successful formation of mixed-matrix gas separation materials. Industrial & Engineering Chemistry Research. https://doi.org/10.1021/ie990799r.

Mao, G., Liu, T., Chen, Y., Gao, X., Qin, J., Zhou, H., Jin, W., 2022. Polyamide@GO microporous membrane with enhanced permeability for the molecular sieving of nitrogen over VOC. Journal of Membrane Science 652, 120443. https://doi.org/10.1016/j.memsci.2022.120443.

Mashhadikhan, S., Moghadassi, A., Ebadi Amooghin, A., Sanaeepur, H., 2020. Interlocking a synthesized polymer and bifunctional filler containing the same polymer's monomer for conformable hybrid membrane systems. Journal of Materials Chemistry A. https://doi.org/10.1039/c9ta13375e.

Matsumoto, M., Inomoto, Y., Kondo, K., 2005. Selective separation of aromatic hydrocarbons through supported liquid membranes based on ionic liquids. Journal of Membrane Science. https://doi.org/10.1016/j.memsci.2004.08.013.

Matsumoto, M., Ueba, K., Kondo, K., 2009. Vapor permeation of hydrocarbons through supported liquid membranes based on ionic liquids. Desalination. https://doi.org/10.1016/j.desal.2007.11.090.

Mc Keown, N.B., Budd, P.M., 2006. Polymers of intrinsic microporosity (PIMs): organic materials for membrane separations, heterogeneous catalysis and hydrogen storage. Chemical Society Reviews. https://doi.org/10.1039/b600349d.

Merkel, T.C., Bondar, V., Nagai, K., Freeman, B.D., 2000. Sorption and transport of hydrocarbon and perfluorocarbon gases in poly(1-trimethylsilyl-1-propyne). Journal of Polymer Science Part B: Polymer Physics. https://doi.org/10.1002/(SICI)1099-0488(20000115)38:2<273::AID-POLB1>3.0.CO;2-X.

Mulder, M., 1991. Basic Principles of Membrane Technology. https://doi.org/10.1524/zpch.1998.203.part_1_2.263.

Obuskovic, G., Majumdar, S., Sirkar, K.K., 2003. Highly VOC-selective hollow fiber membranes for separation by vapor permeation. Journal of Membrane Science. https://doi.org/10.1016/S0376-7388(03)00079-6.

Ordoñez, M.J.C., Balkus, K.J., Ferraris, J.P., Musselman, I.H., 2010. Molecular sieving realized with ZIF-8/Matrimid® mixed-matrix membranes. Journal of Membrane Science. https://doi.org/10.1016/j.memsci.2010.06.017.

Ortiz-Martínez, V.M., Salar-García, M.J., 2019. Separation of Volatile Organic Compounds by Using Immobilized Ionic Liquids, Green Sustainable Process for Chemical and Environmental Engineering and Science: Ionic Liquids as Green Solvents. Elsevier Inc. https://doi.org/10.1016/B978-0-12-817386-2.00004-4.

Ozturk, B., Kuru, C., Aykac, H., Kaya, S., 2015. VOC separation using immobilized liquid membranes impregnated with oils. Separation and Purification Technology 153, 1–6. https://doi.org/10.1016/j.seppur.2015.08.032.

Park, H.B., Lee, Y.M., 2002. Separation of toluene/nitrogen through segmented polyurethane and polyurethane urea membranes with different soft segments. Journal of Membrane Science. https://doi.org/10.1016/S0376-7388(01)00663-9.

Park, S., Jeong, H.K., 2020. In-situ linker doping as an effective means to tune zeolitic-imidazolate framework-8 (ZIF-8) fillers in mixed-matrix membranes for propylene/propane separation. Journal of Membrane Science. https://doi.org/10.1016/j.memsci.2019.117689.

Petrusová, Z., Machanová, K., Stanovský, P., Izák, P., 2019. Separation of organic compounds from gaseous mixtures by vapor permeation. Separation and Purification Technology 217, 95–107. https://doi.org/10.1016/j.seppur.2019.02.028.

Pinnau, I., Toy, L.G., 1996. Transport of organic vapors through poly(1-trimethylsilyl-1-propyne). Journal of Membrane Science. https://doi.org/10.1016/0376-7388(96)00041-5.

Qian, Q., Asinger, P.A., Lee, M.J., Han, G., Mizrahi Rodriguez, K., Lin, S., Benedetti, F.M., Wu, A.X., Chi, W.S., Smith, Z.P., 2020. MOF-based membranes for gas separations. Chemical Reviews 120, 8161–8266. https://doi.org/10.1021/acs.chemrev.0c00119.

Ren, X., Ren, J., Li, H., Deng, M., 2013. Permeation characteristics of light hydrocarbons through poly(amide-6-β-ethylene oxide) multilayer composite membranes. Chinese Journal of Chemical Engineering. https://doi.org/10.1016/S1004-9541(13)60462-0.

Richard, W.B., 2004. Membrane Technology and Applications.Pdf.

Robeson, L.M., 2008. The upper bound revisited. Journal of Membrane Science 320, 390–400. https://doi.org/10.1016/j.memsci.2008.04.030.

Robeson, L.M., 1991. Correlation of separation factor versus permeability for polymeric membranes. Journal of Membrane Science 62, 165–185. https://doi.org/10.1016/0376-7388(91)80060-J.

Sadrzadeh, M., Amirilargani, M., Shahidi, K., Mohammadi, T., 2009. Gas permeation through a synthesized composite PDMS/PES membrane. Journal of Membrane Science 342, 236–250. https://doi.org/10.1016/j.memsci.2009.06.047.

Sakai, H., Tomita, T., Takahashi, T., 2001. p-Xylene separation with MFI-type zeolite membrane. Separation and Purification Technology 25, 297–306.

Sedláková, Z., Clarizia, G., Bernardo, P., Jansen, J.C., Slobodian, P., Svoboda, P., Kárászová, M., Friess, K., Izak, P., 2014. Carbon nanotube- and carbon fiber-reinforcement of ethylene-octene copolymer membranes for gas and vapor separation. Membranes. https://doi.org/10.3390/membranes4010020.

Shen, B., Zhao, S., Yang, X., Carta, M., Zhou, H., Jin, W., 2022. Relation between permeate pressure and operational parameters in VOC/nitrogen separation by a PDMS composite membrane. Separation and Purification Technology 280, 119974. https://doi.org/10.1016/j.seppur.2021.119974.

Shi, Y., Burns, C.M., Feng, X., 2006. Poly(dimethyl siloxane) thin film composite membranes for propylene separation from nitrogen. Journal of Membrane Science 282, 115–123. https://doi.org/10.1016/j.memsci.2006.05.011.

Shimoyama, Y., Ito, A., 2011. Calculation of permeabilities and numerical simulation of separations for volatile organic compound vapor through triethylene glycol derivative liquid membranes. Journal of Membrane Science. https://doi.org/10.1016/j.memsci.2010.12.019.

Sui, G., Li, J., Ito, A., 2009. Separation of VOC vapor from air by a surface-soaked liquid membrane module using triethylene glycol. Separation and Purification Technology. https://doi.org/10.1016/j.seppur.2009.05.006.

Suleman, M.S., Lau, K.K., Yeong, Y.F., 2016. Plasticization and swelling in polymeric membranes in CO2 removal from natural gas. Chemical Engineering & Technology. https://doi.org/10.1002/ceat.201500495.

Sun, J., Yu, C., Jeong, H.K., 2018. Propylene-selective thin zeolitic imidazolate framework membranes on ceramic tubes by microwave seeding and solvothermal secondary growth. Crystals. https://doi.org/10.3390/cryst8100373.

Tanaka, K., Taguchi, A., Hao, J., Kita, H., Okamoto, K., 1996. Permeation and separation properties of polyimide membranes to olefins and paraffins. Journal of Membrane Science. https://doi.org/10.1016/S0376-7388(96)00182-2.

Tien-Binh, N., Rodrigue, D., Kaliaguine, S., 2018. In-situ cross interface linking of PIM-1 polymer and UiO-66-NH2 for outstanding gas separation and physical aging control. Journal of Membrane Science. https://doi.org/10.1016/j.memsci.2017.11.054.

Tirouni, I., Sadeghi, M., Pakizeh, M., 2015. Separation of C3H8 and C2H6 from CH4 in polyurethane-zeolite 4Å and ZSM-5 mixed matrix membranes. Separation and Purification Technology. https://doi.org/10.1016/j.seppur.2014.12.012.

Vanherck, K., Koeckelberghs, G., Vankelecom, I.F.J., 2013. Crosslinking polyimides for membrane applications: a review. Progress in Polymer Science. https://doi.org/10.1016/j.progpolymsci.2012.11.001.

Vaughn, J.T., Harrigan, D.J., Sundell, B.J., Lawrence, J.A., Yang, J., 2017. Reverse selective glassy polymers for C3+ hydrocarbon recovery from natural gas. Journal of Membrane Science. https://doi.org/10.1016/j.memsci.2016.09.003.

Vergadou, N., Theodorou, D.N., 2019. Molecular modeling investigations of sorption and diffusion of small molecules in Glassy polymers. Membranes. https://doi.org/10.3390/membranes9080098.

Vu, D.Q., Koros, W.J., Miller, S.J., 2003. Effect of condensable impurity in CO2/CH4 gas feeds on performance of mixed matrix membranes using carbon molecular sieves. Journal of Membrane Science. https://doi.org/10.1016/S0376-7388(03)00245-X.

Wang, D., Teo, W.K., Li, K., 2002. Permeation of H2, N2, CH4, C2H6, and C3H8 through asymmetric polyetherimide hollow-fiber membranes. Journal of Applied Polymer Science 86, 698–702. https://doi.org/10.1002/app.10966.

Wang, J., Luo, J., Feng, S., Li, H., Wan, Y., Zhang, X., 2016. Recent development of ionic liquid membranes. Green Energy & Environment 1, 43–61. https://doi.org/10.1016/j.gee.2016.05.002.

Wang, Y., Goh, T.Y., Goodrich, P., Atilhan, M., Khraisheh, M., Rooney, D., Thompson, J., Jacquemin, J., 2020. Impact of ionic liquids on silver thermoplastic polyurethane composite membranes for propane/propylene separation. Arabian Journal of Chemistry. https://doi.org/10.1016/j.arabjc.2017.05.008.

Wang, Y., Su, Y., Yang, L., Su, M., Niu, Y., Liu, Y., Sun, H., Zhu, Z., Liang, W., Li, A., 2022. Highly efficient removal of PM and VOCs from air by a self-supporting bifunctional conjugated microporous polymers membrane. Journal of Membrane Science 659. https://doi.org/10.1016/j.memsci.2022.120728.

Wang, Z., Ren, H., Zhang, S., Zhang, F., Jin, J., 2017. Polymers of intrinsic microporosity/metal-organic framework hybrid membranes with improved interfacial interaction for high-performance CO2 separation. Journal of Materials Chemistry A. https://doi.org/10.1039/c7ta01773a.

Yan, X., Anguille, S., Bendahan, M., Moulin, P., 2019. Ionic liquids combined with membrane separation processes: a review. Separation and Purification Technology. https://doi.org/10.1016/j.seppur.2019.03.103.

Yang, W., Zhou, H., Zong, C., Li, Y., Jin, W., 2018. Study on membrane performance in vapor permeation of VOC/N2 mixtures via modified constant volume/variable pressure method. Separation and Purification Technology 200, 273–283. https://doi.org/10.1016/j.seppur.2018.02.044.

Yeom, C.K., Lee, S.H., Song, H.Y., Lee, J.M., 2002a. A characterization of concentration polarization in a boundary layer in the permeation of VOCs/N2 mixtures through PDMS membrane. Journal of Membrane Science. https://doi.org/10.1016/S0376-7388(02)00075-3.

Yeom, C.K., Lee, S.H., Song, H.Y., Lee, J.M., 2002b. Vapor permeations of a series of VOCs/N2 mixtures through PDMS membrane. Journal of Membrane Science 198, 129–143. https://doi.org/10.1016/S0376-7388(01)00657-3.

Yu, L., Grahn, M., Ye, P., Hedlund, J., 2017. Ultra-thin MFI membranes for olefin/nitrogen separation. Journal of Membrane Science. https://doi.org/10.1016/j.memsci.2016.11.077.

Žák, M., Klepic, M., Štastná, L.Č., Sedláková, Z., Vychodilová, H., Hovorka, Š., Friess, K., Randová, A., Brožová, L., Jansen, J.C., Khdhayyer, M.R., Budd, P.M., Izák, P., 2015. Selective removal of butanol from aqueous solution by pervaporation with a PIM-1 membrane and membrane aging. Separation and Purification Technology. https://doi.org/10.1016/j.seppur.2015.07.041.

Zhang, C., Dai, Y., Johnson, J.R., Karvan, O., Koros, W.J., 2012. High performance ZIF-8/6FDA-DAM mixed matrix membrane for propylene/propane separations. Journal of Membrane Science. https://doi.org/10.1016/j.memsci.2011.10.003.

Zhang, C., Gao, X., Qin, J., Guo, Q., Zhou, H., Jin, W., 2021. Microporous polyimide VOC-rejective membrane for the separation of nitrogen/VOC mixture. Journal of Hazardous Materials 402, 123817. https://doi.org/10.1016/j.jhazmat.2020.123817.

Zhang, F., Sun, W., Liu, J., Zhang, W., Zhongqi, R., 2014. Extraction separation of tolueneclohexane with hollow fiber supported ionic liquid membrane.pdf. Korean Journal of Chemical Engineering 31, 1049–1056.

Zhang, J., Xin, Q., Li, X., Yun, M., Xu, R., Wang, S., Li, Y., Lin, L., Ding, X., Ye, H., Zhang, Y., 2019. Mixed matrix membranes comprising aminosilane-functionalized graphene oxide for enhanced CO2 separation. Journal of Membrane Science. https://doi.org/10.1016/j.memsci.2018.10.075.

Zhang, Z., Wu, Q., Xu, S., Yue, Z., Zhou, H., Jin, W., 2023. Ultra-stable fully-aromatic microporous polyamide membrane for molecular sieving of nitrogen over volatile organic compound. Journal of Hazardous Materials 459, 132151. https://doi.org/10.1016/j.jhazmat.2023.132151.

Zhou, H., Su, Y., Chen, X., Yi, S., Wan, Y., 2010. Modification of silicalite-1 by vinyltrimethoxysilane (VTMS) and preparation of silicalite-1 filled polydimethylsiloxane (PDMS) hybrid pervaporation membranes. Separation and Purification Technology. https://doi.org/10.1016/j.seppur.2010.08.017.

Zong, C., Yang, X., Chen, D., Chen, Y., Zhou, H., Jin, W., 2021. Rational tuning of the viscosity of membrane solution for the preparation of sub-micron thick PDMS composite membrane for pervaporation of ethanol-water solution. Separation and Purification Technology. https://doi.org/10.1016/j.seppur.2020.117729.

CHAPTER

5

Adsorption of VOCs

Xueyang Zhang

School of Environmental Engineering, Xuzhou University of Technology, Xuzhou, China

5.1 Overview of adsorption technology for VOC abatement

The World Health Organization (WHO) defines VOCs as organic compounds with melting points below room temperature and boiling points between 50C and 260°C (Makoś-Chełstowska, 2023). There are many types of VOCs, including nonmethane hydrocarbons, oxygenated organic compounds, halogenated hydrocarbons, nitrogen-containing organic compounds, and sulfur-containing organic compounds. They come from a variety of sources, such as petrochemical, electronic, pharmaceutical, and transportation emissions. VOCs not only endanger people's health but also cause serious environmental pollution because they are toxic and carcinogenic and can be the precursors of organic aerosols, haze, as well as the culprit of photochemical smog (Li et al., 2020a).

Numerous technologies have been developed to eliminate VOCs, such as direct combustion, thermal combustion, catalytic combustion, biological degradation, photocatalytic oxidation, ozone catalytic oxidation, plasma decomposition, absorption, adsorption, condensation, and membrane separation technology (Zhang et al., 2017). These aforementioned techniques can be categorized as decomposition technology and recovery technology. As an important industrial raw material, VOCs are very valuable, so recovery technology should be prioritized for the treatment of exhaust gas with high-concentration VOCs. However, the concentration of VOCs in most exhaust gases is low, so it is necessary to condense the VOCs before recycling. The recovery technology represented by the adsorption method, thus, has an important position in VOC recovery and has been widely used.

In adsorption, an adsorbent with a large specific surface area and abundant pores is used to adsorb, concentrate, and separate VOC molecules. When the waste gas passes through the adsorbent, the VOCs in the waste gas are adsorbed on the surface of the adsorbent by physical and/or chemical adsorption, so that the waste gas is purified. After adsorption saturation, the VOCs can be desorbed by heating or other manners, and the adsorbent is regenerated. Because it can realize the recovery of VOCs and the regeneration of adsorbents at the same time, the adsorption method is a green and sustainable method which greatly reduces the cost of waste gas treatment.

Volatile Organic Compound Removal
https://doi.org/10.1016/B978-0-443-15674-8.00007-X

There are many factors affecting adsorption, including the characteristics of the adsorbent (specific surface area, pore volume, pore-size distribution, surface functional groups, particle size, etc.), the nature of VOC molecules (molecule structure, polarity, boiling point, etc.), and the operation conditions (adsorption temperature, humidity, gas velocity, etc.) (Zhang et al., 2017; Zhu et al., 2020). For a specific waste gas, the adsorbate and adsorption conditions are fixed; thus, the selection of the adsorbent becomes the main factor affecting the treatment efficiency. Almost all common adsorbents, such as activated carbon (AC), activated carbon fibers (ACFs), biochar, carbon nanotubes (CNTs), graphene, zeolite molecular sieves, metal–organic frameworks (MOFs), clay, silica, and organic polymers and composites, can be used for VOC adsorption (Ye et al., 2023; Zhang et al., 2019a; Yang et al., 2019). The physical and chemical properties of these adsorbents are different, which determine whether they are suitable for the removal of VOCs or not. Generally, carbonaceous adsorbents, such as AC and ACFs, have excellent aromatics and hydrophobicity, which are conducive to the adsorption of nonpolar VOCs, such as benzene and cyclohexane. However, hydrophilic adsorbents, such as a zeolite molecular sieve, clay, and silica gel, have a large number of hydrophilic groups on their surface, inhibiting the adsorption of nonpolar VOCs. Although CNTs, graphene, and other nanomaterials have great potential for the adsorption of VOCs, their adverse characteristics, such as small particle size, easy aggregation, and difficult formation, have limited their practical applications in VOC adsorption (David and Niculescu, 2021). Because the composition of a VOC waste gas is complex and its amount of humidity is large in practical engineering, it is difficult to meet the actual needs using a single type of adsorbent. Therefore, the development of composite materials, combining the advantages of various adsorbents, is expected to fulfill the need for real exhaust gas adsorption.

In summary, as a promising VOC treatment technology, adsorption technology has many advantages, such as its simple structure, stable operation, wide application range, and the fact that it is renewable. It is suitable for the treatment of a large volume of waste gas with a low VOC concentration. This technology not only solves the problem of environmental pollution, but also creates economic benefits by recovering valuable VOCs. In practical applications, adsorption technology has been widely applied, providing us with more possibilities for the effective management and control of VOCs.

5.2 VOC adsorption mechanisms

5.2.1 Solid surface characteristics

Adsorption occurs on the surface of a solid adsorbent; thus, we need to understand the surface properties of the solid before studying the adsorption mechanism. One of the special properties of solid surfaces is that their molecules (atoms) are bound and cannot move freely as the liquid molecules do. Molecules on a solid surface are subjected to the same asymmetric forces as those on a liquid surface. However, the shape of a solid material is already determined during the forming process, which cannot be changed like a liquid. Therefore, a solid tends to retain its shape when it is formed, while a liquid spontaneously forms a ball under the action of surface tension (Rouquerol et al., 2013).

The molecule inside the solid is in the bulk phase, and the force between it and its surrounding molecules is balanced. For molecules on the outermost surface of the solid, the force is unbalanced, that is, the molecules on the outer surface have surface stress. The molecules on the surface of solids have a residual force field and cannot reduce the surface stress by motion like that in liquids. Therefore, it can only reduce its surface free energy by adsorbing external substances, so that the surface is in a more stable state. This is the fundamental reason why solid surfaces have the adsorption ability.

5.2.2 Adsorption mechanisms

The unsaturated force field of the solid surface exerts an attraction on the surrounding substances. When a gas or liquid molecule is close to the solid surface, it is drawn to the solid surface and is enriched, so that the concentration on the solid surface is greater than that of the surrounding environment. This phenomenon is called adsorption.

According to the nature of the adsorption force, adsorption on the solid surface can be divided into physical adsorption and chemical adsorption. There is no electron transfer in physical adsorption, except for the change of the position of the electron cloud's center. In chemical adsorption, there is electron transfer between the adsorbate and the adsorbent, and a chemical bonding force is generated between them. The main physical adsorption force is the van der Waals force. The physical adsorption layer can be regarded as a liquid film formed by vapor condensation, so the value of physical adsorption heat is similar to that of liquid heat, generally below 40 kJ/mol. Because the van der Waals force exists between any molecules, physical adsorption is not selective. As long as the conditions are suitable, any gas can be adsorbed by any solid, and the amount of adsorption varies with the type of adsorbent and adsorbate. Physical adsorption is a multimolecular layer adsorption, where the adsorption rate is fast, and the adsorption is reversible (Toth, 2002).

However, chemisorption refers to the adsorption caused by the formation of a chemical bond between the adsorbent and the adsorbent surface. Chemisorption heat is similar to chemical reaction heat, generally ranging between 80 and 400 kJ/mol. Chemisorption can occur only between specific substances and is a selective and monolayer adsorption. Because chemisorption is similar to a chemical reaction and requires activation energy, the adsorption process is relatively slow and irreversible (Rodrigues et al., 2012). The difference between physical and chemical adsorption is shown in Table 5.1.

The adsorption properties of solids are closely related to their surface energy. Crushing a bulk solid into fine particles can increase its total surface area, thus improving the adsorption capacity of the solid. Another way to improve the adsorption capacity of solids is to increase their specific surface area by creating more apertures within the adsorbent. The pore size and distribution of porous adsorbents are important factors controlling their adsorption properties. The pore size of the adsorbent that is suitable for the size of the adsorbate molecule is most favorable for adsorption. In addition to changing the morphology of the solid adsorbent, improving its surface chemical properties is also an effective method to enhance the adsorption of VOCs.

TABLE 5.1 Physical versus chemical adsorption (Kondo et al., 2006).

Adsorption type	Physical adsorption	Chemical adsorption
Adsorption power	van der Waals forces	Chemical bonding force
Heat of adsorption	Similar to liquid heat, below 40 kJ/mol	Similar to chemical reaction heat, ranging from 80 to 400 kJ/mol.
Selectivity	No	Yes
Adsorption layer	Mono- or multimolecular layer adsorption	Monomolecular layer adsorption
Adsorption velocity	Fast	Slow
Reversibility	Reversible	Irreversible
Adsorption temperature	Below the critical temperature of the adsorbent	Much higher than the boiling point of the adsorbent

5.2.3 Factors affecting adsorption

5.2.3.1 Adsorbent properties

Specific surface area, pore structure, particle size, and the surface chemical functional groups of an adsorbent are the four crucial factors determining VOC adsorption (Zhang et al., 2017). The surface of the adsorbent is where the adsorption site is located; thus, a large specific surface area can provide more adsorption sites. Li et al. (2020a) summarized VOC adsorption on four types of adsorbents, i.e., MOFs, AC, hypercrosslinked polymeric resin, and zeolites. They found a positive correlation between the specific surface area (2653, 1404, 1225, and 800 m^2/g, respectively) of the four adsorbents and their VOC adsorption capacities (796.2, 460, 384.1, and 137.2 mg/g, respectively). Therefore, a high specific surface area of an adsorbent means a high VOC adsorption capacity. In order to improve the adsorption ability of an adsorbent, numerous studies have been carried out to develop methods to enlarge the surface area of adsorbents. The commonly used enlargement methods include acid, base, CO_2, water vapor, ball milling, impregnation, and heat treatment. Zhang et al. (2019b) studied the treatment of hydrochar with H_3PO_4 and KOH. They found that, while the surface area of pristine hydrochar was very low (7–8 m^2/g), it was increased to 222–571 m^2/g after KOH activation and 1091–1436 m^2/g after H_3PO_4 activation. With the increasing surface area, the adsorption amount of acetone and cyclohexane on hydrochar increased from 15.98 to 25.36 mg/g to 50.57–159.66 mg/g. Both CO_2 and water vapor are commonly used as activation agents in AC production. Zhang et al. (2020) found that CO_2 activation increased hydrochar's surface area from <8 to 1308 m^2/g, so the uptake of acetone increased from 19.19 to 121.74 mg/g. Ball milling is an efficient mechanical method that is used to improve the surface of an adsorbent by grinding it into powder. Xiang et al. (2020) studied the adsorption performance of ball-milled biochar on five different VOCs (i.e., acetone, ethanol, chloroform, cyclohexane, and toluene). They found that biochar's specific surface area (SSA) increased

1.38–29.06 times after ball milling because it not only increased the external surface area by reducing the particle size, but also increased the internal surface area by opening the internal pore. As a result, the adsorption of VOCs by ball-milled biochar increased by 1.3–13.0 folds, and the maximum adsorption capacity of acetone was up to 103.4 mg/g. Subsequently, Zhang et al. (2021) and Qi et al. (2022) confirmed that ball milling can effectively improve the specific surface area of biochar. Ogungbenro et al. (2020) found that a high activated temperature is conducive to enhancing biochar's surface area; the AC prepared at 900°C has a higher specific surface area of 189.18 m^2/g than that prepared at 600°C.

In addition to the specific surface area, the pore size of an adsorbent also plays an important role in VOC adsorption. The pore size of an adsorbent can be classified into a micropore (pore diameter <2 nm), mesopore (2 nm $<$pore diameter <50 nm), and macropore (pore diameter >50 nm). They play different roles in VOC adsorption. Generally, the micropore provides principal adsorption sites, whereas the mesopore enhances the intraparticle diffusion and shortens the adsorption time. It has been widely acknowledged that VOC adsorption on a porous adsorbent is dominated by the micropore, particularly the narrow micropore. Zhang et al. (2022a) prepared a series of lignin-impregnated biochars and studied their adsorption ability for acetone. A linear analysis of acetone uptake and the pore structure parameters show that the micropore volume had a stronger linear correlation ($R^2 = 0.7861$) than the total pore volume ($R^2 = 0.6066$) and the mesopore volume ($R^2 <0.0237$) to acetone uptake, suggesting that a narrow aperture is beneficial for VOC adsorption. Additionally, Yu et al. (2023) prepared a novel silica-based adsorbent containing abundant micropores with pore sizes ranging from 0.5 to 2.0 nm. They found that this adsorbent showed good adsorption performance for small-molecule VOCs, such as acetone, isopropylacetone, toluene, and styrene, demonstrating the importance of micropores. It should be noted that although the micropore is beneficial for VOC adsorption, the adsorption rates may decrease in the narrow pore. Kim and Ahn (Kim and Ahn, 2012) carried out batch adsorption experiments and found that the narrower the pore size, the lower the adsorption kinetic rate, which was subsequently confirmed by Lung et al. (Tsai et al., 2008). Many VOC molecules prefer to be adsorbed in micropores because they commensurate with their molecular size (Carrasco-Marín et al., 1991). However, not all VOCs favor micropores. For example, adsorbates with a larger pore size prefer a mesopore. Dai et al. (2021a) prepared a meso–microporous ZSM-5 containing a large number of uniform mesopores, which showed excellent adsorption and desorption performance for toluene. This is in good agreement with the fact that micropores facilitate the adsorption of small-sized VOCs while large VOCs prefer mesopores (Kim et al., 2006).

The surface chemical functional groups of an adsorbent also influence the adsorption of VOCs. Li et al. (2020a) studied the relationship between the surface chemical functional groups of AC and its VOC adsorption capacity. They found that the adsorption capacity of benzene by AC with the functional group of 1.0 mmol/g, which was approximately 70 mg/g higher than that of 0.2 mmol/g, showing that the VOC adsorption capacity was positively correlated with the amount of chemical functional groups. The surface functional groups of adsorbents are associated with both the nature of raw materials and activation or modification methods, such as heating, chemical, and electrochemical treatments. Janu et al. (2021) suggested that the pyrolysis temperature is the key factor determining the surface functional groups of biochar. They found that the pyrolysis temperature of 600°C resulted in a partial loss of functional groups on the surface of biochar, while 750°C resulted in an almost complete loss of functional groups. The heteroatoms, like oxygen, nitrogen, halogen, and hydrogen, of surface functional groups governed their surface chemistry. The functional groups containing oxygen and nitrogen on porous carbonaceous adsorbents have been recognized as the most important species for adsorption (Shen et al., 2008; Li et al., 2002).

The particle size of a solid adsorbent is closely related to its specific surface area. The smaller the particles, the larger the specific surface area, which can increase the adsorption sites for VOCs (Lyu et al., 2017). Ball milling is an effective method that is used to reduce the particle size by grinding the bulk material into powder (Lyu et al., 2018; Zhang et al., 2019c). The studies on the adsorption of VOCs by zeolites have also shown that changing the particle size of a molecular sieve is an effective way to adjust its morphology to improve its adsorption ability. Song et al. (2004) synthesized nanocrystalline ZSM-5 with the particle size of 15 and 60 nm and found that the adsorption amount of toluene on ZSM-5 with a particle size of 15 nm was twice of that with a larger particle size of 60 nm.

5.2.3.2 VOC properties

The characteristics of VOCs, such as the molecular structure, polarity, and boiling point, also impact adsorption. Because of the size exclusion effect of porous adsorbents, the molecular structure of VOCs directly affects their adsorption. Zhang et al. (2022b) prepared a series of microwave biochars and evaluated their adsorption ability for benzene series VOCs of different molecular sizes. Compared with benzene and toluene, o-xylene had a larger kinetic diameter and showed lower adsorption on the biochars. In addition, Zhang et al. (2021) studied the adsorption of benzene VOCs on ball-milled biochars and also found that the molecular size and structure of VOCs affected their adsorption. Although p-xylene, m-xylene, and o-xylene have the same molecular weight (106 g/mol) and similar boiling points (138.36−144.43°C), the adsorption of p-xylene on biochar was higher than that of m-xylene and o-xylene. This is because the kinetic diameter of p-xylene (0.58 nm) is the smallest among them, making it more easy to enter the pore channels to reach the adsorption sites.

The polarity of VOC molecules directly affects their adsorption on porous materials, which depends on the polarities of both the VOCs and the surface of the adsorbents. Generally, polar VOCs prefer to be adsorbed by adsorbents with polar surface groups, and vice versa. The surface of AC is primarily nonpolar, thus it prefers to adsorb nonpolar or weakly polar VOCs. Ma et al. (2021) studied the adsorption of VOCs on AC and found that the stronger the polarity of VOCs, the less the adsorption amount. Meng et al. (2019) explored the contribution of oxygen groups of ACFs to the adsorption of different VOCs. They found that the adsorption capacity of toluene (169.41 mg/g) was greater than that of methanol (133.06 mg/g) and acetone (105.48 mg/g). This suggests that the presence of oxygen groups may inhibit the specific interaction between nonpolar VOCs and carbon electron-rich regions, while contributing to the adhesion of polar VOCs. Zhang et al. (2019b) studied VOC adsorption on hydrochar and found that the attractive forces between VOC molecules and the carbon surface were affected by polarity. In terms of the polar molecules of acetone, there were three types of intermolecular forces, while there was only the induced dipole−induced dipole force for the nonpolar molecules of cyclohexane. The reinforced molecular forces of induced dipole−dipole and dipole−dipole on polar molecules are comparable to medium-strength hydrogen bonds, which are stronger than the induced dipole−induced dipole force. Thus, acetone showed higher adsorption than nonpolar cyclohexane on the hydrochar with a polar surface.

As a nonpolar molecule, benzene with the dipole moments of 0 D is more readily adsorbed by non- or weakly polar surfaces. Thus, the adsorption capacity of benzene on the nonpolar microwave biochar WB500 reached a maximum of 53.9 mg/g (Zhang et al., 2022b). However, some carbonaceous materials may exhibit strong polarity natively or after modification, such as oxidization by nitric acid or ozone. Consequently, they tend to adsorb polar VOCs. Hou et al. (2021) prepared a coconut shell-based AC and modified it with glycine impregnation. They found that the breakthrough adsorption capacity of toluene on 10 wt% glycine-modified ACs was approximately 147% higher than that of the original ACs, and 139% higher than that of 5 wt% ammonia-modified ACs.

The physical adsorption of VOCs on a porous material is a vapor—liquid phase transitions process. The VOCs with higher boiling points would be preferentially adsorbed than those with a lower boiling point, due to the stronger intermolecular forces and stronger affinity with adsorbents (Guo et al., 2013). Similarly, Chiang et al. (2001) suggest that liquid-like condensation plays an important role in VOC adsorption; thus, the boiling point of VOCs is a crucial factor for adsorption. Both Li et al. (2012) and Zhang et al. (2019a) confirmed this conclusion by evaluating the influence of the physicochemical properties of adsorbents on VOC adsorption.

5.2.3.3 Adsorption conditions

In general, the adsorption of VOCs on porous adsorbents is mainly controlled by physical exothermic interactions, in which the adsorption temperature plays an important role (Tsai et al., 2008; Dou et al., 2011). Zhang et al. (2019b) compared the adsorption of acetone and cyclohexane on biochar at 20, 40, and 60°C. They found that the adsorption amount of VOCs on biochar decreases by 9%—64% with the increase of the temperature. However, not every increase in temperature resulted in a decrease in the adsorption capacity. Xiang et al. (2022) investigated the adsorption of benzene series VOCs on six types of microwave biochar. They found that the adsorption of benzene increased slightly with the increase of the temperature, indicating a different adsorption mechanism. They suggested that a slightly high temperature would accelerate the diffusion process of benzene entering the biochar interior pores.

The moisture and gas velocity can also influence VOC adsorption. The presence of water vapor is a deathblow to VOC adsorption because the water molecules would occupy the sites and pores competitively. The gas velocity has an evident influence on the performance of VOC adsorbents. It affects the accumulation of VOCs on the exterior surface of the adsorbents. A high gas velocity means a short residence time, which causes the VOCs to pass through the adsorbent before being adsorbed, decreasing the adsorption efficiency.

5.3 VOC adsorption systems

5.3.1 Selection and regeneration of adsorbents

5.3.1.1 Adsorbent types

The selection of adsorbent is critical for VOC adsorption systems. A good absorbent should have the advantages of a large specific surface area, abundant micropore structure, a large number of chemical functional groups, good thermal stability, strong hydrophobicity, low price, easy regeneration, etc (Ye et al., 2023). A large number of porous materials can be used for the adsorption of VOCs. Among them, the most researched and applied porous adsorbents include AC, ACFs, biochar, CNTs, graphene, zeolite molecular sieves, MOFs, clay, silica, organic polymers, and composite materials (Zhu et al., 2020). As discussed previously, the adsorption capacities of the adsorbents for VOCs are varied with their physicochemical properties. The characteristics of the commonly studied VOC adsorbents are described below.

5.3.1.1.1 AC

AC is an excellent and most widely used adsorbent with a large specific surface area, well-developed pore structure, abundant surface functional groups, stable structure, high VOC adsorption capacity, and renewability (Ye et al., 2023). Commercial AC is mainly prepared from coal, petroleum asphalt, wood, coconut shells, nutshells, and other carbon-containing materials by pyrolysis and activation under oxygen-starving conditions (Cheng et al., 2021). The shapes of commercial ACs are generally granular, powdery, spherical, and columnar. Compared to other adsorbents, AC is more widely used in practical VOC adsorption engineering. It can adsorb most of the VOCs, such as alkanes, alcohols, ethers, aldehydes, ketones, esters, and aromatic hydrocarbons (Wang et al., 2023a; Zhang et al., 2022c). However, AC also has some drawbacks in practical VOC adsorption applications, such as (1) AC is generally a nonpolar adsorbent, thus having limited adsorption capacity for hydrophilic VOCs (Yan et al., 2023); (2) AC is primarily a microporous material. Therefore, it usually has a low adsorption kinetic rate (Shen, 2023); (3) due to irreversible adsorption, such as chemical adsorption, the desorption of VOCs in AC may be incomplete, affecting its regeneration and increasing the cost (Cai et al., 2022); and (4) physical adsorption is mainly an exothermic process, so it may release a lot of heat during the VOC adsorption of AC, resulting in an increase in the temperature to impede further adsorption (Wang et al., 2021).

5.3.1.1.2 ACFs

ACF is a fibrous carbon material, often produced with organic fibers as precursor and carbonized at 700–1000°C followed by activation using water vapor or CO_2 (Hassan et al., 2020a). Compared with AC, ACFs may have a larger specific surface area (800–2000 m^2/g) and pore volume (0.36–0.92 cm^3/g), and it is a microporous material with its pore sizes mainly in the range of 0.5–1.0 nm (Zhang et al., 2010). In addition, the pores of an ACF are shorter and straighter, which means that it would adsorb VOCs with less mass-transfer resistance, higher adsorption rate, lower pressure loss, and higher adsorption capacity (Wang et al., 2023b). Furthermore, as reported by Lillo-Ródenas et al. (2011), the surface oxygen-containing groups of ACFs are in the range of 735–865 μmol/g, which is much smaller than that of AC (1570–4289 μmol/g). As a result, an ACF has better hydrophobicity and much higher adsorption capacity for nonpolar VOCs than that for polar ones (Kim et al., 2012). An ACF is a promising VOC adsorbent, but it also has some shortcomings, such as the expensive raw materials, complex fiber spinning processes, and high mass loss during high-temperature carbonization. These shortcomings of the ACF have limited its applications to some extent; however, it has been used in many practical engineering purposes for VOC adsorption.

5.3.1.1.3 Biochar

Biochar is a low-cost porous carbon material with a complex pore structure and rich functionalities. It is mainly produced by the pyrolysis of biomass at a low temperature (<700°C). Biochar feedstock has a wide amount of sources, including timber, forest residues, agricultural straws, fruit by-products, and almost all kinds of biomasses (Feng et al., 2021; Rong et al., 2023). Long-time pyrolysis at a high temperature may facilitate the development of biochar's pore structures, while insufficient pyrolysis may introduce surface functional groups (e.g., hydroxyl, carboxyl, carbonyl, phenolic hydroxyl, and methylene) (Hassan et al., 2020b). Therefore, the biochars produced under different conditions have different pore structures and chemical properties, which would further influence their adsorption of VOCs (Dhar et al., 2022). Compared with AC, biochar can be prepared at lower temperatures and does not require activation; thus, it has a lower price tag. The estimated price of biochar is about 350−500 euros/ton, which is much cheaper than that of AC (1200−10,000 euros/ton) (Tibor and Grande, 2022). Biochar has demonstrated great application potential in environmental pollution control (Ponnuchamy et al., 2023). However, the pore structure of biochar is not fully developed due to the low preparation temperature. Its specific surface area is often less than 400 m^2/g, which is much smaller than that of commercial AC. The physical and chemical properties of biochars are greatly affected by their raw materials, and the VOC adsorption capacity of biochars prepared from different raw materials vary greatly (Sadegh et al., 2024). In terms of VOC adsorption, micropores play an important role, but the number of micropores in biochar is relatively small, limiting its adsorption ability. Thus, increasing the number of micropores of biochar is beneficial to increase its VOC adsorption capacity. Acid and base activation (Zhang et al., 2019b), CO_2 activation (Zhang et al., 2020), ball milling (Xiang et al., 2020), and impregnation (Zhang et al., 2022a) are all effective methods to improve the porosity of biochar. Moreover, surface chemical modification can also improve the chemical properties of biochar to further enhance its VOC adsorption capacity. Cheng et al. (2022) investigated alkylation-modified pistachio-based biochar and found that toluene adsorption increased by 48−80 mg/g after modification. Similar to AC, biochar has some shortcomings in VOC adsorption, such as hygroscopicity and flammability (Zhu et al., 2020). Future research should focus on the functional modification of biochar to improve its adsorption capacity and promote its application in practical engineering for VOC adsorption.

5.3.1.1.4 CNTs

CNTs are a nanomaterial in which graphene sheets are rolled into a cylindrical structure (Slepičková Kasálková et al., 2021). Based on the arrangement of graphene cylinders, CNTs can be classified into single-walled CNTs and multi-walled CNTs. They are usually prepared by arc discharge, laser ablation, and chemical vapor deposition. As a new type of carbon nanomaterial, CNTs have excellent VOC adsorption ability due to their properties of a large specific surface area, controllable structure, and hydrophobic tube wall (Barrejón and Prato, 2022; Aslam et al., 2021). Gauden et al. (2006) investigated the adsorption of benzene on multi-walled CNTs and found that the adsorption capacity reached 1.5 cm^3/g, which is higher than that of AC (1.13 cm^3/g). Furthermore, Dai et al. (2021b) explored the adsorption of 20 types of VOCs (including propane, ethylene, formaldehyde, acetone, and benzene) on single-walled CNTs through theoretical chemical calculations. They found that the adsorption energies of these VOCs on CNTs were higher than that of water (3.66 kcal/mol), indicating the excellent selectivity of CNTs to VOCs over water vapor. CNTs have a great potential for VOC adsorption; however, they have the common problems of nanoadsorbents, such as a small particle size, easy aggregation, and difficulty in formation, which seriously hinder their application in practical engineering.

5.3.1.1.5 Graphene

Graphene is a carbon nanomaterial with a monolayer two-dimensional cellular lattice structure connected by the sp^2 hybridization of carbon atoms (Salahdin et al., 2022). It can be prepared by mechanical exfoliation, hydrothermal self-assembly, and chemical vapor deposition (Asiya et al., 2021). Graphene oxide (GO) is a derivative of graphene after oxidation treatment to increase the oxygen-containing functional group, while reduced graphene oxide (RGO) is the product of GO after reduction treatment to remove the oxygen-containing functional group (Hidayat et al., 2020). Graphene and its derivatives have excellent physicochemical properties and can be used for the adsorption of hazardous substances in wastewater and toxic gases, such as VOCs in exhaust gases. Graphene-based materials have excellent adsorption properties, but their manufacturing costs are high. In addition, they are easy to agglomerate, resulting in a decrease in the adsorption performance. Therefore, graphene-based materials are often doped with other materials to improve their VOC adsorption performances (Kong et al., 2021). Szcześniak et al. (2018) prepared MIL-101/GO by doping graphene with MOFs (MIL-101). They found that the methane adsorption amount of the doped absorbent reached 7.5 mmol/g at 25°C and 25 bar, which is higher than that of the precursor. In the future, optimizing the production process of graphene-based materials and developing assembly molding technologies are important ways to promote their practical applications in VOC adsorption.

5.3.1.1.6 Zeolite molecular sieve

Zeolite molecular sieve is a silica—aluminate material containing alkali metals or alkaline earth metal oxides with an empty skeleton and a more regular pore cage structure (Mekki and Boukoussa, 2019). It has a regular micropore structure with a specific surface area of about 500—1000 m^2/g and a pore size of 0.4—1 nm (Cejka et al., 2007; Chen et al., 2022). The pore size of a zeolite is close to the kinetic size of most VOC molecules and its molecular sieve skeleton is rich in the cage structure, which also provides space for the storage of VOC molecules; thus, a zeolite molecular sieve often demonstrates high VOC adsorption capacity (Zhang et al., 2023). Hunter-Sellars et al. (2020) investigated the adsorption of toluene on molecular sieve 13X and found that the adsorption amount reached 657.2 mg/g at 25°C. The adsorption of VOCs by a molecular sieve mainly depends on its pore structure and the Coulomb force field between the equilibrium cation and the molecular sieve skeleton. Therefore, the VOC adsorption performance of a molecular sieve can be changed by adjusting its pore structure, elemental composition, and the type of equilibrium cation to achieve selective adsorption (Zhu et al., 2020). On the one hand, adjusting the pore size of a zeolite can selectively adsorb VOCs with different molecular sizes. On the other hand, the surface hydrophobicity of the zeolite increases with the increase of the Si—Al ratio, which can effectively avoid the competitive adsorption of water molecules.

Compared with AC, a zeolite molecular sieve has a more uniform pore structure, which is conducive to adsorbing VOC molecules smaller than its pore diameter. In addition, the intercrystallite pores formed between zeolite molecular sieve particles can prioritize the adsorption of different polarity and unsaturated molecules, so VOCs are sieved by zeolites according to their properties, such as polarity, saturation, molecular size, and boiling point (Shen et al., 2021). A zeolite molecular sieve has good regeneration performance, and most VOCs could be completely desorbed at 150°C, which is much lower than that of other carbon-based materials (300°C). As an inorganic silico—aluminate material, the thermal stability of a zeolite is better than that of AC, which can prevent spontaneous combustion caused by the exothermic adsorption of VOCs (Calabrese et al., 2019). However, the surface of a molecular sieve has a large number of oxygen-containing hydrophilic groups (e.g., silicon hydroxyl groups) and cations, which easily combine with water molecules via hydrogen bonding or electrostatic interactions. Therefore, the presence of water molecules in the exhaust gas may seriously affect the adsorption of VOCs by zeolite molecular sieves (Wu et al., 2021). In addition, compared to AC, the feedstock prices of molecular sieves are higher, and the synthesis process is also more complex. Therefore, the production cost a zeolite molecular sieve is relatively high, which creates certain obstacles in its commercial applications.

5.3.1.1.7 MOFs

A MOF is a complex polymer with a multidimensional periodic framework structure, which is composed of metal ions or metal—oxygen clusters and organic ligands through coordination self-assembly (Liu et al., 2021; Chen et al., 2020). MOFs have the properties of high porosity, huge internal surface area, abundant organic functional groups, and a flexible and adjustable pore structure, so they have great advantages in VOC adsorption (Siu et al., 2023). The VOC adsorption capacity of MOFs is much larger than that of traditional adsorption materials, such as AC and zeolite molecular sieves. Li et al. (2020a) summarized the adsorption of VOCs on MOFs and found that the adsorption amount of VOCs on MOFs ranged from 240 to 1375 mg/g with an average amount of 796.2 mg/g, which is higher than those of AC (460 mg/g), zeolite molecular sieves (137.2 mg/g), and hypercrosslinked polymeric resin (384.1 mg/g). Similarly, Siu et al. (2023) compared the adsorption of VOCs on various adsorbents and confirmed that MOFs had a higher VOC adsorption capacity than AC and hypercrosslinked polymeric resin. MOFs have different skeleton structures and pore sizes due to the different structures of ligands and the binding ways of ligands and metals. In addition, the modification of organic ligands in the skeleton structure of MOFs can enhance their hydrophobicity and selectivity in VOC adsorption. Zhu et al. (2019) succeeded in introducing alkyl chains into the organic linkers of UiO-67 and enhance their hydrophobicity by decreasing the surface energy of the internal pores and by generating surface roughness to hinder water adsorption. Although MOFs exhibit excellent adsorption capabilities for VOCs, their industrial application is hindered by several drawbacks, including high production costs, poor hydrothermal stability, and low mechanical strength (Li et al., 2020a; Gaikwad et al., 2022).

5.3.1.1.8 Clay minerals

Clay minerals are the most common natural adsorbents. Clay mineral adsorbents usually have a large specific surface area, unique combination of micropores and mesoporous pores, and a fast mass transfer rate, so they have good adsorption capacity for environmental pollutants (Barakan and Aghazadeh, 2021). The commonly used clay adsorbents for VOCs are kaolinite, polywater kaolinite, montmorillonite, and so on (Brigatti et al., 2006). Deng et al. (2017) studied the adsorption of benzene on clay and found that its maximum adsorption capacity reached 141.2 mg/g at 30°C. Although clay minerals have the advantages of low cost and good thermal stability, their adsorption of VOCs is limited by the undeveloped pore structure and the presence of silanol surface groups (Si—OH) (Li et al., 2021). Acid treatment and organic modification are effective measures to improve the adsorption capacity and hydrophobicity of different types of clays. Sun et al. (2022) used tetramethylammonium bromide to modify montmorillonite and found that the modification increased its toluene adsorption capacity from 8.8 to 55.9 mg/g.

5.3.1.1.9 Silica gel

Silica gel is an amorphous inorganic material with a three-dimensional tetrahedral structure; it has the characteristics of good stability, low density, high proportion of micropores, and an abundant number of functional groups (Sui et al., 2019). Sui et al. (2019) studied the adsorption of acetone on silica gel and found that the adsorption amount reached 375 mg/g at 25°C and 30 kPa. Similar to clay materials, the hydrophilic Si—OH on the surface of silica gel impairs its VOC adsorption performance under high humidity conditions (He et al., 2024). Coating the surface of silica gel materials with organic modifiers, such as trimethylchlorosilane, can effectively improve their hydrophobicity. Huang et al. (2017) modified silica gel using trimethylchlorosilane and found that its adsorption of water vapor decreased from 0.154 to 0.022 g/g, suggesting that the modification enhanced the hydrophobicity of silica gel.

5.3.1.1.10 Hypercrosslinked polymers

Hypercrosslinked polymers (HCPs) are porous organic polymers produced by the Friedel—Crafts alkylation reaction (Huang and Turner, 2018). The high degree of crosslinking and rigidity prevent the tight packing of polymer chains, so HCPs have rich pores (0.5—2 nm) (Radjabian and Abetz, 2020). HCPs are generally composed of nonmetallic light elements, such as C, H, O, N, and B (Masoumi et al., 2021). They can be used in the field of VOC abatement, for their advantages of having a low density skeleton, designable structure, easy functional group modification, and high stability (Liu et al., 2022a). Ahmadi et al. (Ahmadi and Kim, 2022) found that HCPs could adsorb benzene with the adsorption amount of 17.85 mmol/g. In addition, HCPs showed a high VOC adsorption capacity in both dry and humid environments due to their large specific surface area and superhydrophobicity. HCPs effectively adsorbed toluene under high humidity conditions, and their good adsorption performances were maintained even after four cycles of reuse (Wang et al., 2013). However, the complex synthesis process and high production cost of HCPs limit their large-scale production and application.

5.3.1.1.11 Composite materials

The VOC gases to be treated in practical applications are generally more complex and are often accompanied by water vapor. Therefore, it is often difficult for a single adsorbent to cope with such complex conditions to meet the actual engineering needs. Composite materials can integrate the advantages of different adsorbents, allowing them to adapt to more complex application conditions. They often have a multistage pore structure, promoting

VOC abatement. Both zeolite molecular sieves and MOFs have controllable pore sizes and large specific surface areas. When they are combined, the diffusion rate and the adsorption of VOCs can be improved. Saini and Pires (2017) prepared ZMF (a composite structure of a zeolite and metal–organic skeleton), a composite structure of a zeolite and metal–organic skeleton by incorporating 59.8% of fine MOF-199 into the foamy porous zeolite ZSM-5. They found that the combination improved the pore structure of ZMF, increasing the specific surface area by 1635%, the total pore volume by 155%, and the microporous volume by 160%. As a result, the adsorption amount of n-hexane, benzene, and cyclohexane on ZMF increased by 150%, 283%, and 468%, respectively. Li et al. (2020b) successfully prepared Cu–BTC@ZSM-5 composites by compositing MOFs with microporous molecular sieves. They found that the specific surface area of Cu–BTC@ZSM-5 increased by 232.26 m^2/g, and its dispersibility and hydrophobicity were also significantly improved. The adsorption of toluene on Cu–BTC@ZSM-5 was 2.53 times higher than that of the precursor Cu–BTC. The excellent adsorption performance of the composite is mainly attributed to the synergistic effect of the precursor adsorbents, which promotes the easy entry of VOC molecules into the adsorbent surface. Other composites, such as AC/ACFs, GO/CNTs, and GO/SiC, also have their potential applications in VOC adsorption (Zhu et al., 2021).

Combining hydrophobic materials with molecular sieves can change their pore structure and hydrophilicity to efficaciously improve their adsorption of VOCs under humid conditions. Li et al. (2020c) successfully combined ZSM-5 with MCM-41 (SBA-15), which increased the adsorption of toluene significantly. When the microporosity of ZSM-5/MCM-41 reached 75%, the adsorption penetration time of toluene molecules under the same humidity adsorption condition increased 2.4 times compared with that of pure ZSM-5. In addition, Li et al. (2018) prepared a zeolite–LDH core-shell structure composite by combining ZSM-5 with Mg–Al-layered double hydroxides. They found that the prepared composite had strong hydrophobicity, similar to the surface of lotus leaves, making it have excellent adsorption ability for toluene under the high humidity of 50%. In addition, the adsorption penetration time was extended by 1.2–2.1 times.

It is worth noting that the preparation of composites with a multistage pore structure generally requires presynthesis or pretreatment methods, which may lead to additional production costs and increased pollutant emissions. Therefore, the development of simple, flexible, and environmentally friendly synthesis methods remains the focus of the research on composites adsorbents for VOC adsorption.

5.3.1.2 Selection principles

The selection of the adsorbent is very important in the design of VOC adsorption technology. The most important criterion to select a VOC adsorbent is good selectivity and high adsorption capacity for the target waste gas. This depends mainly on the physical and chemical structure of the adsorbent itself and the properties of VOCs (such as polarity, molecular size, boiling point, and concentration). Generally, for polar VOC molecules, molecular sieves and silica gel are preferred, while AC is suitable for nonpolar VOCs. For VOCs with large molecular sizes, silica gel with a large pore size is suitable, whereas for the adsorption of small VOCs, a molecular sieve should be used for its regular micropores. When the concentration of VOCs in the exhaust gas is high, but the purification requirement is low, the adsorbent with a moderate adsorption capacity and cheap price can be selected. However, when the concentration of VOCs and the purification requirements are both high, it is difficult to meet the requirements with a single adsorbent. In such cases, a two-stage adsorption treatment using different adsorbents can be considered.

In principle, a good VOC adsorbent should have the following advantages: (1) abundant pores and a large specific surface area. For any adsorbents, abundant pores and a large specific surface area usually mean superior adsorption performance because they provide the sufficient sites for adsorbates. (2) Good selectivity. There are often impurities, such as water vapor, in the exhaust gas. Good selectivity can improve the separation efficiency for specific VOCs. (3) Easy regeneration. The regeneration of the adsorbent directly affects the operating cost, so an adsorbent that is easily regenerated and has stable regeneration performance is preferred. (4) Good mechanical strength and thermal and chemical stability. (5) Abundant feedstock materials, low production cost, and low price.

In practical engineering, before deciding to use an adsorbent, the following evaluation should be carried out to ensure that it is the most suitable for the targeted VOC elimination.

(1) Activity and life experiment: The activity and life tests of one or more selected adsorbents should be carried out. The activity experiment is generally carried out on a laboratory scale, and the life experiment of the adsorbent with good activity should be carried out in the pilot test.
(2) Economic analysis: The activity, service life, desorption performance, price, and other aspects of the preliminary selected adsorbents should be compared in a cost–benefit analysis.
(3) Reusability assessment. The adsorbent needs to be reused. After the adsorption mechanism reaches saturation, the adsorbent should be regenerated, and its adsorption performance should be tested again after recycling to evaluate its reusability.

5.3.1.3 Regeneration methods

The adsorption capacity of an adsorbent is limited, commonly ranging from 1% to 40% (mass fraction). In order to improve the processing ability of the adsorption device and reduce the operation cost, it is generally necessary to regenerate the adsorbent after use. When the adsorbent reaches saturation or is close to saturation, the regeneration process is started to desorb the VOCs for regeneration. There are many kinds of regeneration technologies, including thermal regeneration, reduced pressure regeneration, replacement regeneration, chemical oxidation regeneration, microwave-assisted regeneration, electrochemical regeneration, photocatalytic regeneration, extraction regeneration, and supercritical CO_2 regeneration. The characteristics of the most used regeneration technologies are as follows:

(1) Thermal regeneration: By directly heating the adsorption bed with a hot airflow (e.g., steam, hot air, or inert gas), the adsorbate can be desorbed and released and the adsorbent can restore its adsorption performance. The determination of the regeneration temperature is related to the type of the adsorbent and the nature of the adsorbate. A regeneration temperature that is too low may not be enough to regenerate the adsorbent completely. VOC residues may remain in the adsorbent, inevitably reducing its cyclic adsorption performance. A desorption temperature that is too high, however, may bring many risks, such as inactivation, burning, and the structure destruction of the adsorbent. Thereby, the bed temperature should be strictly controlled to minimize the aforementioned risks. The heating methods of the adsorbent bed in the desorption stage can be the superheated steam method, flue gas method, inductance heating, and microwave heating.

(2) Reduced pressure regeneration: The adsorption capacity of an adsorbent decreases when the pressure decreases. Therefore, the adsorbed VOCs can be desorbed by lowering the pressure. For some VOC adsorbents, the greater the reduction in pressure, the more complete the desorption. However, the pressure change of the system should be strictly controlled to reduce power consumption and avoid the violent relative movement of adsorbent particles. Different from thermal regeneration, the regeneration temperature of reduced pressure regeneration can be the same as the adsorption temperature, and thus, the structural damage or even the combustion of the adsorbent caused by a high desorption temperature can be avoided.

(3) Replacement regeneration: In this method, an adsorbable purge gas is used to replace the adsorbed VOCs to regenerate the adsorbent. The stronger the adsorption of the purge gas by the adsorbent, the better the desorption efficiency. Replacement regeneration is particularly suitable for adsorbents with a strong thermal sensitivity. It has a high desorption efficiency and little residual adsorbate. Water vapor is a common purge gas for replacement regeneration. Since the desorption product contains both an adsorbate and a displacer, it needs to be separated in posttreatment. In order to achieve a better separation effect, the difference in the boiling point between the adsorbate and the displacer should be large.

(4) There are several other regeneration methods, such as chemical oxidation regeneration, microwave-assisted regeneration, electrochemical regeneration, photocatalytic regeneration, extraction regeneration, supercritical CO_2 regeneration, and microbial regeneration. In practical engineering, the most economical and effective regeneration methods should be selected according to the specific operation conditions. Sometimes, it is necessary to combine different regeneration methods to achieve the best regeneration effect.

5.3.1.4 Decrease of the adsorption capacity

Regeneration can realize the reuse of the adsorbent, but with the increase in the number of regeneration times, the adsorption capacity of the adsorbent will gradually decrease. For example, after thermal regeneration, the internal structure of AC may collapse, resulting in a decrease in the specific surface area and pore volume and, thus, decrease the adsorption performance. The thermal stability and regeneration performance of different adsorbents are different. A zeolite molecular sieve has a rigid skeleton structure; heat treatment has little effect on its crystal structure and porosity. Thus, after repeated regeneration, the adsorption performance of a zeolite molecular sieve remains almost unchanged. For most VOC adsorbents, however, their adsorption capacity decreases more or less after repeated regeneration. The reasons for this are summarized as follows:

(1) Regeneration is incomplete. A portion of the adsorbate always remains in the adsorbent after regeneration. The amount of residual VOCs is determined by the regeneration conditions, such as regenerant concentration and temperature. Even under the same regeneration condition, the residual amount of VOCs on different adsorbents is different.

(2) The adsorbent may be covered by polymers, compounds, and other substances, which reduce its surface area and adsorption site. In practical engineering, it is common for the adsorbent to be contaminated by other substances. Some adsorbates with higher boiling points solidify more easily on the surface of the adsorbent, forming a polymer to block the pores.
(3) At a high regeneration temperature, the state of the adsorbent changes and becomes semimolten, resulting in the blockage and disappearance of some pores. For example, the semimelting phenomenon of Si and Al adsorbents occurs at about 320°C.
(4) The original structure of the adsorbent may be destroyed by the chemical reaction during regeneration. For example, if the purge gas contains acidic or alkaline components, it will damage the pore structure of some adsorbents, such as zeolites and activated alumina, resulting in a decrease in the adsorption performance.

5.3.2 Adsorption configurations

5.3.2.1 Fixed-bed adsorber

According to the filling method of the adsorbent, the fixed-bed adsorber can be divided into vertical, horizontal, and ring adsorbers.

5.3.2.1.1 Vertical adsorber

A vertical adsorber is usually cylindrical, with both the bottom and top covers usually being conical or elliptical. It is suitable for small gas volumes and high concentrations of exhaust gases. According to the flow direction of the waste gas, it can be divided into upflow and downflow adsorbers. For the upflow adsorber, the exhaust gas enters from the lower inlet, and the purified gas is discharged from the upper outlet. It is the opposite for the downflow adsorber. The adsorbent is filled on a grid plate at a height of 0.5–2 m. In order to prevent the adsorbent from leaking out, a layer of stainless-steel mesh or blocky gravel is laid out at the bottom, and the top of the adsorbent is also covered with stainless-steel mesh. It should be noted that when the exhaust gas contains corrosive gases, it is necessary to use an anticorrosion material lining, such as a refractory brick and ceramic plate.

5.3.2.1.2 Horizontal adsorber

A horizontal adsorber is a cylindrical device placed horizontally, and the adsorbent filling height is 0.5–1.0 m. In comparison to vertical adsorbents, horizontal adsorbents have the advantages of a large cross-sectional area, low passing velocity, and low resistance. Therefore, the horizontal adsorber is suitable for treating waste gases with a large gas volume at low concentration. However, because of the large cross-sectional area of the adsorption bed, it may have an uneven airflow distribution.

5.3.2.1.3 Ring adsorber

The structure of a ring adsorber is more complex than that of vertical and horizontal adsorbers. The adsorbent is filled between two concentric cylinders and, therefore, has a large adsorption cross-sectional area. The exhaust gas enters from the outer cylinder, passes through the adsorption layer along the radial direction and is discharged from the central cylinder after purification. The ring adsorber has many advantages. For example, the adsorber is relatively compact, the adsorption cross-sectional area is large, the resistance is small, and the processing capacity is large.

5.3.2.2 *Moving-bed adsorber*

A moving-bed adsorber can realize continuous cyclic adsorption separation. The adsorbent is continuously added to the top of the adsorber, gradually moved down, and finally discharged continuously from the bottom. The exhaust gas enters the moving adsorption bed from the bottom, and the purified gas is discharged from the top. Its main components include an adsorbent cooler, adsorbent charging device, adsorbent discharge device, adsorbent distribution board, and adsorbent desorption device.

The cooler is a vertical tubular heat exchanger located above the adsorption section to cool the adsorbent. The adsorbent enters the cooler from the top of the adsorber and falls into the adsorption section after cooling. The function of the adsorbent dispensing plate is to distribute the adsorbent particles evenly along the cross-section of the device. The adsorbent desorption device is an upright tube bundle connected to two tube plates. The adsorbent moves along the pipe, and the heating medium passes between the gaps. The selection of the heating medium is mainly based on the desired desorption temperature, and the commonly used heating medium is water vapor.

5.3.2.3 *Fluidized-bed adsorber*

Adsorbents in a fluidized-bed adsorber are in a fluid state because of the high gas velocity. With the increasing gas velocity, the gaps between the adsorbent particles increase, and the particles are suspended in the flow. At this time, the state of the adsorbent is similar to that of the liquid, so it is called a fluidized bed.

The fluidized-bed reactor has many advantages. For example, the contact area between the fine particle adsorbent and the waste gas is large in the suspension state, which is conducive to the heterogeneous reaction and improves the adsorption efficiency. In addition, the adsorbent in the fluidized bed has fluid-like properties and can be easily added and removed, facilitating the regeneration operation. At the same time, it also has some shortcomings. For example, the violent collision of the adsorbent in the fluidized-bed reactor may not only cause the crushing of the adsorbent but also seriously wear the reactor.

5.3.3 Selection of the adsorption process

5.3.3.1 *Classification*

Depending on whether the exhaust gas can be treated continuously or not, the VOC adsorption process can be classified into three categories, which are intermittent, semicontinuous, and continuous. Their respective characteristics as well as advantages and disadvantages are summarized in Table 5.2.

5.3.3.2 *Typical posttreatment process*

According to the methods of adsorbent regeneration and desorbed gas treatment, there are four typical posttreatment processes (M.o.E.P. Department of Science and Technology Standards and China, 2013).

TABLE 5.2 Comparison of VOC adsorption processes.

Adsorption process	Characteristics	Advantages and disadvantages
Intermittent adsorption	Adsorption device is simple without the regeneration unit. Adsorbent is removed after saturation. Suitable for exhaust gas treatment with a small volume and low VOC concentration.	Simple equipment structure. High adsorption efficiency. Less adsorbent wear. Low operation cost Large pressure loss. Easy to clog.
Semicontinuous adsorption	Two or more adsorbers are used for alternating between adsorption and regeneration. Gas can pass through the adsorber continuously. Each adsorber performs intermittent adsorption or regeneration. Used for continuous exhaust gas discharge or when waste gas pollutants need to be recovered. Used when the gas volume is large or small and the VOC concentration is high or low.	High adsorption efficiency. Less adsorbent wear. Adsorbate can be recovered. Large pressure loss. Easy to clog.
Continuous adsorption	Adsorption in some parts, desorption and regeneration in others. Adsorption and regeneration are carried out continuously.	High adsorption efficiency. Adsorbate can be recovered. Complicated operation. Severe adsorbent wear. Large energy consumption.

5.3.3.2.1 Water vapor regeneration and condensation recovery

First, the collected exhaust gas enters the pretreatment equipment, where impurities, such as particles in the exhaust gas, are removed. Then, the pretreated exhaust gas is transported to the adsorption system for further purification. In order to ensure the continuous operation of the system, the adsorption system is usually composed of two or more parallel adsorbers. This ensures that, while one adsorber is being regenerated, another is still available for adsorption. The purified exhaust gas is then finally discharged through the exhaust funnel.

When the adsorbent in the adsorption system approaches adsorption saturation, it enters the desorption regeneration stage. Water vapor is introduced to the saturated adsorber to desorb the VOCs. The desorbed VOCs are then condensed through a condenser. Subsequently, the condensed VOCs and the water vapor mixture are separated by a solvent separator. The separated VOCs are fed into the reservoir, and the wastewater is treated and then discharged. In the condenser, a small amount of uncondensed gas is transported back to the absorber for close-loop adsorption. A fixed-bed adsorber is suitable for this water vapor regeneration and condensation recovery process. The process flow diagram is shown in Fig. 5.1.

5.3.3.2.2 Hot gas flow regeneration and condensation recovery

The process of hot gas flow regeneration and condensation recovery is very similar to the water vapor one, except that the desorption medium is changed from water vapor to hot air or an inert gas. Apart from this difference, the other components are identical. In addition, the operation flow of the process is basically the same. What is slightly different is that the

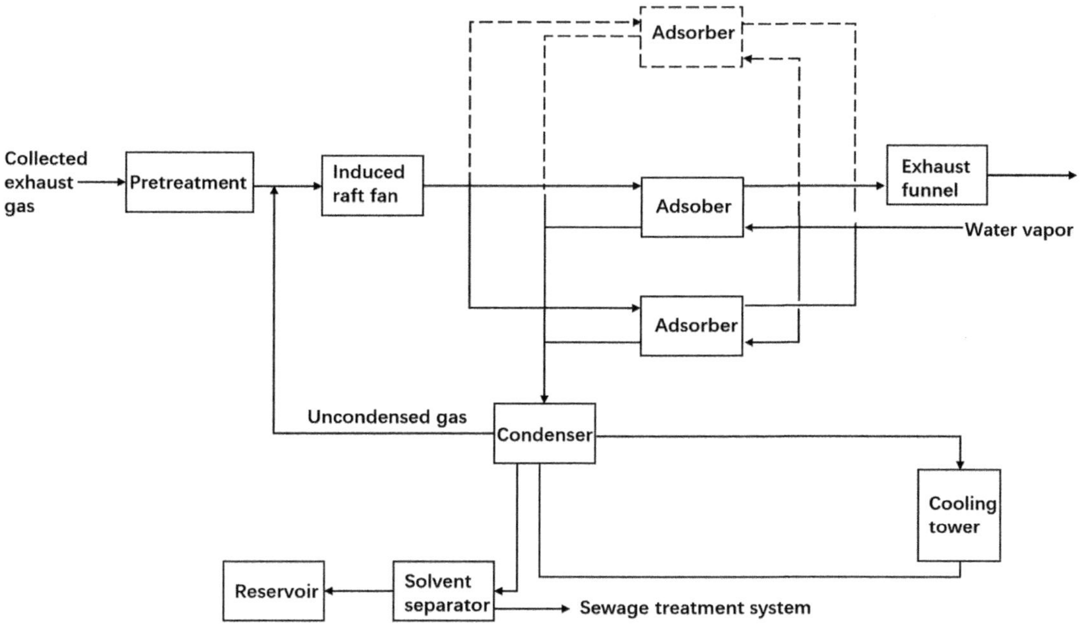

FIGURE 5.1 Process flow diagram of the water vapor regeneration of adsorbents and the condensation recovery of desorbed VOCs.

desorption gas from the atmosphere needs to be purified using a dust collector and then blown into the heater by an induced draft fan to raise the gas temperature. After the temperature is increased, the hot gas enters the adsorber for the desorption of the saturated adsorbent. Both fixed-bed and moving-bed adsorbers can be used in the hot gas flow regeneration and condensation recovery process. The process flow diagram is shown in Fig. 5.2.

5.3.3.2.3 Hot air flow regeneration and catalytic combustion

The process flow diagram is shown in Fig. 5.3. In the pretreatment process, it is necessary to exclude impurities, such as dust and aerosol, in the exhaust gas that cause the blockage and poisoning of the adsorbent. In addition, when catalytic combustion is adopted in the subsequent process, special attention needs to be paid to the removal of substances that can cause catalyst poisoning. The purification efficiency of the adsorber should not be less than 90%, and the pressure loss of the adsorber should be less than 4 kPa. The design space velocity of the catalytic combustion device should be between 4000 and 10,000 h^{-1}, and the pressure loss should be less than 2 kPa. When using hot air to regenerate AC and ACF adsorbents, the desorption temperature should be below 120°C. In addition, the VOC concentration of the desorption gas should be controlled to avoid explosions.

When the organic matter in the exhaust gas contains heteroatoms (e.g., S, N, and Cl), secondary pollutants, such as SO_2, HCl, and NO_x, will be produced in catalytic combustion. As a result, the combustion exhaust gas must be further treated before it can be discharged. The residual heat of the high-temperature flue gas produced in catalytic combustion can be used for adsorbent regeneration.

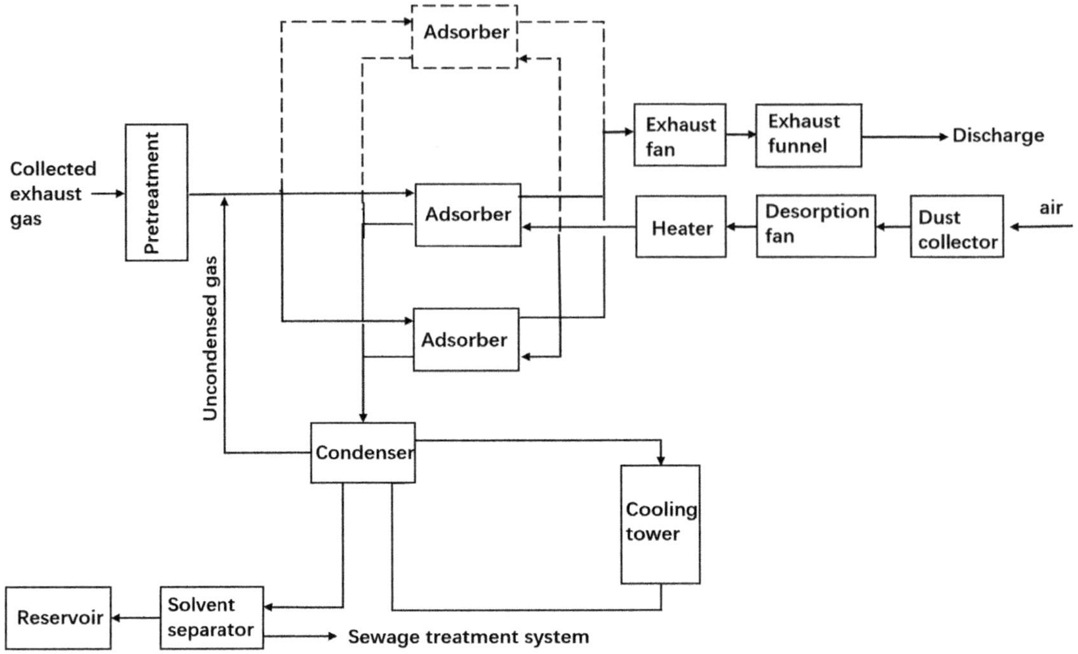

FIGURE 5.2 Process flow diagram of the hot gas flow regeneration of adsorbents and the condensation recovery of desorbed VOCs.

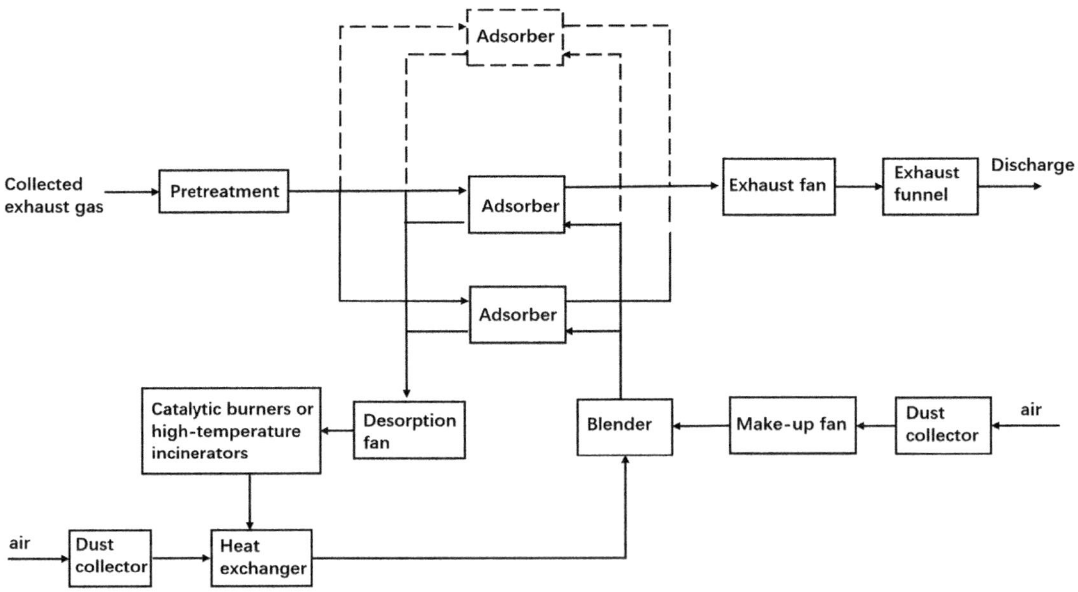

FIGURE 5.3 Process flow diagram of the hot air flow regeneration of adsorbents and the catalytic combustion of desorbed VOCs.

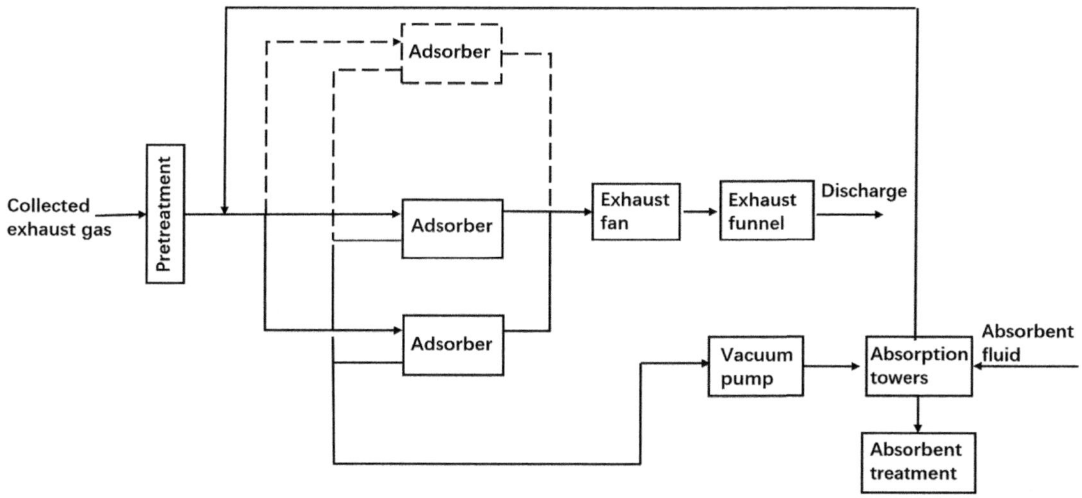

FIGURE 5.4 Process flow diagram of the decompression desorption regeneration of adsorbents and the liquid absorption of desorbed VOCs.

There are two ways to use the waste heat of the high-temperature flue gas for adsorbent regeneration. First, fresh air is added to the high-temperature flue gas by the fan, so that the mixed gas temperature is reduced to the required temperature for adsorbent regeneration. The second is the use of a heat exchanger to heat the fresh air to the temperature required for adsorbent regeneration. Both fixed-bed and moving-bed adsorbers are suitable for the process of hot air flow regeneration and catalytic combustion.

5.3.3.2.4 Decompression desorption regeneration and liquid absorption

In the adsorption stage, the process is similar to the aforementioned process, that is, after pretreatment, the exhaust gas is input into the adsorber for purification. However, its desorption method is different, which uses vacuum desorption. In addition, the desorbed gas treatment system is also different. It uses the liquid absorption method in an absorption tower. A small amount of gas that is not absorbed is sent back to the adsorber inlet again for closed-loop adsorption. Fixed-bed adsorbers are suitable for this process of decompression desorption regeneration and liquid absorption. The process flow diagram is shown in Fig. 5.4.

5.4 Case studies

5.4.1 Expanded polystyrene production line

Chemical enterprises are one of the main sources of VOC emissions. In this case, an ACF-based combined purification process was adopted to treat the VOC waste gas produced in the expanded polystyrene (EPS) production process (Liu, 2023). After treatment, the exhaust VOC emission concentration was lower than 20 mg/m^3 and the particulate matter emission concentration was below 10 mg/m^3, meeting the local exhaust emission standards. This case shows that the application of ACFs in styrene waste gas treatment is feasible. Compared with the traditional honeycomb AC, the ACF has the advantages of a high VOC removal efficiency, fast adsorption and desorption rate, and high flame retardancy. Under the same working conditions, ACFs showed a longer replacement cycle and lower operating cost than AC.

5.4.1.1 Project background

The main components of waste gas from the EPS production line were toluene, styrene, pentane, etc. The exhaust gas flow was 35,000 m^3/h, and the inlet exhaust gas concentration was 500 mg/m^3, and the temperature was 35–40°C. A combined process of three-stage dry filtration, fixed-bed adsorption–desorption, and combustion was used to treat this exhaust gas. A self-made ACF was chosen as the adsorbent, which was prepared by mixing AC, montmorillonite, special fiber, flame retardant, and adhesive with a fibrillated fiber as a carrier. This ACF contained a high flame retardant, in-line with the United States UL94-V-0 highest flame-retardant standard. It showed the advantages of a strong adsorption capacity, fast adsorption and desorption rate, small pressure drop, long service life, low operation and maintenance cost, etc.

5.4.1.2 Process flow

Adsorption: As shown in Fig. 5.5, the exhaust gas produced by polystyrene production was first purified by a three-stage dry filter (primary, medium, and high efficiency), where the particles were removed. The concentration of the particles in the exhaust gas after filtration was reduced to less than 1 mg/m^3. The filtered VOC waste gas was then adsorbed by a fixed bed, where the ACF in the fixed bed efficiently adsorbed toluene, styrene, and pentane. Finally, the purified exhaust gas was sent to the exhaust chimney by the fan, and the exhaust gas was then discharged.

Desorption: The air extracted using the desorption fan was heated to 100°C by the heat exchanger. Hot air was then delivered to a fixed-bed adsorber, where the ACFs filled in the adsorber were heated. The adsorbed VOCs on the ACF were desorbed and transported to the boiler for incineration. The main parameters of VOC waste gas treatment are shown in Table 5.3.

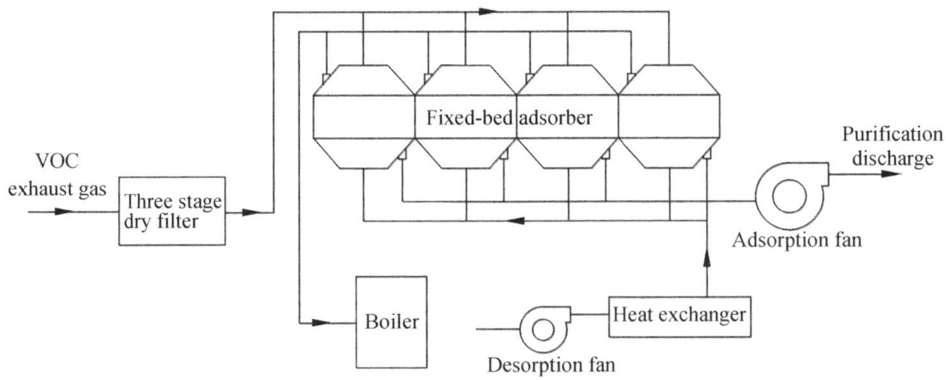

FIGURE 5.5 ACF-based combined purification process for VOC exhaust gas from an EPS production line.

TABLE 5.3 Main device configurations and parameter settings (Liu, 2023).

Equipment	Project	Value
Dry filter	Equipment size (m)	4 × 1.8 × 2.4
Three-stage filtration (primary, medium, and high efficiency)	Handling air volume (m^3/h)	35,000
Fixed-bed adsorber	Equipment size (m)	1.8 × 1.8 × 2.4
(four units, three adsorptions, and one desorption)	Single unit handles air volume (m^3/h)	12,000
	Air velocity of an empty bed (m/s)	1.1
	Filling amount of ACFs per unit (m^3)	1.95
	Filling height (m)	0.6
	Equipment resistance (Pa)	<600
Adsorption fan	Fan pressure (Pa)	3000
	Power (kW)	55
	Air volume (m^3/h)	35,000
Desorption fan	Fan pressure (Pa)	2500
	Power (kW)	5.5
	Air volume (m^3/h)	3000
Heat exchanger	Equipment size (m)	1.06 × 0.67 × 0.36
	Heat transfer area (m^2)	89

5.4.1.3 Performance and costs

After 15 days of operation, the concentrations of pollutants, such as total nonmethane hydrocarbon and particulate matter, were tested. VOC emission and particulate matter concentrations were below 20 and 10 mg/m^3, respectively, much lower than the emission regulation standards.

After the equipment was running for a long time, the operation parameters, such as temperature, desorption/regeneration time, desorption gas concentration, and adsorption efficiency, were measured. The results showed that the combined purification process ran stably.

In addition, the performance parameters and operating costs of ACFs and traditional honeycomb AC were compared (Table 5.4). Under the same operating conditions, ACFs had higher removal efficiency, faster adsorption and desorption rate, higher flame-retardant level, longer replacement cycle, lower replacement material cost, and lower operating costs than AC.

5.4.2 Tank farm

In this case, a multilayer fluidized-bed process was used to treat the VOC waste gas and the air exchange in the pumping room space of a storage tank area (Liu et al., 2022b). High wear resistance and large adsorption capacity resins were used for VOC adsorption. The saturated adsorption resins were regenerated online, realizing the integrated continuous operation of adsorption and regeneration. The system covered a small area and operated

TABLE 5.4 Performance parameters and operation costs of ACFs and honeycomb AC (Liu, 2023).

		ACF	Honeycomb AC
Performance parameter	Toluene removal efficiency (%)	≥80	≤80
	Material desorption regeneration time (h)	≤1.5	≥2
	Desorption temperature (°C)	≤150	≤120
	Cooling time to room temperature (min)	≤2	≥30
Operating cost	Replacement cycle (a)	1	0.5
	Single change dosage (m^3)	7.8	7.8
	Unit price of materials (10,000 yuan/m^3)	1.1	0.7
	Material replacement cost (10,000 yuan/a)	8.58	10.92

safely. The treatment of VOC exhaust gas ran stably and efficiently. After the treatment, the emission concentration met the environmental protection emission standards.

5.4.2.1 *Project overview*

A pharmaceutical factory had two storage tank areas. The no. 1 storage tank area had 17 storage tanks, a single tank reserve of 400 m^3, while the no. 2 storage tank area had 16 storage tanks, a single tank maximum reserve of 100 m^3. The tanks stored raw materials for drug production, of which alcohols (methanol and isopropyl alcohol) accounted for the majority (79%). There were obvious odors around the pump house and tank farm, which seriously affected people's normal work and life. These odors came from the exhaled air of the storage tank and the air exchange from the pump room. Their main components were toluene, methanol, isopropyl alcohol, and ketones. Based on the information, the adsorption process was selected to purify the emitted VOC gas. A resin with an average pore size of 2.6 nm was selected as the adsorbent because its abundant mesoporous pores could effectively remove VOCs.

The comprehensive economic and environmental analysis and evaluation of waste gas treatment projects were carried out. Considering the emission requirement, investment, and operating cost, a combined process of "multilayer fluidized-bed resin adsorption + condensation" was selected. The process flowchart of VOC adsorption and recollection is shown in Fig. 5.6.

As mentioned before, the multilayer fluidized-bed structure has the advantages of a large driving force for mass transfer and strong load-impact resistance, which is conducive to improving the utilization level of the adsorbent. It can also realize online continuous displacement discharge, the step-less adjustment of the gas—solid ratio during the operation, and is suitable for various complex working conditions. This process uses a cascade condensing unit to directly condense and recover the high-concentration VOC gas.

5.4.2.2 *Process flow*

Condensing system: The condensing section treatment was used for the high concentration of a tank-exhaled gas and a desorbed organic waste gas. These two gases contained high-concentration VOCs but the flow rate was low. In order to satisfy the gas emission

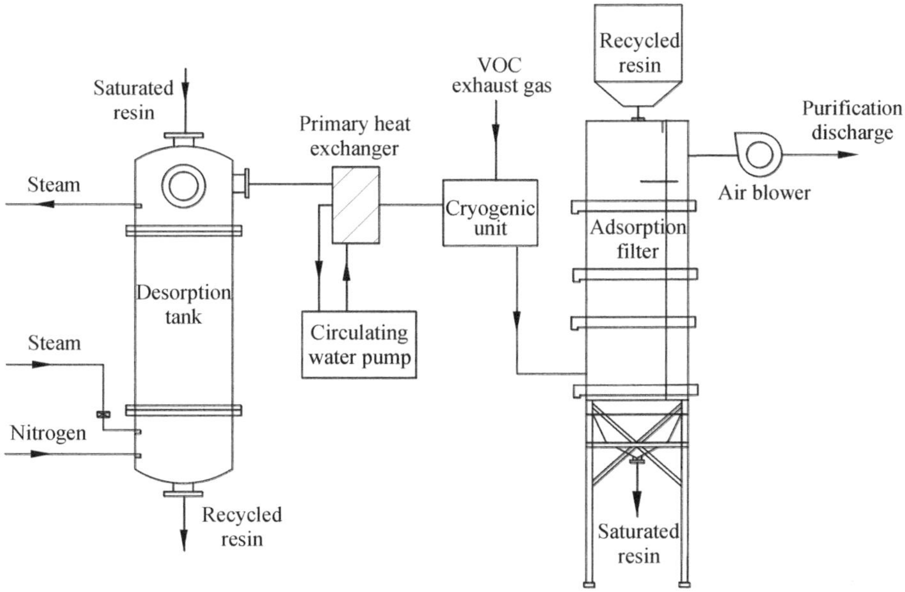

FIGURE 5.6 Multilayer fluidized-bed with a resin adsorbent for the VOC exhaust gas from a tank farm.

standards, a water-cooled two-stage cascade refrigeration unit was used to provide a cold source and so that the emission temperature of the VOC exhaust gas could be reduced to −35°C. After the exhaust gas was cooled, the resulting condensate was collected, while nitrogen and the noncondensing VOC gas were further treated by the adsorption system.

Adsorption system: After condensation, the main components of the exhaust gas are nitrogen and a small amount of VOCs, which were sent into the fluidized-bed adsorption system together with a low concentration of the workshop exhaust gas. The main body of the fluidized-bed adsorption tank was made of stainless steel. The multilayer resin fluidized-bed adsorption system was equipped with six fluidized beds. Fresh or recycled resin was placed on the top layer of the storage tank, while the resin in the adsorption tank fell to the next layer through the automatic turnover mechanism. The saturated resin fell to the lowest layer, so as to complete the replenishment and layer-by-layer replacement of the resin in the fluidized bed. The adsorbent used in this process was styrene–divinyl polymeric resin, which is a yellow–brown ball. It has the characteristics of a large specific surface area and good fluidity, which is conducive to the full contact between the exhaust gas and the resin. The VOC pollutants in the waste gas were adsorbed and the purified gas was discharged directly. The saturated resin that fell to the lowest layer was fed into the desorption tank for regeneration.

Regeneration system: The heat used for resin recycling was 0.3 MPa of low pressure steam at 150°C. In order to ensure the fluidity of the resin after desorption, the desorption steam and the saturated resin were used in the indirect heat exchange. In addition, intermittent pulses of nitrogen were sprayed into the resin, on the one hand to avoid

TABLE 5.5 Parameters of the treatment facility for tank farm VOCs (Liu et al., 2022b).

Processing phase	Project	Parameters
Adsorption section	Waste gas flow (m^3/h)	3500
	Gas speed (m/s)	0.43
	Adsorbent amount (kg)	500
	Bed number	6
	Pressure drop (Pa)	1500
Desorption section	Adsorbent amount (kg)	350
	Regeneration temperature (°C)	135
	Regenerative pressure (kPa)	−95 ~ −20
	Waste gas flow (m^3/h)	<500
	Condensing temperature (°C)	−35

the local overheating of the resin and on the other hand to carry the desorbed VOC exhaust gas away from the regeneration section. The high-concentration VOC waste gas produced by desorption was pumped using a vacuum pump into the refrigeration unit for condensation. The condensate was collected and stored in the liquid division tank. The residual pollutants in waste gas flowed into the fluidized-bed adsorber for further purification.

Exhaust piping system: The exhaust gas collection system consisted of fiberglass pipes and stainless-steel pipes. The VOC gas exhaled by the high-concentration storage tank was collected and transported by stainless-steel pipes. The VOC exhaust gas generated by the ventilation of the pump room was collected and transported by glass fiber-reinforced plastic pipes. Considering the micronegative pressure of the collection system, the design air volume was 3500 m^3/h to prevent exhaust gas leakage.

The main parameters of the waste gas treatment system are shown in Table 5.5.

5.4.2.3 Performance and costs

The total investment of this project was 1.76 million Chinese yuan and the annual operating time was 7920 h. The operating costs mainly included electricity, steam, and resin costs of 150,400, 73,600, and 50,000 yuan/year, respectively. The total annual operating cost was 274,000 yuan. The system could realize the online continuous operation of the adsorption regeneration system. Saturated resin replacement, resin regeneration, fresh resin replenishment, etc., were automated, thus reducing labor costs and operating costs. After the formal operation of the treatment facility, the concentration of the exhaust gas was tested. The emission indicators of various pollutants were lower than the limits required by the local environmental protection agency (Table 5.6). The air quality of the tank farm region has been significantly improved.

TABLE 5.6 Tank farm VOC concentrations before and after treatment (Liu et al., 2022b).

VOCs	Standard limit (mg/m^3)	Original emission value (mg/m^3)	Posttreatment emission value (mg/m^3)
Total nonmethane hydrocarbons	100	30	3
Toluene	60	15	<0.04
Isopropyl alcohol	700	80	<0.02
Carbinol	50	30	1.2
Acetone	450	80	<0.01

5.4.3 Automotive painting workshop

5.4.3.1 Project overview

There are four major processes in automobile manufacturing, i.e., stamping, welding, painting, and assembly. Among them, painting releases the maximum amount of VOC pollutants. This case introduces the treatment of VOCs from an automobile production company's painting workshop (Mu and Feng, 2019). This automobile manufacturing company used water-based paint for spraying to avoid some VOC production at the source. Nevertheless, a large amount of VOC exhaust gas was still generated during the painting process. The painting workshop of this automobile production company had two production lines with a total exhaust gas flow of 96,000 m^3/h (standard state), exhaust gas temperature of 23°C, relative humidity of 65%, and VOC concentration of 1000 mg/m^3 (standard state). The exhaust gas had the characteristics of a large flow and low concentration of VOCs, so it is suitable to be treated by the adsorption concentration and then incineration. This project adopted the combined process of zeolite wheel and inline burner to treat the VOC waste gas and finally realized a VOC treatment efficiency of more than 95%. It reduced the VOC emission concentration to less than 60 mg/m^3, passing the emission standard.

5.4.3.2 Process flow

As shown in Fig. 5.7, the exhaust gas treatment system consisted of five parts, namely, ① two-stage filter (pretreatment module), ② zeolite wheel (VOC concentration module), ③ VOC incineration module (exhaust gas incineration-centralized heating device), ④ flue gas heat exchanger (flue gas heat exchange module), and ⑤ gas–water heat exchanger (waste heat utilization module). The exhaust gas treatment process is as follows: first, the exhaust gas discharged from the painting workshop entered the pretreatment module for coarse and fine filtration. Then, the filtered exhaust gas entered the VOC concentration module for adsorption treatment, and the treated clean gas was directly discharged into the atmosphere. At the same time, a small amount of filtered exhaust gas was purged into the cooling zone of the zeolite wheel. Then, the gas that passed through the cooling zone entered the flue gas heat exchanger, where the temperature rose to about 180°C and became a desorption gas. The desorption gas entered the regeneration zone of the zeolite wheel to desorb the adsorbed VOCs on the zeolite. Finally, the gas containing high-concentration VOCs entered the exhaust gas incineration unit and was incinerated at 730°C. The high-temperature flue gas after

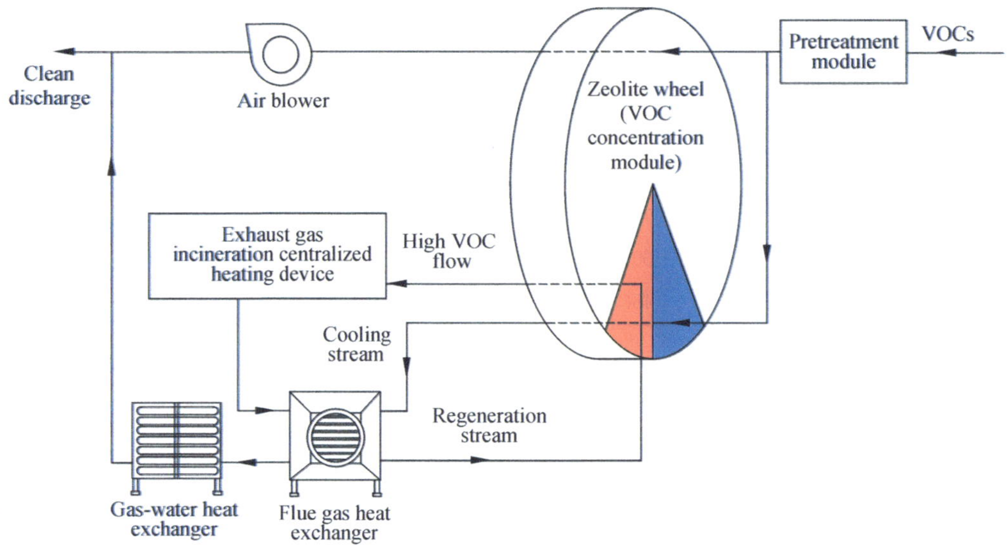

FIGURE 5.7 Zeolite wheel-based exhaust gas treatment system for an automotive painting workshop.

incineration entered the flue gas heat exchange device for heating the desorbed gas. Finally, the flue gas was fed into a gas–water heat exchange module, where waste heat was recovered and then discharged into the atmosphere.

5.4.3.3 Main technical parameters

The core of this exhaust gas treatment system is the zeolite wheel (VOC concentration module). The core material of the VOC concentration wheel is zeolite. As a porous framework-type aluminosilicate, its internal pores can selectively adsorb molecules of different sizes, resulting in a very high VOC adsorption efficiency. The actual operation parameter set is shown in Table 5.7.

The concentration ratio, wheel speed, desorption zone temperature, and sealing are the key factors affecting the operation efficiency of the zeolite wheel. Taking the concentration ratio of the wheel as an example, it is 8:1:1, that is, the adsorption area of the entire zeolite wheel accounts for 4/5 of the total area, and the desorption area and the cooling area each account for 1/10 of the total area. The speed of the wheel affects the time of adsorption and desorption and then affects the adsorption and concentration ability of the wheel.

TABLE 5.7 Main technical parameters of the zeolite wheel (Mu and Feng, 2019).

Project	Concentration ratio	Wheel speed	Desorption zone temperature (°C)	Sealing
Parameter setting	8:1:1	8 r/h	180	Replace ring and strip seals annually

TABLE 5.8 Zeolite wheel and inline burner annual operating costs (Mu and Feng, 2019).

Project	Dosage	Cost	Total cost
Natural gas	595,350 m^3	2.3 ¥/m^3	1.394 million yuan
Electric energy	1,814,400 kW h	0.6 ¥/kW·h	1.134 million yuan
Equipment maintenance	–	–	0.100 million yuan

Optimal speeds cause the times of adsorption and desorption to be in the best balanced state, achieving the highest VOC removal efficiency.

At different temperatures, the adsorption ability of the zeolite material to VOC molecules can be significantly different. At room temperature, the zeolite wheel effectively adsorbed VOC molecules from the exhaust gas passing through it. As the temperature rose, its adsorption efficiency gradually decreased. When it reached about 180°C, the zeolite wheel not only lost its ability of adsorbing VOC molecules from the exhaust gas, but also released some of the adsorbed VOCs, thereby achieving the purpose of regeneration and concentrating the exhaust gas. It is worth noting that some VOCs have a higher boiling point and need a higher temperature to desorb, so high-temperature cleaning may be needed in actual applications.

Insufficient sealing can lead to the direct discharge of the unabsorbed exhaust gas into the atmosphere and cause the exhaust gas to cross among the three areas of the wheel, affecting the purification efficiency. There is a ring seal between the wheel and the chamber, and there is a strip seal between the high-temperature area and the low-temperature area of the wheel. The over-temperature alarm device set behind the wheel can identify the state of the seal ring. In general, replacing the ring and strip seals annually can ensure the efficiency of the wheel.

5.4.3.4 *Performance and costs*

As an advanced VOC treatment technology, the zeolite wheel and inline burner device achieved a treatment efficiency of over 95% and reduced the VOC concentration from 1000 mg/m^3 to less than 60 mg/m^3. Nevertheless, its energy consumption was very high, and the annual operating cost was a total of 2.628 million yuan. Detailed numbers are shown in Table 5.8.

5.4.4 Thin-plate coating line

5.4.4.1 *Project overview*

During the painting process, a large amount of organic waste gas is produced. The main components of the waste gas are alcohols, esters, ethers, and so on. If the exhaust gas is not effectively treated and directly discharged, it will affect the surrounding environment and cause damage to people's health. This case introduces the treatment of the VOC exhaust gas released from a thin-plate coating line of an enterprise (Wang, 2023). The production line had a total exhaust gas amount of 170,000 m^3/h at a low VOC concentration. The combined process of AC adsorption and catalytic combustion was used to purify the VOC exhaust gas.

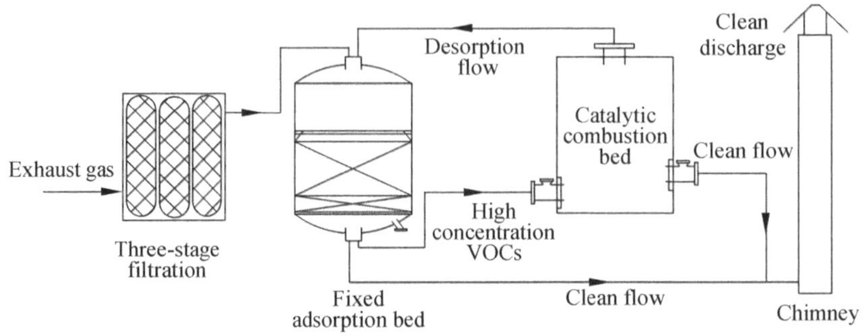

FIGURE 5.8 AC adsorption and catalytic combustion system for the VOC exhaust gas from a thin-plate coating line.

5.4.4.2 *Process flow*

The exhaust gas purification process flow for the thin-plate coating line is shown in Fig. 5.8. There were three main components: the pretreatment unit composed of a dry filter, the adsorption unit with honeycomb AC as the main component, and the catalytic combustion unit. The waste gas produced in the production workshop was collected and transported to this AC-based waste gas purification system by the waste gas pipe. Because the exhaust gas contained impurities, such as paint fog and dust, a three-stage filter was set at the entrance to pretreat it. The pretreated exhaust gas flowed into the adsorption unit, where AC adsorbed the low-concentration VOCs. The purified exhaust gas was directly discharged from the chimney. When the adsorbent was close to saturation, hot air was introduced to heat the adsorbent, causing the VOCs to be desorbed and for them to enter the catalytic combustion unit, where the desorbed VOCs were catalyzed to be burned. At last, the exhaust gas from catalytic combustion was discharged from the chimney together with the purified gas after AC adsorption.

In this process, the low-concentration VOC waste gas was concentrated into a high-concentration one by the adsorption–desorption of AC and then completely eliminated by catalytic combustion. This method combines the advantages of adsorption and catalytic combustion and effectively solves the problems existing in the purification treatment of a large volume of the low-concentration VOC waste gas. This combination method is mature and practical in VOC treatment.

5.4.4.3 *Process parameters*

The process parameters of the pretreatment equipment, adsorption equipment, and catalytic combustion equipment are shown in Table 5.9. A three-stage bag filter was set at the front of the adsorption bed to treat the exhaust gas discharged from the painting workshop to avoid its direct entry into the adsorption equipment. The filter cotton used in the filter has the advantages of a large dust capacity, high efficiency, low pressure loss, and low cost. As a result, it showed a good effect on painting waste gas treatment. The bag filter frame was produced by a 20 × 20 mm galvanized pipe.

The waste gas after pretreatment was evenly distributed and passed through the AC layer in the fixed-adsorption bed. The adsorption system consisted of five parallel-adsorption

TABLE 5.9 Process parameters of the main treatment equipment for the thin-plate coating line (Wang, 2023).

Processing module	Name	Parameter
Pretreatment equipment	Three-stage bag filter	Frame: 20 × 20 mm Size: 592 × 592 × 400 mm
Adsorption equipment	Adsorbent	Honeycomb AC
	Single bed handling air volume	45,000 m^3/h
	Processing efficiency	Organic removal rate ≥80% Overall purification efficiency ≥80%
	Air speed	Velocity of filtering: 0.9–1.2 m/s Residence time: ≥0.5 s
	AC-bed desorption temperature	80–120°C
Catalytic combustion equipment	Texture	Carbon steel, thickness ≥4 mm
	Design flow rate	5000 m^3/h
	Combustion chamber operating temperature	250–300°C
	Catalytic combustion efficiency	≥97%
	Catalyst lifetime	Pt/Pd honeycomb ceramic type ≥12,000 h

equipment, each of which had a processing air volume of 45,000 m^3/h. After a certain residence time (≥0.5 s), AC adsorbed the VOC molecules to purify the waste gas. The concentrated waste gas entered the catalytic combustion equipment and was catalyzed by noble-metal catalysts. The supporter of the noble-metal catalyst was cordierite honeycomb ceramic and γ-Al_2O_3, and the main active components were Pd and Pt.

5.4.4.4 *Performance*

The performance of the exhaust gas treatment system is shown in Table 5.10. The total removal efficiency of the system's exhaust gas reached more than 85%. It can stably meet

TABLE 5.10 Performance of the AC-based exhaust gas treatment system for the thin-plate coating line (Wang, 2023).

	Exhaust gas concentration/(mg/m^3)		
Detect location	**Detect 1**	**Detect 2**	**Detect 3**
Before AC adsorption	76	58	65
After AC adsorption	8.96	7.37	8.68
Before catalytic combustion	1652	2115	2632
After catalytic combustion	118.52	96.29	85.28
Chimney	10.45	9.23	10.89

the emission standards. In summary, the "pretreatment (dry filtration) + AC adsorption + desorption catalytic combustion" process is efficient and feasible for the water-based paint exhaust gas.

5.4.5 Wastewater treatment plant

During the operation of a wastewater treatment plant, there are a lot of foul odors, which contain many VOCs. This is worse in chemical wastewater treatment plants. In recent years, the treatment of VOC waste gas in wastewater treatment plants has become the focus of attention. It is urgent to develop effective technology to solve this problem.

5.4.5.1 Project overview

The wastewater of a petroleum refining plant contained a large amount of VOCs, such as benzene, toluene, hexyl hydride, and pinene (Cheng, 2023). When this wastewater was treated, the VOC gas was generated in multiple treatment units, including the scum removal tank, accident tank, sequencing batch reactor (SBR) regulating tank, and oil separation tank. The VOC gas from this wastewater treatment plant had the characteristics of a high concentration and large amount. The exhaust gas generation from main wastewater treatment units is shown in Table 5.11.

5.4.5.2 Process flow and operating parameters

A "biodegradation and ACF adsorption" process was used to treat the VOC-containing exhaust gas produced by this wastewater treatment plant. The process flow diagram is shown in Fig. 5.9, and the operating parameters are shown in Table 5.12. First, the exhaust gas was collected and transported to the biological treatment equipment, where the pollutants, such as hydrocarbon, hydrogen sulfide, and ammonia, were degraded by the metabolism of microorganisms. Then, the exhaust gas entered an ACF adsorption box. At this stage, the ACF efficiently adsorbed the recalcitrant benzene series, nonmethane total hydrocarbons, and other VOC pollutants. At last, the purified gas was sent to the exhaust pipe using the induced draft fan for discharge.

The ACF in the adsorption box was regenerated with 0.1 MPa of steam after running for a period of time. The desorbed VOCs entered the condenser with the steam to be condensed

TABLE 5.11 VOC gas generation from the main wastewater treatment units of a petroleum refining plant (Cheng, 2023).

Structures	Benzene (mg/m^3)	Toluene (mg/m^3)	Hexyl hydride (mg/m^3)	Pinene (mg/m^3)	Nonmethane total alkane (mg/m^3)
Scum removal tank	376	84	79	40	531
Accident tank	1264	175	125	89	1358
SBR regulating tank	187	56	47	24	483
Oil separation tank	512	97	83	31	567

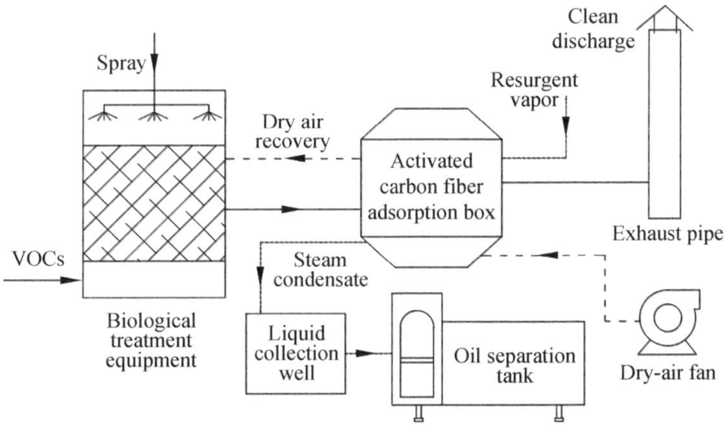

FIGURE 5.9 Combined biodegradation and ACF adsorption system for the VOC waste gas from a wastewater treatment plant.

TABLE 5.12 Operational parameters of the combined biodegradation and ACF adsorption system (Cheng, 2023).

Sections	Projects	Units	Minimum	Maximum
Biological treatment section	Flow of waste gas	m^3/h	19,066	23,000
	Volume of the biological box	m^3	330	330
	Residence time	s	60	60
	Temperature	°C	27	39
	pH	–	7	9
Sprinkler system	Duration of the spray (upper layer)	min	2	2
	Spray interval time (upper layer)	min	1440	1440
	Duration of the spray (lower layer)	min	2	2
	Spray interval time (lower layer)	min	120	120
Adsorption section	Adsorption temperature	°C	11	63
	Desorption temperature	°C	100	136
	Temperature of dry air	°C	25	66
	Water vapor flow	t/h	0.001	0.286
	Steam pressure of desorption	MPa	0.10	0.13

into a liquid. At last, they were collected and sent to the grease trap to achieve a closed cycle of the whole treatment process. The regenerated ACF was dried by dry air from the outside, which was filtered, heated, and, finally, blown into the adsorption box using a fan. The exhaust gas discharged from the drying process entered the air intake of biological treatment for circulation treatment.

TABLE 5.13 VOC concentration changes before and after combined treatment (Cheng, 2023).

Types of VOCs	Inlet concentration (mg/m^3)	Maximum outlet concentration (mg/m^3)	Emission standard (mg/m^3)
Benzene	45–100	1.434	4
Methylbenzene	15–25	0.405	15
Hexyl hydride	15–50	3.54	100
Pinene	15–30	1.87	–
Nonmethane total alkane	100–250	9.12	1210

5.4.5.3 *Performance*

The performance of the combined treatment system is summarized in Table 5.13. The total exhaust gas volume was 21,549 m^3/h. After treatment, the main pollutants in the exhaust gas were removed with the total removal rate of over 90%. Among them, the removal effects of benzene and toluene were better than those of others. The concentration of benzene dropped from 100 mg/m^3 to 1.434 mg/m^3 with a removal rate of 98%. The concentration of toluene dropped from 25 mg/m^3 to 0.405 mg/m^3 with a removal rate of nearly 99%. In summary, the concentration of each VOC in the waste gas satisfies the local standards for discharge.

5.5 Summary and perspectives

The massive release of VOCs from various industrial production processes poses a serious threat to human health and the environment. Therefore, the development of effective adsorbents and economical technologies for VOC elimination has been an important research hotspot. Among various VOC abatement technologies, adsorption is still by far the most widely used method in practical applications. VOC adsorption can be categorized into physical adsorption and chemical adsorption. Although chemisorption has the advantage of high selectivity, physical adsorption characterized by a fast kinetic rate, reversibility, and multilayer adsorption is still dominant in engineering applications.

The choice of the adsorbent is crucial for VOC adsorption. Numerous porous adsorbents, including AC, ACFs, biochar, CNTs, graphene, zeolite molecular sieves, metal–organic frameworks, clay, silica, organic polymers, and composite materials, can be used as VOC adsorbents. Among them, AC, ACFs, zeolite molecular sieves, and resin are commonly used in practical engineering. When selecting the adsorbent, not only VOC properties and operation conditions, but also the properties of the adsorbent itself should be considered. The surface area, pore volume, pore size, and surface chemistry of the VOC adsorbent are the key factors affecting the adsorption performance. To enhance the VOC adsorption capacity of the adsorbent, functional modification and activation are effective. Moreover, the development of hierarchical porous materials and composite materials can enhance their VOC adsorption capacity and kinetic rate simultaneously. To promote the application of adsorption in VOC treatment, future research can focus on the following aspects:

(1) Optimizing physicochemical methods, such as acid treatment, alkali treatment, nitrogen doping, metal doping, organic polymer coating, and impregnation, to fine tune the microstructure and surface chemistry of adsorbents to further enhance their ability to adsorb VOCs.
(2) Water vapor and other gas molecules present in real exhaust gases can compete with VOC molecules to reduce their adoption. To address this, research on the hydrophobic surface enhancement of adsorbents should be promoted to enhance their practical applications in actual exhaust gas treatment.
(3) Improving the mechanical strength and heat resistance of VOC adsorbents to increase their service life, improve the desorption efficiency, and enhance the regenerative performance.
(4) Compared with the physical and chemical production and modification methods, biotechnology is a more economical and promising method, which has not been fully explored in VOC adsorbent development. Future research can pay attention to this direction to apply biotechnology in the development of VOC adsorbents.
(5) For nanomaterial adsorbents, it is necessary to conduct research to reduce their preparation cost, increase their yield, and develop novel derivatives, including composites for practical VOC removal applications.
(6) Composite materials have shown better VOC adsorption performance and practical application potential than single-component adsorbents. Therefore, research on the preparation and application of composite VOC adsorbents should be strengthened.
(7) Molecular simulations and quantum chemical calculations can provide insights into the VOC adsorption processes and mechanisms. These types of theoretical studies should be promoted to better explain the VOC adsorption phenomenon and reveal the governing mechanisms.

In practical applications, the composition of the VOC waste gas is complex, and the properties of each component are different. Therefore, to achieve high treatment efficiency and reduce cost, adsorption can be combined with other treatment technologies, such as thermal combustion, catalysis decomposition, and biological degradation, absorption, and condensation. In the future, we should explore the optimal combinations of these technologies, comprehensively analyze the composition of industrial VOC emissions, optimize operating conditions, understand the standard restrictions, and maximize economic benefits, which will lead to selecting the best VOC treatment route for practical applications.

References

Ahmadi, Y., Kim, K.-H., 2022. Recent progress in the development of hyper-cross-linked polymers for adsorption of gaseous volatile organic compounds. Polymer Reviews 63, 365–393.

Asiya, S., Kyzas, G.Z., Pal, K., de Souza Jr., F.G.J.J.o.M.S., 2021. Graphene functionalized hybrid nanomaterials for industrial-scale applications: a systematic review. Journal of Molecular Structure 1239, 130518.

Aslam, M.M.-A., Kuo, H.-W., Den, W., Usman, M., Sultan, M., Ashraf, H., 2021. Functionalized carbon nanotubes (CNTs) for water and wastewater treatment: preparation to application. Sustainability 13, 5717.

Barakan, S., Aghazadeh, V., 2021. The advantages of clay mineral modification methods for enhancing adsorption efficiency in wastewater treatment: a review. Environmental Science and Pollution Research 28, 2572–2599.

Barrejón, M., Prato, M., 2022. Carbon nanotube membranes in water treatment applications. Advanced Materials Interfaces 9, 2101260.

Brigatti, M.F., Galan, E., Theng, B., 2006. Structures and mineralogy of clay minerals. Developments in Clay Science 1, 19–86.

Cai, J., Wang, D., Zhang, M., Sui, H., Li, X., He, L., 2022. Deactivation mechanisms of engineering adsorbents for VOCs adsorption and the lifetime-prolonging strategy. Industrial & Engineering Chemistry Research 61, 5952–5962.

Calabrese, L., Bonaccorsi, L., Bruzzaniti, P., Proverbio, E., Freni, A., 2019. SAPO-34 based zeolite coatings for adsorption heat pumps. Energy 187, 115981.

Carrasco-Marín, F., Domingo-Garcia, M., Fernandez-Morales, I., López-Garzón, F.J., 1991. Dynamic adsorption of methyl iodide on activated carbons. Carbon 29, 629–634.

Cejka, J., van Bekkum, H., Corma, A., Schueth, F., 2007. Introduction to Zeolite Molecular Sieves. Elsevier.

Chen, R., Yao, Z., Han, N., Ma, X., Li, L., Liu, S., Sun, H., Wang, S., 2020. Insights into the adsorption of VOCs on a cobalt-adeninate metal–organic framework (Bio-MOF-11). ACS Omega 5, 15402–15408.

Chen, L., Deng, Y., Han, W., Jiaqiang, E., Wang, C., Han, D., Feng, C., 2022. Effects of zeolite molecular sieve on the hydrocarbon adsorbent and diffusion performance of gasoline engine during cold start. Fuel 310, 122427.

Cheng, X., 2023. Optimization and Research of "Biodegradation + Activated Carbon Adsorption" for the Treatment of Waste Gases, in, Northeast Petroleum University.

Cheng, T., Li, J., Ma, X., Zhou, L., Wu, H., Yang, L., 2021. The adsorption properties of microporous activated carbon prepared from pistachio nut shell for low-concentration VOCs under low-medium temperatures. Environmental Science Pollution Research 28, 65216–65228.

Cheng, T., Li, J., Ma, X., Zhou, L., Wu, H., Yang, L., 2022. Alkylation modified pistachio shell-based biochar to promote the adsorption of VOCs in high humidity environment. Environmental Pollution 295, 118714.

Chiang, Y.-C., Chiang, P.-C., Huang, C.-P., 2001. Effects of pore structure and temperature on VOC adsorption on activated carbon. Carbon 39, 523–534.

Dai, J., Zhao, C., Hu, X., Chen, D., Yun, J., Liu, C., Zhong, M., Huang, D., Cen, C., Chen, Z., 2021a. One-pot synthesis of meso-microporous ZSM-5 and its excellent performance in VOCs adsorption/desorption. Journal of Chemical Technology Biotechnology 96, 78–87.

Dai, Z., Li, D., Ao, Z., Wang, S., An, T., 2021b. Theoretical exploration of VOCs removal mechanism by carbon nanotubes through persulfate-based advanced oxidation processes: adsorption and catalytic oxidation. Journal of Hazardous Materials 405, 124684.

David, E., Niculescu, V.-C., 2021. Volatile organic compounds (VOCs) as environmental pollutants: occurrence and mitigation using nanomaterials. International Journal of Environmental Research and Public Health 18, 13147.

Deng, L., Yuan, P., Liu, D., Annabi-Bergaya, F., Zhou, J., Chen, F., Liu, Z., 2017. Effects of microstructure of clay minerals, montmorillonite, kaolinite and halloysite, on their benzene adsorption behaviors. Applied Clay Science 143, 184–191.

Dhar, S.A., Sakib, T.U., Hilary, L.N.J.B.C., 2022. Effects of pyrolysis temperature on production and physicochemical characterization of biochar derived from coconut fiber biomass through slow pyrolysis process. Biorefinery 12, 2631–2647.

Dou, B., Hu, Q., Li, J., Qiao, S., Hao, Z., 2011. Adsorption performance of VOCs in ordered mesoporous silicas with different pore structures and surface chemistry. Journal of Hazardous Materials 186, 1615–1624.

Feng, D., Guo, D., Zhang, Y., Sun, S., Zhao, Y., Shang, Q., Sun, H., Wu, J., Tan, H., 2021. Functionalized construction of biochar with hierarchical pore structures and surface O-/N-containing groups for phenol adsorption. Chemical Engineering Journal 410, 127707.

Gaikwad, S., Kim, Y., Gaikwad, R., Han, S., 2022. Enhanced VOC adsorption capacity on MOF thin layer with reduced particle size by cryogrinding and microwave method. Journal of Environmental Chemical Engineering 10, 107567.

Gauden, P.A., Terzyk, A.P., Rychlicki, G., Kowalczyk, P., Lota, K., Raymundo-Pinero, E., Frackowiak, E., Béguin, F., 2006. Thermodynamic properties of benzene adsorbed in activated carbons and multi-walled carbon nanotubes. Chemical Physics Letters 421, 409–414.

Guo, Y., Li, Y., Zhu, T., Ye, M., 2013. Effects of concentration and adsorption product on the adsorption of SO_2 and NO on activated carbon. Energy & Fuels 27, 360–366.

Hassan, M.F., Sabri, M.A., Fazal, H., Hafeez, A., Shezad, N., Hussain, M., 2020a. Recent trends in activated carbon fibers production from various precursors and applications—a comparative review. Journal of Analytical and Applied Pyrolysis 145, 104715.

Hassan, M., Liu, Y., Naidu, R., Parikh, S.J., Du, J., Qi, F., Willett, I.R., 2020b. Influences of feedstock sources and pyrolysis temperature on the properties of biochar and functionality as adsorbents: a meta-analysis. Science of The Total Environment 744, 140714.

He, S., Zhu, R., Chen, Q., Tang, N., Ji, S., Wei, H., Du, J., Yang, Y., Zhu, J., 2024. Development of a novel hierarchical porous and hydrophobic silica from montmorillonite for benzene adsorption. Separation and Purification Technology 329, 125031.

Hidayat, R., Wahyuningsih, S., Ramelan, A., 2020. Simple synthesis of rGO (reduced graphene oxide) by thermal reduction of GO (graphene oxide). IOP Conference Series: Materials Science and Engineering 858, 012009.

Hou, B., Zhao, Y., Sun, W., Lu, S., Chen, S., 2021. Glycine based modification of activated carbons for VOCs adsorption. Chemical Engineering Journal Advances 7, 100126.

Huang, J., Turner, S.R., 2018. Hypercrosslinked polymers: a review. Polymer Reviews 58, 1–41.

Huang, W., Xu, J., Tang, B., Wang, H., Tan, X., Lv, A., 2017. Adsorption performance of hydrophobically modified silica gel for the vapors of n-hexane and water. Adsorption Science and Technology 36, 888–903.

Hunter-Sellars, E., Tee, J.J., Parkin, I.P., Williams, D.R., 2020. Adsorption of volatile organic compounds by industrial porous materials: impact of relative humidity. Microporous and Mesoporous Materials 298, 110090.

Janu, R., Mrlik, V., Ribitsch, D., Hofman, J., Sedláček, P., Bielská, L., Soja, G., 2021. Biochar surface functional groups as affected by biomass feedstock, biochar composition and pyrolysis temperature. Carbon Resources Conversion 4, 36–46.

Kim, K.-J., Ahn, H.-G., 2012. The effect of pore structure of zeolite on the adsorption of VOCs and their desorption properties by microwave heating. Microporous and Mesoporous Materials 152, 78–83.

Kim, K.-J., Kang, C.-S., You, Y.-J., Chung, M.-C., Woo, M.-W., Jeong, W.-J., Park, N.-C., Ahn, H.-G., 2006. Adsorption–desorption characteristics of VOCs over impregnated activated carbons. Catalysis Today 111, 223–228.

Kim, K.-D., Park, E.J., Seo, H.O., Jeong, M.-G., Kim, Y.D., Lim, D.C., 2012. Effect of thin hydrophobic films for toluene adsorption and desorption behavior on activated carbon fiber under dry and humid conditions. Chemical Engineering Journal 200–202, 133–139.

Kondo, S., Ishikawa, T., Abe, I., 2006. Adsorption Science, second ed. Chemical Industry Press.

Kong, Q., Shi, X., Ma, W., Zhang, F., Yu, T., Zhao, F., Zhao, D., Wei, C., 2021. Strategies to improve the adsorption properties of graphene-based adsorbent towards heavy metal ions and their compound pollutants: a review. Journal of Hazardous Materials 415, 125690.

Li, Y., Lee, C., Gullett, B., 2002. The effect of activated carbon surface moisture on low temperature mercury adsorption. Carbon 40, 65–72.

Li, L., Sun, Z., Li, H., Keener, T.C., 2012. Effects of activated carbon surface properties on the adsorption of volatile organic compounds. Journal of the Air & Waste Management Association 62, 1196–1202.

Li, R., Xue, T., Bingre, R., Gao, Y., Louis, B., Wang, Q., 2018. Microporous Zeolite@ Vertically Aligned Mg–Al layered double hydroxide core@ shell structures with improved hydrophobicity and toluene adsorption capacity under wet conditions. ACS Applied Materials Interfaces 10, 34834–34839.

Li, X., Zhang, L., Yang, Z., Wang, P., Yan, Y., Ran, J., 2020a. Adsorption materials for volatile organic compounds (VOCs) and the key factors for VOCs adsorption process: a review. Separation and Purification Technology 235, 116213.

Li, M., Li, Y., Li, W., Liu, F., Qi, X., Xue, M., Wang, Y., Zhao, C., 2020b. Synthesis and application of Cu-BTC@ ZSM-5 composites as effective adsorbents for removal of toluene gas under moist ambience: kinetics, thermodynamics, and mechanism studies. Environmental Science Pollution Research 27, 6052–6065.

Li, R., Xue, T., Li, Z., Wang, Q., 2020c. Hierarchical structure ZSM-5/SBA-15 composite with improved hydrophobicity for adsorption-desorption behavior of toluene. Chemical Engineering Journal 392, 124861.

Li, Z., Jin, Y., Chen, T., Tang, F., Cai, J., Ma, J., 2021. Trimethylchlorosilane modified activated carbon for the adsorption of VOCs at high humidity. Separation and Purification Technology 272, 118659.

Lillo-Ródenas, M.A., Cazorla-Amorós, D., Linares-Solano, A., 2011. Benzene and toluene adsorption at low concentration on activated carbon fibres. Adsorption 17, 473–481.

Liu, L., 2023. Engineering practice of activated carbon fiber for VOCs waste gas treatment. Environmental Science and Technology 36, 36–38.

Liu, C., Wang, J., Wan, J., Yu, C., 2021. MOF-on-MOF hybrids: synthesis and applications. Coordination Chemistry Reviews 432, 213743.

Liu, Z., Liu, H., Wang, Y., Yu, H., Wang, J., 2022a. Preparation of hypercrosslinked polymers with cashew nut shell liquid for removal of volatile organic compounds. Polymer Engineering Science 62, 1823–1832.

Liu, L., Wang, B., Deng, S., 2022b. Engineering application of fluidized bed for VOCs adsorption in tank area. Technology & Development of Chemical Industry 51, 62–64.

Lyu, H., Gao, B., He, F., Ding, C., Tang, J., Crittenden, J.C., 2017. Ball-milled carbon nanomaterials for energy and environmental applications. ACS Sustainable Chemistry & Engineering 5, 9568–9585.

Lyu, H., Gao, B., He, F., Zimmerman, A.R., Ding, C., Tang, J., Crittenden, J.C., 2018. Experimental and modeling investigations of ball-milled biochar for the removal of aqueous methylene blue. Chemical Engineering Journal 335, 110–119.

Ma, X., Yang, L., Wu, H., 2021. Removal of volatile organic compounds from the coal-fired flue gas by adsorption on activated carbon. Journal of Cleaner Production 302, 126925.

Makoś-Chełstowska, P., 2023. VOCs absorption from gas streams using deep eutectic solvents – a review. Journal of Hazardous Materials 448, 130957.

Masoumi, H., Ghaemi, A., Gilani, H.G., 2021. Evaluation of hyper-cross-linked polymers performances in the removal of hazardous heavy metal ions: a review. Separation and Purification Technology 260, 118221.

Mekki, A., Boukoussa, B., 2019. Structural, textural and toluene adsorption properties of microporous–mesoporous zeolite omega synthesized by different methods. Journal of Materials Science 54, 8096–8107.

Meng, F., Song, M., Wei, Y., Wang, Y., 2019. The contribution of oxygen-containing functional groups to the gas-phase adsorption of volatile organic compounds with different polarities onto lignin-derived activated carbon fibers. Environmental Science and Pollution Research 26, 7195–7204.

M.o.E.P. Department of Science and Technology Standards, China, 2013. Technical Specifications of Adsorption Method for Industrial Organic Emissions Treatment Project. China Environment Press, Beijing, pp. 1–16.

Mu, H., Feng, Y., 2019. Application of zeolite rotor and waste gas combustor in automobile paint shop. Electroplating & Finishing 38, 497–503.

Ogungbenro, A.E., Quang, D.V., Al-Ali, K.A., Vega, L.F., Abu-Zahra, M.R.M., 2020. Synthesis and characterization of activated carbon from biomass date seeds for carbon dioxide adsorption. Journal of Environmental Chemical Engineering 8, 104257.

Ponnuchamy, M., Kapoor, A., Jacob, M.M., Awasthi, A., Mukhopadhyay, M., Nandagobu, S., Raghav, A., Arvind, D., Chakraborty, P., Prabhakar, S., 2023. Adsorptive removal of endocrine disruptor bisphenol A from aqueous environment using sugarcane bagasse derived biochar. Journal of the Taiwan Institute of Chemical Engineers 105216.

Qi, G., Pan, Z., Zhang, X., Miao, X., Xiang, W., Gao, B., 2022. Effect of ball milling with hydrogen peroxide or ammonia hydroxide on sorption performance of volatile organic compounds by biochar from different pyrolysis temperatures. Chemical Engineering Journal 450, 138027.

Radjabian, M., Abetz, V., 2020. Advanced porous polymer membranes from self-assembling block copolymers. Progress in Polymer Science 102, 101219.

Rodrigues, A.E., LeVan, M.D., Tondeur, D., 2012. Adsorption: Science and Technology. Springer Science & Business Media.

Rong, Y., Pan, C., Song, K., Nam, J.C., Wu, F., You, Z., Hao, Z., Li, J., Zhang, Z., 2023. Bamboo-derived hydrophobic porous graphitized carbon for adsorption of volatile organic compounds. Chemical Engineering Journal 461, 141979.

Rouquerol, J., Rouquerol, F., Llewellyn, P., Maurin, G., Sing, K., 2013. Adsorption by Powders and Porous Solids: Principles, Methodology and Applications. Academic Press.

Sadegh, F., Sadegh, N., Wongniramaikul, W., Apiratikul, R., Choodum, A., 2024. Adsorption of volatile organic compounds on biochar: a review. Process Safety and Environmental Protection 182, 559–578.

Saini, V.K., Pires, J., 2017. Development of metal organic framework-199 immobilized zeolite foam for adsorption of common indoor VOCs. Journal of Environmental Sciences 55, 321–330.

Salahdin, O.D., Sayadi, H., Solanki, R., Parra, R.M.R., Al-Thamir, M., Jalil, A.T., Izzat, S.E., Hammid, A.T., Arenas, L.A.B., Kianfar, E., 2022. Graphene and carbon structures and nanomaterials for energy storage. Applied Physics A 128, 703.

Shen, Y., 2023. Biomass-derived porous carbons for sorption of Volatile organic compounds (VOCs). Fuel 336, 126801.

Shen, W., Li, Z., Liu, Y., 2008. Surface chemical functional groups modification of porous carbon. Recent Patents on Chemical Engineering 1, 27–40.

Shen, X., Du, X., Yang, D., Ran, J., Yang, Z., Chen, Y., 2021. Influence of physical structures and chemical modification on VOCs adsorption characteristics of molecular sieves. Journal of Environmental Chemical Engineering 9, 106729.

Siu, B., Chowdhury, A.R., Yan, Z., Humphrey, S.M., Hutter, T.J.C.C.R., 2023. Selective adsorption of volatile organic compounds in metal-organic frameworks (MOFs). oordination Chemistry Reviews 485, 215119.

Slepičková Kasálková, N., Slepička, P., Švorčík, V., 2021. Carbon nanostructures, nanolayers, and their composites. Nanomaterials 11, 2368.

Song, W., Justice, R.E., Jones, C.A., Grassian, V.H., Larsen, S.C., 2004. Synthesis, characterization, and adsorption properties of Nanocrystalline ZSM-5. Langmuir 20, 8301–8306.

Sui, H., Liu, J., He, L., Li, X., Jani, A., 2019. Adsorption and desorption of binary mixture of acetone and ethyl acetate on silica gel. Chemical Engineering Science 197, 185–194.

Sun, W., Li, J., Li, H., Jin, B., Li, Z., Zhang, T., Zhu, X., 2022. Mechanistic insights into ball milling enhanced montmorillonite modification with tetramethylammonium for adsorption of gaseous toluene. Chemosphere 296, 133962.

Szcześniak, B., Choma, J., Jaroniec, M., 2018. Gas adsorption properties of hybrid graphene-MOF materials. Journal of Colloid and Interface Science 514, 801–813.

Tibor, S.T., Grande, C.A., 2022. Industrial production of activated carbon using circular bioeconomy principles: case study from a Romanian company. Cleaner Engineering Technology 7, 100443.

Toth, J., 2002. Adsorption. CRC Press.

Tsai, J.-H., Chiang, H.-M., Huang, G.-Y., Chiang, H.-L., 2008. Adsorption characteristics of acetone, chloroform and acetonitrile on sludge-derived adsorbent, commercial granular activated carbon and activated carbon fibers. Journal of Hazardous Materials 154, 1183–1191.

Wang, T., 2023. Application of activated carbon adsorption catalytic combustion technology in waste gas from waterborne paint painting. Value Engineering 42, 145–147.

Wang, G., Dou, B., Wang, J., Wang, W., Hao, Z., 2013. Adsorption properties of benzene and water vapor on hypercross-linked polymers. RSC Advances 3, 20523–20531.

Wang, J., Wu, Z., Niu, Q., Liu, L., Yang, L., Fu, M., Ye, D., Chen, P., 2021. Highly efficient adsorptive removal of toluene using silicon-modified activated carbon with improved fire resistance. Journal of Hazardous Materials 415, 125753.

Wang, B., Lan, J., Bo, C., Gong, B., Ou, J., 2023a. Adsorption of heavy metal onto biomass-derived activated carbon. RSC Advances 13, 4275–4302.

Wang, H., Zu, D., Jiang, X., Xu, Y., Cui, Z., Du, P., Cheng, Z., Li, Z., Li, L., Yang, C., Bai, X., Zhang, B., Zhou, Y., Wang, K., Li, B., Huang, Z., Zhao, L., Li, B., Wu, H., 2023b. Bifunctional activated carbon ultrathin fibers: combining the removal of VOCs and PM in one material. Advanced Fiber Materials 5, 1934–1948.

Wu, S., Wang, Y., Sun, C., Zhao, T., Zhao, J., Wang, Z., Liu, W., Lu, J., Shi, M., Zhao, A., 2021. Novel preparation of binder-free Y/ZSM-5 zeolite composites for VOCs adsorption. Chemical Engineering Journal 417, 129172.

Xiang, W., Zhang, X., Chen, K., Fang, J., He, F., Hu, X., Tsang, D.C.W., Ok, Y.S., Gao, B., 2020. Enhanced adsorption performance and governing mechanisms of ball-milled biochar for the removal of volatile organic compounds (VOCs). Chemical Engineering Journal 385, 123842.

Xiang, W., Zhang, X., Cao, C., Quan, G., Wang, M., Zimmerman, A.R., Gao, B., 2022. Microwave-assisted pyrolysis derived biochar for volatile organic compounds treatment: characteristics and adsorption performance. Bioresource Technology 355, 127274.

Yan, M., Rong, Y., Wu, F., You, Z., Wang, D., Yang, X., Hao, Z., Li, J., Zhang, Z., 2023. Micro-mesoporous graphitized carbon fiber as hydrophobic adsorbent that removes volatile organic compounds from air. Chemical Engineering Journal 452, 139184.

Yang, C., Miao, G., Pi, Y., Xia, Q., Wu, J., Li, Z., Xiao, J., 2019. Abatement of various types of VOCs by adsorption/catalytic oxidation: a review. Chemical Engineering Journal 370, 1128–1153.

Ye, Q., Chen, Y., Li, Y., Jin, R., Geng, Q., Chen, S., 2023. Management of typical VOCs in air with adsorbents: status and challenges. Dalton Transactions 52, 12169–12184.

Yu, Q., Li, J., Liu, W., Wang, Y., Chu, W., Zhang, X., Xu, L., Zhu, X., Li, X., 2023. Rational design of a novel Silica-Based material with abundant open micropores for efficient VOC removal. Chemical Engineering Journal 454, 140077.

Zhang, S., Shao, T., Kose, H.S., Karanfil, T., 2010. Adsorption of aromatic compounds by carbonaceous adsorbents: a comparative study on granular activated carbon, activated carbon fiber, and carbon nanotubes. Environmental Science & Technology 44, 6377–6383.

Zhang, X., Gao, B., Creamer, A.E., Cao, C., Li, Y., 2017. Adsorption of VOCs onto engineered carbon materials: a review. Journal of Hazardous Materials 338, 102–123.

Zhang, G., Feizbakhshan, M., Zheng, S., Hashisho, Z., Sun, Z., Liu, Y., 2019a. Effects of properties of minerals adsorbents for the adsorption and desorption of volatile organic compounds (VOC). Applied Clay Science 173, 88–96.

Zhang, X., Gao, B., Fang, J., Zou, W., Dong, L., Cao, C., Zhang, J., Li, Y., Wang, H., 2019b. Chemically activated hydrochar as an effective adsorbent for volatile organic compounds (VOCs). Chemosphere 218, 680–686.

Zhang, Q., Wang, J., Lyu, H., Zhao, Q., Jiang, L., Liu, L., 2019c. Ball-milled biochar for galaxolide removal: sorption performance and governing mechanisms. Science of the Total Environment 659, 1537–1545.

Zhang, X., Xiang, W., Wang, B., Fang, J., Zou, W., He, F., Li, Y., Tsang, D.C.W., Ok, Y.S., Gao, B., 2020. Adsorption of acetone and cyclohexane onto CO_2 activated hydrochars. Chemosphere 245, 125664.

Zhang, X., Miao, X., Xiang, W., Zhang, J., Cao, C., Wang, H., Hu, X., Gao, B., 2021. Ball milling biochar with ammonia hydroxide or hydrogen peroxide enhances its adsorption of phenyl volatile organic compounds (VOCs). Journal of Hazardous Materials 403, 123540.

Zhang, X., Cao, L., Xiang, W., Xu, Y., Gao, B., 2022a. Preparation and evaluation of fine-tuned micropore biochar by lignin impregnation for CO_2 and VOCs adsorption. Separation and Purification Technology 295, 121295.

Zhang, X., Xiang, W., Miao, X., Li, F., Qi, G., Cao, C., Ma, X., Chen, S., Zimmerman, A.R., Gao, B., 2022b. Microwave biochars produced with activated carbon catalyst: characterization and sorption of volatile organic compounds (VOCs). Science of the Total Environment 827, 153996.

Zhang, W., Li, G., Yin, H., Zhao, K., Zhao, H., An, T., 2022c. Adsorption and desorption mechanism of aromatic VOCs onto porous carbon adsorbents for emission control and resource recovery: recent progress and challenges. Environmental Science: Nano 9, 81–104.

Zhang, W., Fang, D., Huang, G., Li, D., Zheng, Y., 2023. Research and application development of catalytic redox technology for zeolite-based catalysts. Catalysts 13, 1197.

Zhu, N.X., Wei, Z.W., Chen, C.X., Wang, D., Cao, C.C., Qiu, Q.F., Jiang, J.J., Wang, H.P., Su, C.Y., 2019. Self-generation of surface roughness by low-surface-energy alkyl chains for highly stable superhydrophobic/superoleophilic MOFs with multiple functionalities. Angewandte Chemie International Edition 58, 17033–17040.

Zhu, L., Shen, D., Luo, K.H., 2020. A critical review on VOCs adsorption by different porous materials: species, mechanisms and modification methods. Journal of Hazardous Materials 389, 122102.

Zhu, J., Huang, W., Fu, L., Zhu, B., Li, X., Wang, X., Wang, Y., Chen, W., 2021. Nanoporous asphalt-based activated carbon prepared from emulsified asphalt and graphene oxide as high-thermal-conducting adsorbers for n-hexane vapor recovery. ACS Applied Nano Materials 4, 12453–12460.

CHAPTER

6

Absorption and condensation of VOCs

Bing Wang[1,2], *Tao Jiang*[2] *and Nana Wang*[2]

[1]College of Resources and Environmental Engineering, Guizhou University, Guiyang, China; [2]Key Laboratory of Karst Georesources and Environment (Guizhou University), Ministry of Education, Guiyang, China

6.1 Overview of absorption and condensation technologies

Volatile organic compounds (VOCs) usually refer to organic compounds with a saturated vapor pressure higher than 70 Pa at room temperature and a boiling point below 260°C at atmospheric pressure (Dwivedi et al., 2004). VOCs possess the characteristics of diffusivity, toxicity, and volatility. They can undergo photooxidation reactions to form secondary organic aerosols under strong light or react with nitrogen oxides to form O_3, which may cause harm to environment and human health. Therefore, it is of great significance to explore effective and applicable VOC treatment technologies. Currently, VOC emission-reduction technologies are divided into two categories, destruction technology and recycling technology, according to their processes and principles. Destruction technology mainly includes biodegradation, incineration, photocatalytic oxidation, and catalytic oxidation. These technologies can ultimately convert VOCs into CO_2, H_2O, and nontoxic or less toxic compounds. However, destruction technology may result in excessive CO_2 emission, exacerbating the greenhouse effect. Therefore, VOC recycling is consistent with the current concept of a circular economy. Recycling technology, such as condensation, absorption, and membrane separation, tends to recover VOCs and offers a higher economic value than that of destruction technology.

Among all the recycling technologies, condensation and absorption technologies are greatly attractive due to their simple operation, high recovery rate, and wide application range. Condensation technology utilizes the different saturated vapor pressures of VOCs and converts them into liquid by step cooling or pressurization. In the 1970s, Edwards (1978) applied for a relatively mature patent for the condensate recovery of VOCs. The method employed a multilevel cooling cascade refrigeration system that took into account

Volatile Organic Compound Removal
https://doi.org/10.1016/B978-0-443-15674-8.00011-1

the defrost problem and provided a reference for the subsequent work. Unlike condensation technology, absorption technology used low-volatility or nonvolatile liquids as absorbents to separate the soluble and insoluble components using absorption devices, such as packed towers and spray towers. VOCs are separated from waste gas according to the difference in their solubility or chemical reaction characteristics in the absorbent, and adsorption includes physical and chemical absorption. Physical absorption depends on the solubility of VOCs in the absorbents. The chemical reaction in chemical absorption increases the mass transfer coefficient and absorption driving force. When waste gas is characterized as having a large volume and low concentration of harmful components, the performance of chemical absorption may be better than that of physical absorption.

Condensation and absorption technologies have their respective characteristics in the recovery of VOCs. In comparison to absorption technology, condensation technology is less affected by the external temperature and pressure. Its recovery performance is determined by the vapor pressure and boiling point of VOCs, and it is suitable for recycling VOCs with low flow rates, high concentration, and single components. Absorption technology can treat organic waste gas with a high concentration, high pressure, and low temperature. The selection and design of the adsorbent and device are the keys to recycling VOCs by absorption technology. The energy consumption and cost are relatively high by using condensation technology or absorption technology alone, thus they usually couple with other recycling technologies to recover VOCs. For example, condensation technology is combined with adsorption technology or adsorption technology is coupled with membrane separation. The following sections discuss the application of absorption technology and condensation technology in VOC recovery in detail.

6.2 Absorption of VOCs

6.2.1 Types of absorbent

Absorption is a method of purifying waste gas that uses liquid absorbents to absorb VOCs. The treatment of VOCs by absorption technology consists of two steps: first, VOCs in the waste gas are transferred to the liquid phase, and then, the VOCs are recovered or removed from the liquid phase. The type of absorbent is a primary factor to limit the treatment effect of VOCs. Therefore, the following characteristics should be considered when selecting absorbents (Darracq et al., 2010): (1) high solubility, large absorption capacity, and fast absorption rate; (2) low volatility or nonvolatile; (3) no corrosivity or low corrosivity; and (4) low toxicity, nonflammable, and low cost. In practical applications, the selection of absorbents should be considered based on the nature of the waste gas, the production process conditions, and the expected treatment effect. At present, absorbents can be divided into four categories, which are organic solvents, surfactants, ionic liquids, and deep eutectic solvents.

Organic solvent absorbents mainly select substances with higher boiling points, such as waste engine oil, lubricating oil, washing oil, and biodiesel (Ozturk and Yilmaz, 2006). There are also some pure organic solvents, such as phthalate esters and adipate esters (Heymes et al., 2006a). Their history can be traced back to the 1970s using these absorbents. Lalanne et al. (2008) used a mixture of water and lubricating oil as a detergent to purify VOC exhaust gases in a bio-washing device. The results showed that the absorption efficiency of aromatic and chloride VOCs was increased. Heymes et al. (2006b) compared the viscosity, equilibrium constant, and diffusion coefficient of polyethylene glycol, dimethyl phthalate, adipate ester, and silicone oil as absorbents for toluene. They found that adipate ester was the most suitable absorbent for toluene. It is worth noting that although organic solvents have a high absorption capacity for VOCs, most organic solvent components are complex and volatile, and there are certain safety risks and secondary pollution associated with them. In addition, some organic solvents have high costs and viscosity, which make them difficult to separate from VOCs and can easily lead to uneven liquid distribution in the absorption equipment along with an increased pressure drop. These have limited their industrial applications (Biard et al., 2016; Metzger, 2001).

A surfactant is an amphiphilic compound with both hydrophilic and hydrophobic groups. In the liquid phase, when the concentration of the surfactant reaches a certain level, it can spontaneously form nonpolar micelles, significantly increasing the solubility of insoluble or poorly soluble organic substances (Fourmentin et al., 2007). Blach et al. (2008) used cyclodextrin as an absorbent to purify toluene waste gas. The results showed that β-cyclodextrin had the strongest absorption capacity for toluene, about 250 times that of water. Li et al. (2018) removed xylene from the gas stream through sodium hypochlorite (NaClO) and sodium dodecylbenzenesulfonate (SDBS). Furthermore, the effects of pH, the concentration of NaClO and SDBS solutions, liquid and gas flow rates, the rotational speed of xylene removal efficiency, and the total mass transfer coefficient were investigated. The results showed that SDBS could reduce the surface tension of the NaClO solution and enhance the removal ability of organic compounds. In recent years, surfactants have gradually become the focus of scientific researchers. Mixing water and surfactants is an effective way to improve the absorption rate of VOCs by reducing the surface tension and changing the physicochemical properties of the solution. The hydrophilic–lipophilic properties of surfactants also enable them to increase the dispersion, emulsification, and solubility of VOCs in water. However, surfactants have low selectivity for specific VOCs and may form emulsions with VOCs, making it challenging to separate VOCs from the surfactant water mixture (Shen et al., 2022).

Ionic liquids (ILs) are ionic compounds of liquid at low temperatures (<100°C) that are composed of organic cations and organic or inorganic anions (Fahri et al., 2020). At present, the common cations include imidazole, quaternary ammonium, and pyridine. The common anions include $[Tf_2N]^-$, $[BF_4]^-$, $[PF_6]^-$, $[Ac]^-$, and $[CH_3SO_3]^-$ (Salar García et al., 2017). ILs, as "designable" green solvents, can selectively absorb specific VOCs by introducing specific functional groups and have received widespread attention in recent years. Cichowska Kopczyńska et al. (2018) studied the application of the supported ionic liquid membrane for the separation of toluene in the gas phase. The results showed that the permeation rate of toluene was significantly higher than that of pure nitrogen, and the selectivity range of an ionic liquid for toluene/N_2 was 17–341, which fully met the separation requirements. Moufawad et al. (2022) selected a suitable absorber for chlorobenzene and o-dichlorobenzene from the 289 ILs based on the COSMO-RS model, which was 1-ethyl-3-methylimidazolium tetrafluoroborate $[EMIM][BF_4]$. The results show that the chlorobenzene/o-dichlorobenzene content of the purge gas was reduced to 13.6 mg L^{-1}/9.9 mg L^{-1} (molar fraction). As a new type of absorbent, ILs have shown great potential in recycling VOCs. However, there are still some urgent issues that need to be addressed: (1) the high viscosity of ILs often limits the mass transfer between the gas and liquid phases; (2) ILs have a relatively high cost and poor economy (Rodriguez Castillo et al., 2019).

Many researchers have been continuously designing and optimizing absorbents to overcome the drawbacks of the aforementioned absorbents and extend their industrial application range. Deep eutectic solvents (DESs) have recently become green substitute for ionic liquid absorbents and have emerged as a new frontier for absorbents (Perna et al., 2020). A DES is an organic mixture composed of different ratios of organic halides or hydrogen-bond donors and metal salts or hydrogen-bond acceptors. Its internal components can freely connect through hydrogen bonds to form a deep eutectic mixture with a melting point lower than each component (Paiva et al., 2014). Moufawad et al. (2022) prepared eight types of hydrophilic and hydrophobic DESs to absorb nine types of VOCs and their mixtures. The results showed that the DESs exhibited high absorption capacity for various VOCs and did not saturate, even at high VOC concentrations. Among them, the DES based on a mixture of tetrabutylammonium bromide and decanoic acid enhanced the maximum absorption capacity over water up to 60,000 times. Villarim et al. (2022) compared the absorption performance of hydrophobic DESs and conventional solvents for three types of VOCs (toluene, citric acid, and siloxane D4) in biogas. The results showed that all DESs had significantly lower K-values compared to water and other absorbents, indicating that a significant amount of VOCs was absorbed in this type of solvent. Particularly for limonene, the K-values were reduced up to 20,000 times. Although DESs and ILs have similar physicochemical properties, such as low volatility, wide liquid-phase line range, and nonflammability (Chen et al., 2021), a DES has superior performance, mainly due to its low production cost, simple preparation process, minor by-products, good biocompatibility, and biodegradability (Moura et al., 2017). These advantages promise to make DESs a successful alternative to ILs and conventional organic solvent absorbents in many industrial applications.

6.2.2 Types of absorption equipment

The absorption equipment is another key factor that constrains the efficiency of VOC treatment using the absorption method. To achieve the absorption of VOCs' effective components, the most important requirement is to closely contact the VOCs and the absorbing liquid and minimize resistance while maximizing the driving force. Therefore, the selection of absorption equipment should focus on the following basic requirements (Cypes and Engstrom, 2004; Monnier et al., 2010): (1) a large contact area and a certain contact time between the gas and the liquid; (2) a strong disturbance between the gas and the liquid, along with low absorption resistance and high absorption efficiency; (3) stable operation and appropriate flexibility; (4) a simple structure, being easy to manufacture and maintain. A tower is the most commonly used absorption device. According to the absorption principle, absorption towers can be divided into three categories (their characteristics are listed in Table 6.1): packed tower, bubble tower, and spray tower.

Packed towers have been around for over 100 years. They are the most typical types of absorption towers, which are equipped with various forms of solid fillers inside (Ghoreyshi et al., 2014). In a packed tower, the liquid absorbent enters from the top of the tower and is distributed onto the packing layer using the spraying device at the top of the tower, and it then flows down along the surface of the packing using gravity. VOCs enter from the

TABLE 6.1 Characteristics of the commonly used absorption towers.

Absorption equipment	Application scope	Advantages	Disadvantages	References
Packed tower	Fast and instantaneous reaction processes	Simple structure, small pressure drop, can adapt to various corrosive media, and not easy to cause solution foaming	Short liquid residence time, uneven liquid distribution, and wall flow conditions	(Ghoreyshi et al., 2014; Lin and Chien, 2008)
Bubble tower	Medium and slow reaction processes and processes with a high amount of heat release	Good gas–liquid mixing, good reaction, high mass, and high heat transfer efficiency	Serious remixing of the liquid phase in the axial direction, large pressure drop, and easy bubble formation	(Lima et al., 2018; Lv et al., 2019)
Spray tower	Instantaneous, interfacial, and fast reaction processes, and it is also applicable to the system of solid product formation	Low construction cost, high flue gas flow rate, and high absorption efficiency	Low storage capacity, low mass transfer coefficient in the liquid phase, and easy to cause the remixing of the gas –liquid phase	(De Paepe et al., 2014; Kumar et al., 2011)

bottom of the tower and pass through the gap in the packing from the bottom to the top under the driving force of the pressure difference and countercurrently come in contact with the absorbent being sprayed down. The gas and liquid phases are in close contact with the wetted packing surface and fully transfer the mass to absorb and purify the pollutants in the VOC exhaust gas through various physicochemical interactions (Li et al., 2021b; Wu et al., 2010). Its structure is shown in Fig. 6.1.

As the carrier for gas–liquid contact and mass transfer in the packed tower, the performance of fillers directly determines the gas–liquid mass transfer efficiency of the packed tower. Common fillers can be divided into two categories: bulk fillers and structured fillers (Aroonwilas et al., 1999). Bulk packing consists of particles with a certain external structure, mainly including ring packing, saddle packing, ring saddle packing, and spherical packing. Structured packing, represented by metal wire mesh corrugated packing and metal plate corrugated packing, is a type of filler with regular geometric shapes and relatively neat stacking (Hsu and Lin, 2012). Guillerm et al. (2016) compared the absorption and purification efficiency of silicone oil (polydimethylsiloxane, PDMS) in a lacy ring bulk stack packed tower, an intalox metal tower packing (IMTP) bulk stack packed tower, and a Flexipac regular packed tower for toluene waste gas. The results showed that the mass transfer performance of Flexipac structured packing was better than that of IMTP, and its gas–liquid mass transfer efficiency was the highest. In addition, the absorption and purification efficiency of the two PDMS absorbents for toluene exhaust gas could reach 100%.

A bubble tower is a type of multiphase reaction equipment. This tower is filled with liquid. Then, gas is passed through the gas distributor installed at the bottom of the tower, where it is dispersed into bubbles that rise along the liquid and react with the liquid while agitating it

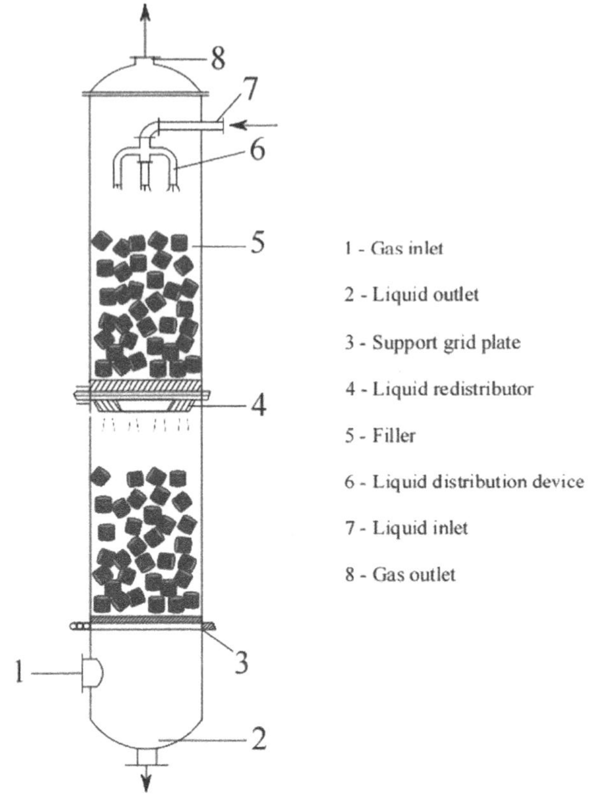

FIGURE 6.1 Structure of the packed tower for VOC absorption.

to increase the mass transfer rate (Chen et al., 2018). According to the structural characteristics, the bubble towers can be categorized into the hollow type, the multi-stage type, and the stripping type. Their structures are shown in Fig. 6.2.

A hollow bubble tower is most suitable for the slow chemical reaction system or reaction systems with significant thermal effects in the liquid phase. When the thermal effect is large, the heat exchange unit can be installed inside or outside the tower to transform it into a bubble tower with a heat exchange unit. To avoid liquid back-mixing in the tower, a multi-stage bubble tower is often used to ensure the reaction efficiency when the height-to-diameter ratio is large. To meet the requirements of a high gas–liquid flow rate or to reduce bubble condensation for a high-viscosity liquid, a stripping bubble tower has been applied. It has the advantages of a uniform radial gas–liquid flow velocity, low axial dispersion coefficient, high heat-transfer coefficient, and an adjustable liquid circulation velocity (Tarelho et al., 2011; Zhou et al., 2021). Li et al. (2021a) studied the absorption and desorption of toluene using a switchable hydrophilic solvent (N, N-dimethylcyclohexylamine ($CyNMe_2$)) and a bubbling tower. The results showed that $CyNMe_2$ exhibited excellent toluene absorption performance compared to the common absorbents. At 60°C, 90% absorption efficiency was achieved by bubbling N_2 through $CyNMe_2$, and at 25°C, 94% of toluene was desorbed from $CyNMe_2$

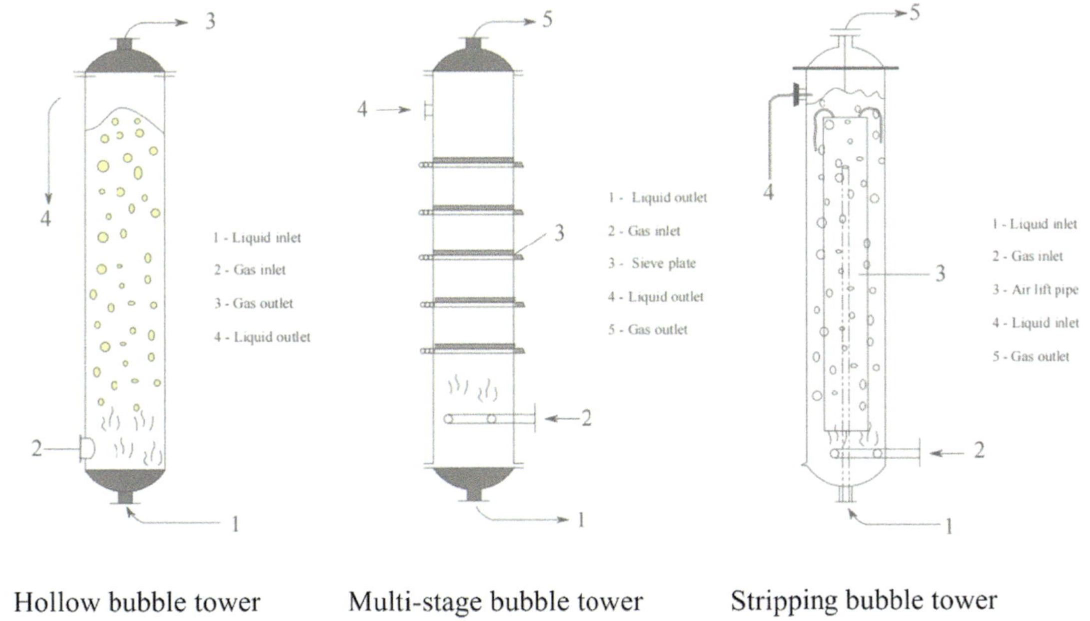

FIGURE 6.2 Structures of the bubble tower for VOC absorption.

by bubbling CO_2. Even after five cycles of absorption and desorption, its absorption capacity was still comparable to that of a new absorbent.

A spray tower is one of the earliest gas–liquid mass transfer equipment in the tower device. Its structure is simple. There are only a few nozzles inside the tower. The gas enters from the bottom and the liquid absorbent is sprayed from the top. A demister is installed at the top of the tower (Mohan et al., 2008). Currently, the application of spray towers in the treatment of hazardous gases has attracted great attention, and many new structural forms have emerged (Cui et al., 2018). The structure of the spray tower is shown in Fig. 6.3.

Liao et al. (2021) accurately identified 60 types of VOCs with their total concentrations ranging from 3.54–26.42 mg L^{-1} in leachate treatment by proton transfer reaction time-of-flight mass spectrometry. A spray tower and a biotrickling filter were used as VOC treatment devices. The results showed that the devices had good removal efficiency. The VOC, which had a concentration as high as 3539.27 μg L^{-1} in the first sampling, was significantly reduced to 635.02 μg L^{-1} after treatment, with a removal efficiency of 82.21%. In addition, spray towers are not only used for absorption in VOC treatment, but also as pretreatment and for the cooling of exhaust gases during adsorption and catalytic combustion (Liu et al., 2022).

6.2.3 Factors affecting VOC absorption

The selection of absorbents and the design of the absorption equipments are the key factors in treating VOCs by absorption. Various parameters act as important factors that affect absorption performance and can be optimized to improve the treatment efficiency. The common

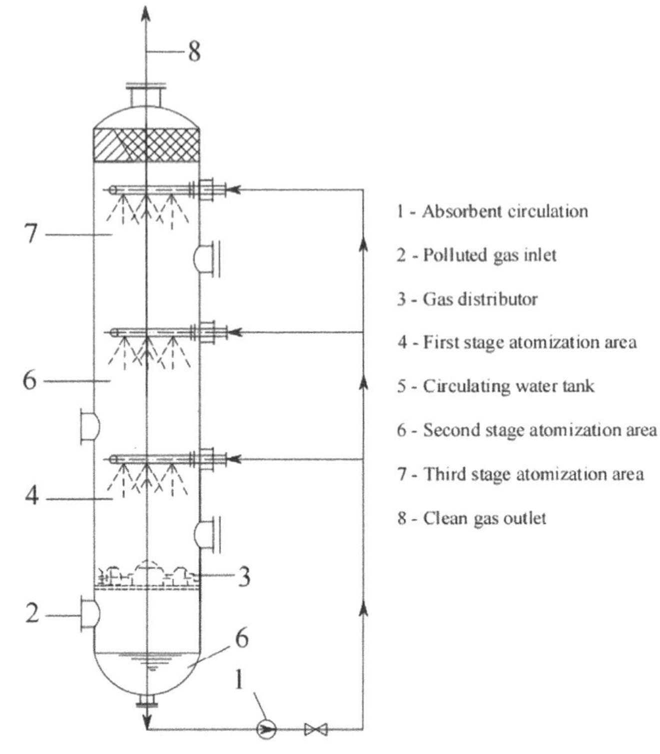

FIGURE 6.3 Structure of the packed tower for VOC absorption.

influencing factors include the air flow rate, liquid-to-gas ratio, absorbent concentration, and pH. First, selecting and controlling the appropriate empty tower gas velocity is beneficial to improve the absorption rate (Liu and Zhu, 2019). When the air velocity is too high, the packed towers are prone to liquid flooding (the absorption solution does not flow down smoothly and accumulates in the tower) (Wu et al., 2010). Bubble towers are prone to spraying (the bubbling layer does not form on the tower plate and the gas flows directly through the plate). In addition, high air velocity in the tower can cause severe mist entrainment, which increases the burden on the demister and poses a risk of exceeding the emission standard (Tyagi and Buwa, 2017). Of course, the higher the air velocity, the greater the treatment capacity, but the higher the tower height must be, considering the gas–liquid contact time. Second, controlling the liquid-to-gas ratio can also improve the absorption efficiency accordingly. The liquid-to-gas ratio is the volume of the absorbent required to treat 1 m^3 of gas. Increasing the liquid-to-gas ratio increases the gas–liquid mass transfer rate, thereby increasing the removal rate of pollutants (Sarhan et al., 2018). In engineering applications, the minimum allowed liquid-to-gas ratio is determined by the operating characteristics of the absorber tower and can be calculated based on the material balance and the operating line equation (Li et al., 2018). Finally, the absorption of VOCs can also be improved by adjusting the concentration of the absorbent solution (pH) and by paying attention to factors such as system fouling and blockage, temperature, pressure, and leakage. Energy consumption should also be considered, and variable frequency motors can also be considered where conditions permit (Lin et al., 2006).

6.3 Condensation of VOCs

6.3.1 Principle of condensation technology

Condensation technology is one of the most conventional methods to recover VOCs. When the components of VOCs in the waste gas are relatively simple, condensation technology becomes attractive. Currently, condensation technology can be used alone or in combination with other recovery techniques, such as adsorption and membrane separation. In case it is used alone, condensation technology has strict requirements for equipment. Generally, a low temperature is achieved using a precooler, mechanical refrigeration, or liquid nitrogen. The operating temperature of the precooler is above the freezing point of each component. After the waste gas enters the precooler, the temperature of the waste gas drops to 4°C, and most of the water vapor is removed by liquefying it into water. However, it is impossible to altogether remove the water vapor of the waste gas using the precooler, and residual water vapor could be frosted in the subsequent processes. Frosting significantly impacts the heat exchange performance of the heat exchanger, which may block the gas flow channel and increase flow resistance. Most of the VOCs in waste gas can be recovered into liquid by mechanical refrigeration. Nevertheless, mechanical refrigeration is limited by the temperature that is difficult to reach, which is below −120°C. If a lower temperature is required to meet a higher recovery rate, liquid nitrogen can be connected after mechanical refrigeration. The refrigeration temperature of liquid nitrogen can be as low as −180°C, which can meet the increasingly stringent emission standards.

Condensers can be categorized into the direct contact condenser and the surface condenser according to the way of contact between the waste gas and the coolant. Waste gas can directly contact the coolant in the direct contact condenser, and the heat quantity of VOCs is transmitted to the coolant through convection and heat conduction. The heat quantity is transferred through the parietal wall in the latter one, and a tubular heat exchanger is a typical representative of the surface condenser. Besides, the selection of the refrigerant is also a key factor affecting the recycling efficiency of VOCs, it needs to meet the characteristics of environmental protection, safety, and economy. Currently, there are mainly two types of refrigerants, hydrofluorocarbons (HFCs) and natural refrigerants. HFCs stay in the atmosphere for a long time and have a high greenhouse potential, their extensive use would cause global warming. At present, many countries are also gradually reducing the use of HFCs. According to the Montreal Protocol, most countries are striving to reduce HFC use to more than 80% before 2045. Based on this, alternatives to HFCs are also being vigorously developed, which should be characterized by the zero ozone depletion potential, low greenhouse effect potential, low costs, and good safety, such as ammonia, carbon dioxide, liquid nitrogen, and hydrocarbon (e.g., propane, butane, and isobutane).

6.3.2 Factors affecting VOC condensation

The condensation recovery efficiency of VOCs can be affected by several aspects, such as the inlet air pressure, temperature, and mixed gas composition. Currently, the condensation system is designed as a three-stage process, that is, the precondensation stage, the high-temperature condensation stage, and the low-temperature stage. Most of the VOCs are not recovered in the high-temperature stage, they need to be recovered in the low-temperature stage, resulting in high energy consumption. Therefore, the pressure of VOCs is appropriately increased and can improve the condensation temperature to improve the aforementioned situation. On the one hand, the removal rate of water vapor is greatly improved after increasing the pressure of VOCs, minimizing the frost formation of water vapor. On the other hand, the condensation temperature of VOCs would be increased with the enhancement of the pressure. It is worth noting that the pressure increase is accompanied by the temperature rise of the waste gas, and in the case of ensuring safety, an appropriate pressure increase can effectively reduce the requirements of the emission concentration on the cooling temperature. The standard of China for oil and gas recovery is that the recovery rate of non-methane hydrocarbons should be above 95% and the emission concentration should be below 25 g m^{-3}. If oil and gas are to meet the recovery requirements, the cooling temperature must be at least −80°C. Furthermore, in case the requirements of the emission concentration are met, it must be at least −100°C. Providing that the VOC pressure is increased from the ordinary pressure to 0.5 MPa, the temperature would decrease from −100 to −80°C when meeting the emission concentration. On this basis, the pressure continues to rise to 0.7 MPa, the requirements can be achieved at −75°C, and the recovery rate reaches 98.5%. The large increase in pressure does not bring a significant rise in the recovery rate because it may increase the refrigeration load and cause a sharp increase in the compressor power consumption and exhaust temperature.

The formation of various waste gases has different environments and processes, and the components are not the same. More carbon-containing components are recovered simply by condensation technology, specifically for components with a carbon atom number greater than or equal to four. It is more suitable for VOCs with heavier hydrocarbon components to use condensation technology for recovery. Therefore, the recovery of VOCs by the condensation technique should be optimized by adjusting the inlet pressure and temperature based on the waste gas composition to minimize operating costs.

6.3.3 Coupling of condensation technology

The recovery of VOCs consumes much energy using condensation technology alone. The condensation system must work at a low temperature, if the waste gas possesses high recovery requirements. Therefore, to improve this situation, condensation technology can be combined with other recovery methods to reduce energy consumption and increase the recovery rate. The combination of membrane separation and condensation is a typical mixed system used to recover VOCs. Membrane separation acts as a pretreatment device to improve the operating temperature of condensation (Song et al., 2021a). Picaud-Vannereux et al. (2020) studied the energy efficiency of a mixed membrane/condensation system for the recovery of propane. The dense polymer membrane and the porous inorganic membrane (zeolite, metal-organic framework with a carbon molecular sieve) were selected as the target permeation membranes. The N_2 and propane mixture was recovered through five processes, respectively. When the propane content in the feed was about 5%, the recovered purity of propane was 99.17% by the Pebax 2533 membrane hybrid process. It is worth noting that its energy demand was degraded by 74% compared with the condensation technique alone. In another work, Belaissaoui et al. (2016) explored a hybrid membrane/condensation process to recover VOCs from air waste gas. The pretreatment of the membrane separation system significantly improved the concentration efficiency of VOCs. The result showed that VOCs with high boiling points, such as toluene, octane, and acetone, were more energy conserved using the standalone condensation technique. The hybrid process obtained a higher energy efficiency for the intermediate to low boiling temperature VOCs, such as propane, ethane, and ethylene.

The condensation-coupled adsorption technology is also a common hybrid system that fully utilizes its characteristics. The recovery of high-concentration VOCs has the advantages of high stability and efficiency by using condensation technology. Adsorption technology can control the emission concentration of VOCs in a low range, especially low-concentration VOCs. For example, Gautham and Mahmoud (2000) studied the treatment of dichloromethane and toluene by coupling condensation and adsorption technology. A mathematical model was established to predict the mass concentration range of VOC recovery. Simulation results showed that the flow rate of the coolant and the size of the condenser limited the recovery rate. The inlet temperature of the cryogen was appropriately reduced, which was beneficial for the removal of VOCs, while resulting in water vapor forming frost in the condenser tube. An integrated treatment process of condensation and adsorption was used to remove VOCs from laundry dry-cleaning. The mixed system reduced the generation of ozone and secondary organic aerosols by 90.4% and 95.9%, respectively. The concentration of the total amount of VOCs was decreased from 4099 to 58 ppmv, and the removal rate was as high as 98.5%. In addition, the regular regeneration of adsorbers could keep it continuous working, this mixed system still efficiently removed VOCs after 30 cycles (Song et al., 2021b).

The traditional refrigeration system mostly adopts cascade compression refrigeration, and the minimum refrigeration temperature is generally not lower than −80°C due to the limitation of the refrigerant and compressor performance. However, the refrigeration system must be improved with the continuous improvement of the VOC emission standards and the limitation of a low temperature. Li et al. (2020) proposed a new deep condensation recovery system (VOC-DCR), which used turbo expansion refrigeration to replace traditional cascade compression refrigeration. The steady-state simulation showed that the recovery rate of VOCs was 99.97% and the emission concentration was 45.17 mg Nm^{-3} after treatment, below the emission limit. Xu et al. (2021) also proposed a novel cryogenic condensation system based on a heat-driven refrigerator without a power input for oil gas VOC recovery. A thermodynamic calculation model was established to quantitatively calculate the recovery efficiency. The results indicated that the remaining 91.0% of VOCs could be recovered by burning 9.0% of VOCs when the refrigeration temperature was 200 K, the heat source temperature was 873 K, and the concentration was 33.5% at 100 kPa. It has great potential for multiple application scenarios, such as ocean tankers and crude oil storage tank areas. The recent technical advantages of VOC recovery by condensation technology can be attributed to the rapid progress of refrigeration technology. The refrigeration compressor improves the energy efficiency ratio, volumetric efficiency, compressor displacement, indication efficiency, and reliability. Developing and applying new refrigerants and heat transfer technologies have improved the refrigeration capacity per unit mass and effectively reduced the unit volume and irreversible heat transfer loss of the refrigeration system. At present, the application of computer technology and intelligent controllers makes the condensing system more automatic. The refrigeration process parameters allow it to be accurately controlled and optimized, improving the stability and safety of condensation technology.

6.4 Case studies

Conventional packed beds have gradually been eliminated due to their poor gas–liquid mass transfer, large volume, and high investment and operating costs. Therefore, this case introduces an alternative method to enhance mass transfer between a gas and a liquid, the rotating packed bed. High-speed rotating packing can disperse liquid into thin films or tiny droplets, which reduce the thickness of the liquid film and increase the gas–liquid contact area, thus improving the mass transfer process due to its strong shearing effect. The schematic diagram of a water absorber + rotating packed bed is shown in Fig. 6.4. It contains two

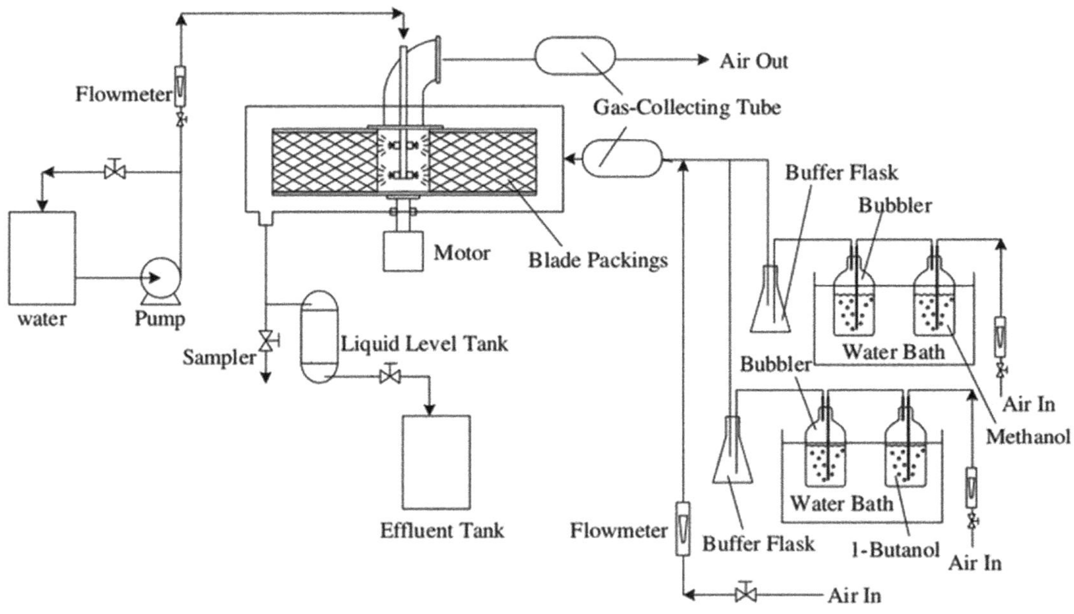

FIGURE 6.4 Schematic diagram of a water absorbent and rotating packed bed device. *Reproduced with permission from Hsu, L.J., Lin, C.C., 2012. Binary VOCs absorption in a rotating packed bed with blade packings. Journal of Environmental Management 98, 175–182.*

subsystems producing a binary gas stream containing methanol and 1-butanol. The clean gas stream is discharged from the top of the rotating packed bed, which is filled with blade-shaped packing. Water enters the inner edge of the packed bed through a liquid distributor and then flows out from the outer edge due to centrifugal acceleration. Finally, the water containing methanol and 1-butanol is discharged from the bottom of the rotating packed bed with vane packing. The blade-shaped packing is made of wire mesh of stainless steel, with an inner diameter of 1.95 cm, an outer diameter of 6.25 cm, and an axial height of 2.95 cm. It has a specific surface area of 93 $m^2\ m^{-3}$ and a porosity of 0.99. The rotation speed can reach 600–1800 rpm, the gas flow rate ranges from 5–55 L min^{-1}, and the liquid flow rate ranges from 0.15–0.65 L min^{-1}. The effect of the methanol and 1-butanol concentration at the inlet on the total gas-phase mass transfer coefficient (K_{Ga}) of methanol and 1-butanol can be ignored. The K_{Ga} values of methanol and 1-butanol increase with the increasing rotation speed, gas flow rate, and liquid flow rate. The dependence of K_{Ga} on the gas flow rate is higher than that on the liquid flow rate, indicating that mass transfer in binary VOC absorption may be largely controlled by gas-phase mass transfer. Due to the centrifugal force caused by the rotating annular packing, the contact area between the gas and the liquid is increased, reducing the total resistance of mass transfer. Compared with traditional packed beds, rotating packed beds have a smaller capacity, lower investment, and lower operating costs.

The energy consumption of VOC recovery by condensation technology is a key factor in determining its operating cost, and reducing energy consumption as much as possible is a bottleneck that needs to be broken. Therefore, an example introduces a new low-temperature condensing system combined with gas turbines to contribute to low carbon emissions and energy conservation. The schematic of the VOC cryogenic condensation recovery system is shown in Fig. 6.5. Part of the exhaust gas is combusted to drive the gas turbine for the generation of electric energy, which provides power for a *Stirling* refrigerator or directly drives the absorption refrigerator. The *Stirling* refrigerator adopts pollution-free natural helium as a refrigerant and has no greenhouse effect, and the cooling temperature is comparable to or even lower than liquid nitrogen. In process 1, the refrigerants are different in the precooling stage and the low-temperature condensation stage, the former is an absorption refrigerator and the latter is a *Stirling* refrigerator. First, the exhaust gas is cooled using cryogenic purified gas in a recuperator. Then, it enters the absorption refrigerator for precooling and is condensed using the *Stirling* refrigerator. The heat source of the absorption refrigerator is provided by the combustion of the exhaust gas. For the *Stirling* refrigerator, part of the exhaust gas is burned to drive the gas turbine for the generation of electric energy, and the output power drives the compressor for refrigeration. The exhaust gas temperature of gas turbines is usually high; the residual heat carried by the exhaust gas can be used as another heat source for an absorption refrigerator, reducing the combustion of the exhaust gas. The cooled exhaust gas is, in turn, separated using the three-stage gas–liquid separator, and the condensate is collected in the storage tank. Compared with process 1, the *Stirling* refrigerator replaces the absorption refrigerator in the precooling stage of process 2. All the imported exhaust gas enters the gas turbine for combustion to generate electric energy, thereby causing the *Stirling* refrigerator to obtain the corresponding cooling temperature and capacity in the precooling stage and the cryogenic stage. This system can achieve near-zero VOC emission and a high VOC recovery efficiency.

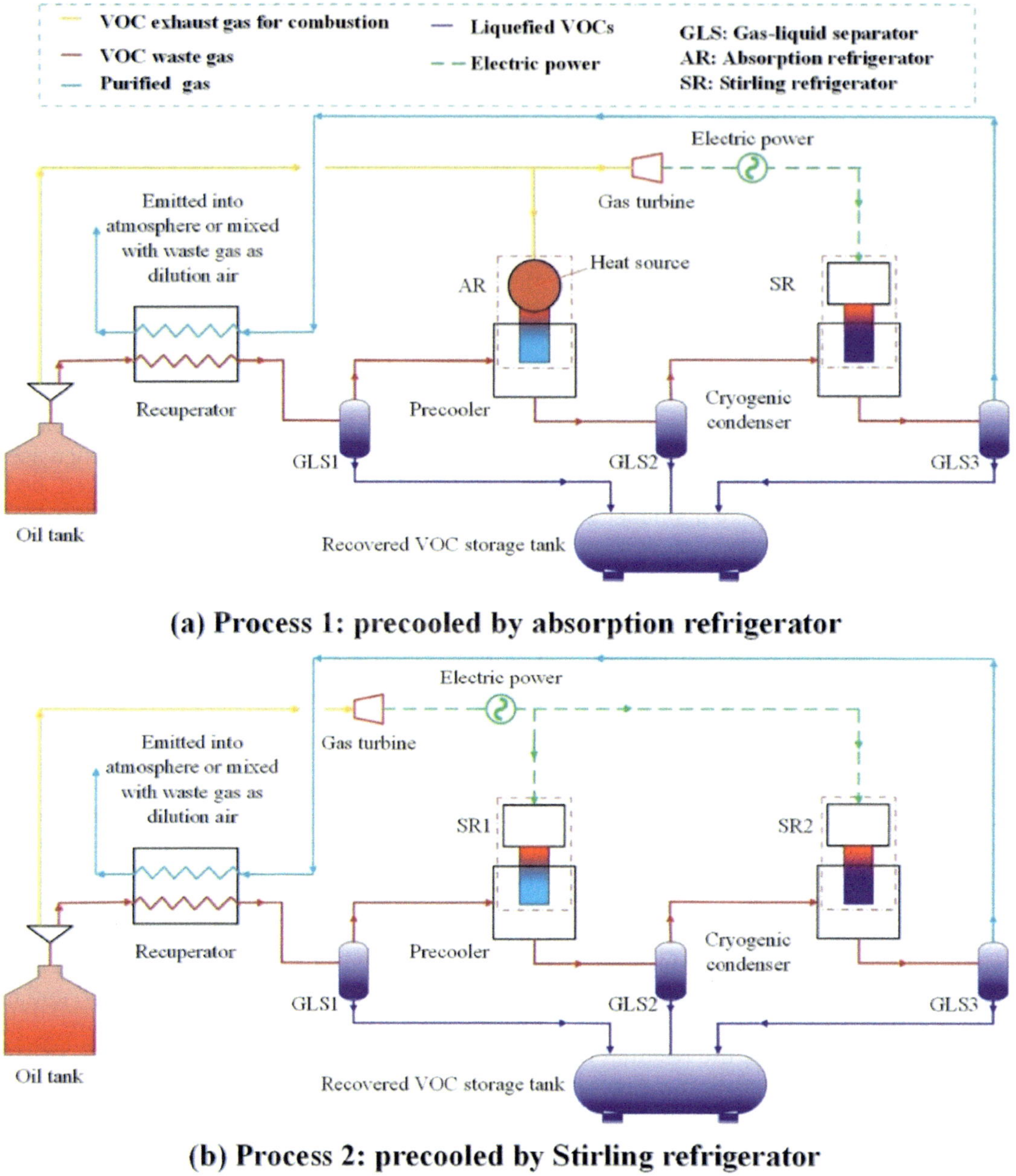

FIGURE 6.5 Schematic of the VOC cryogenic condensation recovery system. *Reproduced with permission from Xu, H., Xu, X., Chen, L., Guo, J., Wang, J., 2022. A novel cryogenic condensation system combined with gas turbine with low carbon emission for volatile compounds recovery. Energy 248, 123604.*

6.5 Summary and perspectives

Absorption and condensation are economical and efficient due to their simple processes and low operating costs, which are suitable for treating medium to high concentrations of VOCs. Many new organic compounds, such as solvents, surfactants, ionic liquids, and deep eutectic solvents, have been widely used in VOC absorption technology. Absorption equipments, such as packing towers, bubble towers, and spray towers, have also aroused great interest. Meanwhile, the three-stage condensation process is mostly used for VOC recovery, the requirements of condensation are strict regarding equipment materials and the heat insulation system. Different factors can affect the efficiency of VOC recovery in condensation and absorption systems, such as the flow rate, absorbent concentration, pressure, and components of VOCs. Coupling condensation with other recovery techniques, such as membrane separation and adsorption, has become a developing trend. Both absorption and condensation technologies for VOC recovery have seen rapid progresses in theoretical and experimental research.

In the practical applications of VOC recovery, there are still some urgent problems to be solved in absorption and condensation technologies: (1) many studies have reported that VOC absorbents possess good absorption effects when treating small gas volumes with relatively simple compositions in the bench scale experiment. However, the adsorption efficiency is relatively poor in medium-scale or industrial-scale experiments. (2) Current absorbents have good performance, but the price is relatively expensive, limiting their industrial applications. It should focus on developing absorbents with the characteristics of strong adaptability and low cost. In addition, more efforts should be made to optimize absorption equipment to further improve the gas–liquid mass transfer efficiency, reduce operating energy consumption, and improve the level of automation. (3) For VOCs with different concentrations and compositions, coupling condensation and other recovery technologies can achieve high recovery efficiency. The selection and development of specific condensates and the optimization of integrated processes and structures are the key points to be focused on in the future. (4) The secondary volatilization of the recovered VOCs is also a complex problem because liquid-phase VOCs would be changed to the gas phase along with the temperature rise or pressure drop. (5) Frosting and blocking in heat exchangers are intractable problems, defrosting devices should be developed to reduce the resistance of heat exchangers. (6) Thermoelectric refrigeration is an emerging refrigeration technology with good application prospects in recent years, it relies on the direct propagation of energy by electrons and holes in semiconductor devices to achieve refrigeration. (7) The development of renewable energy should be vigorously advocated, and energy-saving technologies for solar photovoltaic power generation should be widely studied and applied.

References

Aroonwilas, A., Veawab, A., Tontiwachwuthikul, P., 1999. Behavior of the mass-transfer coefficient of structured packings in CO_2 absorbers with chemical reactions. Industrial and Engineering Chemistry Research 38 (5), 2044–2050.

Belaissaoui, B., Le Moullec, Y., Favre, E., 2016. Energy efficiency of a hybrid membrane/condensation process for VOC (Volatile Organic Compounds) recovery from air: a generic approach. Energy 95, 291–302.

Biard, P.F., Coudon, A., Couvert, A., Giraudet, S., 2016. A simple and timesaving method for the mass-transfer assessment of solvents used in physical absorption. Chemical Engineering Journal 290, 302–311.

Blach, P., Fourmentin, S., Landy, D., Cazier, F., Surpateanu, G., 2008. Cyclodextrins: a new efficient absorbent to treat waste gas streams. Chemosphere 70 (3), 374–380.

Chen, C.C., Huang, Y.H., Hung, S.M., Chen, C.Y., Lin, C.W., Yang, H.H., 2021. Hydrophobic deep eutectic solvents as attractive media for low-concentration hydrophobic VOC capture. Chemical Engineering Journal 424, 130420.

Chen, H.Y., Liu, J.M., Pei, Y.P., Zhao, P., Zhang, Y., Yan, L.C., Zhang, T.W., Li, W.H., Wu, C.D., Hussain, M., 2018. Study on the synergistic effect of UV/Fenton oxidation and mass transfer enhancement with addition of activated carbon in the bubble column reactor. Chemical Engineering Journal 336, 82–91.

Cichowska Kopczyńska, I., Joskowska, M., Debski, B., Aranowski, R., Hupka, J., 2018. Separation of toluene from gas phase using supported imidazolium ionic liquid membrane. Journal of Membrane Science 566, 367–373.

Cui, H., Li, N., Peng, J., Yin, R., Li, J., Wu, Z., 2018. Investigation on the thermal performance of a novel spray tower with upward spraying and downward gas flow. Applied Energy 231, 12–21.

Cypes, S.H., Engstrom, J.R., 2004. Analysis of a toluene stripping process: a comparison between a microfabricated stripping column and a conventional packed tower. Chemical Engineering Journal 101 (1), 49–56.

Darracq, G., Couvert, A., Couriol, C., Amrane, A., Le Cloirec, P., 2010. Integrated process for hydrophobic VOC treatment—solvent choice. Canadian Journal of Chemical Engineering 88 (4), 655–660.

De Paepe, W., Contino, F., Delattin, F., Bram, S., De Ruyck, J., 2014. New concept of spray saturation tower for micro humid air turbine applications. Applied Energy 130, 723–737.

Dwivedi, P., Gaur, V., Sharma, A., Verma, N., 2004. Comparative study of removal of volatile organic compounds by cryogenic condensation and adsorption by activated carbon fiber. Separation and Purification Technology 39 (1–2), 23–37.

Edwards, R.C., 1978. Vapor Condensing and Recovery Assembly Including a Decanter for Vapor Recovery System. Google Patents.

Fahri, F., Bacha, K., Chiki, F.F., Mbakidi, J.P., Panda, S., Bouquillon, S., Fourmentin, S., 2020. Air pollution: new bio-based ionic liquids absorb both hydrophobic and hydrophilic volatile organic compounds with high efficiency. Environmental Chemistry Letters 18 (4), 1403–1411.

Fourmentin, S., Outirite, M., Blach, P., Landy, D., Ponchel, A., Monflier, E., Surpateanu, G., 2007. Solubilization of chlorinated solvents by cyclodextrin derivatives: a study by static headspace gas chromatography and molecular modelling. Journal of Hazardous Materials 141 (1), 92–97.

Gautham, P., Mahmoud, M.E.H., 2000. Optimum mass integration strategies for condensation and allocation of multicomponent VOCs. Chemical Engineering Science 55 (5), 881–895.

Ghoreyshi, A.A., Sadeghifar, H., Entezarion, F., 2014. Efficiency assessment of air stripping packed towers for removal of VOCs (volatile organic compounds) from industrial and drinking waters. Energy 73, 838–843.

Guillerm, M., Couvert, A., Amrane, A., Norrant, E., Lesage, N., Dumont, É., 2016. Absorption of toluene in silicone oil: effect of the solvent viscosity on hydrodynamics and mass transfer. Chemical Engineering Research and Design 109, 32–40.

Heymes, F., Demoustier, P.M., Charbit, F., Fanlo, J.L., Moulin, P., 2006a. Recovery of toluene from high temperature boiling absorbents by pervaporation. Journal of Membrane Science 284 (1), 145–154.

Heymes, F., Manno Demoustier, P., Charbit, F., Fanlo, J.L., Moulin, P., 2006b. A new efficient absorption liquid to treat exhaust air loaded with toluene. Chemical Engineering Journal 115 (3), 225–231.

Hsu, L.J., Lin, C.C., 2012. Binary VOCs absorption in a rotating packed bed with blade packings. Journal of Environmental Management 98, 175–182.

Kumar, R., Dhar, P.L., Jain, S., 2011. Development of new wire mesh packings for improving the performance of zero carryover spray tower. Energy 36 (2), 1362–1374.

Lalanne, F., Malhautier, L., Roux, J.C., Fanlo, J.L., 2008. Absorption of a mixture of volatile organic compounds (VOCs) in aqueous solutions of soluble cutting oil. Bioresource Technology 99 (6), 1699–1707.

Li, W., Liu, H., Li, S., Li, W., Chen, Y., Gao, J., Cao, Y., 2018. Removal of hydrophobic volatile organic compounds with sodium hypochlorite and surfactant in a co-current rotating packed bed. Journal of Environmental Sciences 64, 190–196.

Li, X., Ma, J., Ling, X., 2020. Design and dynamic behaviour investigation of a novel VOC recovery system based on a deep condensation process. Cryogenics 107, 103060.

Li, Y., Chang, H., Yan, H., Tian, S., Jessop, P.G., 2021a. Reversible absorption of volatile organic compounds by switchable-hydrophilicity solvents: a case study of toluene with N,N-Dimethylcyclohexylamine. ACS Omega 6 (1), 253–264.

Li, Y.B., Liu, W., Zhang, X., Sun, B.C., Luo, Y., Chu, G.W., 2021b. Enhancement of devolatilization performance in a rotating packed bed with different packing structures. Separation and Purification Technology 278, 119527.

Liao, W., Liang, Z., Yu, Y., Li, G., Li, Y., An, T., 2021. Pollution profiles, removal performance and health risk reduction of malodorous volatile organic compounds emitted from municipal leachate treating process. Journal of Cleaner Production 315, 128141.

Lima, V.N., Rodrigues, C.S.D., Borges, R.A.C., Madeira, L.M., 2018. Gaseous and liquid effluents treatment in bubble column reactors by advanced oxidation processes: a review. Critical Reviews in Environmental Science and Technology 48 (16–18), 949–996.

Lin, C.C., Chien, K.S., 2008. Mass-transfer performance of rotating packed beds equipped with blade packings in VOCs absorption into water. Separation and Purification Technology 63 (1), 138–144.

Lin, C.C., Wei, T.Y., Hsu, S.K., Liu, W.T., 2006. Performance of a pilot-scale cross-flow rotating packed bed in removing VOCs from waste gas streams. Separation and Purification Technology 52 (2), 274–279.

Liu, B., Ji, J., Zhang, B., Huang, W., Gan, Y., Leung, D.Y.C., Huang, H., 2022. Catalytic ozonation of VOCs at low temperature: a comprehensive review. Journal of Hazardous Materials 422, 126847.

Liu, Y., Zhu, J., 2019. Hydrodynamics of a bubble-driven liquid-solid fluidized bed. Chemical Engineering Science 195, 730–736.

Lv, X., Zhang, Y., Li, L., Zhang, W., Liang, P., 2019. Experimental study of ethanol adsorption using a multistage bubbling fluidized bed. Process Safety and Environmental Protection 125, 64–70.

Mohan, B.R., Biswas, S., Meikap, B.C., 2008. Performance characteristics of the particulates scrubbing in a counter-current spray-column. Separation and Purification Technology 61 (1), 96–102.

Monnier, H., Falk, L., Lapicque, F., Hadjoudj, R., Roizard, C., 2010. Intensification of G/L absorption in microstructured falling film. Application to the treatment of chlorinated VOC's – part I: comparison between structured and microstructured packings in absorption devices. Chemical Engineering Science 65 (24), 6425–6434.

Moufawad, T., Costa Gomes, M., Fourmentin, S., 2022. Deep eutectic solvents as absorbents for VOC and VOC mixtures in static and dynamic processes. Chemical Engineering Journal 448, 137619.

Moura, L., Moufawad, T., Ferreira, M., Bricout, H., Tilloy, S., Monflier, E., Costa Gomes, M.F., Landy, D., Fourmentin, S., 2017. Deep eutectic solvents as green absorbents of volatile organic pollutants. Environmental Chemistry Letters 15 (4), 747–753.

O. Metzger, J., 2001. Organic reactions without organic solvents and oils and fats as renewable raw materials for the chemical industry. Chemosphere 43 (1), 83–87.

Ozturk, B., Yilmaz, D., 2006. Absorptive removal of volatile organic compounds from flue gas streams. Process Safety and Environmental Protection 84 (5), 391–398.

Paiva, A., Craveiro, R., Aroso, I., Martins, M., Reis, R.L., Duarte, A.R.C., 2014. Natural deep eutectic solvents – solvents for the 21st century. ACS Sustainable Chemistry and Engineering 2 (5), 1063–1071.

Perna, F.M., Vitale, P., Capriati, V., 2020. Deep eutectic solvents and their applications as green solvents. Current Opinion in Green and Sustainable Chemistry 21, 27–33.

Picaud-Vannereux, S., Lutin, F., Favre, E., Roizard, D., 2020. Energy efficiency of membrane vs hybrid membrane/cryogenic processes for propane recovery from nitrogen purging vents: a simulation study. Separation and Purification Technology 240, 116613.

Rodriguez Castillo, A.S., Biard, P.F., Guihéneuf, S., Paquin, L., Amrane, A., Couvert, A., 2019. Assessment of VOC absorption in hydrophobic ionic liquids: measurement of partition and diffusion coefficients and simulation of a packed column. Chemical Engineering Journal 360, 1416–1426.

Salar García, M.J., Ortiz Martínez, V.M., Hernández Fernández, F.J., de los Ríos, A.P., Quesada Medina, J., 2017. Ionic liquid technology to recover volatile organic compounds (VOCs). Journal of Hazardous Materials 321, 484–499.

Sarhan, A.R., Naser, J., Brooks, G., 2018. CFD modeling of bubble column: influence of physico-chemical properties of the gas/liquid phases properties on bubble formation. Separation and Purification Technology 201, 130–138.

Shen, Y., Ma, C., Li, Z., Ji, W., Zhu, Y., Zhang, S., 2022. Enhanced removal of VOCs from wood by coupling extraction with surfactant and, oxidation using ultrasound-activated persulfate. Industrial Crops and Products 189, 115818.

Song, C., Zhang, Z., Li, R., Lian, S., Guo, H., Jia, C., Liu, Q., 2021a. Optimization of membrane-cryogenic hybrid propane recovery process: from molecular to process simulation. Journal of Cleaner Production 321, 129049.

Song, M., Kim, K., Cho, C., Kim, D., 2021b. Reduction of volatile organic compounds (VOCs) emissions from laundry dry-cleaning by an integrated treatment process of condensation and adsorption. Processes 9 (9), 1658.

Tarelho, L.A.C., Neves, D.S.F., Matos, M.A.A., 2011. Forest biomass waste combustion in a pilot-scale bubbling fluidised bed combustor. Biomass and Bioenergy 35 (4), 1511−1523.

Tyagi, P., Buwa, V.V., 2017. Experimental characterization of dense gas-liquid flow in a bubble column using voidage probes. Chemical Engineering Journal 308, 912−928.

Villarim, P., Genty, E., Zemmouri, J., Fourmentin, S., 2022. Deep eutectic solvents and conventional solvents as VOC absorbents for biogas upgrading: a comparative study. Chemical Engineering Journal 446, 136875.

Wu, H., Feng, T.-C., Chung, T.-W., 2010. Studies of VOCs removed from packed-bed absorber by experimental design methodology and analysis of variance. Chemical Engineering Journal 157 (1), 1−17.

Xu, H., Xu, X.F., Chen, L.B., Guo, J., Wang, J.J., 2021. A novel cryogenic condensation system based on heat-driven refrigerator without power input for volatile organic compounds recovery. Energy Conversion and Management 238, 114157.

Zhou, R., Zhou, R., Alam, D., Zhang, T., Li, W., Xia, Y., Mai-Prochnow, A., An, H., Lovell, E.C., Masood, H., Amal, R., Ostrikov, K., Cullen, P.J., 2021. Plasmacatalytic bubbles using CeO_2 for organic pollutant degradation. Chemical Engineering Journal 403, 126413.

CHAPTER

7

Traditional functional materials for VOC removal

Xirong Huang, Xuqiong Zha, Chuangye Yang, Jiawen Ding and Zhuhong Ding

School of Environmental Science & Engineering, Nanjing Tech University, Nanjing, China

7.1 Introduction

Various volatile organic compound (VOC) purification methods have been introduced in previous chapters, including catalytic degradation, thermal oxidation, membrane separation, adsorption, absorption and condensation. Among those methods, catalytic oxidation technology is regarded as an efficient method due to its high economic feasibility, low cost and low levels of the generation of secondary pollutants, and it can be operated at a lower temperature with the controlled selectivity of the by-products. Of the catalytic oxidation methods, the VOC reaction rate is highly depended on the catalytic materials during the degradation process. Therefore, the fabrication of highly active and cost-effective catalytic materials for the elimination of VOCs is critically important in realistic applications. Similarly, adsorption method is one of the popular technologies to control VOCs. It is especially suitable for treating low concentrations of VOCs. Adsorption materials are the key factor in VOC adsorption technology, which directly affects investment cost, operating cost, and safety.

In actual use, the properties of VOCs and their emission characteristics should be fully considered to choose appropriate functional materials for treatment process. Taking all above factors into account, this chapter mainly focuses on the traditional functional materials for VOC removal through adsorption and catalytic oxidation. In addition, other functional materials are referred briefly. Each section is devoted to discussing material preparation, mechanism of action, catalysis or adsorption process, and other aspects.

Volatile Organic Compound Removal
https://doi.org/10.1016/B978-0-443-15674-8.00006-8

7.2 Adsorbents

There are many kinds of traditional functional materials that can be used to adsorb VOCs (Bhargavi et al., 2011). They can be divided into three simple categories: carbon-based adsorbents, inorganic materials, and organic polymers.

7.2.1 Carbon-based adsorbents

7.2.1.1 Activated carbon

Adsorption is a simple and safe method that separates mixtures and is highly efficient in removing low concentrations of toxic and hazardous substances. Activated carbon is currently the adsorbent of choice for the treatment of VOC pollution. It has the advantages of well-developed pore structure, large specific surface area, rich surface chemical groups, good adsorption performance, cheap and sufficient raw materials, and sample preparation process. At present, activated carbon for VOC adsorption needs to undergo two major treatment measures, namely activation and modification.

7.2.1.1.1 Activation treatment of activated carbon

There are three main types of activated carbon activation treatment methods: chemical activation, physical activation and physical-chemical activation.

1) Chemical activation method

Chemical activation method is to prepare activated carbon by uniformly mixing various raw materials containing carbon with chemicals and then, at a certain temperature, going through the processes of carbonization, activation, the recovery of chemicals, rinsing and drying. The most commonly used activators are phosphoric acid, zinc chloride, and potassium hydroxide. Chemical activators can inhibit the formation of tar by-products and prevent tar from blocking the fine pores generated during pyrolysis, which can improve the yield of activated carbon.

2) Physical activation method

Physical activation method is a process in which the charred raw material is brought into contact with activating gases such as water vapor (Li et al., 2015), flue gas (a mixture of water vapor, CO_2, N_2, etc.) (Ch et al., 2015), CO_2 (Bahrim et al., 2016), flue gas or air at a high temperature of 800 to 1000°C, thus carrying out an activation reaction.

3) Physical-chemical activation method

Physical-chemical activation method is an association of physical and chemical activation, i.e., the char is first treated chemically and then further activated by physical methods.

7.2.1.1.2 Surface modification of activated carbon

Activated carbon is modified to optimize their pore structure and surface chemistry for improved adsorption of VOCs (Fig. 7.1). The principal methods to adjust the surface chemistry of activated carbon are oxidation treatment, reduction treatment, metal loading, microwave treatment and organic modification.

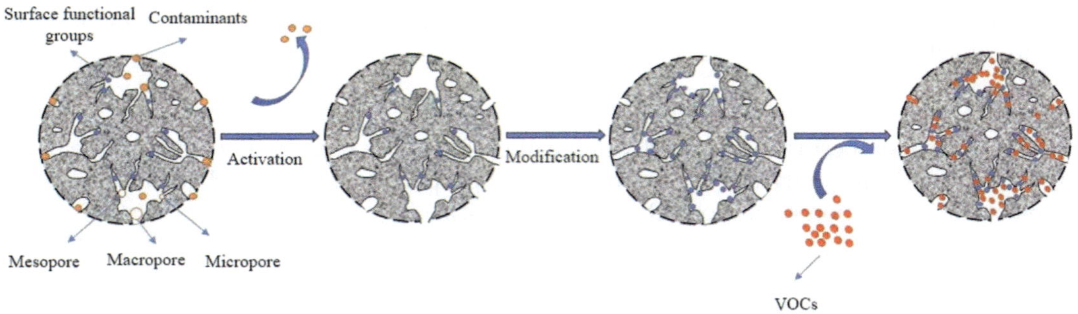

FIGURE 7.1 Scheme of activated carbon modification for adsorption of VOCs.

1) Surface oxidation modifications

It is now generally accepted that nitrogen-containing functional groups and oxygen-containing functional groups have strong impact on the adsorption performance of activated carbon. Different surface modification methods including surface oxidation modification have been developed to introduction functional groups on activated carbon surface. Surface oxidation uses oxidants to treat activated carbon to increase the number of acidic oxygen-containing groups, enhance the surface polarity, and increase the affinity for polar substances. The commonly used oxidants for oxidation treatment are HNO_3, HCl, H_2O_2, HCl, H_2PO_4, H_2SO_4, etc.

2) Surface reduction modifications

Surface reduction treatment is the reduction of oxygen-containing functional groups on the surface of activated carbon using a reducing agent at a suitable temperature. It can increase the number of oxygen-containing basic groups on the surface of activated carbon and weaken the surface polarity, thus providing better adsorption performance for nonpolar VOCs. Li et al. (2011) treated coconut shell activated carbon with NH_3 and NaOH and compared it with acid treatment methods such as nitric acid, sulfuric acid and phosphoric acid. The alkali treatment reduced the oxygenated acidic groups on the surface of the carbon, which facilitated the adsorption of hydrophobic VOCs on the modified activated carbon, resulting in a significant increase in the adsorption of o-xylene, while the caustic treatment had the opposite effect.

3) Metal-impregnated modifications

Surface loaded metal is often used in the modification of activated carbon. The principle is that the metal ions are first adsorbed on the surface of the activated carbon by physical adsorption, and then decreased to monomers or low valence ions by means of reduction. This enhances the adsorption and separation performance of the activated carbon by increasing the binding force of the low valent or precious metal to the adsorbent.

4) Organic modification

The principle of organic modification is to modify the activated carbon with organic solvents to improve the adsorption capacity of activated carbon for VOCs. Monsieur et al. (Monser and Adhoum, 2002) used sodium diethyldithiocarbamate and tetrabutylammonium

(TBA) to impregnate activated carbon and found organic modification improved adsorption ability of activated carbon to various contaminants.

7.2.1.1.3 Activated carbon adsorption of VOCs

The adsorption performance of activated carbon for VOCs is not only related to the nature of the activated carbon itself, but also the physical properties of the adsorbate and the conditions of the adsorption operation (Dobre et al., 2014).

1) Pressure swing adsorption

Pressure swing adsorption (PSA) is a cyclic process of adsorption and desorption of adsorbent at different pressures by periodically changing the system pressure at constant temperature or without a heat source.

2) Influence of adsorbate properties

The ability of adsorbate molecules to enter the pores of activated carbon is related to their individual kinetic diameters. According to the size exclusion theory, adsorbate molecules can only enter the pores of activated carbon if the pore diameter is greater than the diameter of the adsorbate molecules.

3) Influence of operating conditions

The temperature, inlet concentration, gas flow rate, pressure, moisture and gas composition during the adsorption operation all affect the adsorption performance of the activated carbon. It is important to select the appropriate operating conditions for targeted VOCs to maximize activate carbon performance.

As the most widely used adsorbent for VOC adsorption in industry, activated carbon may have some shortcomings in practical application, such as low adsorption capacity, poor regeneration ability, and the adsorption performance is greatly affected by environmental factors such as water and gas. Based on a comprehensive consideration of the influencing factors of VOC treatment by adsorption on activated carbon, the recovery and comprehensive utilization equipment of VOCs should be improved and developed, and the best process operating conditions should be optimized, so that activated carbon can be more widely used in the treatment of VOCs.

7.2.1.2 Activated carbon fibers (ACF)

Activated carbon fiber is a new type of charcoal multifunctional adsorbent developed in the 1960s and expanded in the 1970s. Activated carbon fibers are a promising carbon adsorbent for the removal of volatile organic compounds due to their developed pore structure, larger specific surface area, better regeneration and operability, faster adsorption and desorption kinetics, more uniform micro porous structure, 1−10 times the adsorption capacity and higher mass transfer rate than ordinary activated carbon (Fig. 7.2) (Miura et al., 2000; Balanay Jo et al., 2015).

7.2.1.2.1 Activated carbon fiber preparation process

The usual preparation of activated carbon fibers consists of three stages: pretreatment, carbonization, and activation. (1) Pretreatment: The purpose of pretreatment is to protect the fibers from being destroyed during the subsequent carbonization process, thus ensuring the successful preparation of activated carbon fibers. (2) Carbonization: The purpose of the

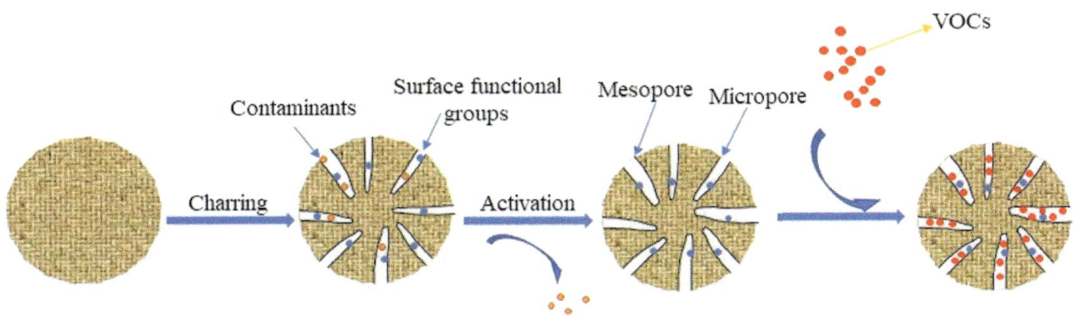

FIGURE 7.2 VOC adsorption by activated carbon fibers.

carbonization process is to allow the activated carbon fibers to form a fibrous structure and at the same time form pores. During the charring process, the organic components of the raw material are destroyed, and the carbon content rises. (3) Activation Treatment: The purpose of activation treatment of activated carbon fibers is to amplify the pore size and specific surface area of the pores formed during the charring stage. Activation methods mainly include physical activation and chemical activation. Physical activation is the treatment of activated carbon fibers with gas at high temperatures, while chemical activation is the treatment of activated carbon fibers with chemical reagents such as potassium hydroxide, sodium hydroxide and aluminum chloride.

7.2.1.2.2 The nature of activated carbon fibers

Activated carbon fiber adsorbs organic waste gas because the organic matters in the gas can be adsorbed and retained by the activated carbon fibers, thus achieving a purification effect. Compared to other adsorbents, activated carbon fibers have a faster adsorption and desorption rate and a longer service life.

7.2.1.2.3 Application of activated carbon fiber adsorption technology

The adsorption of VOCs using activated carbon fibers has been widely reported in the literature, and many scholars have studied the adsorption performance of dissimilar materials of activated carbon fibers (Zhang et al., 2017; Tang et al., 2007; Chiang et al., 2009). Joel Guticérrez-Martinez et al. (Gutiérrez-Martínez et al., 2020) prepared a new large pore size carbon material by chemical vapor deposition and investigated its adsorption effect on the volatile organic compound benzene. The experimental results showed that the surface area of carbon fibers increased from 120 m^2/g to 185 m^2/g and the benzene removal rate increased by 1.5 times with a maximum adsorption capacity of 55.53 mg/g in a static adsorption system at 25°C for 26 h. Lin et al. (2013) investigated the adsorption and removal of toluene from activated carbon fibers with a concentration of 4936.07 mg/m^3 and the adsorption capacity was up to 569 mg/g. Wang et al. (2020a) prepared activated carbon fibers by reductive carbonization of cellulose fibers with a nickel nanocatalyst in the presence of hydrogen.

This section highlights the principles of VOC adsorption activated carbon fibers, the preparation of activated carbon fibers and the current research on activated carbon fibers. Activated carbon fibers are operative in adsorption of a wide range of pollutants and have a very promising application in the field of adsorption. However, the research on the adsorption mechanism of activated carbon fibers is still not deep enough, and most of the current research still focuses on the adsorption of single VOC pollutant under ideal conditions. It is thus important to future research to further improve activated carbon fibers' ability to adsorb different types of VOCs, especially their mixtures, for real life applications.

7.2.1.3 Graphitized carbon black

Various carbon sorbents can be used to enrich and separate VOCs and graphitized carbon black (GCB) is one of them. After calcination at high temperature and high pressure, the heteroatoms on the surface of carbon black are removed and the dense arrangement and rigid structure are formed on the surface. Compared with activated carbon, GCB has obvious advantages in the field of chromatography. The adsorption and elution rules of compounds on GCB are easy to grasp, so GCB is used as a conventional adsorbent to adsorb VOCs.

7.2.1.3.1 Preparation of GCB

High temperature solid carbonization method is one of the most common methods used at present. It refers to the method of carbonizing solid precursors directly at high temperature under inert atmosphere conditions. GCB is made from soot in an inert atmosphere at 2700°C. Soot is usually produced from oil or natural gas. GCB is carbon black made from hot carbon black (soot pretreated at 3000°C) and is classified as hot GCB (Dettmer and Engewald, 2002). By adding catalyst, the reaction temperature can be greatly reduced. In addition, the process equipment is simple, the yield is large, and the product purity is high.

The degree of graphitization of any carbon after high temperature treatment depends largely on its initial structure. Thus, graphitization of some activated carbons at temperatures higher than 1200°C can greatly reduce or even completely eliminate their porosity (Matisová and Škrabáková, 1995). In addition, hydrogen gas assisted graphitization process at 1000°C can remove polar groups on the carbon surface. Washing with acid, such as perchloric acid or phosphoric acid, can remove the basic carbonic complexes and sulfides from the surface (Dettmer and Engewald, 2002).

7.2.1.3.2 Characteristics of GCB

GCB is composed of loosely assembled colloid-sized particles and bound by a weak van der Waals force. In its crystalline form, the carbon atoms are located at the apex of a regular hexagon. Three adjacent carbon atoms are covalently bonded together to form a two-dimensional planar layer called a basal plane. Adjacent basal surfaces are staggered, and a carbon atom in the upper layer is in the hexagonal grid of the lower layer. Every other layer, the position of carbon atoms is the same, which becomes a three-dimensional arrangement (Xu et al., 1996). GCB is a hydrophobic material. The affinity of water vapor depends on the chemical structure of the surface and can be enhanced by oxidation or hydration.

7.2.1.3.3 Adsorption properties of GCB

With uniform surface and no micropores, GCB can adsorb various compounds on its outer surface. It is prepared at high temperature, so its surface is almost completely free of unsaturated bonds, unshared electron pairs, free radicals and ions. The adsorption of compounds on GCB is controlled by the availability of two types of adsorption sites. The vast majority of GCB surface sites are nonpolar, corresponding to graphite-like arrays of carbon atoms. These sites do not show a tendency to preferentially interact with molecules carrying functional groups, and dispersive interactions dominate the adsorption behavior. The number of polar adsorption sites is small, but they can establish specific strong interactions with polar compounds. The gas adsorption isotherm of GCB is a reversible type II isotherm, which indicates that the adsorption is mainly single-multilayer adsorption on nonporous particles (Loi et al., 2021). GCB is the mode of monolayer adsorption (Fig. 7.3). When the pressure increases to a certain value and the temperature is constant, the monolayer adsorption saturates, and then the multilayer absorption begins. Corcia et al. (Dicorcia et al., 1993) found an oxygen complex with a chromium-like structure in the residue of combustion during GCB production, which in the presence of water rearranged this surface group into a benzopyrene salt. This allowed GCB to act as an anion exchanger in solid phase extraction (SPE). GCB can be used for volatile organic compounds sampling in extremely humid atmospheres because of its extremely high hydrophobicity.

GCB has many active sites and complex interaction forces. The application of GCB in more fields is also on the agenda.

7.2.2 Zeolite

Zeolite is a porous skeletal silica-aluminate molecular sieve that can be used as an ion exchanger, adsorbent separator, catalyst, etc (Zhu et al., 2020). Zeolite is an excellent material for the removal of hydrocarbons, radioactive cations, ammonia and heavy metal cations in industrial environments as an efficient absorber and adsorbent. Essentially, zeolite acts as a molecular sieve that absorbs, binds and immobilizes solid particles, liquids and gases in its honeycomb structure. Compared with other popular adsorbents, zeolite has a smaller specific surface area and thus might have a lower adsorption capacity for VOCs, however, it is easier to be regenerated by heat treatment (Fig. 7.4).

7.2.2.1 Preparation of zeolite

There are three main methods of zeolite synthesis: hydrothermal synthesis, template synthesis, and template-free synthesis. There are increasing concerns over hydrothermal and

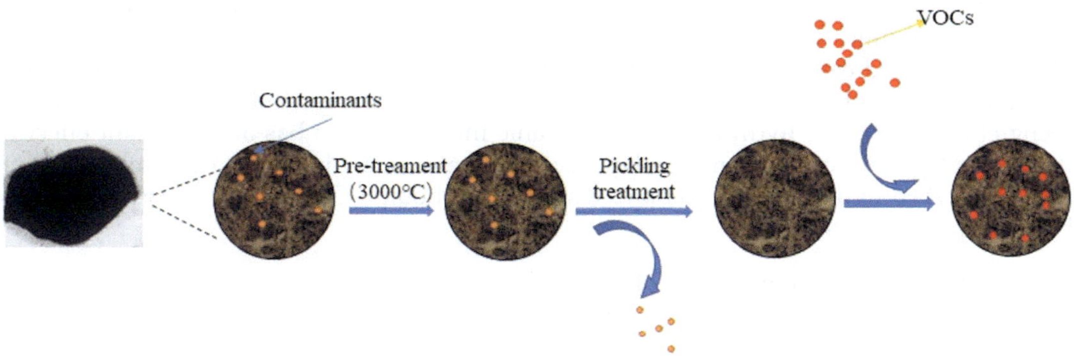

FIGURE 7.3 Adsorption of VOCs by carbon black.

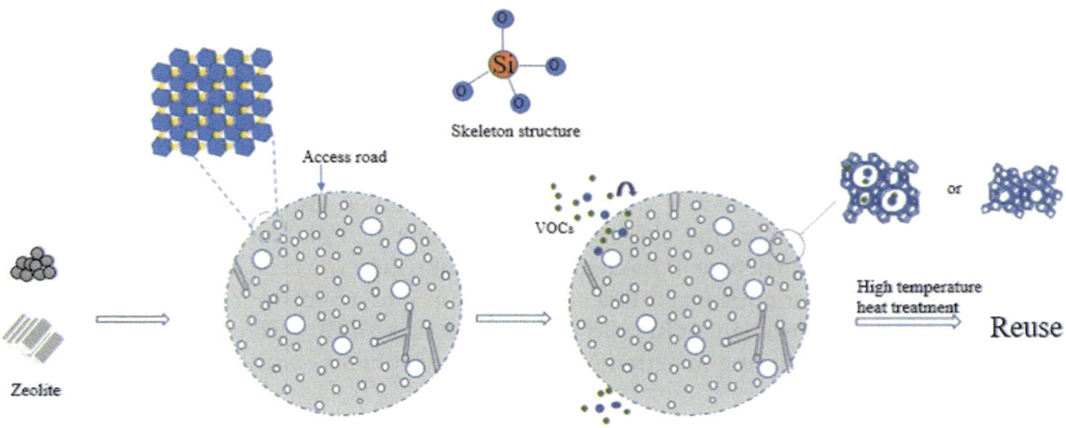

FIGURE 7.4 Application of zeolite in VOC adsorption.

organic template syntheses because of their potential toxicity and high cost. Initially, synthetic zeolite molecular sieves and natural zeolites were synthesized under nonorganic template conditions in the presence of inorganic cations. Considering environmental protection and efficient use of energy, researchers have explored inorganic/organic/no template conditions to synthesize zeolite molecular sieve materials. Currently, there are three types: modulation of the initial gel ratio; addition of molecular sieve guide; crystal species guide method.

7.2.2.2 *Adsorptive properties of zeolite*

The pore size in zeolite skeleton is uniform due to the crystalline nature of the pore, and the pores can selectively adsorb molecules of different sizes. As a result, zeolite has been widely used in VOC adsorption field. Zeolite molecular sieves contain more acid sites and have certain catalytic activity, so they are very suitable as catalyst support materials. Zeolite-based catalysts have been developed and applied in the catalytic oxidation of VOCs. The adsorption characteristics of different zeolites are different due to their different internal pore structures, so the adsorption of VOCs with different molecular sizes can be effectively carried out by adjusting the pore size of zeolites (Cundy and Cox, 2003). The surface hydrophilicity of zeolites can be controlled by changing the ratio of silicon to aluminum and surface modification to improve their adsorptive properties to VOCs. Modification of zeolites by balanced cation exchange and doping can modulate their electric field, pore structure and surface physical chemistry properties, and thus enhance their adsorption ability (Feng et al., 2018). The hydrophilicity of organic molecules also has an important effect on the adsorptive properties of molecular sieves. In general, hydrophilic VOC molecules are easily adsorbed by hydrophilic zeolite, while hydrophobic VOCs are more easily adsorbed by hydrophobic ones.

7.2.2.3 *Adsorption performance of zeolite*

According to the literature, the adsorption ability of zeolite can be improved by changing the ratio of Si to Al and loading metals and metal oxides. First, the ratio of silicon to aluminum is the most important parameter of zeolite, which has a great influence on the adsorption performance of zeolite. Zeolite can be divided into three classes according to the different ratio of silicon to aluminum (Kim and Ahn, 2011). With the increase of Si/Al ratio, the hydrophobicity of zeolite increases, which is beneficial to the adsorption of organic molecules including VOCs in water vapor (Li et al., 2021). The regeneration of adsorbents should be considered in the treatment of VOC pollutants. The most commonly used industrial regeneration methods are high temperature water vapor desorption, pressure drop, and vacuum desorption. Microwave heating is another effective method for VOC desorption, especially for polar ones. The dielectric property of zeolite has a great influence on microwave desorption. Zeolite often contains abundant compensating cations and has good dielectric property, which is beneficial to microwave desorption.

The cations in the zeolite skeleton have a strong interaction with VOC molecules. The metal ions can enhance the adsorption of VOCs on the zeolite, but they may reduce the specific surface area and block some pores to increase pore diffusion resistance. Therefore, the adsorption of VOCs on zeolite can only be improved by choosing suitable metal ions. For example, the adsorption of ethylene on zeolite was enhanced by the presence of Ag^+. Because of the charge-compensated strong interaction between Ag^+ and ethylene double bonds in the zeolite backbone, the adsorption was increased by two orders of magnitude (Veerapandian et al., 2019). More toluene can be adsorbed by Ag-loaded HY due to the π complex bond between AG/HY and toluene (Zhang et al., 2016a). Metal oxides can be used as active VOC adsorption sites, effectively improving the selective adsorption capacity of specific VOCs. Metal oxides are appropriately encapsulated in zeolites and can improve the adsorption capacity for VOCs at high relative humidity (Yin et al., 2022). Adsorption capacities for benzene and toluene molecules can be significantly increased after the oxygen substitution of Ni and Co atoms.

7.2.3 Silica gel

Silica gel is commonly used by industry as an adsorbent become of its low cost. Most silica gels are translucent or milky white with spherical morphology and porous structure. Silica gel also has a large specific surface area and stable physical and chemical properties. It has strong binding affinity to polar molecules, which depends on its pore size distribution. Microporous silica gels have excellent adsorption ability to VOCs at low concentrations. On the contrary, mesoporous silica are more suitable for high concentration VOCs (Wang et al., 2015).

The basic structural unit of silica gel is the silicon oxygen tetrahedron, in which the silicon atom is located in the center of the tetrahedron (Yang et al., 2019). Its chemical composition is $mSiO_2 \cdot nH_2O$. The four oxygen atoms are distributed in the four corners of the tetrahedron, and each oxygen atom is shared by two adjacent silicon atoms. Tetrahedra are connected to each other in different ways and accumulate to form SiO_2 particles, which condense to form the skeleton structure of silica gel.

7.2.3.1 Adsorptive properties of silica gel

The arrangement and connection of silica dioxide in silica gel are irregular. Due to the presence of unsaturated valence bonds, atoms in these positions have strong binding affinity in trapping foreign molecules. The gas adsorption on the surface of silica gel is mainly through physical adsorption, but there may be some chemical adsorption. Compared with the monolayer chemical adsorption, physical adsorption (physisorption) can be either monolayer or multilayer. The N_2 adsorption isotherms of silica gel are mainly type IV, reflecting its mesoporous nature. Some polar molecules and aromatic gas molecules can form hydrogen bonds with hydroxyl groups on the surface of silica gel (Fig. 7.5). After the adsorption of the first layer on the surface is completed, the second layer will be formed with the increase of relative pressure. For some nonpolar molecules, however, they may not be able to form hydrogen bonds and are mainly adsorbed on silica gel through physisorption (Fig. 7.5). Based on the static and dynamic adsorption-desorption tests, Sui concluded that silica gel had higher adsorption capacity to toluene and much more complete desorption compared with activated carbons (Sui et al., 2017). There was almost no accumulation of toluene on the silica gel after several times of reuse. In general, silica gel has a fast VOC desorption rate, leading to low energy consumption during regeneration (Sui et al., 2019).

7.2.3.2 *Preparation of silica gel*

The silica gels are synthesized by three main steps, namely gel preparation, aging, and drying (Sui et al., 2019). Dilute sodium silicate with a silica content of about 16% and dilute sulfuric acid with a concentration of 20% are mixed under a certain pressure, and the gel particles are formed by air granulation reaction at a certain pH value. Then the gel particles enter

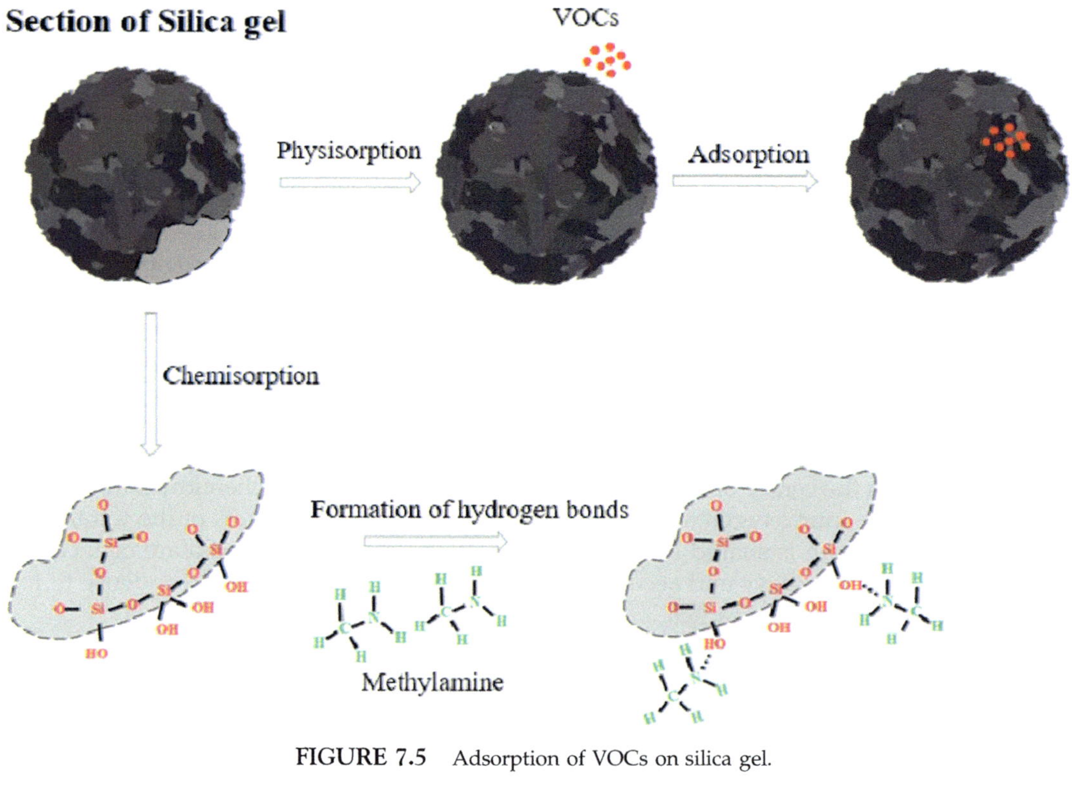

FIGURE 7.5 Adsorption of VOCs on silica gel.

the gel particle acceptor, the circulating water in the acceptor is neutral, aging for about 10 min. Through the conveyor belt into the washing tank, the dilute sulfuric acid with an acid content of about 2.5% is used to exchange for washing for more than 5 h, until the pH value of the glue block is greater than 5. After washing, coarse pore silica gel is soaked at 20–30°C temperature for 16 h by using 0.13%–0.18% dilute ammonia and fine pore silica gel directly entered into the tunnel drying.

7.2.3.3 Modification of silica gel

Silica gel has certain adsorption ability, which can be enhanced by chemical modifications (Price et al., 2000). (1) Silicon hydroxyl group modified by chlorination: The silicon hydroxyl group on the surface of silica gel can be transformed into a reactive silicon chloride group through chlorination reaction. Silicon hydroxyl group can react with Grignard reagent and organic amine derivatives, so the corresponding alkyl or derivative amine groups can be introduced. (2) Silicon hydroxyl group modified by the reaction of chlorosilane with silicon hydroxyl group: Various alkyl chains can be introduced to the surface of silica gel by chlorosilane, especially long alkyl or aromatic groups. In the anhydrous condition of the bonding reaction, organic amine catalyst is sometimes added to absorb HCl in order to promote the progress of the reaction. (3) Silanol group coupling method: Hydrophobic modification is achieved by this method (Haq et al., 2017). The conjugate reagent is generally expressed as X_3SiR, where X is the hydrolyzable alkoxy group and R represents the organic functionalized group. The commonly used coupling reagents are MTMS, DMDMS, and TMCS. Hydrophobic modified silica gel has better selectivity for VOCs due to the change of surface properties.

In view of the complexity of VOCs and the inevitable coexistence of water vapor, the development of efficient and low-cost adsorption materials has become a key technology. With excellent thermal, mechanical, and chemical stability, silica gel becomes an excellent alternative adsorbent for VOC removal. Therefore, in-depth research on adsorption of VOCs on silica gel is imperative.

7.2.4 Activated alumina

Activated alumina is made by heating and dehydrating aluminum hydrate, the nature of activated alumina depends on the initial structural state of hydroxide. It is generally not pure Al_2O_3 but partially hydrated porous structured material. In activated alumina, there is not only amorphous gel, but also crystals of hydroxide. Because of its high activity on the surface of capillary channels, it is called activated alumina. Activated alumina refers to intermediate alumina and can be divided into three categories including γ series of low temperature alumina (Fig. 7.6), δ series of high temperature alumina, and ρ-Al_2O_3 (Bénard et al., 2014). The most referred activated alumina in the literature is γ-alumina. Qin (Qin et al., 2016) used Ag–Mn/γ-Al_2O_3 to adsorb toluene and found that the loading sequence of Ag and Mn affected the adsorption performance. A large amount of Ag^+ existed on the surface of γ-Al_2O_3 after loading Mn and Ag, and Ag^+ as a new adsorption active site was beneficial to the adsorption of toluene. Ankur et al. (Ankur and Vimal, 2009) found that the carbon-oxygen functional groups present on the surface of alumina were effective for adsorbing DBT onto alumina.

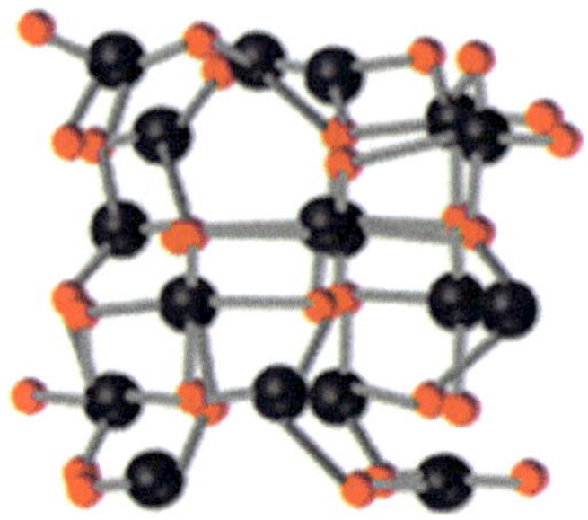

FIGURE 7.6 Structural unit of γ-Al_2O_3.

7.2.4.1 *Adsorption mechanisms*

Physical adsorption of VOCs on activated alumina is mainly through van der Waals force, which is formed by the instantaneous asymmetric dipoles of electrons in atoms or molecules. The appearance and adsorption ability of activated alumina adsorbent do not change during physical adsorption (Fig. 7.7). Under the condition of high temperature gas purification, the adsorbed gas can be completely desorbed without changing the properties of sorbent and sorbate, and the water vapor absorbed inside activated alumina can be discharged to restore the adsorption capacity.

Chemical adsorption of VOCs on activated alumina relies on chemical interactions between the adsorbents and the adsorbate such as electron exchange, transfer, or combination between the two, leading to the formation and destruction of chemical bonds and composition. Chemical adsorption usually occurs on active sites, such as marginal unsaturated carbohydrates (Burgos et al., 2002).

7.2.4.2 *Preparation of activated alumina*

Several synthesis methods have been developed for the preparation of activated alumina. (1) Dehydration method: Aluminum hydroxide is heated and dehydrated to obtain γ-Al_2O_3. Methods such as acid precipitation, alkali precipitation, alcohol-based alumina hydrolysis are commonly used to synthesize aluminum hydroxide feedstock (Abyzov, 2019). (2) Sol-gel method: This method is to boil metal aluminum in hydrochloric acid or aluminum chloride solution to get clear and colorless aluminum sol. Then the aluminum sol is mixed with cycloheximide tetraamine solution. The mixture is dropped into a hot oil column, and then the activated

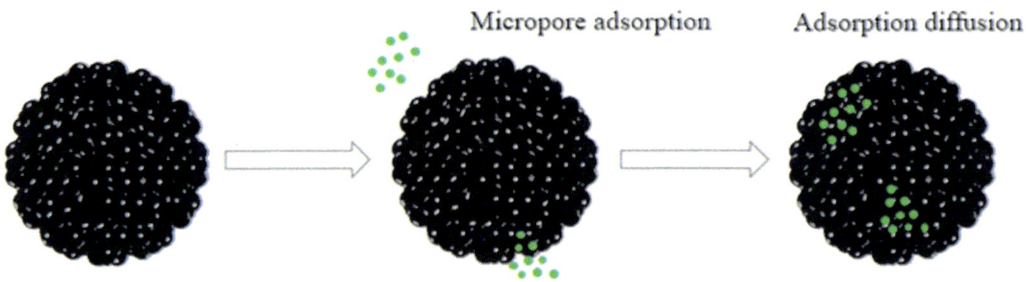

FIGURE 7.7 Physical adsorption of VOCs on γ-Al_2O_3.

alumina is produced by aging, washing, drying and calcination. The resultant activated alumina has low density and good mechanical strength. The method also has advantages such as convenient washing, simple operation, and easy to realize continuous production. (3) Citric acid method: Compared with the conventional methods, citric acid method for activated alumina production has shorter preparation cycle and the process is easier to control.

7.2.5 Diatomite

Diatomite, also known as diatomite or kieselguhr, is a kind of siliceous sedimentary rock resulting from fossilizing of diatom frustules, which consists of porous siliceous algae skeletons with Si—O tetrahedral interconnection nets. It mainly contains amorphous hydrated or opaline silica ($SiO_2 \cdot nH_2O$) accompanied by amounts of clay minerals, silica sand, carbonate minerals, iron oxides, and organic matter. Diatomite has high porosity, low density, and high specific surface area because of the large amount of diatom frustule at microscale and pores at nanoscale (Zhao et al., 2019). It thus has been used as an adsorbent for VOCs.

7.2.5.1 Adsorption of VOCs

The VOC adsorption capacity of diatomite is related to its surface functional group, surface charge, and surface structure. There are a large number of hydroxyl groups on the surface of diatomite. The more they are, the better the adsorption performance of diatomite is. Change of surface hydroxyl group of diatomite may have a strong effect on its adsorption of VOCs. These hydroxyl groups are active, so they may interact with other substances, which would change the adsorption of VOCs on diatomite.

H_2O and VOCs in the air are constantly moving, constantly colliding with the surface of the diatomite during adsorption. When encountering large pores, most of them return to the air (Fig. 7.8). When the pore size is about twice the diameter of a VOC molecule, it enters

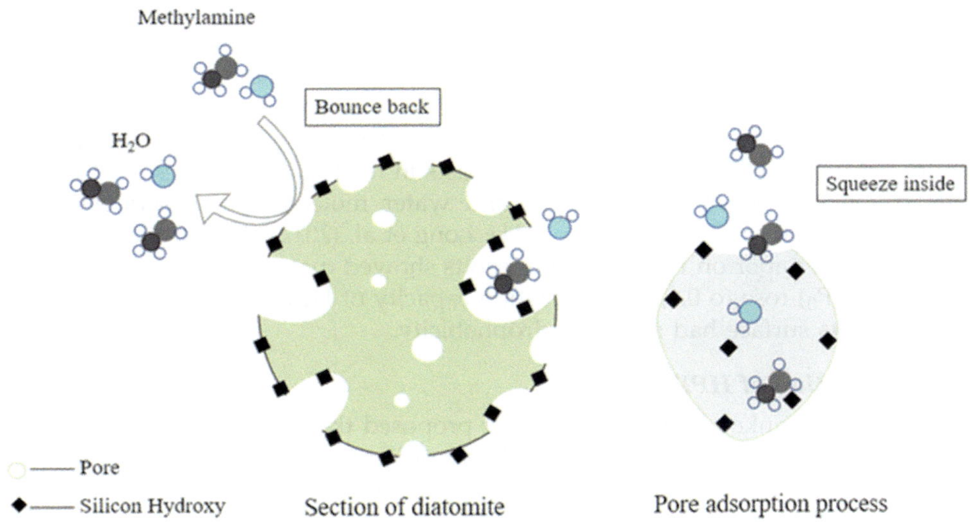

FIGURE 7.8 Adsorption of VOCs on diatomite.

the pore and is likely to be trapped (Fig. 7.8). In addition, it may collide with the next molecule and move deeper until the pore is filled.

7.2.5.2 *Modification of diatomite*

Diatomite is a biosiliceous rock with porous structure formed by preliminary diagenesis after the accumulation of the remains of ancient single-celled and low plant diatom. The raw diatomite generally contains impurities and also has obvious structural defects, which greatly limit its adsorption of VOCs. Therefore, diatomite needs to be modified to improve its VOC adsorption capacity (Zhang et al., 2013). For example, acid treatment can enhance diatomite's adsorption capacity due to an increase of the mineral specific surface and porosity (Jin et al., 2014). The adsorption capacity of acid-thermal activated diatomite (A-diatomite) to VOCs can reach 0.833 wt at 2.63 MPa and 298 K, much higher than that of many known adsorbents (Zaitan et al., 2006).

7.2.6 Hypercrosslinked polymeric resin

Hypercrosslinked polymeric resin (HPR) is listed as one of the main adsorption materials for controlling VOCs by the US EPA. HPR has many advantages such as the rich pore structure, stable physical and chemical properties, and adjustable surface functional groups (Li et al., 2020). It has been proved to be a promising alternative adsorbent to activated carbons for the adsorption of VOCs. The appearance and morphology of HPR are characterized by porous spongy three-dimensional structure (Huang and Turner, 2018). Some HPRs appear milky white, brick red, and other colors because of the differences in the structure and properties of the feedstock materials.

7.2.6.1 *Adsorption of VOCs*

HPR has high specific surface, large micropore volume, and excellent adsorption ability to VOCs, especially small molecular ones. The sorption of VOCs by unfunctionalized HPR is mainly based on hydrophobic and p-p interactions, but interaction with functional groups can also boost VOC adsorption (Fig. 7.9). Shi et al. (2018) showed that VOC adsorption capacity of HPRs with different pore sizes was related to pressure. At relative low pressure ($P/P_0 = 0.04$), the equilibrium adsorption capacity of VOCs was positively correlated with the micropore volume of HPR. Water inevitably exists in many VOC sources, and the competitive adsorption between VOC molecules and water molecules may seriously affect the adsorption performance (Águeda et al., 2011). Long et al. (2012) studied the adsorption performance of water vapor on HPR, and the results showed that when the relative pressure of water vapor (P/P_0) rose to 0.6, the adsorption capacity of HPR to water was the minimum, indicating that its surface had strong hydrophobicity.

7.2.6.2 *Preparation of HPR*

In the 1970s, Davankov and Tsyurupa first proposed the third generation of polystyrene mesh polymer - HPR. In a typical synthesis, linear or light cross-linking precursors are stirred in a solvent (usually 1, 2-dichloroethane) until expansion to create a free volume between the polymer chains. Then catalyst (e.g., $SnCl_4$ and $FeCl_3$) and external cross-linker are added to create cross-links that lock the polymer chains in the expansion state (Huang and Turner,

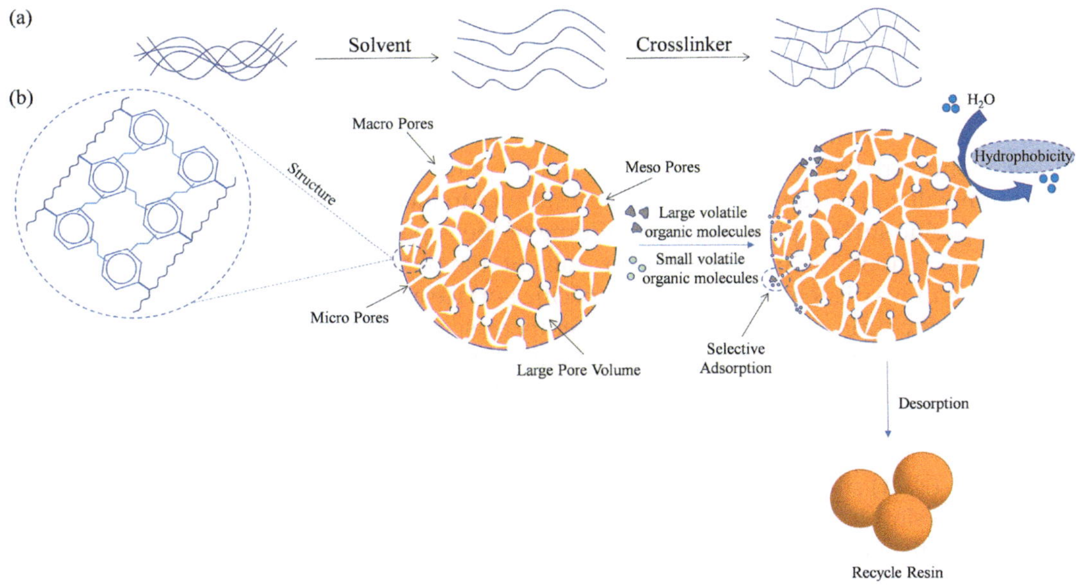

FIGURE 7.9 Preparation of HPR (a) for VOC adsorption (b).

2018). After the solvent is removed in vacuum, the space left by the solvent forms micropores, forming HPRs. Three commonly used HPR synthesis routes are summarized below.

1) Friedel–Crafts crosslinking method

Linear polystyrene is sufficiently swollen in the organic solvent cyclohexane or nitrobenzene. Then the catalyst tin tetrachloride ($SnCl_4$) and the cross-linking agents trichloromethyl benzene (CMM), 1,2,4,5-phenyltetrachloric anhydride, p-dibenzyl chloride (XDC), chloromethyl ether (MCDE) and 4, 4-dichloromethyl biphenyl (CMB), are used for cross-linking.

2) Post cross-linking of copolymers

Styrene - divinylbenzene is swelled in organic solvent under the catalyst Lewis acid, and the crosslinker chloromethyl ether is added. After a series of reactions, HPR can be synthesized.

3) Post-crosslinking copolymer with polar monomer

The copolymer is chloromethylated in chloromethyl ether and then swelled in an organic solvent. Using acrylic monomer as raw material, under the catalysis of Lewis acid, Friedel–Crafts crosslinking reaction after the resin is prepared. Common organic solvents are cyclohexane, dichloroethane, and nitrobenzene. The commonly used Lewis acid catalysts are anhydrous $SnCl_4$, anhydrous $ZnCl_2$, anhydrous $FeCl_3$ and anhydrous $AlCl_3$.

The specific surface area and pore volume of HPR determine its adsorption capacity of VOCs. By changing the reaction conditions during crosslinking process, such as the type of crosslinking agent, reaction temperature and reaction time, HPR structure can be regulated to target specific VOC compounds.

4) Pore-forming agent

According to different mechanisms of polymer pore formation, pore-forming agents can be generally divided into good solvent (toluene) and poor solvent (liquid wax). For good solvent, when the crosslinking degree is 56% and the amount of toluene is 18 times of the weight of the monomer, the specific surface area and pore volume of the polymer reaches the maximum value. If the amount of toluene continues to increase, the copolymer would shrink, and the specific surface area and pore volume of the copolymer would decrease with the increase of the amount of toluene. For poor solvent, the phase separation is difficult to occur if the amount of wax is too small, and the specific surface area of the copolymer is 0. The pores and the specific surface area of the copolymer increase with the increase of liquid wax dosage. The pore volume and pore size increase rapidly when the amount of liquid paraffin is more than 0.5 times that of monomer, resulting in the decrease of copolymer specific surface area.

5) Crosslinking agent

Liu et al. (2021) studied the effect of the amount of divinylbenzene cross-linker on the pore volume of resin by using liquid paraffin and benzene as pore-forming agents. The results showed that the effect of the amount of cross-linker on pore volume has a mutational range. In the range of 24%–28% of monomer content, the pore volume increases with the increase of cross-linker content but remains almost unchanged outside this range.

6) Reaction time and temperature

Reaction time and reaction temperature determine the degree of crosslinking of HPR. Generally, the reaction time and temperature are inversely proportional to the chlorine content and directly proportional to the specific surface area of the resin. In addition, higher reaction temperature leads to higher adsorption rate. Under certain reaction conditions, high reaction temperature may oxidize chloromethyl to formaldehyde carbonyl and ketone carbonyl. These groups can improve the hydrophilicity of HPR to enhance its adsorption of polar VOCs (Meng et al., 2007).

Modification or functionalization is commonly applied to improve the adsorption ability of HPR to VOCs, which also relies on enhancing the physical and chemistry characteristics of the sorbents. Hybridizing inorganic or organic components to into HPRs is the major routine for the improvement of their physicochemical and sorptive properties for VOC removal.

7.3 Catalysts

Catalytic oxidation is considered one of the most promising technologies for removing VOCs for its low energy consumption, high efficiency, and no secondary pollution in the process of treatment. Catalyst is the key substance for catalytic oxidation and directly determines the removal efficiency of VOCs. This section mainly focuses on the preparation methods, properties and catalytic performance of conventional catalysts for VOC removal. Besides, the characteristics and the factors affecting catalytic performance of these catalysts are also discussed. Precious metals and metal oxides are most commonly used traditional catalysts used for the oxidation of VOCs.

7.3.1 Precious metal catalysts

Precious metal catalysts can change the rate of a chemical reaction without being involved in the end product. Precious metal catalytic oxidation and decomposition of VOCs uses catalysts to reduce the activation energy and to enrich the VOCs on the catalyst surface, resulting in their oxidation and decomposition into CO_2 and H_2O. Precious metal catalysts have an irreplaceable catalytic activity and selectivity and are of great importance in the treatment of VOCs. In addition, they have been widely researched because of their high catalytic activity, superior thermal stability, low reaction temperature, and universality. Owing to their easy adsorption of reactants on their surfaces and their moderate strength, precious metal catalysts such as Au, Pt, Pd and Ru facilitate the formation of intermediate "active compounds" with high catalytic activity. Examples of precious metal catalysts for VOC removal are summarized below.

7.3.1.1 Pt-based catalysts

Pt-based catalysts often use carriers such as γ-Al_2O_3 (Liang et al., 2019), TiO_2, ZrO_2 and zeolite molecular sieves for their extraordinary activity in the catalytic oxidation of VOCs (Fig. 7.10). For example, Ruud et al. (van den Brink et al., 1998) found 2% Pt/γ-Al_2O_3 is an excellent catalyst for the oxidation of perchlorobenzene.

7.3.1.2 Pd-based catalysts

Pd is in the fifth period, subgroup VIII. It is in the same group as Pt, so Pd is chemically similar to Pt and can be used as a catalysis for VOC removal. Tidahy et al. (2006) prepared three catalysts using Pd loaded onto TiO_2, ZrO_2, and ZrO_2-TiO_2 for the degradation of toluene and chlorobenzene, respectively. The results showed that 0.5% Pd/TiO_2 gave the best catalytic performance for both toluene and chlorobenzene. Barakat et al. (2014) showed that the catalytic activity of the Pd/TiO_2 catalysts for toluene increased with the increase of Ce content. The increase in Ce not only increased the surface area of the catalyst, but also enhanced the synergistic effect between the carrier and the active components.

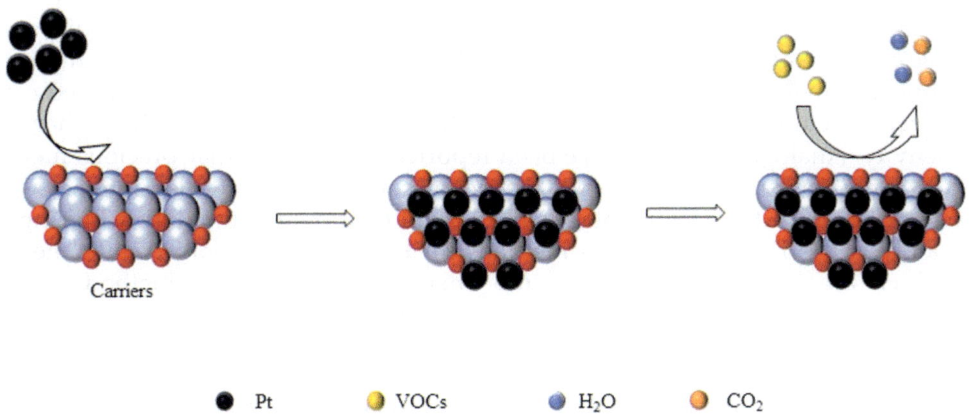

FIGURE 7.10 Pt catalyst for VOC removal.

7.3.1.3 *Au-based catalysts*

Au has been taken into account a catalytically inert precious metal for a long time. However, Japanese scientists Harute et al. (Harute et al. (1989) synthesized nano-Au loaded on metal oxides in 1989, showing excellent catalytic activity at low temperatures. Scirè et al. (2003) prepared Au/CeO_2 catalysts and obtained AuCeDP and AuCeCP catalysts by deposition-precipitation and coprecipitation methods. They were used for the catalytic oxidation of 2-propanol, methanol and toluene and exhibited excellent catalytic activity. The Au (6 nm) particles in AuCeDP were smaller than those in AuCeCP (8 nm) as characterized by XRD and TEM. XPS showed that more Au particles were exposed on the surface of AuCeDP than AuCeCP. In addition, H_2-TPR shows that the AuCeDP surface was covered with more oxygen than AuCeCP.

7.3.1.4 *Ru-based catalysts*

Ru-based catalysts were developed later than Au, Pt and Pd based ones for catalytic oxidation of VOCs. In recent years, it has been found that the catalytic activity of Ru-based catalysts is not much different from that of Au, Pt and Pd; however, they are much cheaper. Okai et al. (Okai and Zawadzki, 2009) used $RuCl_3$ as a precursor to synthesize $Ru/\gamma\text{-}Al_2O_3$ for the catalytic oxidation of butane by primary wet impregnation. The $Ru/\gamma\text{-}Al_2O_3$ catalyst showed good catalytic activity due to the decomposition of Ru_xO_y to produce reactive oxygen species. Mitsui et al. (2008) investigated the effect of carriers on Ru-based catalysts. While Ru was loaded onto CeO_2, ZrO_2, $\gamma\text{-}Al_2O_3$ and SnO_2 by impregnation, Ru/CeO_2 showed the best activity in the catalytic oxidation of ethyl acetate, acetaldehyde and toluene because of the good redox properties of CeO_2 at low temperatures. The catalytic activity was significantly increased if they were treated with H_2 reduction. TEM and XPS showed that smaller Ru particles were produced, enhancing the low temperature reduction.

7.3.1.5 *Influencing factors*

The catalytic performance of precious metal catalysts can be influenced by many factors including dispersion, particle size, choice of the carrier, and preparation method. (1) Dispersion is a factor that has a strong impact on catalysts. In general, the higher the dispersion of the active component, the better the catalysis. (2) Particle size is another very significant factor for precious metal catalysts. Scirè et al. (2003) prepared a variety of Au-based catalysts using diverse methods and found that the smaller the Au particle size, the higher the catalytic activity. Analogous results have been reported for Pt and other precious metal catalysts. (3) Some carriers not only carry functional components, but also interact with the active components or have some catalytic activity of their own, ultimately achieving synergistic catalysis. (4) There are various methods for the preparation of precious metal catalysts, which are broadly subdivided into chemical and physical methods. Chemical methods include impregnation, ion exchange, hydrolysis, photochemical reduction, etc. Physical methods include plasma, microwave synthesis, inert gas evaporation, ultrasonic crushing, etc. Different preparation methods have a significant impact on the structure of the catalyst, the state of action of the active components, and the nature of the catalytic reaction.

7.3.2 Metal oxide catalysts

Metal oxide catalysts are inexpensive and environmentally friendly but are not durable and inefficient. They are usually loaded on a suitable carrier with high thermal stability and a high specific surface area. The most commonly used metal-oxide catalysts include cobalt oxide, cerium dioxide, cupric oxide, titanium oxide, manganese oxide, and chromium oxide.

7.3.2.1 Cobalt oxide catalyst

Co_3O_4 materials have been synthesized by a variety of methods, including (1) Hydrothermal method: The synthesis process is mainly to dissolve $Co(NO_3)_2 \cdot 6H_2O$ in deionized water and stirring for 30 min. A certain amount of urea is then added to the above solution. After stirring for another 30 min, the resulting uniform solution is transferred to a stainless-steel autoclave for high-temperature and high-pressure treatment (Ren et al., 2018). Co_3O_4 with different morphologies and exposed crystal surfaces can be synthesized by controlling the test dose and different hydrothermal conditions during the hydrothermal process. (2) Precipitation method: The precipitant is added to the metal salt solution to generate insoluble metal salt or metal hydration oxide, and then precipitated from the solution. The catalytic material is prepared by aging, filtering, washing, drying, roasting, molding, activation, and other processes. (3) Impregnation method: The catalyst is obtained by soaking the carrier in a salt solution (nitrate, acetate, ammonium salt, etc.) with soluble and thermal decomposition characteristics, followed by drying and roasting.

Co_3O_4 exhibits spinel structure with a cell length of 0.8084 nm. The lattice oxygen is packed in a cell, in which Co^{2+} occupies 1/4 of the tetrahedral site and Co^{3+} occupies 1/2 of the octahedral site (Fig. 7.11a). This excellent morphology and structural stability ensure its catalytic stability in the process of VOC oxidation (Huang et al., 2015). There are mobile oxygen species in spinel structure of Co_3O_4, which has oxygen vacancies and excellent reducing power, as well as high concentration of electrophilic oxides.

Taking methylbenzene as an example, its oxidation over Co_3O_4 usually follows the Mars-van Krevelen (MVK) mechanism, in which the formation of oxygen vacancy is theoretically proposed to be the rate-determining step (Fig. 7.11b). In the reaction, Co^{2+} sites are more easily oxidized to Co^{3+} species in comparison to tetrahedrally coordinated sites, and Co^{3+} is responsible for the oxidative breakage of the benzene rings to generate the intermediate species (carboxylates), giving rise to good catalytic activity (Wang et al., 2017).

The oxidation of VOC over Co_3O_4 involves both surface and lattice oxygen species, which is more strongly dependent on the shape of the catalyst particle. Previous studies have found that as the shape of Co_3O_4 nanoparticles changes from spherical nanoparticles to nanorods, the catalytic activity and persistence during VOC oxidation are greatly improved due to the higher surface area of the material and the larger proportion of oxygen vacancies (He et al., 2019).

7.3.2.2 Cerium oxide catalyst

Cerium is the most abundant of the rare earth elements. CeO_2 has been applied in the oxidation and purification of VOCs. It is synthesized mainly using the redox reaction method (Abbasi et al., 2011). Briefly, Ce $(NO_3)_3 \cdot 6H_2O$ solution is mixed with 30% hydrogen peroxide

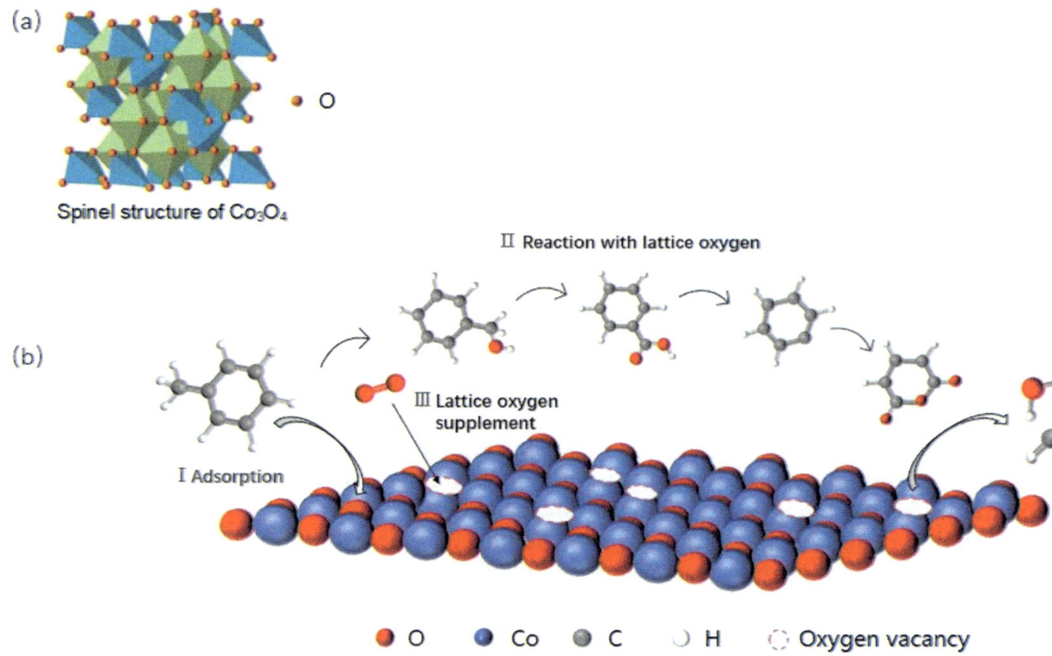

FIGURE 7.11 (a) Co_3O_4 unit cell; (b) mechanism of methylbenzene degradation over Co_3O_4.

at a volume ratio of 3:1. The mixture is then heated to boiling point and kept at boiling temperature for 4 h. The samples are then dried in an oven for 24 h at 110°C. Next, the solids are calcined in air for 4 h at 500°C to obtain the CeO_2 catalyst.

CeO_2 has a cubic fluorite crystal structure, consisting of a cubic array of four times coordinated oxygen ions and cerium ions, occupying half of the eight times coordinated cation gap (Huang et al., 2015). It is often used as a key component or structure of heterogeneous catalysts and electronic promoters because of its excellent oxygen storage capacity. Various studies have shown that the surface alkalinity, high oxygen mobility and oxygen supply capacity of CeO_2 catalyst are the main reasons for its good catalytic performance.

Catalytic oxidation of VOCs by CeO_2 also follows MVK mechanism. Its essence is that the lattice oxygen in the catalyst oxidizes VOCs. In this reaction process, firstly, VOCs react with lattice oxygen and generate oxygen vacancies, and the metal oxides are reduced consequently. Secondly, oxygen vacancies are filled by oxygen in the air. The adsorption capacity and oxygen transfer capacity are the key contributing factors of the MVK mechanism. For example, in the process of CeO_2 catalyzing methylbenzene, the coexistence of Ce^{3+}/Ce^{4+} pairs give the catalyst good redox ability, and the catalytic reaction is in accordance with the MVK mechanism (Fig. 7.12).

CeO_2 shows great potential in the deep oxidation of VOCs due to its many unique properties such as high oxygen storage capacity and the ability to undergo redox shuttling between Ce^{3+} and Ce^{4+} (He et al., 2019). CeO_2 and CeO_2-based catalysts are inexpensive, environmentally friendly, and have an excellent oxidation effect on nonchlorinated VOCs

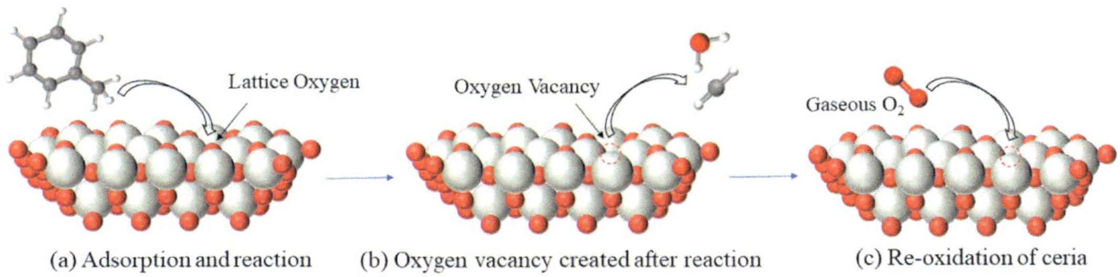

FIGURE 7.12 Catalytic mechanism of methylbenzene degradation over Co_3O_4.

such as methane, methanol, and propane. Cerium based catalysts are more suitable for non-chlorinated VOCs because they will be deactivated by adsorption of HCl or Cl_2 (Kamal et al., 2016).

7.3.2.3 Cupric oxide catalyst

CuO has a lower cost compared with other metal oxide catalysts. It is widely used because of its ability to interact with many other metals. CuO is often used in the deep oxidation of methane, methanol, ethanol, and acetaldehyde. It is also an excellent catalytic material for the selective oxidation of methanol and propylene. It is often synthesized through the following method. Anhydrous copper acetate is dissolved in ethanol and NaOH solution (0.5 mol/L) is then added. The obtained solution is stirred at room temperature for 30 min, then transferred into a stainless-steel autoclave and heated with water at 100°C for 6 h. The precipitate obtained from the reaction is collected, washed to neutral with deionized water and ethanol, and dried at 80°C for 12 h (Hermawan et al., 2020).

The catalytic activity of CuO is affected by many factors, such as morphology, valence of copper ions and the mobility of oxygen species. The morphology and structure of CuO are mainly affected by the preparation method. The valence of copper ion affects the mechanism of oxidation reaction and plays an important role in catalytic reaction. Lattice oxygen in CuO also plays an active role in the oxidation reaction, and the oxidation rate is limited by the diffusion rate of lattice oxygen to the surface. VOCs can extract almost all lattice oxygen from CuO (Hermawan et al., 2020). In recent years, researchers have found that it is helpful to improve the catalytic activity when CuO is combined with other transition metals to form composite metal oxides for catalytic reaction, because of the good synergistic effect between Cu and other metals. For example, HCHO oxidation over CuO/MnO_2 usually follows the MVK mechanism.

7.3.2.4 Titanium oxide catalyst

Titanium oxide is available in a variety of valence states, the most commonly used of which is the titanium dioxide photocatalyst. TiO_2 can be synthesized using the following method. Glacial acetic acid is added into anhydrous ethanol containing butyl titanate and mixed to form solution A. In addition, deionized water is added into anhydrous ethanol, after adjusting the pH and stirring, to obtain solution B. Solution B is slowly dropped into solution A, aging in the dark to form a gel and then microwave drying, and calcining in a high temperature furnace to obtain titanium dioxide crystals. Usually there are three types of TiO_2 crystals: anatase, rutile and slab titanite. It is generally believed that anatase is the most active crystalline form, followed by rutile, while plate titanite and amorphous TiO_2 have no significant photocatalytic activity. TiO_2 sols can be synthesized by hydrolysis of titanium tetrachloride ($TiCl_4$). The measured $TiCl_4$, deionized water and dispersant are mixed by carefully adding water ice cubes to the $TiCl_4$ solution at 0°C. The ice absorbs the heat released by the reaction of $TiCl_4$ with water and prevents the formation and precipitation of colloidal $TiO(OH)_2$ particles during the mixing process. The resulting mixture must be a clear solution, which is then heated to 80°C, stirred and refluxed at this temperature for 6 h. TiO_2 sols are then obtained. The TiO_2 gels show a well-defined anatase crystal structure of TiO_2 (Chen, 2002).

Titanium dioxide is nontoxic, chemically stable, and hardly reacts with other substances at room temperature, it is an amphoteric oxide. It does not react with oxygen, hydrogen sulfide, sulfur dioxide, carbon dioxide and ammonia, and is insoluble in water, fatty acids and other organic acids and weak inorganic acids, slightly soluble in alkali and hot nitric acid. TiO_2 has good stability as a stable compound in many inorganic and organic media. Titanium dioxide at high temperatures can be reduced by hydrogen, sodium, magnesium, aluminum, zinc, calcium and certain compounds of variable-valent elements to low-valent titanium compounds, but it is difficult to be reduced to titanium metal. Suspended in some organic media, titanium dioxide in the light and air can lead to the oxidation of the media. This photochemical activity is particularly obvious for anatase in the ultraviolet radiation. This property makes titanium dioxide an effective photocatalyst for VOC removal (Fig. 7.13). For example, formaldehyde

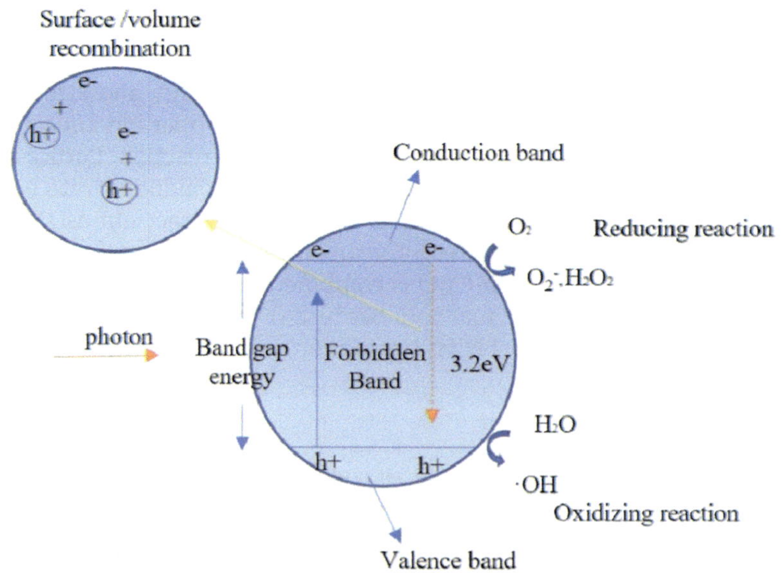

FIGURE 7.13 Reaction mechanism of TiO_2 photocatalytic degradation of VOCs.

with a mass number of 1×10^{-4} or less can be completely decomposed to CO_2 and H_2O_2 by titanium dioxide photocatalysis, while at higher concentrations it is oxidized to formic acid.

Hayashi et al. (2015) found that the titanium oxide based catalysts are promising oxygen reduction reaction catalysts. Harada et al. (1999) found that TiO_2 photocatalysts can be loaded on activated carbon fibers by a combination of the ICB method and calcination at relatively low temperatures. Under UV irradiation, the surface of TiO_2 becomes amphiphilic (hydrophilic and lipophilic), and this surface property is closely related to the photocatalytic reaction (Zhang et al., 2020). The surface characteristics (specific surface area and acidity) of the photocatalyst TiO_2 material play a key role in the overall reaction kinetics and deactivation stability, and a larger surface area results in higher adsorption and higher activity, when other factors such as lattice defects are equal (Pargoletti et al., 2022).

7.3.2.5 Manganese oxide catalyst

Manganese oxide is a grass green or gray–green cubic crystal powder or octahedral crystal. It is insoluble in water and soluble in acid and ammonium chloride solution. When heated in air, it is easily transformed into other high valence manganese oxides, such as Mn_3O_4, MnO_2, Mn_2O_3, which are more difficult to reduce. The most common manganese oxide catalyst is MnO_2.

Manganese dioxide can be prepared by roasting method, hydrothermal method, sol-gel method, chemical precipitation method, solid phase synthesis method, etc. Roasting is a relatively mature method to prepare manganese dioxide, which can be prepared on a large scale; however, it produces low level of optional manganese salts and single raw material. Meanwhile, it is difficult to produce nanoscale manganese dioxide due to the high temperature roasting process, which is prone to sintering, and the particle size distribution is wide. The following is a summary of the advantages and disadvantages of various preparation methods (Table 7.1).

TABLE 7.1 Summary of manganese dioxide preparation methods (Wang et al., 2020b).

Method	Advantage	Disadvantage
Hydrothermal synthesis	Nanoscale products can be produced with homogeneous morphology, enabling controlled product preparation	High energy consumption, small yield, and demanding reaction conditions
Sol-gel method	Low energy consumption and milder conditions	Disadvantages of complex reaction conditions and low yield
Micro-emulsion method	Good dispersion of micro particles	Poor controllability, large particle size
Sedimentation method	Mild preparation conditions and simple operation	Long pre-treatment time
Solid-phase method	Flexible operation and control conditions, low energy consumption, good reaction selectivity, high yield, and low environmental pollution	Uneven powder contact and insufficient reaction
Combustion method	Low starting temperature, rapid combustion reaction, small component segregation, high product purity	–

Manganese dioxide is also known as manganese peroxide, black manganese oxide, black orthogonal crystallization or brown black powder. It is insoluble in water, weak acid, weak base, nitric acid, cold sulfuric acid, and dissolved in concentrated hydrochloric acid to produce chlorine gas under heating. It's very oxidizing. MnO_2 has an octahedral structure, in which oxygen atoms are at the top of the corner of the octahedron and manganese atoms are in the octahedron. The octahedron is jointly connected to form single or double chains, and these chains are co-topped with other chains to form a tunnel structure of gaps. The octahedron is either hexahedral dense packing or square dense packing. MnO_2 shows various forms according to the arrangement of structural units (Fig. 7.14). It can be modified by ion exchange of Li^+, K^+, Rb^+ and other metals. The Cu−Pd doped MnO_2 catalyst showed high catalytic activity for CO oxidation compared to pristine MnO_2 (Kunkalekar and Salker, 2013).

Zhang et al. (2015) investigated the differences in the catalytic oxidation activity of different crystalline MnO_2 catalysts and explored the correlation between the catalyst crystalline structure and the catalytic oxidation activity. The results showed that the catalytic oxidation activity of different crystalline MnO_2 catalysts differed significantly in the order of α-MnO_2 > δ-MnO_2 > γ-MnO_2 > β-MnO_2. The higher lattice oxygen concentration and higher reducibility of α-MnO_2 catalysts are the key factors for their good catalytic activity in the oxidation of ethanol. The Au0.5Pt0.5/MnO_2/cotton catalyst showed excellent catalytic activity for the oxidative decomposition of formaldehyde (HCHO). The valence state of MnO_2, the porous structure of MnO_2, the synergistic effect between Au1-xPtx NPs and MnO_2, and the synergistic interaction between Au0.5Pt0.5/MnO_2 and cotton were considered as the main reasons for the high catalytic activity of Au0.5 Pt 0.5/MnO_2/cotton catalyst (Yu et al., 2012).

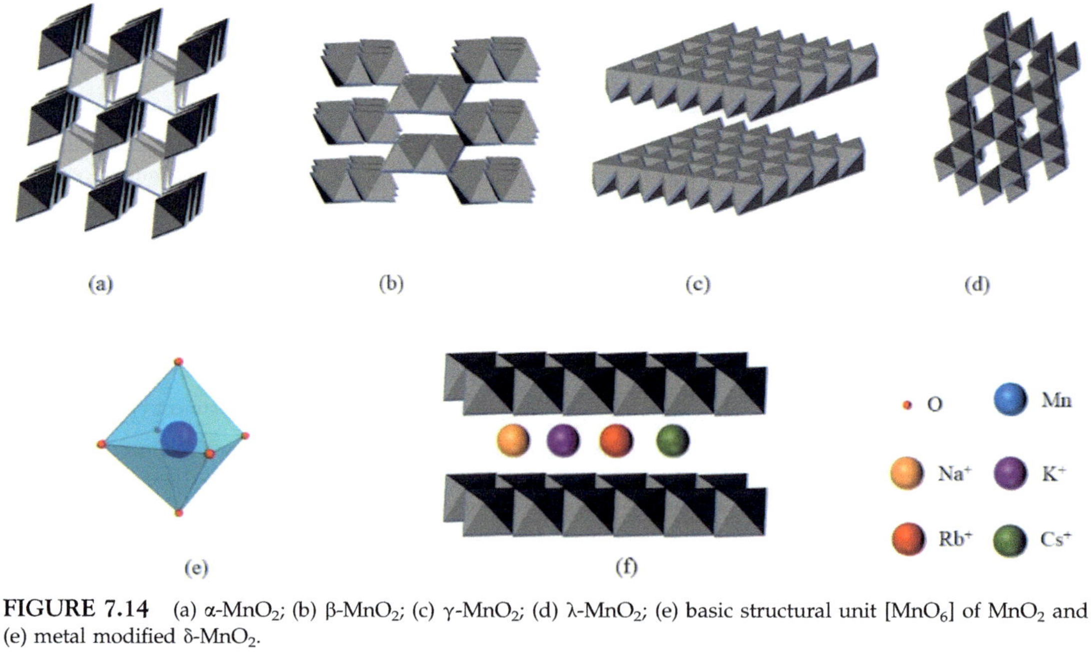

FIGURE 7.14 (a) α-MnO_2; (b) β-MnO_2; (c) γ-MnO_2; (d) λ-MnO_2; (e) basic structural unit [MnO_6] of MnO_2 and (e) metal modified δ-MnO_2.

7.3.2.6 Chromium oxide catalyst

Chromium oxide is light green to dark green fine hexagonal crystals that turn brown when scorched and remain green when cold. The crystals are extremely hard and stable, resistant to acids and bases, stable at high temperatures, with high hardness and good refractory properties, even when red-hot with hydrogen gas, there is no change. Chromium oxide is soluble in heated potassium bromate solution, slightly soluble in acids and bases, almost insoluble in water, ethanol and acetone. In the laboratory, it is often synthesized with the hydrothermal treatment of CrO_3 and HCHO in aqueous solution. In the industry, its general production methods are solid-phase synthesis and calcining chromic hydroxide. Based on the process of Cr_2O_3 preparation by thermal decomposition of chromic anhydride, Cr_2O_3 particles with an average particle size of 200 nm and good dispersibility can be obtained by controlling the conditions. By adding sulfur to sodium chromate solution, the hexavalent chromium is reduced to trivalent chromium, forming chromium hydroxide precipitation, and then calcined chromium hydroxide to produce chromium oxide.

Chromium oxide catalysts have large specific surface area, surface with groups, structure is amorphous structure and so on. Cr_2O_3+x has been studied as an alternative material for TiO_2 for VOC decomposition (Hiramatsu et al., 2010). Pt—Cr_2O_3 can catalyze the combustion of VOCs into CO_2 and H_2O, and the stronger interaction between Pt and Cr_2O_3 inhibits the sintering and deactivation of the catalysts (Fig. 7.15).

Maki et al. (Maki and Mizuguchi, 2010) developed a new catalyst system for VOC removal based on the heater integrated catalyst unit using Cr_2O_3, which showed excellent performance in VOC reduction. Xia et al. (2013) synthesized P4VP-Cr_2O_3, promoted the dispersion of manganese oxide particles and improved the oxidation performance of Mn/P4Vp-Cr_2O_3 catalyst. The interaction between Cr_2O_3, CeO_2 and MnO_x improved the oxidation characteristics of the catalyst. Wang et al. (Wang and Chou, 2000) studied the catalytic oxidation

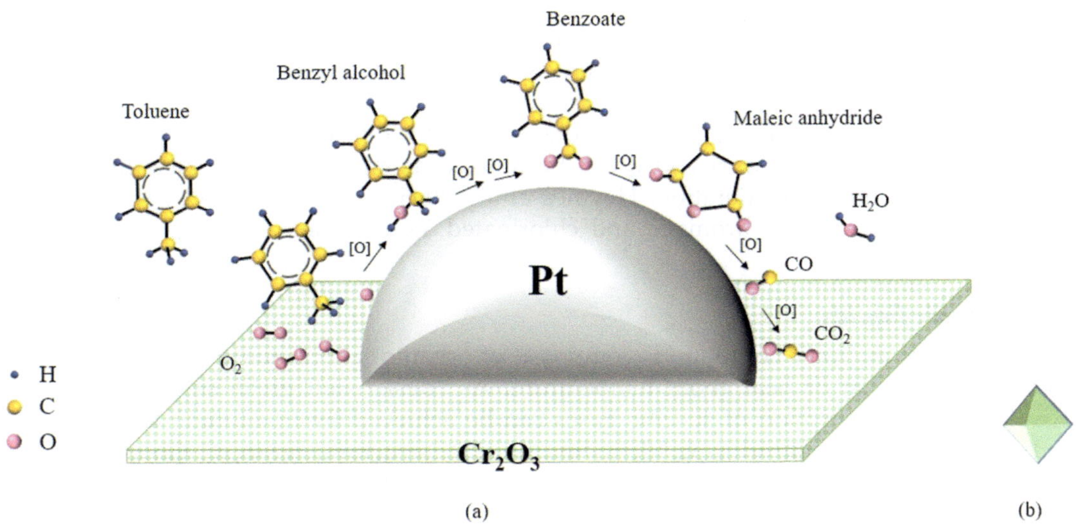

FIGURE 7.15 (a) proposed mechanism of VOC degradation over Pt—Cr_2O_3; (b) basic structural unit of Cr_2O_3.

characteristics of n-hexane, benzene and oil vapor collected during the treatment of oil refining wastewater on Cr_2O_3 catalyst, and concluded that the destruction of volatile organic compounds at low temperature could be achieved by using Cr_2O_3 catalyst.

7.4 Other methods and related materials

7.4.1 Absorption

The absorption method is to purify the VOC exhaust gas by contacting the liquid absorbent with the VOC exhaust gas due to the different solubility of each component of the VOC exhaust gas in the absorbent. Selecting the proper absorbent is crucial for treating VOC exhaust gas (Heymes et al., 2006). At present, the common absorbents in practical applications include mineral oil absorbents, water-based composite absorbents, high boiling point organic solvents, and ionic liquid absorbents.

7.4.1.1 *Mineral oils*

Mineral oil-based absorbents mainly refer to light diesel oil, motor oil, washing oil, lubricating oil, waste vegetable oil, etc. The characteristics of this type of absorbents are high absorption capacity, complex components, easy to cause pollution, and easy to volatilize. Ozturk et al. (Ozturk and Yilmaz, 2006) used waste vegetable oil and lubricating oil to remove benzene, toluene, carbon tetrachloride, and methanol from flue gas, and the results showed that the removal rate of benzene and toluene reached 90%, carbon tetrachloride reached 80%, and methanol reached 70%.

7.4.1.2 *High boiling point organic solvents*

Organic solvent-based absorption fluids are easily accessible and have been explored in initial absorption methods for organic waste gas purification. Silicone oil, diethylhydroxylamine, phthalates and adipates are some of the more common organic solvent-based absorption fluids, all of which have high boiling points and low vapor pressure characteristics. Silicone oil (PDMS) is an effective NAPL because it is insoluble in water, nonbiodegradable, nontoxic to microorganisms, and has a strong affinity for hydrophobic VOCs. Silicone oil is nonbiodegradable and nontoxic, so it can be utilized as an adsorption phase for the integrated process under consideration. Heymes et al. (2006) found that DEHA was the most effective absorber of other aromatic and chlorinated VOCs when the temperature was in the range of 20–70°C.

7.4.1.3 *Ionic liquids*

Ionic liquids are ionic compounds composed of organic cations and organic or inorganic anions in liquids at or near room temperature. As "designated green solvents", selective dissolution and absorption of specific gases can be achieved by fine-tuning their structure and properties and introducing fictitious specific structural groups. Gonzalez et al. (Gonzalez-Miquel et al., 2013) calculated the Henry constants and excess enthalpies of 14 representative VOCs from 135 imidazole ionic liquids in gas-liquid phase by using COSMO-RS code, and studied their dissolution and absorption processes, the cations and anions in ionic liquids play a decisive role in the absorption and purification of VOCs. By replacing the ions in ionic liquids, the interaction between ionic liquids and VOC molecules is enhanced, to improve the efficiency of VOC absorption and purification. As a new type of absorbent, ionic liquids have a good application prospect in the purification and treatment of VOCs, but there are still some problems to be solved and studied: (1) ionic liquids usually have high viscosity, which limits gas-liquid mass transfer; (2) ionic liquids are expensive.

7.4.1.4 Water-based composites

Water is more stable, low viscosity and easy to obtain. Most VOCs have low water solubility, so some scholars have studied water-based composite absorbents to increase the absorption of VOCs. Common composite absorbents are often composed of mineral oil and other liquids with water and surfactants, such as water - wash oil, water - surfactants - additives, etc. Water-based composite absorbents are inexpensive and have no secondary loss, but their absorption effect is poor. Additional research is needed to further improve them.

7.4.2 Condensation method

This method is suitable for high boiling point and high concentration of VOC recovery, but the concentration of VOCs in the condensed gas is high and cannot meet the emission standards. At present, condensation methods are mainly divided into surface condensation and contact condensation. The most important part of the condensation method is the condensate, among which synthetic refrigerants include R134a, R404a, R507, etc., and natural refrigerants include ammonia (R717), carbon dioxide (R744) and hydrocarbons. Aprea et al. (1996) conducted an experimental study on several alternative refrigerants of R502, and the experimental results showed that R402a, R402b, R403b and R408a can be used as short-term alternative refrigerants, while R404a, R407a and FX40 can be used as long-term alternative refrigerants, and R404a has a good application prospect in the long term. Based on the existing refrigerant research, suitable refrigerants should be selected for different components of volatile organic compounds, focusing on the application of R404a, R507, R407c, R410a and other refrigerants in refrigeration systems. This method has a small footprint, the condensed material can be reused, and the economic and environmental benefits are both win-win, but it requires a low-temperature and high-pressure working environment, so the cost is higher.

7.4.3 Membrane separation

Membrane separation is achieved by using the different selective permeability of different VOCs on the separation membrane and the large difference in the rate of each component passing through the separation membrane driven by the pressure difference (Yan et al., 2019). Membrane material performance in the process of membrane separation technology is the key to influence the industrial application of membranes. Membrane separation technology for VOCs can be broadly classified according to the separation mechanism: microporous polymer membranes following the molecular sieving mechanism and dense polymer membranes following the dissolution-diffusion mechanism (Thomas et al., 2009), microporous polymer separation membranes (Zhou et al., 2017), molecular sieve membranes, silicone rubber membranes, etc. Generally, polymer membranes used for gas separation are divided into two types: glassy and rubbery, whose permeation process follows the dissolution-diffusion principle. The permeation coefficient of gases in polymers is determined by the solubility and diffusion coefficients. In glassy polymers such as polysulfone, polyimide, polyetherimide, polyacrylonitrile or polyvinylidene fluoride, the diffusion process of gas is the controlling factor; in rubbery polymers such as silicone rubber and fluoroelastomer, the dissolution process of gas is the controlling factor. The gas molecules first condense or dissolve on the membrane surface, then diffuse from the high concentration side to the low concentration side under the concentration gradient, and finally the gas molecules move to the membrane permeate side (low pressure side) and desorb. For the separation system of VOCs mixture, the rubbery material has both high permeability and high selectivity, and it has preferential permeability to VOCs, especially for VOCs that are easy to condense. The preferential permeability can prevent VOCs from condensing and liquefying on the membrane surface, which can damage the integrity of the membrane.

7.4.4 Combustion method

Combustion method refers to the process of oxidation of VOCs to CO_2 and H_2O by complete combustion under certain conditions. Combustion method can be divided into direct combustion method, catalytic combustion method and thermal storage combustion method.

7.4.4.1 Direct combustion method

The direct combustion method is a method that uses the flammable nature of VOCs and feeds them directly into the incinerator as fuel to make them burn completely and decompose. This method has a small investment and is suitable for treating waste gas with high concentration and small air volume with high removal rate. However, the direct combustion method operates at too high a temperature, usually between 800°C and 1200°C, with high energy consumption and high equipment maintenance costs, and the incineration process of some VOCs is prone to produce harmful by-products such as chlorinated hydrocarbons, nitrogen oxides and dioxins, causing secondary pollution. The combustion temperature, retention time in the combustion chamber and turbulent mixing are usually considered to be the main factors affecting the decomposition and elimination efficiency (DRE) of VOCs.

7.4.4.2 Catalytic combustion method

The catalytic combustion method refers to the addition of a suitable catalyst to the combustion system so that VOCs can be completely catalytically degraded at a lower temperature and is suitable for the treatment of VOCs at concentrations above 5000 mg/m^3. Catalysts are the core of catalytic combustion method, and catalysts used for catalytic combustion of VOCs are mainly divided into precious metal catalysts and nonprecious metal catalysts (Zhang et al., 2016b). Pt, Pd, Rh and other precious metals are commonly used in the catalytic combustion reaction of VOCs (Yang et al., 2020), which have the advantages of good low temperature activity, low ignition temperature, and good product selectivity, but the price is high. The synthesis of low-loading, high-activity precious metal catalysts is of great practical importance. Nonprecious metal catalysts are mainly transition metal oxides such as CuO, MnO_2, NiO, CoO, CeO_2, etc., which have the advantages of low cost, high tolerance and regeneration ability, but the catalytic activity is also relatively low. Therefore, improving the catalytic activity of nonprecious metal catalysts is the focus of future research.

7.4.4.3 *Regenerative combustion method*

The regenerative combustion method is based on the direct combustion method with the addition of a heat accumulator for preheating and reheating, which effectively reduces the use of auxiliary fuels and is suitable for the treatment of VOCs with large air volumes and low to medium concentrations (Vaart et al., 1994). The heat accumulator is one of the key components of RTO devices, and typical heat accumulation materials include ceramic, cordierite, high alumina clay, mullite, alumina titanate, spinel, etc (Amelio and Morrone, 2007). The regenerative combustion method includes the regenerative thermal combustion method and the regenerative catalytic combustion method, and the main structure of both combustion systems consists of a combustion chamber, a heat accumulation chamber and a switching valve, etc. The heat accumulation chamber can be used as a one-chamber, two-chamber or multi-chamber system. At present, most of the industry uses the double chamber system. RCO technology is a combination of catalytic combustion and thermal combustion technology, using the catalytic combustion performance of the catalyst in catalytic combustion technology and the RTO furnace heat recovery capacity, so that VOCs completely degraded at the lower combustion temperature, reducing energy consumption. RCO and RTO have different heat accumulation layer added with catalytic bed layer, and the height of the accumulation layer of RCO is also lower than that of RTO. RCO technology has short preheating time and fast temperature rise, which is especially suitable for intermittent operation and can be turned on and off with intermittent production and save energy consumption. This technology has been developed and applied commercially.

7.4.5 Photocatalytic oxidation method

Photocatalytic oxidation is a new treatment method with excellent performance, first of all, the decomposition of organic waste gases can be carried out at a very fast rate through photolysis. During photolysis, intermediate substances are generally produced, which can be removed by using light and sodium hydroxide solution. Second, photocatalytic decomposition. The application of this technology allows the use of activated UV light or irradiation of TiO_2 to promote the generation of OH-free radicals from H_2O for the effective conversion of organic waste gases into carbon dioxide and water. Chen et al. (2020) found that SnO_2 nanoparticles showed moisture-tolerant photocatalysts with high photocatalytic activity. Qian et al. (2016) found that highly stable carbon quantum dots (CQDs) were combined with Bi_2WO_6 to photocatalyze the removal of gaseous volatile organic compounds, and that fluorescent lamps could be used as light sources in the selection of the corresponding catalysts to effectively remove malodors from organic waste gases and effectively reduce their pollution concentrations. However, from the current experience, it is necessary to pay more attention to improvement of the technology and effectiveness of catalytic decomposition, and to exploring more deeply in this area.

7.4.6 Plasma technology

The application of plasma technology in practice is that under strong electric field, gas discharge will produce some high-energy electrons, ions and free radicals, etc. In the end, oxygen ions and ozone are stimulated at various levels, which react with VOCs and break them down into carbon dioxide, carbon monoxide and water. Kim et al. (2005) has reported that the AG/TiO_2 plasma system was very attractive for the treatment of organic waste gases with low mass concentrations. Subrahmanyam et al. (2007) prepared a quartz tube-based dielectric barrier discharge reactor with sintered metal fibers loaded with transition metals as internal electrodes, and toluene, isopropanol and trichloroethene were degraded by supported different catalysts (Mn, Ti, Co). This technology has a strong applicability in the purification of low concentration, but also can achieve the purification of indoor air. The advantage of this technology is that VOCs can be removed by simple methods and simple equipment, and the safety of emission should be paid attention to in the process, the removal efficiency of VOCs can reach about 50%.

7.4.7 Biodegradation method

The biodegradation method is to use the organic components in VOCs as the carbon source and energy source for microbial metabolism, and by creating a suitable environment for microorganisms to survive, the microorganisms can degrade and convert organic matter into CO_2, H_2O and other inorganic salts in the metabolic process. Microorganisms are the mainstay of biological treatment of organic waste gas, mainly bacteria, fungi and actinomycetes (Oh et al., 2009; Estevez et al., 2005), and the domestication of efficient microorganisms is the focus of research on biological treatment of waste gas. Cheng et al. (2017) degraded CB by inoculation with fungal-bacterial flora, and experiments showed that the removal efficiency of CB by mixed flora (100%, 60 h) was significantly higher than that of single bacteria (100%, 72 h) or fungi (100%, 96 h). Biodegradation methods used for VOC treatment can be further divided into biological trickling, biological filtration and biological scrubbing according to the process (Mudliar et al., 2010). These three biological methods for purification of VOCs are carried out based on the double membrane theory. Generally the biofiltration tower is more often used to treat VOCs, and its packing is composed of a porous packed bed stacked from top to bottom, mainly including compost, peat blocks, soil, plastic filter media, granular activated carbon, ceramic filter media, etc., and the surface is covered by microbial film and water.

References

Abbasi, Z., Haghighi, M., Fatehifar, E., Saedy, S., 2011. Synthesis and physicochemical characterizations of nanostructured Pt/Al_2O_3–CeO_2 catalysts for total oxidation of VOCs. Journal of Hazardous Materials 186 (2–3), 1445–1454.

Abyzov, A.M., 2019. Aluminum Oxide and Alumina Ceramics (review). Part 1. Properties of Al_2O_3 and commercial production of dispersed Al_2O_3. Novye Ogneupory (New RefractoriesE) 1, 16–23.

Águeda, V.I., Crittenden, B.D., Delgado, J.A., Tennison, S.R., 2011. Effect of channel geometry, degree of activation, relative humidity and temperature on the performance of binderless activated carbon monoliths in the removal of dichloromethane from air. Separation and Purification Technology 78, 154–163.

Amelio, M., Morrone, P., 2007. Numerical evaluation of the energetic performances of structured and random packed beds in regenerative thermal oxidizers. Applied Thermal Engineering: Design, Processes, Equipment, Economics 27 (4), 762–770.

Ankur, S., Vimal, C.S., 2009. Adsorptive desulfurization by activated alumina. Journal of Hazardous Materials 170, 1133–1140.

Aprea, C., Mastrullo, R., de Rossi, F., 1996. Behaviour and performances of r502 alternative working fluids in refrigerating plants. International Journal of Refrigeration 19 (4), 257–263.

Bahrim, M. Al, Calvo, L., Gilar, M.A., 2016. Diuron multilayer adsorption on activated carbon from CO_2 activation of grape seeds. Chemical Engineering Communications 203 (1), 103–113.

Balanay Jo, A.G., Floyd, E.L., Lungu, C.T., 2015. Breakthrough curves for toluene adsorption on different types of activated carbon fibers: application in respiratory protection. Annals of Occupational Hygiene 59 (4), 481–490.

Barakat, T., Idakiev, V., Cousin, R., Shao, G.S., Yuan, Z.Y., Tabakova, T., Siffert, S., 2014. Total oxidation of toluene over noble metal based Ce, Fe and Ni doped titanium oxides. Applied Catalysis B: Environmental 146, 138–146.

Bénard, S., Ousmane, M., Retailleau, L., Boréave, A., Vernoux, P., Giroir-Fendler, A., 2014. Catalytic removal of propene and toluene in air over noble metal catalyst. Canadian Journal of Civil Engineering 36 (12), 1935–1945.

Bhargavi, R., Kadirvelu, K., Kumar, N.S., 2011. Vapor phase adsorption of homologous aliphatic ketones on activated spherical carbon. International Journal of Environmental Sciences 1 (5), 938–947.

Burgos, N., Paulis, M., Antxustegi, M.M., Montes, M., 2002. Deep oxidation of VOC mixtures with platinum supported on Al_2O_3/Al monoliths. Applied Catalysis B: Environmental 38 (4), 251–258.

Ch, L.T.L., Chejne, F., Bhatia, S.K., 2015. Effect of activating agents: flue gas and CO_2 on the preparation of activated carbon for methane storage. Energy & Fuels 29 (10), 6296–6305.

Chen, S., Sun, Z., Zhang, L., Xie, H., 2020. Photodegradation of gas phase benzene by SnO_2 nanoparticles by direct hole oxidation mechanism. Catalysts 10 (1), 117-117.

Chen, X., 2002. Preparation and property of TiO_2 nanoparticle dispersed polyvinyl alcohol composite materials. Journal of Materials Science Letters 21 (21), 1637–1639.

Cheng, Z., Li, C., Kennes, C., Ye, J., Chen, D., Zhang, S., Chen, J., Yu, J., 2017. Improved biodegradation potential of chlorobenzene by a mixed fungal-bacterial consortium. International Biodeterioration and Biodegradation 123, 276–285.

Chiang, H., Chen, T., Pan, S., Chiang, H., 2009. Adsorption characteristics of Orange II and Chrysophenine on sludge adsorbent and activated carbon fibers. Journal of Hazardous Materials 161 (2/3), 1384–1390.

Cundy, C.S., Cox, P.A., 2003. The hydrothermal synthesis of zeolites: history and development from the earliest days to the present time. Chemical Reviews 103 (3), 663–701.

Dettmer, K., Engewald, W., 2002. Adsorbent materials commonly used in air analysis for adsorptive enrichment and thermal desorption of volatile organic compounds. Analytical and Bioanalytical Chemistry 373 (6), 490–500.

Dicorcia, A., Marchese, S., Samperi, R., 1993. Evaluation of graphitized carbon-black as a selective adsorbent for extracting acidic organic-compounds from water. Journal of Chromatography A 642 (1–2), 163–174.

Dobre, T., Parvulescu, O.C., Iavorschi, G., Stroescu, M., Stoica, A., 2014. Volatile organic compounds removal from gas streams by adsorption onto activated carbon. Industrial and Engineering Chemistry Research 53 (9), 3622–3628.

Estevez, E., Veiga, M.C., Kennes, C., 2005. Biodegradation of toluene by the new fungal isolates paecilomyces variotii and erophiala oligosperma. Journal of Industrial Microbiology and Biotechnology 32 (1), 33–37.

Feng, A., Yu, Y., Yu, Y., Song, L., 2018. Recent progress in the removal of volatile organic compounds by zeolite and its supported catalysts. Journal of Chemistry 76 (10), 757–773.

Gonzalez-Miquel, M., Palomar, J., Rodriguez, F., 2013. Selection of ionic liquids for enhancing the gas solubility of volatile organic compounds. The Journal of Physical Chemistry B 117, 96–306.

Gutiérrez-Martínez, J., Nieto-Delgado, C., Avalos-Borja, M., Basiuk, E., Rangel-Mendez, J.R., 2020. Fast benzene vapor capture by natural macroporous carbonized fibers improved with carbon nanostructures. Separation and Purification Technology 257, 117956.

Haq, E.U., Zaidi, S.F.A., Zubair, M., Karim, R.M.A., Padmanabhan, S.K., Licciulli, A., 2017. Hydrophobic silica aerogel glass-fiber composite with higher strength and thermal insulation based on methyltrimethoxysilane (MTMS) precursor. Energy and Buildings 151, 494–500.

Harada, M., Tanii, A., Yamashita, H., Anpo, M., 1999. Preparation of titanium oxide photocatalysts loaded on activated carbon by an ionized cluster beam method and their photocatalytic reactivities for the degradation of 2-propanol diluted in water. Zeitschrift für Physikalische Chemie 213, 59–65.

Harute, M., Yamada, N., Kobayashi, T., Iijima, S., 1989. Gold catalysts prepared by coprecipitation for low-temperature oxidation of hydrogen and of carbon monoxide. Journal of Catalysis 115, 301–309.

Hayashi, T., Ishihara, A., Uehara, N., Kohno, Y., Matsuzawa, K., Mitsushima, S., Ota, K., 2015. Kinetic study of oxygen reduction reaction in acid solution using titanium oxide-based catalysts prepared from oxy-titanium tetra-pyrazino-porphyrazine. Electrochemistry 83 (10), 807–809.

He, C., Cheng, J., Zhang, X., Douthwaite, M., Pattisson, S., Hao, Z., 2019. Recent advances in the catalytic oxidation of volatile organic compounds: a review based on pollutant sorts and sources. Chemical Reviews 119 (7), 4471–4568.

Hermawan, A., Zhang, B., Taufik, A., Asakura, Y., Hasegawa, T., Zhu, J., Shi, P., Yin, S., 2020. CuO nanoparticles/ $Ti_3C_2T_x$ MXene hybrid nanocomposites for detection of toluene gas. ACS Applied Nano Materials 3 (5), 4755–4766.

Heymes, F., Manno-Demoustier, P., Charbit, F., Fanlo, J.L., Moulin, P., 2006. A new efficient absorption liquid to treat exhaust air loaded with toluene. Chemical Engineering Journal (115), 225–231.

Hiramatsu, H., Iwamoto, R., Mizuguchi, J., 2010. Thermally activated Cr_2O_{3+x} for elimination of volatile organic compounds. Materials Transactions 51, 1231–1236.

Huang, J., Turner, S.R., 2018. Hypercrosslinked polymers: a review. Polymer Reviews 58 (1), 1–41.

Huang, H., Xu, Y., Feng, Q., Leung, D.Y.C., 2015. Low temperature catalytic oxidation of volatile organic compounds: a review. Catalysis Science and Technology 5 (5), 2649–2669.

Jin, J., Zheng, C., Yang, H., 2014. Natural diatomite modified as novel hydrogen storage material. Functional Materials Letters 7 (3), 1450027.

Kamal, M.S., Razzak, S.A., Hossain, M.M., 2016. Catalytic oxidation of volatile organic compounds (VOCs)-A review. Atmospheric Environment 140, 117–134.

Kim, K.J., Ahn, H.G., 2011. The effect of pore structure of zeolite on the adsorption of VOCs and their desorption properties by microwave heating. Microporous and Mesoporous Materials 152, 78–83.

Kim, H.H., Oh, S.M., Ogata, A., Futamura, S., 2005. Decomposition of gas-phase benzene using plasma-driven catalyst (PDC)reactor packed with Ag/TiO_2 catalyst. Applied Catalysis B: Environmental: An Internationall Journal Devoted to Catalytic Science and Its Applications 56 (3), 213–220.

Kunkalekar, R.K., Salker, A.V., 2013. Low temperature CO oxidation over nano-sized Cu-Pd doped MnO_2 catalysts. Reaction Kinetics, Mechanisms and Catalysis 108 (1), 173–182.

Li, L., Liu, S., Liu, J., 2011. Surface modification of coconut shell based activated carbon for the improvement of hydrophobic VOC removal. Journal of Hazardous Materials 192 (2), 683–690.

Li, G., Tian, F., Zhang, Y., Ding, J., Fu, Y., Wang, Y., Zhang, G., 2015. Bamboo/lignite-based activated carbons produced by steam activation with and without ammonia for SO_2 adsorption. Carbon 85, 448.

Li, X., Zhang, L., Yang, Z., Wang, P., Yan, Y., Ran, J., 2020. Adsorption materials for volatile organic compounds (VOCs) and the key factors for VOC adsorption process: a review. Separation and Purification Technology 235, 116213-1–116213-16.

Li, X., Wang, J., Guo, Y., Zhu, T., Xu, W., 2021. Adsorption and desorption characteristics of hydrophobic hierarchical zeolites for the removal of volatile organic compounds. Chemical Engineering Journal 411, 128558-1–128558-9.

Liang, Y., Liu, Y., Deng, J., Zhang, K., Hou, Z., Zhao, X., Zhang, X., Zhang, K., Wei, R., Dai, H., 2019. Coupled palladium-tungsten bimetallic nanosheets/TiO(2) hybrids with enhanced catalytic activity and stability for the oxidative removal of benzene. Environmental Science and Technology 53 (10), 5926–5935.

Lin, C., Cheng, Y., Liu, Z., Chen, J., 2013. Adsorption and oxidation of high concentration toluene with activated carbon fibers. Journal of Porous Materials 20, 883–889.

Liu, X.L., Zhang, Y., Ju, H., Yang, F., Luo, X., Zhang, L., 2021. Uptake of methylene blue on divinylbenzene cross-linked chitosan/maleic anhydride polymer by adsorption process. Colloids and Surfaces A: Physicochemical and Engineering Aspects 629, 127424.

Loi, Q.K., Phothong, K., Do, D.D., Nicholson, D., 2021. Effects of crystallite stacking on adsorption of ethylene on graphitized thermal carbon black. Microporous and Mesoporous Materials 323, 111202-1–111202-10.

Long, C., Li, Y., Yu, W., Li, A., 2012. Adsorption characteristics of water vapor on the hypercrosslinked polymeric adsorbent. Chemical Engineering Journal 180, 106–112.

Maki, A., Mizuguchi, J., 2010. VOC decomposition system based upon heater-integrated catalyst units using NiO or Cr_2O_3. Materials Transactions 51 (7), 1361–1363.

Matisová, E., Škrabáková, S., 1995. Carbon sorbents and their utilization for the preconcentration of organic pollutants in environmental samples. Journal of Chromatography A 707 (2), 145–179.

Meng, G., Li, A., Yang, W., Liu, F., Yang, X., Zhang, Q., 2007. Mechanism of oxidative reaction in the post crosslinking of hypercrosslinked polymers. European Polymer Journal 43 (6), 2732–2737.

Mitsui, T., Tsutsui, K., Matsui, T., Kikuchi, R., Eguchi, K., 2008. Support effect on complete oxidation of volatile organic compounds over Ru catalysts. Applied Catalysis B: Environmental 81, 56–63.

Miura, K., Nakawaga, H., Okamoto, H., 2000. Production of high density activated carbon fiber by a hot briguetting method. Carbon 38, 119–125.

Monser, L., Adhoum, N., 2002. Modified activated carbon for the removal of copper, zinc, chromium and cyanide from wastewater. Separation and Purification Technology 26 (2–3), 137–146.

Mudliar, S., Giri, B., Padoley, K., Satpute, D., Dixit, R., Bhatt, P., Pandey, R., Juwarkar, A., Vaidya, A., 2010. Bioreactors for treatment of VOCs and odours - a review. Journal of Environmental Management 91 (5), 1039–1054.

Oh, D.I., Song, J.H., Hwang, S.J., Kim, J.Y., 2009. Effects of adsorptive properties of biofilter packing materials on toluene removal. Journal of Hazardous Materials 170 (1), 13–21.

Okai, J., Zawadzki, M., 2009. Catalytic combustion of butane on Ru/γ-Al_2O_3 catalysts. Applied Catalysis B: Environmental 89, 22–32.

Ozturk, B., Yilmaz, D., 2006. Absorptive removal of volatile organic compounds from flue gas streams. Process Safety and Environmental Protection 84 (5), 391–398.

Pargoletti, E., Rimoldi, L., Meroni, D., Cappelletti, G., 2022. Photocatalytic removal of gaseous ethanol, acetaldehyde and acetic acid: from a fundamental approach to real cases. International Materials Reviews 67, 864–897.

Price, P.M., Clark, J.H., Macquarrie, D.J., 2000. Modified silicas for clean technology. Journal of the Chemical Society Dalton Transactions 2, 101–110.

Qian, X., Yue, D., Tian, Z., Reng, M., Zhu, Y., Kan, M., Zhang, T., Zhao, Y., 2016. Carbon quantum dots decorated Bi_2WO_6 nanocomposite with enhanced photocatalytic oxidation activity for VOCs. Applied Catalysis B: Environmental: An International Journal Devoted to Catalytic Science and Its Applications 193, 16–21.

Qin, C., Huang, X., Dang, X., Huang, J., Teng, J., Kang, Z., 2016. Toluene removal by sequential adsorption-plasma catalytic process: effects of Ag and Mn impregnation sequence on Ag-Mn/γ-Al_2O_3. Chemosphere 162, 125–130.

Ren, Q., Mo, S., Peng, R., Feng, Z., Zhang, M., Chen, L., Fu, M., Wu, J., Ye, D., 2018. Controllable synthesis of 3D hierarchical Co_3O_4 nanocatalysts with various morphologies for the catalytic oxidation of toluene. Journal of Materials Chemistry A 6 (2), 498–509.

Scirè, S., Minicò, S., Crisafulli, C., Satriano, C., Piston, A., 2003. Catalytic combustion of volatile organic compounds on gold/cerium oxide catalysts. Applied Catalysis B: Environmental 40, 43–49.

Shi, Q., Yang, X., Wu, L., Liu, H., Zhang, J., Zhang, F., Long, C., 2018. Binary adsorption equilibrium and breakthrough of toluene and cyclohexane on macroporous and hypercrosslinked polymeric resins. Microporous and Mesoporous Materials 271, 73–82.

Subrahmanyam, C., Renken, A., Kiwi-Minsker, L., 2007. Novel catalytic non-thermal plasma reactor for the abatement of VOCs. Chemical Engineering Journal 134 (1/3), 78–83.

Sui, H., Liu, H., An, P., He, L., Li, X., Cong, S., 2017. Application of silica gel in removing high concentrations toluene vapor by adsorption and desorption process. Journal of the Taiwan Institute of Chemical Engineers 74, 218–224.

Sui, H., Liu, J., He, L., Li, X., Jani, A., 2019. Adsorption and desorption of binary mixture of acetone and ethyl acetate on silica gel. Chemical Engineering Science 197, 185–194.

Tang, D., Zheng, Z., Lin, K., Luan, J., Zhang, J., 2007. Adsorption of p-nitrophenol from aqueous solutions onto activated carbon fiber. Journal of Hazardous Materials 143 (1/2), 49–56.

Thomas, S., Pinnau, I., Du, N., Guiver, M.D., 2009. Hydrocarbon/hydrogen mixed-gas permeation properties of PIM-1, an amorphous microporous spirobisindane polymer. Journal of Membrane Science 338 (1/2), 1–4.

Tidahy, H.L., Siffert, S., Lamonier, J.F., Zhilinskaya, E.A., Aboukaïs, A., Yuan, Z.Y., Vantomme, A., Su, B.L., Canet, X., De Weireld, G., Frère, M., N'Guyen, T.B., Giraudon, J.M., Leclercq, G., 2006. New Pd/hierarchical macro-mesoporous ZrO_2, TiO_2 and ZrO_2-TiO_2 catalysts for VOCs total oxidation. Applied Catalysis A: General 310, 61–69.

Vaart, D.R., Marchand, E.G., Bagely-Pride, A., 1994. Thermal and catalytic incineration of volatile organic compounds. Critical Reviews in Environmental Science and Technology 24 (3), 203–236.

van den Brink, R.W., Louw, R., Mulder, P., 1998. Formation of polychlorinated benzenes during the catalytic combustion of chlorobenzene using a Pt/γ-Al_2O_3 catalyst. Applied Catalysis B: Environmental 16, 219–226.

Veerapandian, S.K.P., Geyter, N.D., Giraudon, J.M., Lamonier, J.F., Morent, R., 2019. The use of zeolites for VOCs abatement by combining non-thermal plasma, adsorption, and/or catalysis: a review. Catalysts 9 (1), 98.

Wang, J.B., Chou, M.S., 2000. Kinetics of catalytic oxidation of benzene, n-hexane, and emission gas from a refinery oil/water separator over a chromium oxide catalyst. Journal of the Air and Waste Management Association 50 (2), 227–233.

Wang, H., Rong, X., Han, L., Tang, M., Yu, M., Zhang, J., Huang, W., Chen, R., 2015. Controlled synthesis of hexagonal mesostructure silica and macroporous ordered siliceous foams for VOCs adsorption. RSC Advances 5 (8), 5695–5703.

Wang, X., Liu, Y., Zhang, T., Luo, Y., Lan, Z., Zhang, K., Zuo, J., Jiang, L., Wang, R., 2017. Geometrical-site-dependent catalytic activity of ordered mesoporous Co-based spinel for benzene oxidation: in situ drifts study coupled with Raman and XAFS spectroscopy. ACS Catalysis 7 (3), 1626–1636.

Wang, Y., Bayatpour, S., Qian, X., Vaz, B.F., Wang, P., 2020a. Activated carbon fibers via reductive carbonization of cellulosic biomass for adsorption of nonpolar volatile organic compounds. Colloids and Surfaces A: Physicochemical and Engineering Aspects 612, 125908.

Wang, J., Yu, H., Ma, D., 2020b. Progress in the preparation and application of nanostructured manganese dioxide. Journal of Inorganic Materials 35 (12), 1307–1314.

Xia, P., Zuo, S., Liu, F., Qi, C., 2013. Ceria modified crystalline mesoporous Cr_2O_3 based nanocomposites supported metal oxide for benzene complete oxidation. Catalysis Communications 41, 91–95.

Xu, W., Zerda, T.W., Yang, H., Gerspacher, M., 1996. Surface fractal dimension of graphitized carbon black particles. Carbon 34 (2), 165–171.

Yan, X.R., Anguille, S., Bendahan, M., Moulin, P., 2019. Ionic liquids combined with membrane separation processes: a review. Separation and Purification Technology 222, 230–253.

Yang, C., Miao, G., Pi, Y., Xia, Q., Wu, J., Li, Z., Xiao, J., 2019. Abatement of various types of VOCs by adsorption/catalytic oxidation: a review. Chemical Engineering Journal 370, 1128–1153.

Yang, Y., Wang, G., Fang, D., Han, J., Dang, F., Yang, M., 2020. Study of the use of a Pd–Pt-based catalyst for the catalytic combustion of storage tank VOCs. International Journal of Hydrogen Energy 45 (43), 22732–22743.

Yin, T., Meng, X., Wang, S., Yao, X., Liu, N., Shi, L., 2022. Study on the adsorption of low-concentration VOCs on zeolite composites based on chemisorption of metal-oxides under dry and wet conditions. Separation and Purification Technology 280, 119634-1–119634-14.

Yu, X., He, J., Wang, D., Hu, Y., Tian, H., He, Z., 2012. Facile controlled synthesis of Pt/MnO_2 nanostructured catalysts and their catalytic performance for oxidative decomposition of formaldehyde. Journal of Physical Chemistry C 116 (1), 851–860.

Zaitan, H., Feronnato, C., Bianchi, D., Achak, O., Chafik, T., 2006. Etude des propriétés texturales et adsorbantes dune diatomite marocaine: Application au traitement d'air chargé d'un polluant de type composé organique volatile. Annales de Chimie Science des Matériaux 31, 183–196.

Zhang, G., Cai, D., Wang, M., Zhang, C., Zhang, J., Wu, Z., 2013. Microstructural modification of diatomite by acid treatment, high-speed shear, and ultrasound. Microporous and Mesoporous Materials 165, 106–112.

Zhang, J., Zhang, J., Zhang, C., He, H., 2015. Complete catalytic oxidation of ethanol over MnO_2 with different crystal phase structures. Acta Physico-Chimica Sinica 31, 353–359.

Zhang, L., Peng, Y., Zhang, J., Chen, L., Meng, X., Xiao, F., 2016a. Adsorptive and catalytic properties in the removal of volatile organic compounds over zeolite-based materials. Chinese Journal of Catalysis 37 (6), 800–809.

Zhang, Z., Zheng, J., Shangguan, W., 2016b. Low-temperature catalysis for VOCs removal in technology and application: a state-of-the-art review. Catalysis Today 264, 270–278.

Zhang, X., Gao, B., Creamer, A.E., Cao, C., Li, Y., 2017. Adsorption of VOCs onto engineered carbon materials: a review. Journal of Hazardous Materials 338, 102–123.

Zhang, L., Moralejo, C., Anderson, W.A., 2020. A review of the influence of humidity on photocatalytic decomposition of gaseous pollutants on TiO_2-based catalysts. Canadian Journal of Chemical Engineering 98 (1), 263–273.

Zhao, Y., Tian, G., Duan, X., Liang, X., Meng, J., Liang, J., 2019. Environmental applications of diatomite minerals in removing heavy metals from water. Industrial and Engineering Chemistry Research 58 (27), 11638–11652.

Zhou, H., Tao, F., Quan, L., Zong, C., Yang, W., 2017. Microporous polyamide membranes for molecular sieving of nitrogen from volatile organic compounds. Angewandte Chemie 56 (21), 5755–5759.

Zhu, L., Shen, D., Luo, K., 2020. A critical review on VOC adsorption by different porous materials: species, mechanisms and modification methods. Journal of Hazardous Materials 389, 122102-1–122102-27.

CHAPTER

8

Biochar for VOC removal

Wei Xiang

School of Environmental Engineering, Jiangsu Key Laboratory of Industrial Pollution Control and Resource Reuse, Xuzhou University of Technology, Xuzhou, China

8.1 Overview of biochar for VOC removal

Biochar is a porous carbon-based material with unique physicochemical properties, and is typically produced via thermochemical decomposition of waste biomass (Dissanayake et al., 2020; Xiang et al., 2020a, 2020b). Biochar feedstock can be any organic waste materials which include crop and forest residues, wood chip, algae, sewage sludge, manures, and food wastes (Xie et al., 2022; Xiong et al., 2019). Methods of thermochemical decomposition include pyrolysis, hydrothermal carbonization, gasification, torrefaction, and microwave heating, varying in thermochemical temperature and duration (Abhishek et al., 2022; Xiang et al., 2022; Xu et al., 2021; Zhang et al., 2022a). Interest in biochar is largely based on its two distinct benefits. First, biochar production can offset greenhouse gas emission because it stores carbon in a stable form, preventing feedstock degradation resultant release greenhouse gas into the atmosphere (Cao et al., 2022; Creamer and Gao, 2016; Yang et al., 2018a). Second, biochar is an effective, low-cost, and environment-friendly adsorbent, resulting from its relatively large surface area and abundant surface functional groups (Wang et al., 2017; Zhang et al., 2017a). Biochar can be used to sorb a wide range of contaminants (e.g., heavy metals, antibiotics, and nutrients) from water (Wan et al., 2018; Xiang et al., 2020a; Zheng et al., 2019a), enhance crop productivity (Yoo et al., 2018), and remove gaseous pollutants (e.g., CO_2, H_2S, and VOCs) in waste gases (Cao et al., 2022; Xiang et al., 2020c; Zhang et al., 2021). At present, biochar is regarded as a sustainable alternative to traditional sorbents due to its use of waste feedstock sources and environmental friendliness (Zhu et al., 2020).

Volatile organic compounds (VOCs) usually have toxic, teratogenic, carcinogenic, and irritant odors, which are extremely harmful to human health (Gwenzi et al., 2021; Xiang et al., 2022). In recent years, great efforts have been made to develop efficient VOC abatement techniques, such as catalysis oxidation, condensation, biological degradation, absorption, and adsorption (Zhu et al., 2020). Among them, using biochar as a VOC abatement technique is recognized as one of the most economic and promising control strategies. A great deal

Volatile Organic Compound Removal
https://doi.org/10.1016/B978-0-443-15674-8.00001-9

of research has been conducted to investigate the removal of VOCs by biochar produced with different technologies.

Generally, the purification of VOCs in waste gas by biochar is mainly achieved through the following steps: (1) VOC gas is transported to the biochar surface and further enters the pores; (2) VOC molecules occupy the active site of biochar via surface adsorption or pore filling; and (3) VOCs further react with adsorbed intermediate products (Choudhury and Lansing, 2020). Therefore, biochar with larger specific surface area and pore volume usually has higher VOC adsorption capacity. Hence, enhancing the physical and chemical properties of biochar is conducive to its removal of VOCs.

It has been well documented that feedstock, thermochemical decomposition methods, and their temperature and duration can significantly affect biochar physical and chemical properties. Besides, feedstock pre-treatments and biochar post-treatments can also affect the biochar properties. Pre-treatments vary depending on feedstock and the purposes for biochar use, including physical (dry, crush, sieve, wash, etc.), chemical (treat with chemicals or functional materials, load of precursors and functional agents, etc.), and biological (bacterial treatment, etc.) methods. Post-treatments mainly rely on physical (ball milling, magnetization, etc.) and chemical (corrosive treatment, etc.) methods (Fig. 8.1) (Xiang et al., 2020b). The processed

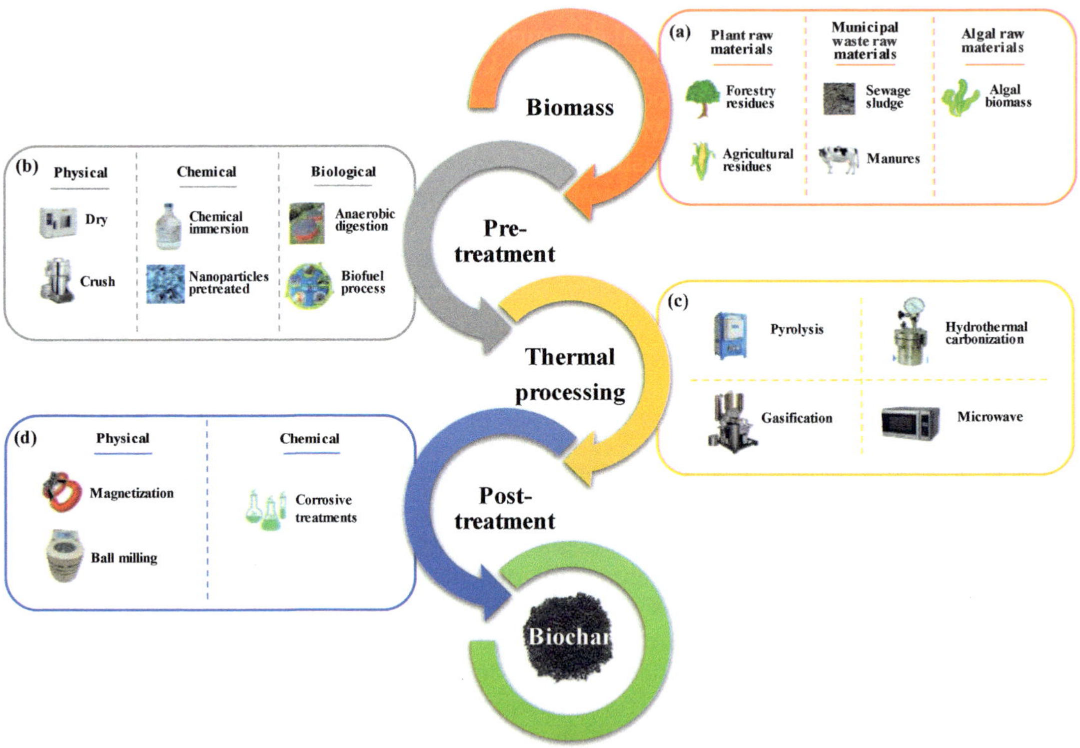

FIGURE 8.1 Biochar production technologies: (a) Biomass. (b) Pre-treatment technologies. (c) Thermal processes. (d) Post-treatment technologies (Xiang et al., 2020b).

biochars are called engineered biochars, which have been applied in VOC removal. For example, biochar prepared from neem was applied to adsorb benzene, toluene, methyl chloride, xylene, chloroform and carbon tetrachloride with the adsorption capacities ranging from 19.7 mg/g to 65.5 mg/g (Kumar et al., 2020a). Biochar derived from sugarcane bagasse could adsorb VOCs effectively and the adsorption capacity (acetone) reached 91.2 mg/g (Zhang et al., 2017b). VOC (acetone) adsorption on ball-milled wood biochar could be up to 304 mg/g (Zhuang et al., 2021). Microwave corn stalk biochar was applied to sorb benzene and o-xylene with adsorption capacities of 6.82–54.75 mg/g and 7.43–48.73 mg/g, respectively (Xiang et al., 2022). The maximum adsorption capacities of microwave wheat straw biochar to benzene, toluene and o-xylene were 53.9 mg/g, 75.8 mg/g and 63.0 mg/g, respectively (Zhang et al., 2022b). These studies have confirmed that the removal of VOCs are related to the pore structure of biochar such as specific surface area and pore volume, and are governed by various mechanisms like surface interaction, partition, and physical pore filling.

This chapter mainly focuses on the utilization of biochar in VOC removal as the adsorbents, catalysts, and others (sections 8.2–8.4). Perspectives and future development of biochar in VOC removal are outlined at the end of the chapter.

8.2 Biochar adsorbents for VOC removal

The adsorption capacity of pristine biochar for pollutants is generally lower than that of engineered biochar. In addition, due to the low density of pristine biochar, it is difficult to be separated from contaminated media, which greatly limits the application. In order to enhance the adsorption ability of the pristine biochar, different modification methods have been developed to improve biochar's physical and chemical properties. As mentioned in section 8.1, the modification methods of biochar mainly include physical methods, chemical methods, and biological methods, which can be conducted before or after biochar production. In the literature, biochar is often post-treated by either physical or chemical modification methods to increase its specific surface area, pore volume, and surface functional groups (Xiang et al., 2020b). This section thus mainly focuses on the application of engineered biochar modified by physical, chemical, and biological technologies for VOC removal, as well as the governing factors and mechanisms.

8.2.1 Engineered biochar for VOC removal

8.2.1.1 Physically modified biochar for VOC removal

Physical modification is a technique to use physical methods to improve or expand the properties of biochar, including gaseous activation, ball milling, magnetic modification, etc (Majumder et al., 2023). It improves the pore structures of biochar to increase the specific surface area and create more micropores and mesopores. For example, gaseous activation and ball milling can greatly enhance biochar's ability to remove VOCs.

Gaseous activation refers to the activation of biochar using steam or carbon dioxide. It requires energy and operates at high temperatures for extended periods (Abuelnoor et al., 2021). At the high temperatures, carbon atoms and activators are exchanged to form surface oxides and hydrogen ($CO + H_2$ or CO), thereby improving the pore structure of biochar (Yaashikaa et al., 2019). the gas activated biochar is widely used for adsorbing small molecule VOCs based on its rich microporous (Li et al., 2020a). Stream activation can improve the amount of surface oxygen containing groups, which attributes to the interaction between water molecules and the biochar under thermal treatment. Stream activated biochar favors the adsorption of polar VOC adsorption. While, carbon dioxide activation mainly develops new narrow micropores and enlarges the original micropores of the biochar, which result in widening the pore size distribution. Then the carbon dioxide activated biochar is beneficial for nonpolar VOC adsorption (Guo et al., 2016; Zhu et al., 2020). For example, carbon dioxide activated biochar adsorbed as high as 121.74 mg/g acetone, much more than the pristine biochar (Zhang et al., 2020). Similarly, carbon dioxide activation increased the adsorption of toluene on biochar from 95.35 to 109.5 mg/g (Qiu et al., 2018).

Ball milling is a simple, efficient and mechanical method using kinetic energy of moving balls to break particles (Lyu et al., 2017). After ball milling, particle size, specific surface area, negative zeta potential, oxygen-containing functional groups, etc. of biochar are improved (Lyu et al., 2018a, 2018b; Xiang et al., 2020c; Zhuang et al., 2021). Actually, these characteristics of biochar are highly associated with the parameters of ball milling process, such as milling type, speed, duration, temperature, and substrate-to-ball ratio (Kumar et al., 2020b). Therefore, ball-milling conditions are often selected based on the requirements of biochar products. Ball-milled biochar has already been applied to remove various pollutants. Its adsorption of VOCs has also received widespread research attention in recent years. Ball-milled hickory wood biochar effectively adsorbed five types of VOCs (acetone, ethanol, chloroform, cyclohexane, and toluene) with adsorption capacities ranged from 23.4 to 103.4 mg/g, which was contributed to the improved structural characteristics (e.g., specific surface area) of biochar by ball milling (Xiang et al., 2020c). Ball-milled pine wood biochar showed enhanced adsorption capacity for acetone (304 mg/g) and toluene (136 mg/g), which mainly contributed to the increased specific surface area and the oxygen functional group content after the ball milling (Zhuang et al., 2021). In addition, ball milling can be combined with chemical modified to control the surface functional groups of biochar, thereby changing the adsorption ability of biochar to different pollutants (Fig. 8.2). Zhang et al. (2021) and Qi et al. (2022) both used the hydrogen peroxide and ammonia hydroxide to modify the ball-milled biochar. The results indicated that the phenyl VOC adsorption capacities by modified ball-milled biochar were significantly enhanced compared to the pristine ball-milled biochar. The main adsorption mechanisms are surface adsorption and partitioning. In a word, ball-milled biochar is a promising adsorbent for VOCs.

8.2.1.2 Chemical modified biochar for VOC removal

Chemical modification can alter the surface chemistry of biochar, which is usually classified as a corrosion treatment. Biochar usually is treated by acidic, alkaline, or oxidizing agents, such as HCl, KOH, $KMnO_4$, and H_2O_2 (Kaikiti et al., 2021; Zheng et al., 2019b). Its surface functional groups, cation exchange capacity and the sorption capacity are all improved after the chemical modification (Wang and Liu, 2018; Yang et al., 2019). This section mainly discusses acid and alkali treated biochar for VOC removal.

Acid modification can remove impurities and introduce acidic functional groups on the surface of biochar. Therefore, the surface area and pore size of biochar are improved

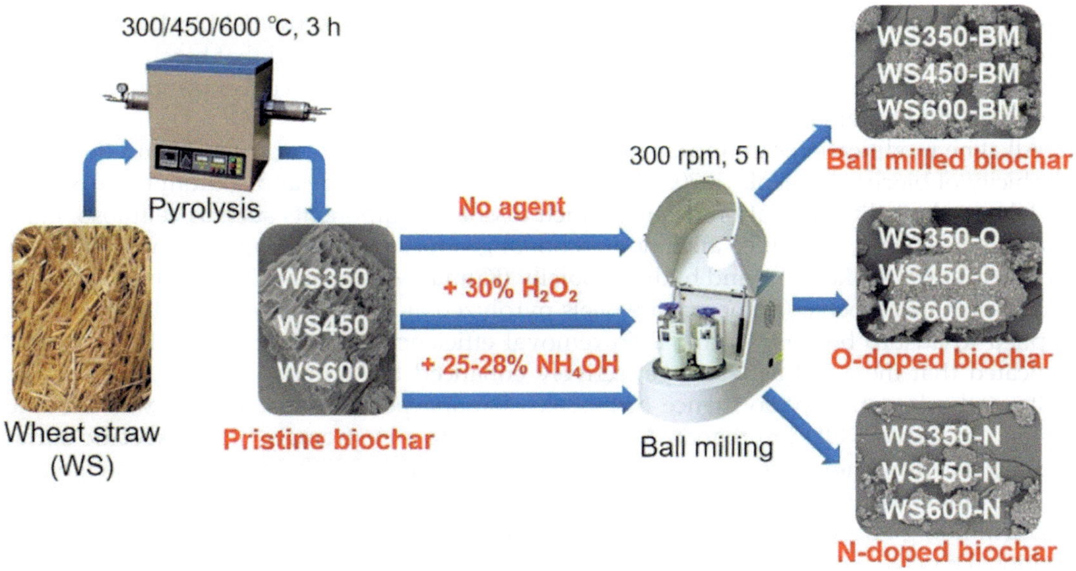

FIGURE 8.2 Ball milling biochar with chemicals (ammonia hydroxide and hydrogen peroxide) for surface functionalization (Qi et al., 2022).

(Wang and Wang, 2019). Hydrochloric acid, phosphoric acid, nitric acid, oxalic acid, etc. are commonly used acid modifiers, which act as both dehydrating agents and oxidants during the modification (Zhu et al., 2020). Among them, phosphoric acid is the most widely used agent for biochar acid modification. Yang et al. (2020) used phosphoric acid to modify waste bovine bone based biochar and obtained higher average pore width and new surface phosphate groups. The phosphoric acid modified biochar had unique organic-inorganic porous structure and surface phosphate groups. As a result, it showed accelerated toluene adsorption (Fig. 8.3). Sessa et al. (2022) found that the specific surface area of biochar obtained

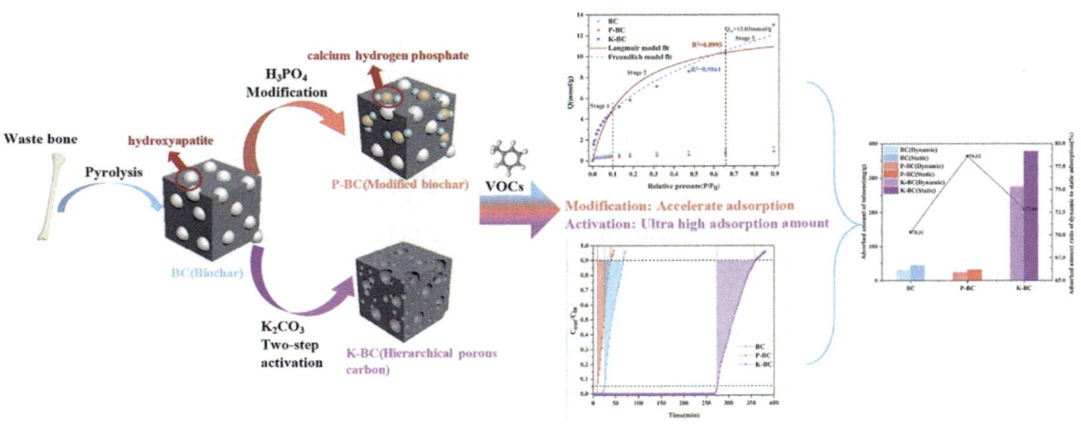

FIGURE 8.3 Acid modified waste bone biochar for VOC adsorption (Yang et al., 2020).

from H_3PO_4 impregnated pine bark was up to 3342 m^2/g. The activated biochar was highly effective for toluene adsorption, up to 5.5 mmol/g. It should be noted that biochar acid modification is a process influenced by multiple factors, including acid concentration, acid types, feedstock, and other conditions.

Alkaline modification can improve the surface alkalinity, porous structure and the hydrophobicity of biochar (Li et al., 2020a, 2020b). Potassium hydroxide and sodium hydroxide are widely used alkaline activation agents, which dissolve components such as cellulose, lignin, and ash in biochar, introduce alkaline functional groups, and thus improve the specific surface area, surface alkalinity, and hydrophobicity (Kumar Mishra et al., 2023; Leng et al., 2021). Fruit-originated biochar (pomegranate peels pyrolyzed at 550°C) was activated by KOH, and applied to adsorb benzene with a 99% removal efficiency (Kaikiti et al., 2021). The results indicated that the KOH activation effectively enhanced the porous structure, surface alkalinity and adsorption performance of biochar. Banana peel-derived biochar was modified via an optimized KOH-impregnated method, showing an extremely higher specific surface area (3746.5 m^2/g) and pore volume (2.5 cm^3/g) (Shen et al., 2020). Meanwhile, the modified biochar demonstrated high adsorption capacities for benzene (27.55 mmol/g) and toluene (23.82 mmol/g). Seven types of biowaste-derived biochar, including paper mill sludge, conventional biochar, composite biochar, gasified biochar from mixed feedstock, gasification biochar from single feedstock, activated biochar with KOH, and activated carbon, were evaluated for their removal of benzene (Khan et al., 2019). Biochar activated with KOH showed that highest benzene adsorption capacity (Fig. 8.4).

8.2.1.3 Biological modified biochar for VOC removal

Biological modification technology refers to the utilization of biological processes to improve biomass feedstock for engineered biochar production (Wang et al., 2017). Bacterial treatment, particularly anaerobic digestion or biofuel processes, of biomass feedstock has been proven to be an effective in producing 'biologically activated' biochar with enhanced properties (Inyang et al., 2010; Yao et al., 2015). The anaerobic digestion pre-treatment makes the obtained biochar have a larger specific surface area and better adsorption performance (Tang et al., 2019). Meanwhile, through microbial decomposition pretreatment, the porous structure of biomass can potentially be enriched and regulated to produce favorable carbonaceous VOC adsorbents with hierarchical porous structures. *Nelumbo nucifera* was decomposed by microbial strains (Phanerochaete chrysosporium, Trichoderma viride, and Aspergillus tubingensis), and the obtained feedstock materials were carbonized and modified by KOH (Zhang et al., 2019b). The resultant biologically modified biochar exhibited high specific surface areas and excellent toluene adsorption performances (maximum: 446 mg/g). In another study, bagasse was decomposed by Bacillus subtilis, mixed with phenolic resin and a low proportion of KOH, and then was pyrolyzed into biochar (Fig. 8.5), which showed excellent adsorption of toluene (635 mg/g) (Wang et al., 2023).

8.2.2 Key factors controlling VOC adsorption

There are several factors controlling the adsorption of VOCs on biochar, including the properties of biochar (specific surface area, pore volume, etc.) and VOCs (boiling point, polarity, etc.), and the environmental conditions (temperature, VOC concentration, etc.). The

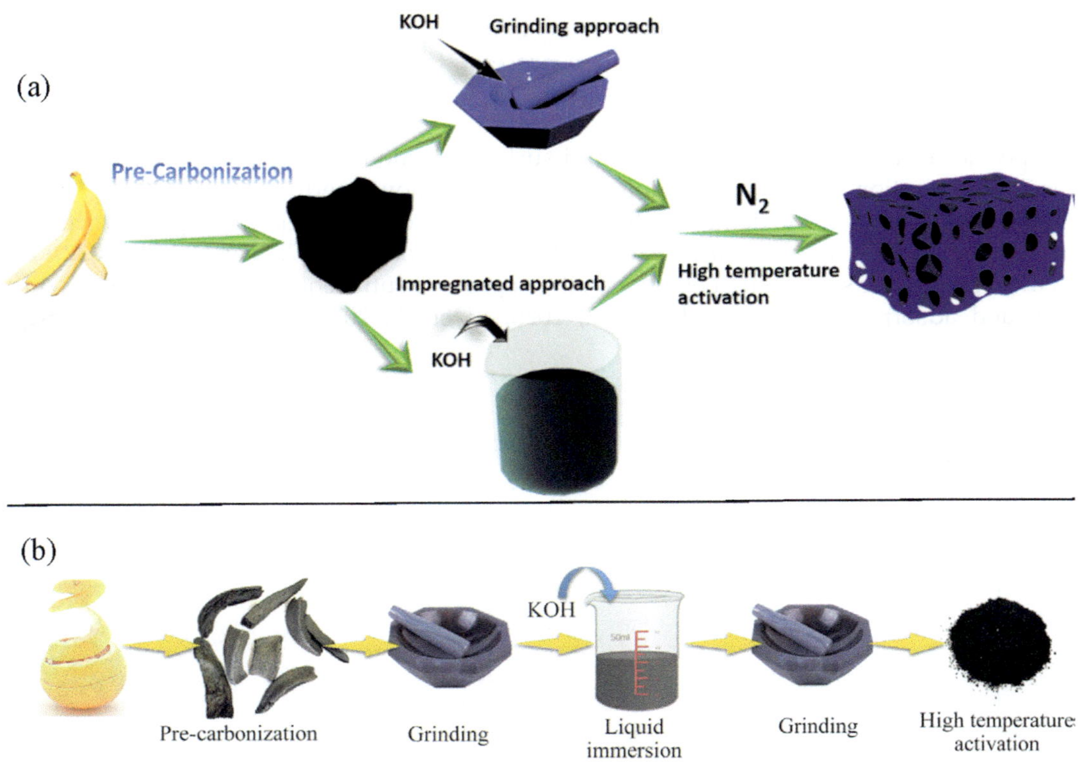

FIGURE 8.4 Alkaline modified biochar from (a) waste banana-peels and (b) shaddock peel (Shen et al., 2020; Yang et al., 2022).

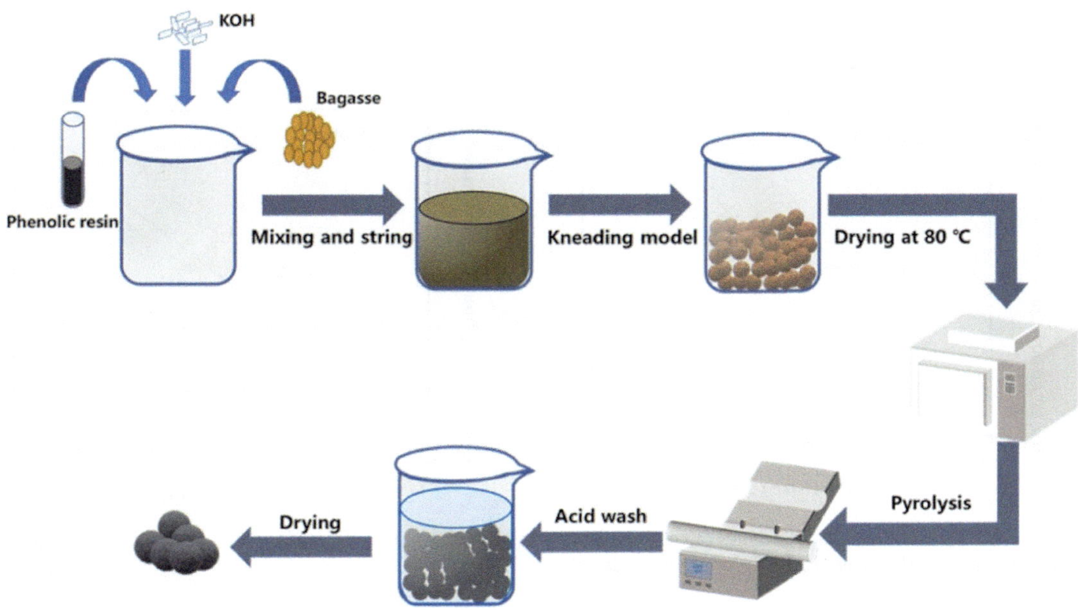

FIGURE 8.5 Production of biochar from bacterial pretreated bagasse (Wang et al., 2023).

comprehensive effect of various factors results in different affinity between biochar and VOC molecules, thereby affecting the adsorption process.

8.2.2.1 *Characteristics of biochar*

Specific surface area, pore structure, and surface functional groups are crucial factors for biochar adsorption, directly determining their VOC removal performance (Fig. 8.6).

Previous studies have demonstrated a good positive linear relationship between VOC adsorption and the specific surface area of biochar (Qi et al., 2022; Xiang et al., 2022). Due to the high specific surface area of biochar, it can provide sufficient active sites for VOC diffusion and adsorption (Dissanayake et al., 2020; Li et al., 2020a). Qi et al. compared the

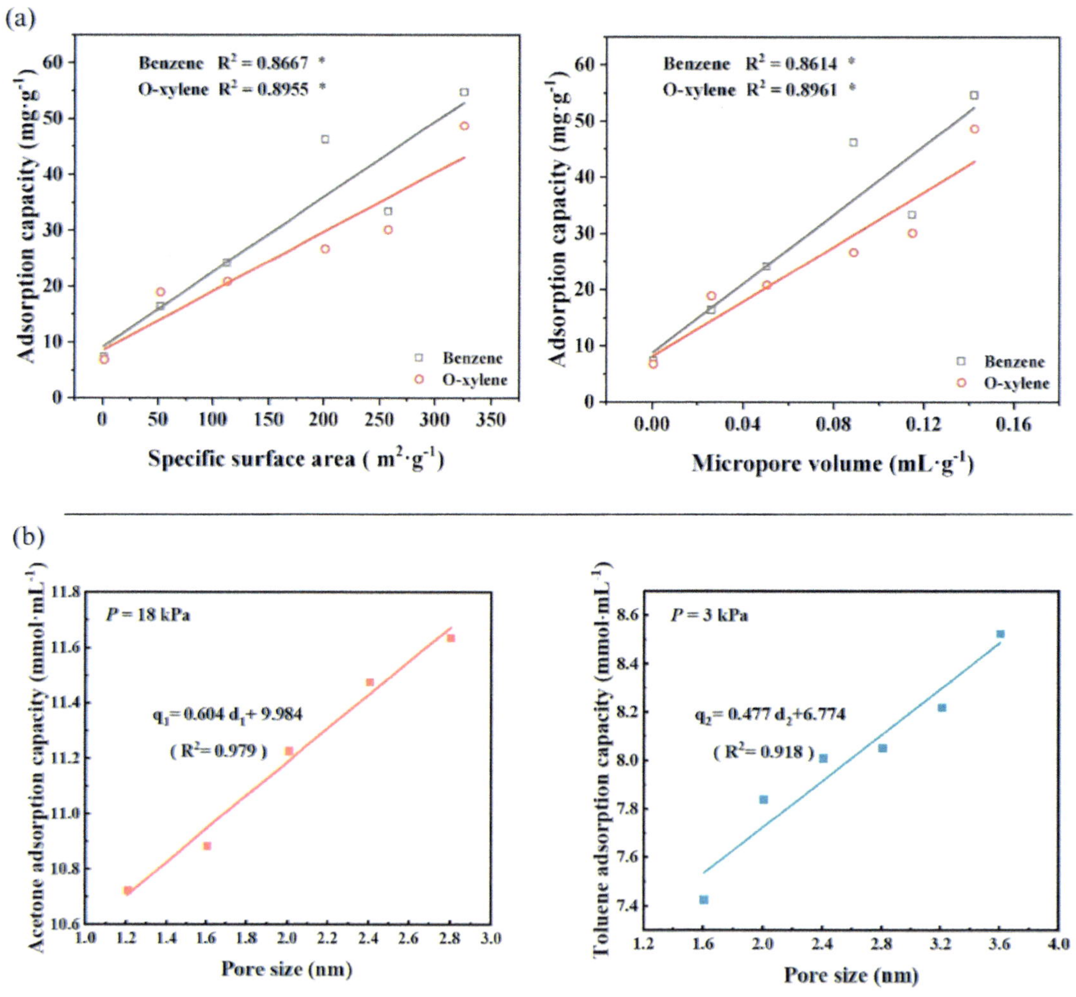

FIGURE 8.6 Biochar characteristics affect the adsorption of VOCs (a) specific surface area and micropore volume and (b) pore size (Jiang et al., 2022; Xiang et al., 2022).

adsorption of benzene, o-xylene, m-xylene, and p-xylene on 12 kinds of biochars, and higher VOC adsorption capacities were observed on the biochar with higher specific surface area (Qi et al., 2022). Opening blocked and creating new pores are two common strategies to increase the specific surface area of biochar. Xiang et al. reported that adsorption capacities of microwave biochar to benzene and o-xylene were positively related with specific surface area, total pore volume, and micropore volume (Xiang et al., 2022).

Pore structure of biochar, especially the distribution of pore size, also strongly affects its ability to adsorb VOCs. The pores can be dived into micropore (pore diameter <2 nm), mesopore (2 nm $<$ pore diameter <50 nm), and macropore (pore diameter >50 nm) (Zhang et al., 2017a). In general, micropores provide principal adsorption sites, while the mesopore and macropore enhance the intra-particle diffusion and shorten the adsorption time, and VOCs would be hardly adsorbed dynamically in mesopore and macropore (Shen, 2023; Wen et al., 2023). As mentioned above, there is a significant linear correlation between the microporous volume or specific surface area of biochar and the adsorption capacity of VOCs (Zhang et al., 2017b, 2021). Zhang et al. applied the fine-tuned micropore biochar to adsorb the benzene and acetone and found that the smaller size aperture of biochar dominated VOC adsorption (Zhang et al., 2022a). Jiang et al. found that the acetone molecules (dynamics diameter is 0.46 nm) were adsorbed not only on the pore wall but also on the adsorbed acetone molecular layer when the pore size was to 1.6–2.8 nm, forming an optimal multilayer adsorption strengthened by an interaction force among the acetone molecules (Jiang et al., 2022). However, this interaction was greatly weakened when the pore size was enlarged.

Surface chemical functional groups of biochar may also be responsible for their adsorption of VOCs, which are associated with both the feedstock characteristic and modification methods such as heating, chemical, and electrochemical treatments (Qiao et al., 2002; Su et al., 2007; Villacañas et al., 2006). The heteroatoms of surface functional groups govern their surface chemistry. They mainly include oxygen, nitrogen, halogen, hydrogen, etc., among which the oxygen and nitrogen groups are recognized as the most important species for VOC adsorption (Li et al., 2002; Shen et al., 2008). Most of the oxygen groups are the source of surface acidity, which contributes to the adhesive of hydrophilic VOCs onto biochar's surface (Li et al., 2011). Carbonyl and carboxyl groups of biochar can react with the hydrogen atoms of xylene methyl groups (mostly p-xylene due to higher polarity) via hydrogen bonding (Rajabi et al., 2021). As mentioned previously, oxidations by acids is among the most efficient methods to introduce surface oxygen groups to biochar. Alkali modification is an efficient method to replace the oxygen groups with alkaline functional groups, improving surface alkalinity and hydrophobicity of biochar. Ammonia hydroxide was used to modify ball-milled biochar, showing improved adsorption performance for VOCs (Qi et al., 2022; Zhang et al., 2021).

8.2.2.2 *Characteristics of VOCs*

Properties of VOC molecules, such as molecular structure, polarity, and boiling point, also play important roles in VOC adsorption onto biochar.

VOC molecular structure is generally measured by molecular dynamic diameter and molecular morphology, which affect their adsorption on biochar (Table 8.1). For example, the higher adsorption of acetone on biochar could be relevant to its smaller kinetic diameter (0.46 nm) than toluene (0.67 nm) and cyclohexane (0.60 nm) because VOCs with smaller

TABLE 8.1 Properties of representative VOCs.

VOCs	Molecular formula	3D model	Molecular weight ($g \cdot mol^{-1}$)	Boiling point (°C)	Dipole moment (D)	Kinetic diameter (nm) (Li et al., 2009)
Acetone	CH_3COCH_3	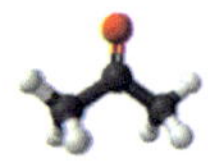	58.08	56.05	2.91	0.46
Benzene	C_6H_6		78.11	80.1	0	0.58
Ethanol	CH_3CH_2OH		46.07	78.24	1.69	0.45
Chloroform	$CHCl_3$		119.37	61.15	1.15	0.54
Cyclohexane	C_6H_{12}		84.16	80.74	0	0.60
Toluene	$C_6H_5CH_3$		92.14	110.60	0.36	0.67

molecule sizes are easy to access the adsorption sites (Xiang et al., 2020c; Zhang et al., 2017b). Benzene has higher diffusion rate than other xylene VOCs, this may be attributed to its small molecular size (0.58 nm), facilitating its diffusion into the micropores and access adsorption sites (Zhang et al., 2021). P-xylene presented better adsorption onto biochar than M-phenyl and O-xylene, attributed to the P-xylene's small kinetic diameter (0.58 nm) (Rajabi et al., 2021). Besides, molecular morphology of benzene and O-xylene also influences their adsorption onto the biochar (Fig. 8.7), especially on narrow micropores (<1 nm). O-xylene with two methyl groups (with a maximum distance of 1.07 nm) has shown lower adsorption than benzene onto biochar because it is difficulty for O-xylene to enter the micropores (Xiang et al., 2022; Yuh-Shan, 2004; Zhu et al., 2020).

The molecules polarity of VOCs is generally determined by dipole moments. For example, the dipole moments of CH_3COCH_3, $CHCl_3$, $C_6H_4(CH_3)_2$, $C_6H_5CH_3$, C_6H_6, and C_6H_{12} are 2.91, 1.15, 0.63, 0.36, 0 and 0 Debyes, respectively. Therefore, C_6H_6, and C_6H_{12} are nonpolar VOCs, while the others are polar ones. Generally, polar VOCs prefer to be adsorbed onto biochars containing rich polar groups, while nonpolar ones are more attractive to biochars without polar groups. The intermolecular forces of polar VOC molecules mainly include dipole-dipole, induced dipole-induced dipole, and dipole-induced dipole, while nonpolar VOC molecules only have induced dipole-induced dipole force (Wade and Simek, 2016). Based on this, polar

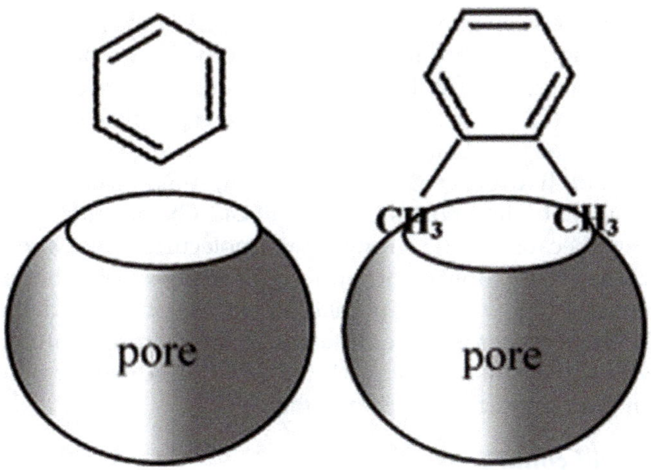

FIGURE 8.7 Morphology of benzene and o-xylene affects their pore entering dynamics (Li et al., 2020a).

acetone was more easily adsorbed than nonpolar cyclohexane onto a hydrochar with polar surface (Fig. 8.8) (Zhang et al., 2020). According to the total acidity of biochar, the surface functional groups increase its surface polarity, thus polar o-xylene is more likely to be adsorbed on biochar than nonpolar benzene (Xiang et al., 2022). The oxygenic functional groups are introduced and the polarity is enhanced during the ball milling of biochar. The adsorption of polar acetone, chloroform, and ethanol on ball-milled biochar can thus be promoted by forces of dipole-dipole and hydrogen bonds (Xiang et al., 2020c). Nonpolar C_6H_6 and CCl_4 show higher adsorption on biochar than other polar VOCs, which is attribute to their zero dipole moment matches well with the biochar (Bansode et al., 2003). In a word, VOCs prefer to be adsorbed onto biochars of the same polarity.

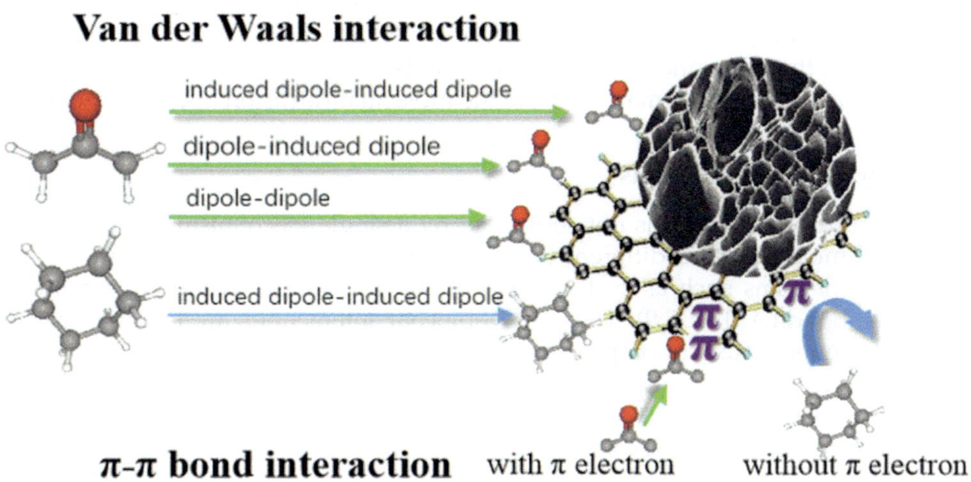

FIGURE 8.8 Adsorption mechanisms of acetone and cyclohexane onto activated hydrochar (Zhang et al., 2020).

The boiling point of VOCs can affect their intermolecular interaction with biochar (Table 8.1). VOCs with higher boiling points are more easily adsorbed onto biochar due to the stronger intermolecular forces (Zhang et al., 2019a). Therefore, ethanol, cyclohexane, and toluene with their higher boiling points (78.2°C, 80.7°C, and 110.6°C, respectively) showed stronger sorption onto ball-milled biochars (Xiang et al., 2020c). Cyclohexane with a boiling point of 80.7°C also had higher adsorption on hydrochar than lower boiling point acetone (56.5°C) (Zhang et al., 2019c). Xylene with higher boiling points (138.46−144.43°C) was more preferentially adsorbed than benzene (80.09°C) because of the stronger intermolecular forces between xylene and biochar (Qi et al., 2022). In general, the adsorption performance of high boiling point VOCs onto biochar is higher than that of low boiling point ones. On the other hand, the desorption of high boiling point VOCs from biochar is more difficult for their stronger affinity with biochar. For example, toluene with higher boiling points (110.6°C) required higher desorption temperature to be desorbed from biochar surface than the other VOCs.

8.2.2.3 *Environmental conditions*

Temperature is an important factor of the biochar adsorption of VOCs. The adsorption of VOCs by biochar mainly involves physical exothermic interactions, thus biochar's VOC adsorption capacity decreases with elevating temperature. For example, chloroform adsorption capacity of biochar reduced by 17%−38% with the adsorption temperature increased from 30°C to 80°C (Tsai et al., 2008). The adsorption of acetone by biochar dropped by 73.1%−98.2% as the adsorption temperature raised from 20°C to 40°C (Zhang et al., 2017b). Benzene uptake and xylenes adsorption by biochar also dropped with enhancement of temperature from 25°C to 45°C (Opatokun et al., 2017; Rajabi et al., 2021). On the other hand, elevating the adsorption temperature can promote the molecular diffusion to enhance the adsorption kinetic rate, facilitating VOCs to enter into the pores of biochar. Therefore, the adsorption of benzene onto the biochar increased slightly with the increase of temperature (Xiang et al., 2022). This may be attributed to the high adsorption rate constant and small kinetic diameter of benzene, which may accelerate the diffusion process to enter the biochar interior pores to increase its adsorption (Fletcher et al., 2006; Wang et al., 2015; Zhang et al., 2017a). In a word, the effect of temperature on VOCs adsorption onto biochar is complicated. Improving the adsorption temperature inhibits the physical adsorption of VOCs onto biochar in general; however, it may promote the molecular diffusion and adsorption kinetics.

8.2.3 Adsorption mechanisms

VOC adsorption on biochar is not controlled by a single mechanism but the interactions between VOC molecules and different biochar aspects, involving a combination of several adsorption mechanisms such as surface adsorption, pore filling, partition, π-π interaction, and hydrogen bonding (Fig. 8.9). These VOC-biochar interactions are largely through both physical and chemical adsorption processes (Rajapaksha et al., 2016).

Physical adsorption is a nonspecific adsorption process mainly controlled by intermolecular attraction, such as van der Waals force, π - π stacking, and hydrophobic interactions. However, these types of attractions between VOCs and biochar are relatively weak, so they do not involve any chemical reactions. Biochar can be easily regenerated and maintain its original structure. The specific surface area and pore structure of biochar are important

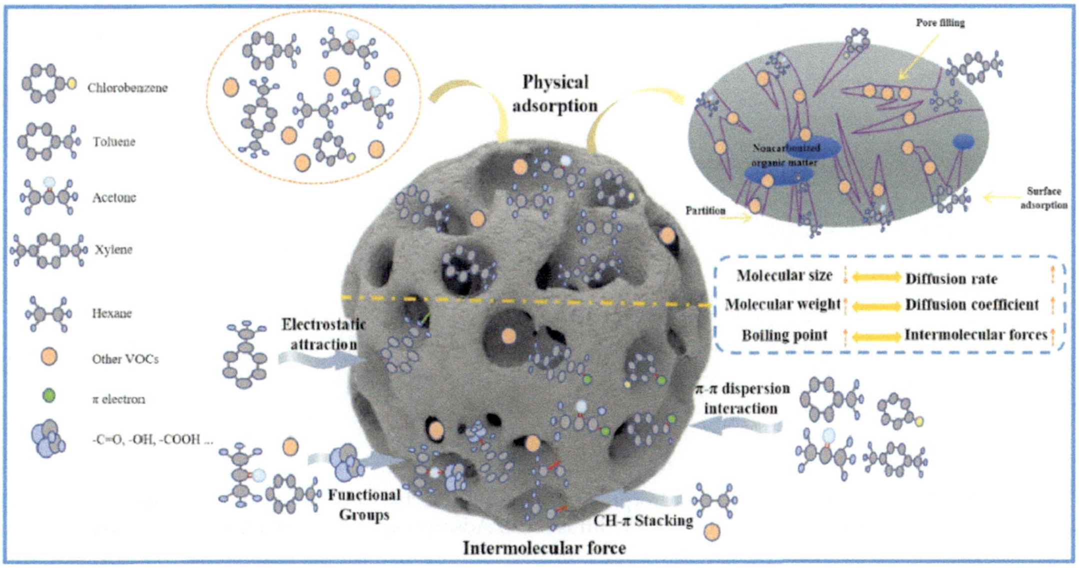

FIGURE 8.9 Summary of adsorption mechanisms of VOCs onto biochar (Wen et al., 2023).

during the physical adsorption process, which usually can be divided into three stages: (1) external surface adsorption, VOCs reach the external surface of biochar via convection and dispersion; (2) internal diffusion, VOC vapors enter into the internal surface of biochar through pore diffusion; and (3) equilibrium phase (Yang et al., 2018b). Micropores offer main adsorptive sites and dominate the physical adsorption of VOCs; macropores are exposed directly to the external surface of biochar; while mesopores are branches of the macropores to provide the transport channels for VOC molecules entering into the micropores (Fig. 8.10). Physical adsorption is a complicated process, which is affected by both biochar and VOC characteristics, as discussed previously.

Chemical adsorption usually refers to the chemical reaction between the special surface sites of adsorbent (biochar) and the adsorbent molecules (VOCs), which may involve electrophilic interactions, ion exchange, surface complexation, and so on (Premarathna et al., 2019). Meanwhile, chemical adsorption is usually irreversible owing to the strong combination of VOCs and biochar surface. Among the common surface sites, the oxygen and nitrogen containing surface functional groups are acknowledged as the most important ones for the chemical adsorption of VOCs. Common oxygen containing groups include carboxyl, hydroxyl, and carbonyl groups, which can promote the surface polarity of biochar to adsorb polar VOCs such as methanol, ethanol, and acetone by forming strong bonds with biochar (Zhu et al., 2020). Nitrogen containing groups on biochar are introduced by doping with ammonium, nitric acid and nitrogen-containing compounds, which can increase the pH value of biochar. Nitrogen functional groups can improve the active sites on the surface of biochar for chemical adsorption of VOCs (Gao et al., 2022; Xu et al., 2019).

Actually, physical adsorption and chemical adsorption may exist simultaneously during VOC adsorption onto biochar. Benzene adsorption on biochar is likely heterogeneous, which

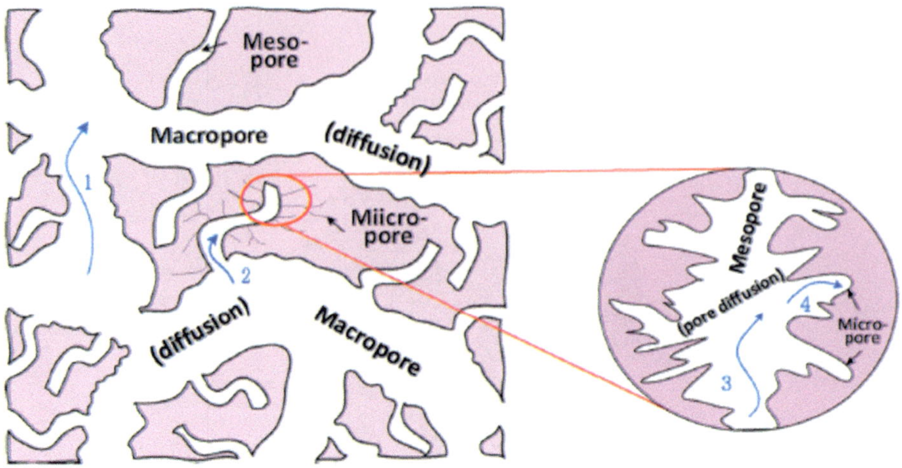

FIGURE 8.10 Illustration of physical adsorption processes of biochar (1. Convection and dispersion in gas; 2. Convective and dispersive mass transfer; 3. Pore diffusion; 4. Adsorption on surfaces.) (Zhu et al., 2020).

may be related to the surface interaction of electronic complexes formatted between carboxyl and benzene ring. However, o-xylene adsorption on biochar is more homogeneous and may be associated with both pore filling and further formed strong hydrogen bond (Rajabi et al., 2021; Xiang et al., 2022; Zhao et al., 2018). Neem-, sugarcane-, and bamboo-based biochars were applied to adsorb benzene, toluene, methyl chloride, xylene, chloroform and carbon tetrachloride (Kumar et al., 2020a). The results indicated that physical adsorption and partitioning were the main mechanisms for the effective removal of the VOCs by these biochars. Meanwhile, feedstock material, pyrolysis temperature, contact time, and specific surface area of the biochars affected their VOC removal efficiencies. Previous studies have shown that adsorption of acetone, ethanol, and chloroform on biochar is mainly controlled by the surface adsorption mechanism due to the high pore diffusion process; while the adsorption of cyclohexane and toluene is governed by both surface adsorption and partition mechanisms (Zhang et al., 2017b, 2019c). Banana peel derived biochar was modified and applied to adsorb benzene and toluene (Shen et al., 2020). The results showed that adsorption of toluene was mainly through π-π superposition with the methyl on toluene acting as an electron donor to strengthen π conjugation. In addition, the adsorption of benzene on the biochar was higher that determined by the pore volume at a higher relative pressure range. Straw-, waste wood-, and pig manure-based biochars showed different VOC adsorption mechanisms with chemical adsorption mechanism for straw and waste wood-based biochar and physical mechanism for pig manure-based biochar (Zhou et al., 2020). Soft wood pellet biochar prepared at 700°C showed excellent adsorption performance for benzene and the adsorption process was largely governed by physical adsorption in combination with surface control and boundary layer effects (Vikrant et al., 2020). In summary, the adsorption mechanism of biochar for VOCs is complex, may involve more than a single mechanism, and is often controlled by multiple factors.

8.3 Biochar catalysts for VOC removal

Based on the fact that the morphology and surface function of biochar can be modified by various physical and chemical treatments (as mentioned in section 8.2), many scholars believe that biochar can mimic traditional catalysts and has great catalysis potential for biodiesel production (İnan et al., 2023), syngas purification (Gaur et al., 2022), and pollution control (Cha et al., 2022). The physical and chemical properties of biochar, such as specific surface area, porosity, and surface chemistry, have important contributions to its application in catalysis. For example, biochar catalytic selectivity and mass transfer can be improved with appropriate pore size. A larger specific surface area of biochar can enable better dispersion of active substances, enhance molecular diffusion of reactants and products, and thus enhance it catalytic activity (Li et al., 2020b). The oxygen-containing or nitrogen-containing functional groups on biochar surface can make it alkaline, acidic, or neutral, thereby exhibiting significant catalytic performance. It has been recognized that biochar can be an effective catalytic material that can be used either directly as a catalyst or as a catalyst carrier to catalyze related reactions.

Biochar-based catalysts have been applied in chemical production, energy production, and material synthesis (Xiong et al., 2017; Yuan et al., 2023). Several previous articles have explored the potential application of biochar-based catalysts in environmental protection, such as degradation of organic contaminants in wastewater, remediation of emerging pollutants in soils, and purification of gaseous pollutants in air (Lee et al., 2019a; Yuan et al., 2023). However, few articles have evaluated the removal of VOCs by biochar-based catalysts, thus this section only briefly discusses biochar catalysts for VOC removal. Table 8.2 summaries some of the biochar-based catalysts that have been used for the removal of VOCs from gaseous or air streams.

8.3.1 Direct catalyst

Coked hardwood biochar as a catalyst was used to reform tar steam that was the mixture of benzene, toluene, and naphthalene (Buentello-Montoya et al., 2020). The removal efficiencies of toluene and naphthalene were more than 90%, while benzene was more than 50%. The porous structure of the hardwood biochar influenced the reform process with micropores contributing to the initially high catalytic reforming activity and mesopores providing surface active sites. Pristine and KOH-modified switchgrass biochar catalysts were synthesis and evaluated for the removal of toluene from biomass-generated producer gas (Bhandari et al., 2014). The results showed that both catalysts were effective in toluene removal with removal efficiency of 81% and 92% at 800°C, respectively. The KOH-modified biochar catalyst exhibited a higher toluene removal because of its higher specific surface area, larger pore diameter, and larger pore volume.

8.3.2 Catalyst carrier

KOH treated sawdust biochar was used as a carrier for MnOx catalyst to remove toluene from chemical industry emissions (Cha et al., 2022). Because biochar carrier enhanced its properties, such as specific surface area, porosity, redox property, (Oads + Ov)/OL ratio,

TABLE 8.2 Catalytic performances of some biochar-based catalysts toward the decomposition of VOCs.

Biochar	Feedstock	Treatment/ Modification	Impregnated catalyst	VOCs	Catalytic activities	Ref.
Direct catalyst						
Hardwood biochar	Hardwood	Physical activated	–	Toluene, naphthalene, and benzene	Removal efficiency of toluene and naphthalene were more than 90%, while benzene was more than 50%	Buentello-Montoya et al. (2020)
Switchgrass biochar	Switchgrass	Pristine	–	Toluene	Removal efficiency was 81%	Bhandari et al. (2014)
	Switchgrass	KOH modified	–	Toluene	Removal efficiency was 92%	
Pine bark biochar	Pine bark	–	–	Toluene	Removal efficiency was 94%	Mani et al. (2013)
Catalyst carrier						
MnOx/ Sawdust biochar	Sawdust	KOH modified	MnOx	Toluene	Removal efficiency was 89.7%	Cha et al. (2022)
Ar/Mallee wood biochar	Mallee wood	Ar activated	–	Toluene	Obviously efficiency, but not given the exact value	Liu et al. (2021)
Perovskite oxide/biochar nanocomposite	Pomegranate peel	Sol-gel method	$La(NO_3)_3 \cdot 6H_2O$ and $Co(NO_3)_3 \cdot 6H_2O$	Toluene	Obviously efficiency, but not given the exact value	Huang et al. (2024)
Fe/pine bark-biochar	Pine bark		$Fe\ (NO_3)_3 \cdot 9H_2O$	Toluene	Removal efficiency was 100%	Kastner et al. (2015)
Biochar/ MnO_2/g-C_3N_4	Cellulose fibers and melamine	Potassium permanganate and ethanol	–	Formaldehyde	Degradation efficiency was 91.78%	Li et al. (2022a)
CuMn/HBC	Pinecones	30% H_2O_2 treatment	12% CuMn	Formaldehyde	Removal efficiency was 89%	Yi et al. (2018)
C/CN-8	Melamine and cellulose fiber	Co-thermal treatment	8% g-C_3N_4	Formaldehyde	Removal efficiency was 84.6%	Li et al. (2019)
TiO2/SWP700	Soft-wood pellets	Ultrasound	TiO_2	Methanol	Removal efficiency was 90%	Paweł Lisowskia et al. (2017)

and Mn^{3+} portion, the catalyst exhibited a high toluene removal efficiency of 89.7%. Perovskite oxide/biochar nanocomposite was applied as an catalyst to photothermic oxidation of toluene and exhibited better catalytic performance than the bare catalyst (Huang et al., 2024). This phenomenon was attributed to three aspects: (1) biochar carrier inhibited the recombination of photogenerated carriers and provided additional active sites for VOCs; (2) large specific surface area and abundant functional groups of biochar enhanced toluene adsorption; and (3) the combined thermal-assisted effect of biochar and the enrichment of oxygen vacancies improved the oxygen mobility and the low-temperature reducibility of the catalyst.

The physical and chemical properties of biochar (e.g., composition, surface function, surface area, porosity, etc.) play an important role in improving the performance of biochar-based catalysts, which present a promising and effective way to reduce the pollution of VOCs. However, the application of biochar-based catalysts to mitigate VOCs is still in its infancy. More research efforts are thus needed to develop biochar-based catalytic applications for mitigating VOCs.

8.4 Other applications of biochar in VOC removal

Biochar has also been used for (1) the removal of VOCs during composting, (2) the adsorption of VOCs in water, and (3) combining with microorganisms and bacteria for VOC removal. This section will briefly discuss these three applications.

8.4.1 Biochar in composting for VOC removal

During composting, VOCs such as hydrocarbons, alcohols, aldehydes, etc. are generated and discharged into the air that may cause environmental risks and health problems. Biochar has been applied as an additive in the composting process to solve this problem. The VOCs generated in the composting process mainly include nitrogen-containing VOCs, oxygen-containing VOCs, and other types of VOCs (Tran et al., 2018). The addition of biochar helps to reduce the emissions of oxygen-containing and nitrogen-containing VOCs during composting (Nguyen et al., 2022). It can affect the internal aeration, temperature, humidity, microbial activity, and other conditions of compost. The high porosity of biochar can promote the aeration rate of compost, strengthen the internal gas exchange of compost, and slow down the production of VOCs. Meanwhile, the pore structure and surface acid functional groups of biochar can intercept toxic emissions of compost, thereby preventing the volatilization of toxic emissions from compost and reducing corresponding pollution (Sanchez-Monedero et al., 2018). The efficiency of biochar in reducing the VOC levels was investigated in a compost mixture containing 90% poultry manure and 10% straw (with 3% biochar addition) (Sánchez-Monedero et al., 2019). The results showed that nitrogen-containing VOCs and oxygen-containing VOCs (ketones, phenols and organic acids) were both significantly reduced during the thermophilic and mesophilic phase of compost with the biochar addition. It was attributed to the large surface area of biochar that lead to the high adsorption of VOCs. Chicken manure and straw mixture with 10% biochar were composted in the composting reactor, 17.55% of total VOCs were adsorbed by the biochar and the removal rate reached 53% following the biotrickling filtration (Li et al., 2022b). These results indicated that biochar can be an important addition to composting in reducing VOC emission. On the other hand, temperature and humidity of the composting process may also affect VOC adsorption on the surface of biochar (Hwang et al., 2018). In summary, the removing VOCs by biochar from composting is complex and controlled by multiple factors and mechanisms (Fig. 8.11).

8.4.2 Biochar for VOC removal in water

VOCs in the water environment are typically released without properly treatment from various commercial, industrial, and household practices, leading to their absorption/dissolution into surface water, groundwater, and sewage systems. This may impose a significant threat on the ecosystems and human health. Biochar presents a new approach to effectively and economically reduce excess VOCs discharged to surface and groundwater worldwide (Muir et al., 2021). The properties of biochar such as specific surface area, pore volume etc. could also affect VOC adsorption in aqueous phase. Biochar prepared from neem, sugarcane, and bamboo (all pyrolyzed at 350°C, 450°C, and 550°C) were used to adsorb six VOCs including benzene, toluene, methyl chloride, xylene, chloroform and carbon tetrachloride from water solutions (Kumar et al., 2020a). Among them, neem-based biochar pyrolyzed at 550°C exhibited the maximum adsorption capacity for toluene (65.5 mg/g). This phenomenon could be attribute to the higher surface area (43.9 mg/g), pore size (1.66 × 101 Å) and carbon content (78.1%) of neem-based biochar. Municipal solid waste biochar was applied to adsorb VOCs (toluene and m-xylene) from aqueous solution (Jayawardhana et al.,

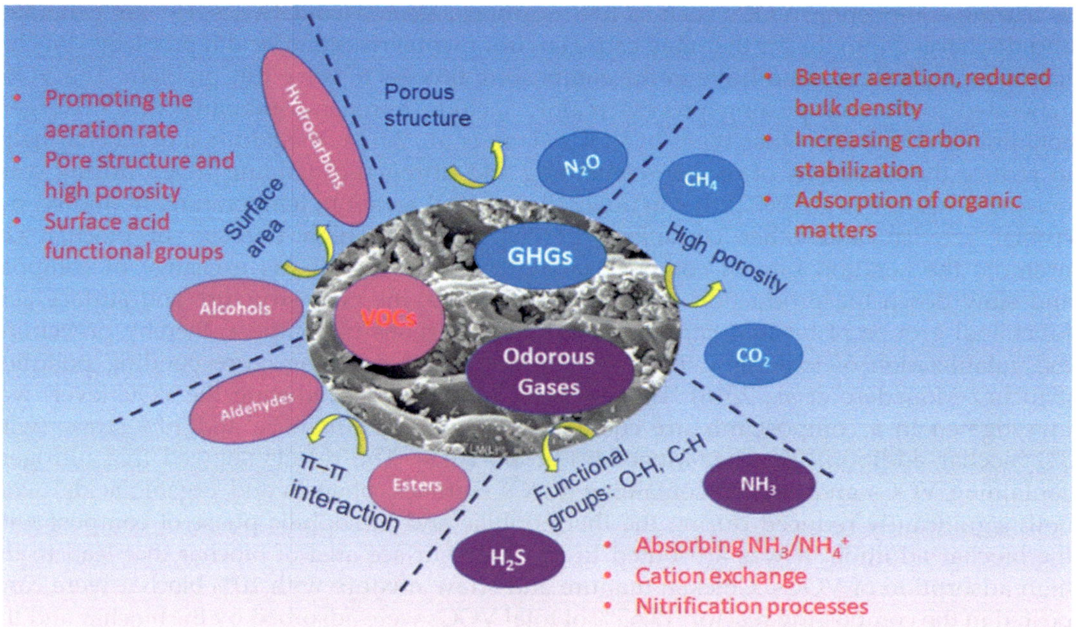

FIGURE 8.11 Effect of biochar on VOC emissions during composting (Nguyen et al., 2022).

2019). The Langmuir adsorption capacities of the biochar were 850 and 550 μg/g for toluene and m-xylene, respectively. The results suggested that the adsorption of toluene and m-xylene were both spontaneous and the adsorption mechanism was a favoring adsorption on a heterogeneous surface with corporative binding, which did not follow an ideal monolayer adsorption. Magnetic biochars were produced by chemical co-precipitation of Fe^{3+}/Fe^{2+} on orange peel powder and subsequently pyrolyzing under 400°C, which had the ability to efficiently remove naphthalene (18 mg/L) from water (Chen et al., 2011). The main mechanism of naphthalene adsorption on the biochars was contributed to the partition and micropores. In a word, biochar has a relatively high potential to adsorb some VOC molecules in the water environment, which is affected by its specific surface area, surface charge, and the ratio of micropores to mesoporous.

8.4.3 Biochar combined with bacteria for VOC removal

Bioremediation utilizes the ability of microorganisms to remove various organic pollutants. VOCs in water or soil can also be degraded by some specific microorganisms, such as *Pseudomonas*, *Rhodococcus*, *Ralstonia*, *Paraburkholderia*, *Rhodanobacter*, *Acinetobacter*, *Bacillus*, *Mycobacterium*, *Pseudoxanthomonas*, *Candida*, and *Penicillum*. These microorganisms are also proven biological resources that degrade benzene, toluene, ethylbenzene, xylene, chlorinated methanes, and ethenes (Lee et al., 2019b; Yoshikawa et al., 2017). Bioremediation discussed in this section is different from the biological modification summarized in section 8.2.1.3. Currently, limited research is available on VOC bioremediation using bacterial-loaded biochar. Chaudhary et al. (2023) used a newly designed diffusion bioreactor to enrich and isolate various VOC degrading bacteria, and then immobilized the most effective ones (*Pseudomonas* sp. DKR-23 and *Rhodococcus* sp. Korf-18) onto rice husk biochar (Fig. 8.12). The bioremediation performance of biochar-immobilized bacteria on benzene, toluene, xylene, and trichloroethane was evaluated by establishing batch microcosm experiments. The bacteria were successfully immobilized on biochar's surface, resulting in uneven surface of biochar and decreased surface

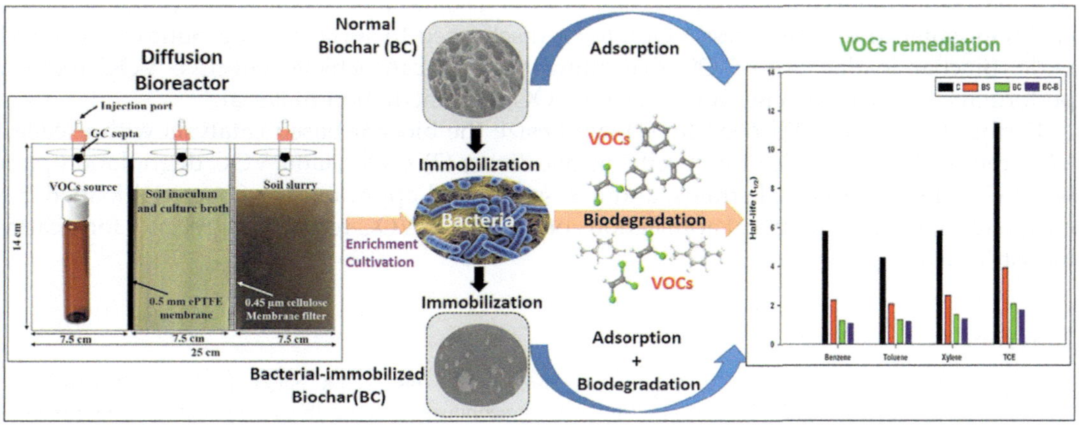

FIGURE 8.12 Enrichment cultivation of VOC-degrading bacteria using diffusion bioreactor and development of bacterial-loaded biochar for VOC bioremediation (Chaudhary et al., 2023).

area, total pore volume, and pore size compared to the pristine biochar. Even though, biochar-immobilized bacteria achieved the highest removal efficiencies (>90%) for benzene, toluene, xylene, and trichloroethane after 7 days of treatment. The results indicated that biochar is an effective support for bacteria in removing VOCs.

8.5 Summary and perspectives

Biochar is a carbon-based and porous material with unique physicochemical properties, which can be produced from various biomass materials including agricultural crop residues, forestry residues, sewage sludge, manures, and solid organic municipal wastes. Therefore, it has been used in VOC removal. This chapter reviews the utilization of engineered biochar modified by physical, chemical and biological technologies for VOC removal, as well as the related key factors and mechanisms. It summarizes the biochar-cased catalysts including direct and composite ones for VOC removal. Finally, it also briefly outlines the removal of VOCs by biochar in three other applications. The main conclusions of this chapter include: (1) Engineering biochar adsorbents can significantly enhance their adsorption performances for VOCs because the different physical, chemical and biological modification technologies can improve their physicochemical properties. (2) Properties of biochar (e.g., specific surface area, pore volume, etc.), properties of VOCs (e.g., boiling point, polarity, etc.), and the adsorption conditions (temperature, VOCs concentration etc.) control the affinity between biochar and VOC molecules, thereby affecting VOC adsorption process. (3) VOC adsorption on biochar involves several mechanisms such as surface adsorption, pore filling, partition, π-π interaction, and hydrogen bonding. (4) Biochar is an effective catalytic material, which can be used as a catalyst directly or a catalyst carrier to catalyze VOC removal. Its catalytic performance is related to its physicochemical properties. Appropriate pore size can improve catalytic selectivity and mass transfer, and large specific surface area can promote material diffusion and improve the catalytic activity of biochar. Surface functional groups can also make biochar exhibit remarkable catalytic performance. (5) Biochar as an additive reduce VOC emission of the composting process. Biochar also has a potential to neutralize VOC molecules in water to reduce excess VOCs to be discharged to surface and groundwater worldwide. Biochar loaded with VOC degrading bacteria can achieve effective VOC removal performance, which becomes a promising VOC bioremediation material.

Further studies are still need to: (1) synthesize the biochar-based catalysts with excellent VOC removal performance for specific applications; (2) understand VOC degradation pathways via effective bacterial strains, and assess the field applicability of bacterial loaded biochar, and (3) increase practical applications of biochar in VOC removal, especially for treating industrial waste gas.

References

Abhishek, K., Shrivastava, A., Vimal, V., Gupta, A.K., Bhujbal, S.K., Biswas, J.K., Singh, L., Ghosh, P., Pandey, A., Sharma, P., Kumar, M., 2022. Biochar application for greenhouse gas mitigation, contaminants immobilization and soil fertility enhancement: a state-of-the-art review. Science of the Total Environment 853, 158562.

Abuelnoor, N., AlHajaj, A., Khaleel, M., Vega, L.F., Abu-Zahra, M.R.M., 2021. Activated carbons from biomass-based sources for CO2 capture applications. Chemosphere 282, 131111.

Bansode, R.R., Losso, J.N., Marshall, W.E., Rao, R.M., Portier, R.J., 2003. Adsorption of volatile organic compounds by pecan shell- and almond shell-based granular activated carbons. Bioresource Technology 90 (2), 175–184.

Bhandari, P.N., Kumar, A., Bellmer, D.D., Huhnke, R.L., 2014. Synthesis and evaluation of biochar-derived catalysts for removal of toluene (model tar) from biomass-generated producer gas. Renewable Energy 66, 346–353.

Buentello-Montoya, D., Zhang, X., Li, J., Ranade, V., Marques, S., Geron, M., 2020. Performance of biochar as a catalyst for tar steam reforming: effect of the porous structure. Applied Energy 259, 114176.

Cao, L., Zhang, X., Xu, Y., Xiang, W., Wang, R., Ding, F., Hong, P., Gao, B., 2022. Straw and wood based biochar for CO2 capture: adsorption performance and governing mechanisms. Separation and Purification Technology 287, 120592.

Cha, J.S., Kim, Y.-M., Lee, I.H., Choi, Y.J., Rhee, G.H., Song, H., Jeon, B.-H., Lam, S.S., Khan, M.A., Andrew Lin, K.-Y., Chen, W.-H., Park, Y.-K., 2022. Mitigation of hazardous toluene via ozone-catalyzed oxidation using MnOx/Sawdust biochar catalyst. Environmental Pollution 312, 119920.

Chaudhary, D.K., Park, J.-H., Kim, P.-G., Ok, Y.S., Hong, Y., 2023. Enrichment cultivation of VOC-degrading bacteria using diffusion bioreactor and development of bacterial-immobilized biochar for VOC bioremediation. Environmental Pollution 320, 121089.

Chen, B., Chen, Z., Lv, S., 2011. A novel magnetic biochar efficiently sorbs organic pollutants and phosphate. Bioresource Technology 102 (2), 716–723.

Choudhury, A., Lansing, S., 2020. Biochar addition with Fe impregnation to reduce H2S production from anaerobic digestion. Bioresource Technology 306, 123121.

Creamer, A.E., Gao, B., 2016. Carbon-based adsorbents for postcombustion CO_2 capture: a critical review. Environmental Science and Technology 50 (14), 7276–7289.

Dissanayake, P.D., You, S., Igalavithana, A.D., Xia, Y., Bhatnagar, A., Gupta, S., Kua, H.W., Kim, S., Kwon, J.-H., Tsang, D.C.W., Ok, Y.S., 2020. Biochar-based adsorbents for carbon dioxide capture: a critical review. Renewable and Sustainable Energy Reviews 119, 109582.

Fletcher, A.J., Yüzak, Y., Thomas, K.M., 2006. Adsorption and desorption kinetics for hydrophilic and hydrophobic vapors on activated carbon. Carbon 44 (5), 989–1004.

Gao, W., Lin, Z., Chen, H., Yan, S., Zhu, H., Zhang, H., Sun, H., Zhang, S., Zhang, S., Wu, Y., 2022. Roles of graphitization degree and surface functional groups of N-doped activated biochar for phenol adsorption. Journal of Analytical and Applied Pyrolysis 167, 105700.

Gaur, V.K., Gautam, K., Sharma, P., Gupta, S., Pandey, A., You, S., Varjani, S., 2022. Carbon-based catalyst for environmental bioremediation and sustainability: updates and perspectives on techno-economics and life cycle assessment. Environmental Research 209, 112793.

Guo, Z., Huang, J., Xue, Z., Wang, X., 2016. Electrospun graphene oxide/carbon composite nanofibers with well-developed mesoporous structure and their adsorption performance for benzene and butanone. Chemical Engineering Journal 306, 99–106.

Gwenzi, W., Chaukura, N., Wenga, T., Mtisi, M., 2021. Biochars as media for air pollution control systems: contaminant removal, applications and future research directions. Science of the Total Environment 753, 142249.

Huang, J., Zhang, Y., Wu, M., Zuo, S., Yao, C., Ni, C., Li, X., 2024. Photothermal catalytic oxidation of toluene by perovskite oxide/biochar nanocomposite: effect of biomass incorporation. Separation and Purification Technology 330, 125316.

Hwang, O., Lee, S.-R., Cho, S., Ro, K.S., Spiehs, M., Woodbury, B., Silva, P.J., Han, D.-W., Choi, H., Kim, K.-Y., Jung, M.-W., 2018. Efficacy of different biochars in removing odorous volatile organic compounds (VOCs) emitted from swine manure. ACS Sustainable Chemistry and Engineering 6 (11), 14239–14247.

İnan, B., Koçer, A.T., Özçimen, D.B., 2023. Valorization of lignocellulosic wastes for low-cost and sustainable algal biodiesel production using biochar-based solid acid catalyst. Journal of Analytical and Applied Pyrolysis 173, 106095.

Inyang, M., Gao, B., Pullammanappallil, P., Ding, W., Zimmerman, A.R., 2010. Biochar from anaerobically digested sugarcane bagasse. Bioresource Technology 101 (22), 8868–8872.

Jayawardhana, Y., Gunatilake, S.R., Mahatantila, K., Ginige, M.P., Vithanage, M., 2019. Sorptive removal of toluene and m-xylene by municipal solid waste biochar: simultaneous municipal solid waste management and remediation of volatile organic compounds. Journal of Environmental Management 238, 323–330.

Jiang, Y., Xu, X., Liu, B., Zhou, C., Wang, H., Qiu, J., Zeng, Z., Ge, Y., Li, L., 2022. Optimal pore size design guided by GCMC molecular simulation for VOCs adsorption. Microporous and Mesoporous Materials 341, 112081.

Kaikiti, K., Stylianou, M., Agapiou, A., 2021. Development of food-origin biochars for the adsorption of selected volatile organic compounds (VOCs) for environmental matrices. Bioresource Technology 342, 125881.

Kastner, J.R., Mani, S., Juneja, A., 2015. Catalytic decomposition of tar using iron supported biochar. Fuel Processing Technology 130, 31–37.

Khan, A., Szulejko, J.E., Samaddar, P., Kim, K.-H., Liu, B., Maitlo, H.A., Yang, X., Ok, Y.S., 2019. The potential of biochar as sorptive media for removal of hazardous benzene in air. Chemical Engineering Journal 361, 1576–1585.

Kumar, A., Singh, E., Khapre, A., Bordoloi, N., Kumar, S., 2020a. Sorption of volatile organic compounds on non-activated biochar. Bioresource Technology 297, 122469.

Kumar, M., Xiong, X., Wan, Z., Sun, Y., Tsang, D.C.W., Gupta, J., Gao, B., Cao, X., Tang, J., Ok, Y.S., 2020b. Ball milling as a mechanochemical technology for fabrication of novel biochar nanomaterials. Bioresource Technology 312, 123613.

Kumar Mishra, R., Jaya Prasanna Kumar, D., Narula, A., Minnat Chistie, S., Ullhas Naik, S., 2023. Production and beneficial impact of biochar for environmental application: a review on types of feedstocks, chemical compositions, operating parameters, techno-economic study, and life cycle assessment. Fuel 343, 127968.

Lee, H.W., Lee, H., Kim, Y.-M., Park, R.-s., Park, Y.-K., 2019a. Recent application of biochar on the catalytic biorefinery and environmental processes. Chinese Chemical Letters 30 (12), 2147–2150.

Lee, Y., Lee, Y., Jeon, C.O., 2019b. Biodegradation of naphthalene, BTEX, and aliphatic hydrocarbons by Paraburkholderia aromaticivorans BN5 isolated from petroleum-contaminated soil. Scientific Reports 9 (1).

Leng, L., Xiong, Q., Yang, L., Li, H., Zhou, Y., Zhang, W., Jiang, S., Li, H., Huang, H., 2021. An overview on engineering the surface area and porosity of biochar. Science of the Total Environment 763, 144204.

Li, J.-R., Kuppler, R.J., Zhou, H.-C., 2009. Selective gas adsorption and separation in metal-organic frameworks. Chemical Society Reviews 38 (5), 1477–1504.

Li, L., Liu, S., Liu, J., 2011. Surface modification of coconut shell based activated carbon for the improvement of hydrophobic VOC removal. Journal of Hazardous Materials 192 (2), 683–690.

Li, X., Fang, G., Qian, X., Tian, Q., 2022a. Z-scheme heterojunction of low conduction band potential MnO2 and biochar-based g-C3N4 for efficient formaldehyde degradation. Chemical Engineering Journal 428, 131052.

Li, X., Qian, X., An, X., Huang, J., 2019. Preparation of a novel composite comprising biochar skeleton and "chrysanthemum" g-C3N4 for enhanced visible light photocatalytic degradation of formaldehyde. Applied Surface Science 487, 1262–1270.

Li, X., Zhang, L., Yang, Z., Wang, P., Yan, Y., Ran, J., 2020a. Adsorption materials for volatile organic compounds (VOCs) and the key factors for VOCs adsorption process: a review. Separation and Purification Technology 235, 116213.

Li, Y., Lee, C., Gullett, B., 2002. The effect of activated carbon surface moisture on low temperature mercury adsorption. Carbon 40 (1), 65–72.

Li, Y., Ma, J., Yong, X., Luo, L., Wong, J.W.C., Zhang, Y., Wu, H., Zhou, J., 2022b. Effect of biochar combined with a biotrickling filter on deodorization, nitrogen retention, and microbial community succession during chicken manure composting. Bioresource Technology 343, 126137.

Li, Y., Xing, B., Ding, Y., Han, X., Wang, S., 2020b. A critical review of the production and advanced utilization of biochar via selective pyrolysis of lignocellulosic biomass. Bioresource Technology 312, 123614.

Liu, Y., Paskevicius, M., Wang, H., Parkinson, G., Wei, J., Asif Akhtar, M., Li, C.-Z., 2021. Insights into the mechanism of tar reforming using biochar as a catalyst. Fuel 296, 120672.

Lyu, H., Gao, B., He, F., Zimmerman, A.R., Ding, C., Tang, J., Crittenden, J.C., 2018a. Experimental and modeling investigations of ball-milled biochar for the removal of aqueous methylene blue. Chemical Engineering Journal 335, 110–119.

Lyu, H.H., Gao, B., He, F., Ding, C., Tang, J.C., Crittenden, J.C., 2017. Ball-milled carbon nanomaterials for energy and environmental applications. ACS Sustainable Chemistry & Engineering 5 (11), 9568–9585.

Lyu, H.H., Gao, B., He, F., Zimmerman, A.R., Ding, C., Huang, H., Tang, J.C., 2018b. Effects of ball milling on the physicochemical and sorptive properties of biochar: experimental observations and governing mechanisms. Environmental Pollution 233, 54–63.

Majumder, S., Sharma, P., Singh, S.P., Nadda, A.K., Sahoo, P.K., Xia, C., Sharma, S., Ganguly, R., Lam, S.S., Kim, K.H., 2023. Engineered biochar for the effective sorption and remediation of emerging pollutants in the environment. Journal of Environmental Chemical Engineering 11 (2), 109590.

Mani, S., Kastner, J.R., Juneja, A., 2013. Catalytic decomposition of toluene using a biomass derived catalyst. Fuel Processing Technology 114, 118–125.

Muir, B., Sobczyk, M., Bajda, T., 2021. Fundamental features of mesoporous functional materials influencing the efficiency of removal of VOCs from aqueous systems: a review. Science of the Total Environment 784, 147121.

Nguyen, M.K., Lin, C., Hoang, H.G., Sanderson, P., Dang, B.T., Bui, X.T., Nguyen, N.S.H., Vo, D.-V.N., Tran, H.T., 2022. Evaluate the role of biochar during the organic waste composting process: a critical review. Chemosphere 299, 134488.

Opatokun, S.A., Prabhu, A., Al Shoaibi, A., Srinivasakannan, C., Strezov, V., 2017. Food wastes derived adsorbents for carbon dioxide and benzene gas sorption. Chemosphere 168, 326–332.

Paweł Lisowskia, J.C.C., Mašekb, O., Lisowskia, W., Lisovytskiya, D., Kamińskaa, A., Łomota, D., 2017. Dual functionality of TiO2/biochar hybrid materials: photocatalytic phenol degradation in liquid phase and selective oxidation of methanol in gas phase. ACS Sustainable Chemistry and Engineering 5, 6274–6287.

Premarathna, K.S.D., Rajapaksha, A.U., Sarkar, B., Kwon, E.E., Bhatnagar, A., Ok, Y.S., Vithanage, M., 2019. Biochar-based engineered composites for sorptive decontamination of water: a review. Chemical Engineering Journal 372, 536–550.

Qi, G., Pan, Z., Zhang, X., Miao, X., Xiang, W., Gao, B., 2022. Effect of ball milling with hydrogen peroxide or ammonia hydroxide on sorption performance of volatile organic compounds by biochar from different pyrolysis temperatures. Chemical Engineering Journal 450, 138027.

Qiao, W., Korai, Y., Mochida, I., Hori, Y., Maeda, T., 2002. Preparation of an activated carbon artifact: oxidative modification of coconut shell-based carbon to improve the strength. Carbon 40 (3), 351–358.

Qiu, W., Dou, K., Zhou, Y., Huang, H., Chen, Y., Lu, H., 2018. Hierarchical pore structure of activated carbon fabricated by CO 2/microwave for volatile organic compounds adsorption. Chinese Journal of Chemical Engineering 26 (1), 81–88.

Rajabi, H., Mosleh, M.H., Mandal, P., Lea-Langton, A., Sedighi, M., 2021. Sorption behaviour of xylene isomers on biochar from a range of feedstock. Chemosphere 268, 129310.

Rajapaksha, A.U., Chen, S.S., Tsang, D.C.W., Zhang, M., Vithanage, M., Mandal, S., Gao, B., Bolan, N.S., Ok, Y.S., 2016. Engineered/designer biochar for contaminant removal/immobilization from soil and water: potential and implication of biochar modification. Chemosphere 148, 276–291.

Sanchez-Monedero, M.A., Cayuela, M.L., Roig, A., Jindo, K., Mondini, C., Bolan, N., 2018. Role of biochar as an additive in organic waste composting. Bioresource Technology 247, 1155–1164.

Sánchez-Monedero, M.A., Sánchez-García, M., Alburquerque, J.A., Cayuela, M.L., 2019. Biochar reduces volatile organic compounds generated during chicken manure composting. Bioresource Technology 288, 121584.

Sessa, F., Merlin, G., Canu, P., 2022. Pine bark valorization by activated carbons production to be used as VOCs adsorbents. Fuel 318, 123346.

Shen, W., Li, Z., Liu, Y., 2008. Surface chemical functional groups modification of porous carbon. Recent Patents on Chemical Engineering 1 (1), 27–40.

Shen, X., Ou, R., Lu, Y., Yuan, A., Liu, J., Gu, J., Hu, X., Yang, Z., Yang, F., 2020. Record-high capture of volatile benzene and toluene enabled by activator implant-optimized banana peel-derived engineering carbonaceous adsorbents. Environment International 143, 105774.

Shen, Y., 2023. Biomass-derived porous carbons for sorption of Volatile organic compounds (VOCs). Fuel 336, 126801.

Su, W., Zhou, Y.-p., Wei, L.-f., Sun, Y., Zhou, L., 2007. Effect of microstructure and surface modification on the hydrogen adsorption capacity of active carbons. New Carbon Materials 22 (2), 135–140.

Tang, Y., Alam, M.S., Konhauser, K.O., Alessi, D.S., Xu, S., Tian, W., Liu, Y., 2019. Influence of pyrolysis temperature on production of digested sludge biochar and its application for ammonium removal from municipal wastewater. Journal of Cleaner Production 209, 927–936.

Tran, H.-T., Vu, C.-T., Lin, C., Bui, X.-T., Huang, W.-Y., Vo, T.-D.-H., Hoang, H.-G., Liu, W.-Y., 2018. Remediation of highly fuel oil-contaminated soil by food waste composting and its volatile organic compound (VOC) emission. Bioresource Technology Reports 4, 145–152.

Tsai, J.-H., Chiang, H.-M., Huang, G.-Y., Chiang, H.-L., 2008. Adsorption characteristics of acetone, chloroform and acetonitrile on sludge-derived adsorbent, commercial granular activated carbon and activated carbon fibers. Journal of Hazardous Materials 154 (1–3), 1183–1191.

Vikrant, K., Kim, K.-H., Peng, W., Ge, S., Sik Ok, Y., 2020. Adsorption performance of standard biochar materials against volatile organic compounds in air: a case study using benzene and methyl ethyl ketone. Chemical Engineering Journal 387, 123943.

Villacañas, F., Pereira, M.F.R., Órfão, J.J.M., Figueiredo, J.L., 2006. Adsorption of simple aromatic compounds on activated carbons. Journal of Colloid and Interface Science 293 (1), 128–136.

Wade, L.G., Simek, J.W., 2016. Organic Chemistry, ninth ed. Pearson Education.

Wan, S., Wu, J., Zhou, S., Wang, R., Gao, B., He, F., 2018. Enhanced lead and cadmium removal using biochar-supported hydrated manganese oxide (HMO) nanoparticles: behavior and mechanism. Science of the Total Environment 616–617, 1298–1306.

Wang, B., Gao, B., Fang, J., 2017. Recent advances in engineered biochar productions and applications. Critical Reviews in Environmental Science and Technology 47 (22), 2158–2207.

Wang, G., Dou, B., Zhang, Z., Wang, J., Liu, H., Hao, Z., 2015. Adsorption of benzene, cyclohexane and hexane on ordered mesoporous carbon. Journal of Environmental Sciences 30, 65–73.

Wang, J., Wang, S., 2019. Preparation, modification and environmental application of biochar: a review. Journal of Cleaner Production 227, 1002–1022.

Wang, Y., Liu, R., 2018. H2O2 treatment enhanced the heavy metals removal by manure biochar in aqueous solutions. Science of the Total Environment 628–629, 1139–1148.

Wang, Y., Zhu, W., Zhao, G., Ye, G., Jiao, Y., Wang, X., Yao, F., Peng, W., Huang, H., Ye, D., 2023. Precise preparation of biomass-based porous carbon with pore structure-dependent VOCs adsorption/desorption performance by bacterial pretreatment and its forming process. Environmental Pollution 322, 121134.

Wen, C., Liu, T., Wang, D., Wang, Y., Chen, H., Luo, G., Zhou, Z., Li, C., Xu, M., 2023. Biochar as the effective adsorbent to combustion gaseous pollutants: preparation, activation, functionalization and the adsorption mechanisms. Progress in Energy and Combustion Science 99, 101098.

Xiang, W., Wan, Y., Zhang, X., Tan, Z., Xia, T., Zheng, Y., Gao, B., 2020a. Adsorption of tetracycline hydrochloride onto ball-milled biochar: governing factors and mechanisms. Chemosphere 255, 127057.

Xiang, W., Zhang, X., Cao, C., Quan, G., Wang, M., Zimmerman, A.R., Gao, B., 2022. Microwave-assisted pyrolysis derived biochar for volatile organic compounds treatment: characteristics and adsorption performance. Bioresource Technology 355, 127274.

Xiang, W., Zhang, X., Chen, J., Zou, W., He, F., Hu, X., Tsang, D.C.W., Ok, Y.S., Gao, B., 2020b. Biochar technology in wastewater treatment: a critical review. Chemosphere 252, 126539.

Xiang, W., Zhang, X., Chen, K., Fang, J., He, F., Hu, X., Tsang, D.C.W., Ok, Y.S., Gao, B., 2020c. Enhanced adsorption performance and governing mechanisms of ball-milled biochar for the removal of volatile organic compounds (VOCs). Chemical Engineering Journal 385, 123842.

Xie, Y., Wang, L., Li, H., Westholm, L.J., Carvalho, L., Thorin, E., Yu, Z., Yu, X., Skreiberg, Ø., 2022. A critical review on production, modification and utilization of biochar. Journal of Analytical and Applied Pyrolysis 161, 105405.

Xiong, X., Yu, I.K.M., Cao, L., Tsang, D.C.W., Zhang, S., Ok, Y.S., 2017. A review of biochar-based catalysts for chemical synthesis, biofuel production, and pollution control. Bioresource Technology 246, 254–270.

Xiong, X., Yu, I.K.M., Tsang, D.C.W., Bolan, N.S., Sik Ok, Y., Igalavithana, A.D., Kirkham, M.B., Kim, K.-H., Vikrant, K., 2019. Value-added chemicals from food supply chain wastes: state-of-the-art review and future prospects. Chemical Engineering Journal 375, 121983.

Xu, S., Chen, J., Peng, H., Leng, S., Li, H., Qu, W., Hu, Y., Li, H., Jiang, S., Zhou, W., Leng, L., 2021. Effect of biomass type and pyrolysis temperature on nitrogen in biochar, and the comparison with hydrochar. Fuel 291, 120128.

Xu, X., Zheng, Y., Gao, B., Cao, X., 2019. N-doped biochar synthesized by a facile ball-milling method for enhanced sorption of CO2 and reactive red. Chemical Engineering Journal 368, 564–572.

Yaashikaa, P.R., Senthil Kumar, P., Varjani, S.J., Saravanan, A., 2019. Advances in production and application of biochar from lignocellulosic feedstocks for remediation of environmental pollutants. Bioresource Technology 292, 122030.

Yang, F., Li, W., Ou, R., Lu, Y., Dong, X., Tu, W., Zhu, W., Wang, X., Li, L., Yuan, A., Pan, J., 2022. Superb VOCs capture engineering carbon adsorbent derived from shaddock peel owning uncompromising thermal-stability and adsorption property. Chinese Journal of Chemical Engineering 47, 120–133.

Yang, F., Xu, Z.B., Yu, L., Gao, B., Xu, X.Y., Zhao, L., Cao, X.D., 2018a. Kaolinite enhances the stability of the dissolvable and undissolvable fractions of biochar via different mechanisms. Environmental Science and Technology 52 (15), 8321–8329.

Yang, X., Yi, H., Tang, X., Zhao, S., Yang, Z., Ma, Y., Feng, T., Cui, X., 2018b. Behaviors and kinetics of toluene adsorption-desorption on activated carbons with varying pore structure. Journal of Environmental Sciences 67, 104–114.

Yang, X.D., Wan, Y.S., Zheng, Y.L., He, F., Yu, Z.B., Huang, J., Wang, H.L., Ok, Y.S., Jiang, Y.S., Gao, B., 2019. Surface functional groups of carbon-based adsorbents and their roles in the removal of heavy metals from aqueous solutions: a critical review. Chemical Engineering Journal 366, 608–621.

Yang, Y., Sun, C., Lin, B., Huang, Q., 2020. Surface modified and activated waste bone char for rapid and efficient VOCs adsorption. Chemosphere 256, 127054.

Yao, Y., Gao, B., Wu, F., Zhang, C., Yang, L., 2015. Engineered biochar from biofuel residue: characterization and its silver removal potential. ACS Applied Materials and Interfaces 7 (19), 10634–10640.

Yi, Y., Li, C., Zhao, L., Du, X., Gao, L., Chen, J., Zhai, Y., Zeng, G., 2018. The synthetic evaluation of CuO-MnOx-modified pinecone biochar for simultaneous removal formaldehyde and elemental mercury from simulated flue gas. Environmental Science and Pollution Research International 25 (5), 4761–4775.

Yoo, J.C., Beiyuan, J., Wang, L., Dcw, T., Baek, K., Bolan, N.S., Ok, Y.S., Li, X.D., 2018. A combination of ferric nitrate/EDDS-enhanced washing and sludge-derived biochar stabilization of metal-contaminated soils. Science of the Total Environment 616.

Yoshikawa, M., Zhang, M., Toyota, K., 2017. Biodegradation of volatile organic compounds and their effects on biodegradability under Co-existing conditions. Microbes and Environments 32 (3), 188–200.

Yuan, X., Cao, Y., Li, J., Patel, A.K., Dong, C.-D., Jin, X., Gu, C., Yip, A.C.K., Tsang, D.C.W., Ok, Y.S., 2023. Recent advancements and challenges in emerging applications of biochar-based catalysts. Biotechnology Advances 67, 108181.

Yuh-Shan, H., 2004. Citation review of Lagergren kinetic rate equation on adsorption reactions. Scientometrics 59 (1), 171–177.

Zhang, G., Liu, Y., Zheng, S., Hashisho, Z., 2019a. Adsorption of volatile organic compounds onto natural porous minerals. Journal of Hazardous Materials 364, 317–324.

Zhang, W., Cheng, H., Niu, Q., Fu, M., Huang, H., Ye, D., 2019b. Microbial targeted degradation pretreatment: a novel approach to preparation of activated carbon with specific hierarchical porous structures, high surface areas, and satisfactory toluene adsorption performance. Environmental Science and Technology 53 (13), 7632–7640.

Zhang, X., Cao, L., Xiang, W., Xu, Y., Gao, B., 2022a. Preparation and evaluation of fine-tuned micropore biochar by lignin impregnation for CO2 and VOCs adsorption. Separation and Purification Technology 295, 121295.

Zhang, X., Gao, B., Creamer, A.E., Cao, C., Li, Y., 2017a. Adsorption of VOCs onto engineered carbon materials: a review. Journal of Hazardous Materials 338, 102–123.

Zhang, X., Gao, B., Fang, J., Zou, W., Dong, L., Cao, C., Zhang, J., Li, Y., Wang, H., 2019c. Chemically activated hydrochar as an effective adsorbent for volatile organic compounds (VOCs). Chemosphere 218, 680–686.

Zhang, X., Gao, B., Zheng, Y., Hu, X., Creamer, A.E., Annable, M.D., Li, Y., 2017b. Biochar for volatile organic compound (VOC) removal: sorption performance and governing mechanisms. Bioresource Technology 245 (Pt A), 606–614.

Zhang, X., Miao, X., Xiang, W., Zhang, J., Cao, C., Wang, H., Hu, X., Gao, B., 2021. Ball milling biochar with ammonia hydroxide or hydrogen peroxide enhances its adsorption of phenyl volatile organic compounds (VOCs). Journal of Hazardous Materials 403, 123540.

Zhang, X., Xiang, W., Miao, X., Li, F., Qi, G., Cao, C., Ma, X., Chen, S., Zimmerman, A.R., Gao, B., 2022b. Microwave biochars produced with activated carbon catalyst: characterization and sorption of volatile organic compounds (VOCs). Science of the Total Environment 153996.

Zhang, X., Xiang, W., Wang, B., Fang, J., Zou, W., He, F., Li, Y., Tsang, D.C.W., Ok, Y.S., Gao, B., 2020. Adsorption of acetone and cyclohexane onto CO2 activated hydrochars. Chemosphere 245, 125664.

Zhao, X., Zeng, X., Qin, Y., Li, X., Zhu, T., Tang, X., 2018. An experimental and theoretical study of the adsorption removal of toluene and chlorobenzene on coconut shell derived carbon. Chemosphere 206 (SEP), 285.

Zheng, Y., Wang, B., Wester, A.E., Chen, J., He, F., Chen, H., Gao, B., 2019a. Reclaiming phosphorus from secondary treated municipal wastewater with engineered biochar. Chemical Engineering Journal 362, 460–468.

Zheng, Y., Yang, Y., Zhang, Y., Zou, W., Luo, Y., Dong, L., Gao, B., 2019b. Facile one-step synthesis of graphitic carbon nitride-modified biochar for the removal of reactive red 120 through adsorption and photocatalytic degradation. Biochar 1, 89–96.

Zhou, X., Moghaddam, T.B., Chen, M., Wu, S., Adhikari, S., 2020. Biochar removes volatile organic compounds generated from asphalt. Science of the Total Environment 745, 141096.

Zhu, L., Shen, D., Luo, K.H., 2020. A critical review on VOCs adsorption by different porous materials: species, mechanisms and modification methods. Journal of Hazardous Materials 389, 122102.

Zhuang, Z., Wang, L., Tang, J., 2021. Efficient removal of volatile organic compound by ball-milled biochars from different preparing conditions. Journal of Hazardous Materials 406, 124676.

CHAPTER

9

Carbon nanomaterials (CNMs) for VOC removal

Zhifei Pan

School of Environmental Engineering, Xuzhou University of Technology, Xuzhou, China

9.1 Overview of CNMs for VOC removal

VOCs represent an important class of chemical pollutants in the environment with high volatility. They have attracted public and academic attentions due to their existence in the atmosphere can cause severe air pollution and health problems (Shih and Li, 2008). Among various treatment technologies, adsorption and catalytic oxidation are two types of promising technologies for VOC removal (Yang et al., 2019). The choice of removal agents (adsorbent or catalyst) is the key factor affecting the efficiency of adsorption and catalytic oxidation process.

Carbon nanomaterials (CNMs) including carbon nanofibers, graphene, carbon nanotubes (CNTs), fullerene, and carbon black nanoparticles (Fig. 9.1) have been utilized in various environmental applications including VOC removal during the last 20 years (Gabal et al., 2019; Pandey et al., 2020). For example, graphene and CNTs are considered as excellent adsorbents for VOCs as their ultra-high theoretical specific surface area and excellent mechanical strength (Kumar et al., 2020). Graphene-based adsorbents are commonly in forms of graphene oxide (GO) and reduced graphene oxide (rGO), while CNTs are usually divided into single-walled carbon nanotubes (SWCNTs) and multi-walled carbon nanotubes (MWCNTs). All these CNMs have been recognized as new generation adsorbents for organic compounds. In addition, graphene and CNTs also exhibit good catalytic efficiency mainly attributed to their enormous specific surface area and high electronic mobility (Kottappara et al., 2019; Lee et al., 2008). Moreover, from the viewpoint of fundamental science, adsorption is the essential and the first-step of catalytic oxidation, followed by catalytic transformation and desorption, thus adsorption and catalytic oxidation are closely correlated (Yang et al., 2019; Zou et al., 2019).

In this chapter, recent literature about utilization of graphene and its derivatives (GO and rGO), and CNTs in VOC adsorption and catalytic oxidation are summarized. Special

https://doi.org/10.1016/B978-0-443-15674-8.00013-5

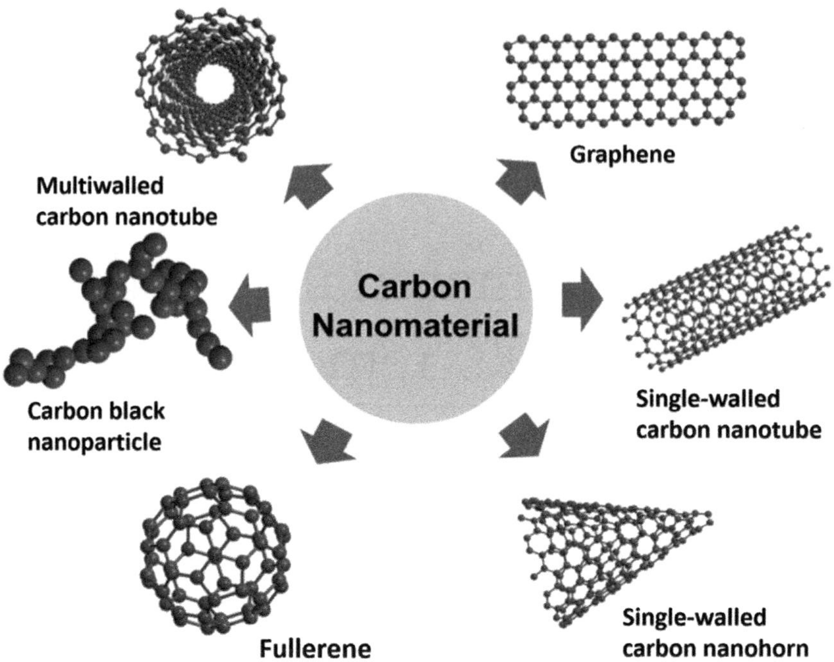

FIGURE 9.1 Examples of commonly used CNMs.

attention is focused on VOC adsorption capacity and oxidation efficiency of CNM-based composites. Meanwhile, corresponding adsorption and catalytic mechanisms of VOC removal are explained.

9.2 CNMs-based adsorbents for VOC removal

Adsorption is extensively applied in the separation and recovery of VOCs, based on the ability of adsorbents to selectively adsorb specific VOCs (Yang et al., 2019). Selecting an appropriate adsorbent is crucial for efficient VOC removal. An ideal VOC adsorbent is expected to possess (a) high adsorption capacity; (b) high hydrophobicity, to overcome the competitive adsorption with water vapor; (c) easy regeneration, high thermal and hydrothermal stability, to ensure the cyclic adsorption-regeneration.

Various carbonaceous materials such as metal organic frameworks (MOFs), activated carbons (ACs), zeolites, porous organic polymers (POPs), and biochar are commonly researched in field of VOC adsorption (Gan et al., 2023; Li et al., 2020; Na et al., 2019). In addition, CNMs such as graphene and CNTs are also considered as promising adsorbents for VOCs (Li et al., 2004; Liu et al., 2007; Na et al., 2019; Rasheed et al., 2020). The characteristics of ultra-high theoretical specific surface area and excellent mechanical strength bring graphene and CNTs strong adsorption ability (Kumar et al., 2020). In this section, recent literature focusing on the adsorption performance of graphene and CNTs, as well as their derivatives including composites are summarized.

9.2.1 Graphene and its derivatives

Graphene, which was firstly discovered by Novoselov et al. (2004) shows a two-dimensional (2D) structure composed of single layer of sp^2-hybridized carbon atoms arranged in hexagons. It can be mainly produced by methods of graphite exfoliation, hydrothermal self-assembly, chemical vapor deposition, nanotube slicing, etc. (Tahriri et al., 2019). GO is the product of graphene oxidation under acidic conditions with different oxygen-containing groups such as carboxylic, hydroxyl, and epoxide. As an adsorbent, GO has advantageous over pristine graphene in the following two aspects: (1) providing hydrophilic functional groups that enable chemical functionalization and dispersion; and (2) featuring wider physical properties than pristine graphene due to structural heterogeneity (Pei and Cheng, 2012; Shin et al., 2009). GO can further be reduced to rGO with reduction processes of eliminating the functional groups of GO by chemical treatment or thermal annealing (Kumar et al., 2018; Pei and Cheng, 2012).

Recent years, graphene and its derivatives (GO and rGO) have been explored in various fields including electronics, sensors, photonics, energy storage, and environment management because of their outstanding physical-chemical characteristics such as huge theoretical surface area, great mechanical strength, conductivity, and chemical functionalization (Loh et al., 2010; Pierson, 1999; Shen et al., 2015). Graphene and its derivatives seem to be attractive alternatives for VOC adsorption. The adsorption of VOCs onto graphene-based nanomaterials involves the transport of VOCs from the gas phase to the surface of adsorbent. Then adsorption occurs on the surface through interactions such as π-π interactions, hydrogen bonding, electrostatic forces of interactions, pore filling, and hydrophobic and hydrophilic interactions. Fig. 9.2 summarizes the main interactions involved in the adsorption of VOCs on the surface of graphene-based nanomaterials. The adsorption process is related to graphene-based nanomaterials' surface characteristics, especially surface area and porosity. Additionally, surface functional groups and VOC properties also affect the adsorption mechanisms.

Polar VOCs usually prefer to interact with polar surfaces, GO is therefore attractive to the adsorption of polar VOCs as there are abundant hydrophilic oxygen-containing groups on its surface (Bao et al., 2016; Dreyer et al., 2010; Petit et al., 2009). On the contrary, rGO exhibits a strong hydrophobicity due to the removal of oxygen-containing groups, which is would be beneficial for adsorption of nonpolar VOCs. Yu et al. (2018b) investigated the adsorption performance of GO and rGO for benzene and toluene at room temperature. The rGO exhibited higher adsorption capacities to benzene and toluene (276.4 and 304.4 mg/g) than GO (216.2 and 240.6 mg/g) at initial concentration of 50 ppm, which can be attributed to the hydrophobic nature and enhanced tendency to form π-π bonds. In addition, larger surface area of rGO (292.6 vs. 236.4 m^2/g) also contributed to its better adsorption performance. The results indicate that rGO is a potential adsorbent for benzene and toluene. In a study by Kim et al. (2018), graphene platelets (GPs), GO with microwave treatment (rGOMW), and rGOMW with KOH treatment (rGOMWKOH) were prepared for the adsorption of toluene and acetaldehyde. Their toluene removal performances followed the order GPs (2 mg/g) $<$ rGOMW (7 mg/g) $<$ rGOMWKOH (14.4 mg/g), in accordance with the order of their specific surface area. As for acetaldehyde, the adsorption capacities increased from 0.8 to 1.23 mg/g after KOH treatment. An rGO/cellulose nano-crystal (CNC) hybrid aerogels were prepared through a mild chemical reduction route, and cellulose fibers (CFs) extracted from waste-

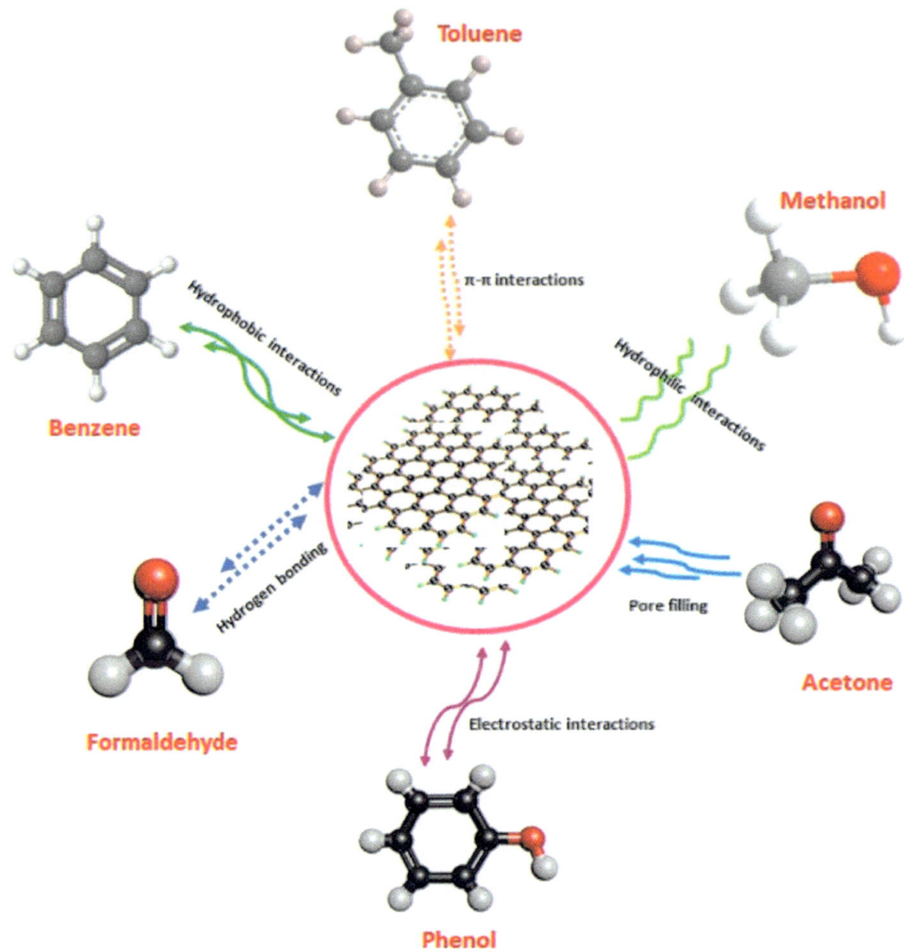

FIGURE 9.2 Proposed adsorption mechanisms of VOCs on graphene-based porous nanohybrids (Bilal et al., 2022).

papers were further incorporated into the pore-structure as reinforcing agents. The synthesized aerogels exhibited an toluene adsorption capacity of 59 mg/g (Shadkam et al., 2021). GO has been researched for selective adsorption of formaldehyde and methanol (Esrafili and Dinparast, 2018). Further, decoration of GO with Al or Si atoms enhanced its adsorption of both VOCs. Compared to formaldehyde, methanol had a larger adsorption energy over the decorated GOs, due to the more favorable orbital interaction as well as electrostatic attraction on the graphene surface. Al-decorated GO were more active in the adsorption of formaldehyde and methanol than Si-decorated GO.

VOC adsorption by porous adsorbents depends on the intermolecular force, and its strength increases with increasing specific surface area (Lizzio and DeBarr, 1996). Graphene based composites usually have high specific surface area. For example, the composite of GO and MIL-101 exhibited an extremely high BET surface area of 3502.2 m^2/g (Sun et al., 2014).

Other materials such as MOF-5 (Liu et al., 2015), Cu-BTC (Li et al., 2016c; Yan et al., 2016) and ZIF-8 (Chu et al., 2018) could also be combined with GO, showing high surface area and promising adsorption of VOCs.

Chu et al. (2018) synthesized ZIF-8 and GO hybrid composites (ZG-X, in which X indicated wt% of GO) to investigate the competitive adsorption behavior of toluene and water vapor. The GO content determined nanoscale morphology and surface area of composites, and the synthetic ZG-X composites showed enhanced toluene adsorption capacity with the GO content below 7 wt%. The ZG-4 with BET surface area of 1112 m^2/g showed the highest toluene adsorption capacity of 116 mg/g at relative humidity (RH) of 55%, increased by 19% compared with the pristine ZIF-8. Due to the competitive adsorption of water molecules on the surfaces of GO, toluene adsorption capacities of ZG-4 decreased from 116 mg/g to 110 mg/g at 80 RH%, while increased to 123 mg/g at 10 RH%. ZG-4 was efficiently regenerated and reused for toluene adsorption over five recycles. These results indicate that the ZIF-8/GO composites are an effective adsorbent for VOCs even in a humid atmosphere. Compared with GO, rGO usually exhibits better VOC adsorption performance in humid atmosphere. For example, Bai et al. (2013) prepared rGO/carbon composite ultrafine fiber, with the GO being reduced to rGO by H_2 during the carbonization. Because of the removal of oxygen-containing groups, more sp^2 carbon atoms emerged on the adsorbent that would weaken the reaction between water vapor and the adsorbent (Asai et al., 2011). Consequently, the resulted composite ultrafine fiber showed higher adsorption ability toward VOCs than water molecular.

ZIF-8/GO composites were also effective in adsorption of chlorinated VOCs such as methylene chloride (Zhou et al., 2016). Hybrid nanocomposites of ZIF-8/GO were prepared at different GO contents of 2, 4, 6, 10, and 15 wt%. The formation of strong interactions between ZIF-8 and GO in the synthesized composites was confirmed. They were composes of aggregated nano-sized particles and exhibited a high methylene chloride adsorption capacity (Fig. 9.3). The increase in GO content resulted in enhancing in adsorption capacity with the maximum ethylene chloride adsorption capacity up to 240 mg/g at GO content of 15 wt%.

The effect of GO on adsorption performance of ordered mesoporous carbons (OMC) for benzene was investigated (Szcześniak et al., 2018a). The incorporation of GO combined with KOH activation led to a surface area of 1370 m^2/g, higher than that of pristine OMC (740 m^2/g). The pore volume also increased from 0.61 to 1.06 cm^3/g after the formation of the GO/OMC composite. As a result, higher benzene adsorption capacities of 750 mg/g

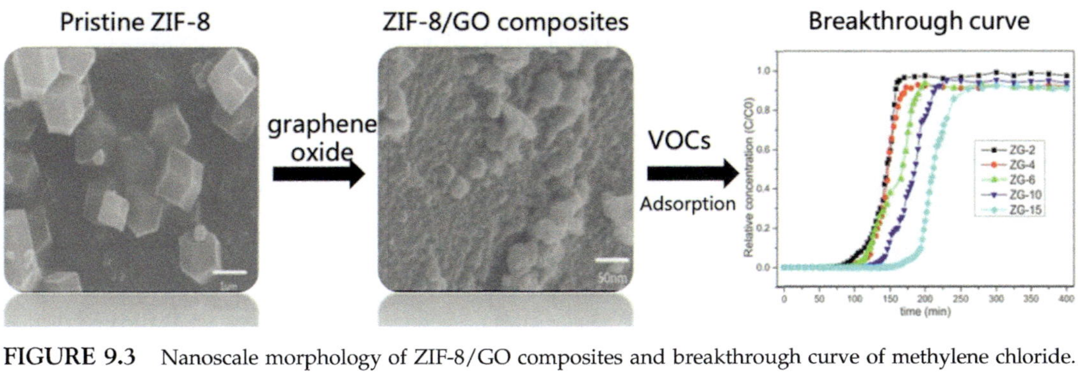

FIGURE 9.3 Nanoscale morphology of ZIF-8/GO composites and breakthrough curve of methylene chloride.

was observed for GO/OMC composite than the 633 mg/g for pristine OMC. Likewise, a series of highly porous carbon materials was prepared from polypyrrole/rGO composite, the composites exhibited high BET surface areas ranging from 1650–2780 m^2/g, large total pore volume in the range of 0.73–1.66 cm^3/g, and high micropore volume in the range of 0.43–0.78 cm^3/g, achieving excellent benzene adsorption of 1234 mg/g (Szcześniak et al., 2018b). Similarly, embedment of GO led to the increased mesopores and high surface oxygen content of carbon composite nanofibers (CNFs) (Guo et al., 2016). The resulting GO/CNF composite was able to adsorb 298.9 mg/g of benzene, which was higher than 255 mg/g of activated carbon fibers (ACF). Under similar conditions, its highest butanone adsorption capacities reached 419.6 mg/g. These demonstrate the great potential of the composite nanofibers in adsorbing VOCs, especially polar ones.

Liu et al. (2015) reported that GO/MOF composites were efficient in the removal of benzene and ethanol. In the research, GO/MOF-5 composites were synthesized with different GO contents (1.75, 3.5, 5.25, and 7 wt%). The composite with 5.25 wt% GO exhibited the highest surface area (727 m^2/g) and pore volume (0.35 cm^3/g), higher than those of MOF-5 (539 m^2/g and 0.27 cm^3/g, respectively). As a result, it showed the best performer in terms of benzene removal capacity (250.8 mg/g), confirming the importance of pore structure and surface properties to its adsorption of VOCs. However, the maximum ethanol adsorption capacity of 158.2 mg/g was achieved by the composite with 3.5 wt% GO due to its abundant oxygen-containing functional groups. Another study also concluded that the combined effect of high specific surface area and rich oxygen functional groups led to a high phenol adsorption capacity of 68 mg/g (Mahmoud, 2020).

Li et al. (2016c) used copper benzene-1,3,5-tricarboxylate (Cu-BTC) and GO (2, 5 and 10 wt%) to synthesize Cu-BTC@GO composites with a solvent-free mechanochemical method (Fig. 9.4). Compared with pristine Cu-BTC with a surface area of 1188.3 m^2/g and a pore

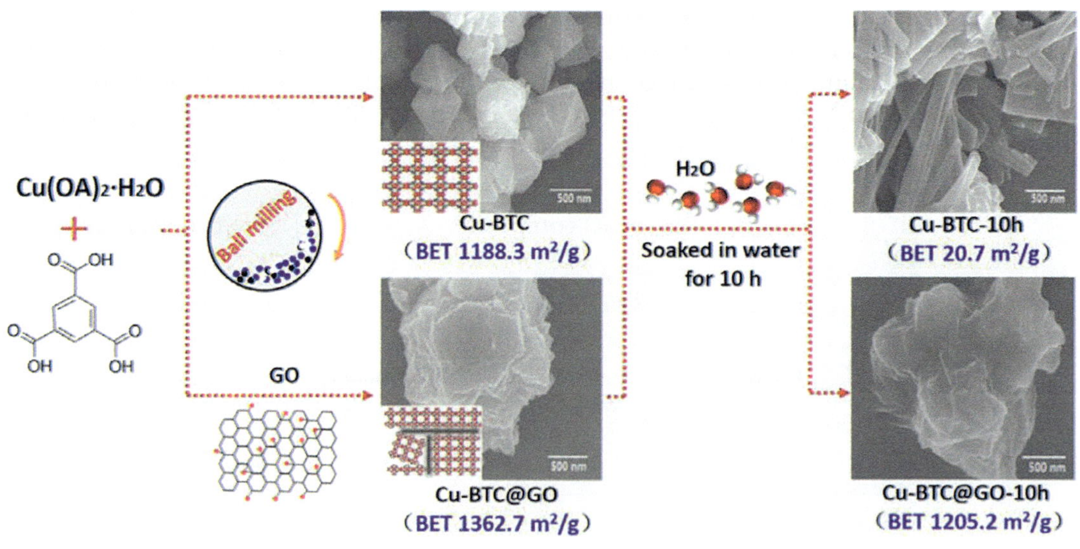

FIGURE 9.4 Preparation methods of Cu-BTC and Cu-BTC@GO composites.

volume of 0.77 cm^3/g, the resulted Cu-BTC@GOs had higher BET surface area and pore volume. Among them, Cu-BTC@GO-5 exhibited the highest surface area of 1362.7 m^2/g and total pore volume of 0.87 cm^3/g. More importantly, its maximum adsorption capacity of toluene was 838.5 mg/g. In another study, Cu-BTC@GO (mass fraction of 20%) composites exhibited high adsorption capacity (183 mg/g) for toluene, which is nearly three times as much as that of Cu-BTC (62.7 mg/g), and the adsorption efficiency can still reach 82.1% after five adsorption–desorption cycles (Dai et al., 2019). Cu-BTC@GO composites with varied GO loading (1, 2, 3, 5 and 10 wt%) prepared by rapid room-temperature synthesis method also showed good potential for ethanol removal (Yan et al., 2016). The Cu-BTC@GO-2 had the highest BET surface area of 1459 m^2/g and pore volume of 0.789 cm^3/g (mesoporous volume of 0.169 cm^3/g and microporous volume of 0.573 cm^3/g), and thus possessed a super high adsorption of ethanol (up to 635 mg/g).

MIL-101(Cr)/GO composites synthesized by a solvothermal synthesis method were confirmed to be efficient adsorbents for the removal of n-hexane from air (Sun et al., 2014). Multiple adsorption–desorption cycles were conducted to evaluate the reversibility of n-hexane on the composites. The adsorption capacities of n-hexane on MIL-101/GO-5 and MIL-101/GO-10 were much higher than that of pure MIL-101. The MIL-101/GO-5 with a surface area of 3502 m^2/g and a pore volume of 1.75 cm^3/g exhibited the maximum n-hexane uptake of 1042.1 mg/g, which is even much higher than those of commercial activated carbons and zeolites. The superior adsorption performance of MIL-101/GO composites is ascribed to their increased specific surface area and enforced surface dispersive forces trigged by the introduction of GO in dense arrays of MOFs. Similarly, MIL-101/GO composites showed high potential for acetone adsorption (Zhou et al., 2014). Characterization results revealed the introduction of GO decrease the crystal size of MIL-101 in the composites. The MIL-101/GO exhibited an acetone adsorption capacity of 1167 mg/g, having an increase of 44.4% in comparison with that of pure MIL-101. Consecutive cycles of adsorption-desorption experiment showed that the desorption efficiency for n-hexane and acetone reached 96.78% and 91.3%, indicating the MIL-101/GO had excellent reversibility for VOC adsorption. In another study, MIL-101/GO (0.25, 0.5, 1.0, and 2.0 wt%) composites exhibited high carbon tetrachloride uptake capacities (Zheng et al., 2018). Among synthesized composites, MIL-101/GO-0.25 was the best performer with a carbon tetrachloride adsorption capacity of 2368.1 mg/g, 16% higher than that of MIL-101 (2044.4 mg/g). Moreover, the MIL-101/GO-0.25 wt% composite exhibited considerably enhanced adsorption performance over most commonly explored conventional adsorbents (e.g., MCM-11, activated carbon, and zeolite Y) by 2.2, 3.9, and 5.5 times, respectively. Several factors were responsible for the improved adsorption performance of GO/MIL-101 for carbon tetrachloride: (1) increased surface area; (2) increased dispersion forces; (3) generation of more defects on the crystal surface; and (4) formation of new pores.

The release of carbonyl VOCs, especially formaldehyde, affects the indoor air quality. Chemical adsorption with amino groups ($-NH_2$) plays an important role in formaldehyde adsorption on CNMs (Matsuo et al., 2008). Graphene based materials including graphene-sponge (G/S) and amino-functional graphene-sponge composites decorated by graphene nanodots (G-GND/S) were used as adsorbents to remove indoor formaldehyde through dynamic adsorption experiment (Wu et al., 2015). Due to the high concentration of amine surface functional groups on its surface, GND/S exhibited a high adsorption capacity (22.8 mg/

g) for formaldehyde, better than G/S (7.5 mg/g). Similarly, a CNT-enhanced graphene aerogel (GN/V) and an amino-functionalized graphene aerogel (GN/E) were synthesized for removing gaseous formaldehyde (Wu et al., 2017). CNTs provided support to graphene layers to reduce the pore diameter of GN/E, leading to the exposure of more amino functionalities for the uptake of formaldehyde. As a result, GN/E exhibited an adsorption capacity of 27.4 mg/g for formaldehyde. The results indicate that chemical adsorption with amino groups plays an important role in formaldehyde removal by graphene-based materials.

As summarized in Table 9.1, the use of graphene and its derivatives exhibited excellent potential in adsorption of VOCs. It has also been proven that graphene-based adsorbents demonstrate better VOC removal performance than conventional adsorbents, e.g., AC and zeolites. However, their adsorption performance is significantly affected by experimental conditions (Szulejko et al., 2019) and aggregation of graphene limits its industrial applications. CNTs have been used to support and connect the graphene layers, thus weakening the aggregation effect and improving the microstructure (Wu et al., 2017). On the other hand, the relative complicated synthesis of graphene materials presents great challenges. Therefore, further studies are encouraged to shorten the fabrication period and optimize production method (Diaz et al., 2007; Li et al., 2012; Koduru et al., 2019).

9.2.2 Carbon nanotubes (CNTs)

CNT is comprised of a graphene sheet rolling up in form of a cylindrical structure (Kauffman and Star, 2008; Vashist et al., 2011). It was discovered by Iijima (1991) in 1991. CNTs are widely researched in many fields due to their unique characteristics like excellent electrical conductivity, optical activity, and mechanical strength (Gulati et al., 2022). In addition, the large surface area, controlled cylindrical hollow structure, natural hydrophobicity as well as thermal stability make CNTs a superior adsorbent for environmental applications (Gupta et al., 2013; Na et al., 2019; Zhang et al., 2019). The mechanisms of VOC adsorption onto CNTs are primary physical adsorption combined with chemical reaction between surface functional groups (e.g., OH, COOH, C=O) and VOC molecules. It has been clarified that CNTs can undergo interactions with organic molecules via noncovalent forces including hydrogen bonding, $\pi-\pi$ stacking, electrostatic forces, van der Waals forces, and hydrophobic interactions (Li et al., 2016b; Rengaraj et al., 2007).

In general, CNTs are divided into single-walled carbon nanotubes (SWCNTs) and multiwalled carbon nanotubes (MWCNTs) according to the number of graphene sheets (Raphey et al., 2019; Yang et al., 2020). The SWCNT indicates a cylinder structure of a single graphene sheet while the MWCNT contains at least two graphene sheets rolling into a stacking of concentric layer cylinders with an interspacing of 0.34 nm (Zare et al., 2015). As for the adsorption sites, adsorption can occur at four sites of SWCNTs: outside the tubes, interstitial channels between the tubes, in the pores of the tubes, and grooves between adjacent tubes (Ren et al., 2011). While three locations, i.e., inside of tubes, external surface, and interstitial space between graphene sheet layers are potentially adsorption sites of MWCNTs. Because the interspace of MWCNTs is generally too small for VOC molecules, MWCNTs are more effective for the adsorption of large molecules; while SWCNTs are highly effective for the adsorption of small molecules. On the other hand, several studies have confirmed that

TABLE 9.1 Performance of graphene-based adsorbents for VOCs (Bilal et al., 2022; Kumar et al., 2020).

Type of graphene material	BET surface area (m^2/g)	Target VOC	Adsorption capacity (given value)	Adsorption capacity (mg/g)	Temperature (K)	Pressure (Pa)	References
GO	236.4	Benzene	216.2 mg/g	216.2	293.15	5.07	Yu et al. (2018b)
GO	236.4	Toluene	240.6 mg/g	240.6	293.15	5.07	Yu et al. (2018b)
rGO	292.6	Benzene	276.4 mg/g	276.4	293.15	5.07	Yu et al. (2018b)
rGO	292.6	Toluene	304.4 mg/g	304.4	293.15	5.07	Yu et al. (2018b)
GO/OMC	1,370	Benzene	9.6 mmol/g	749.9	293.15	10,018	Szcześniak et al. (2018a)
rGO/PPy	2,780	Benzene	15.8 mmol/g	1234.2	293.15	10,018	Szcześniak et al. (2018b)
GO/CNF	473	Benzene	83.2 cm^3/g	289.8	293.15	9,829.4	Guo et al. (2016)
GO/CNF	473	Butanone	130.5 cm^3/g	419.6	293.15	9,270.8	Guo et al. (2016)
GO/MOF-5 (GO 5.25 wt%)	727	Benzene	72 cm^3/g	250.8	303.15	9,829.4	Liu et al. (2015)
GO/MOF-5 (GO 3.5 wt%)	439	Benzene	77 cm^3/g	158.2	303.15	9,808.8	Liu et al. (2015)
GP	54	Toluene	466 mm^3/g	1.9	293.15	3.042	Kim et al. (2018)
GP	54	Acetaldehyde	165 mm^3/g	0.32	293.15	3.042	Kim et al. (2018)
rGOMW	326	Toluene	1710 mm^3/g	7	293.15	3.042	Kim et al. (2018)
rGOMW	326	Acetaldehyde	405 mm^3/g	0.8	293.15	3.042	Kim et al. (2018)
rGOMWKOH	491	Toluene	3510 mm^3/g	14.4	293.15	3.042	Kim et al. (2018)
rGOMWKOH	491	Acetaldehyde	630 mm^3/g	1.23	293.15	3.042	Kim et al. (2018)
Cu-BTC/GO	1,362.7	Toluene	9.1 mmol/g	838.5	298	–	Li et al. (2016c)

(Continued)

TABLE 9.1 Performance of graphene-based adsorbents for VOCs (Bilal et al., 2022; Kumar et al., 2020).—cont'd

Type of graphene material	BET surface area (m^2/g)	Target VOC	Adsorption capacity (given value)	Adsorption capacity (mg/g)	Temperature (K)	Pressure (Pa)	References
Cu-BTC/GO	512.6	Toluene	183 mg/g	183	298	–	Dai et al. (2019)
Cu-BTC/GO	1,459	Ethanol	13.6 mmol/g	635	303	–	Yan et al. (2016)
ZIF-8/GO	1,112	Toluene	123 mg/g	123	303	–	Chu et al. (2018)
ZIF-8/GO	–	CH_2Cl_2	240 mg/g	240	–	28.672	Zhou et al. (2016)
MIL-101/GO	3,502.2	N-hexane	1042.2 mg/g	1042.2	298	–	Sun et al. (2014)
GO/MIL-101	2,928	Acetone	20.1 mmol/g	1167.4	293.15	9,270.8	Zhou et al. (2014)
MIL-101/GO	2,026	CCl_4	2368.1 mg/g	2368.1	303	–	Zheng et al. (2018)
G-GND/S	–	Formaldehyde	22.8 mg/g	22.8	293.15	–	Wu et al. (2015)
GN/E	252	Formaldehyde	27.43 mg/g	27.43	293.15	–	Wu et al. (2017)
Me-GO	621	Phenol	68 mg/g	68	298	–	(Mahmoud, 2020)

VOC molecules are mainly adsorbed on the external surface or internal of the CNTs (Agnihotri et al., 2007; Kondratyuk and Yates, 2005). Therefore, the surface functional groups can also affect adsorption behaviors of VOCs on CNTs. A previous study have shown that carboxylation and nitration modification led to the generation of polar functional groups on MWCNT surface, and the nanotubes showed strong adsorption of polar alcohols, but relatively weaker adsorption for nonpolar and aromatic compounds, such as hexane, benzene, and toluene (Hussain et al., 2009).

CNTs contain both crystalline and amorphous carbon, and both can affect adsorption. Shih and Li (2008) showed that the defects and amorphous carbon on the surface of CNTs greatly affected their adsorption of VOCs. They investigated the adsorption behaviors of n-hexane, benzene, trichlorethylene, and acetone on two MWCNTs (i.e., CNT1 and CNT2). CNT2 had a higher VOC adsorption capacity than CNT1 due to the existence of an exterior amorphous carbon layer as shown in Fig. 9.5 (Chakrapani et al., 2003). The study also concluded that the adsorption of nonpolar VOCs onto CNTs is mainly controlled by physical processes, while the adsorption of polar VOCs onto CNTs may be affected by the combination of

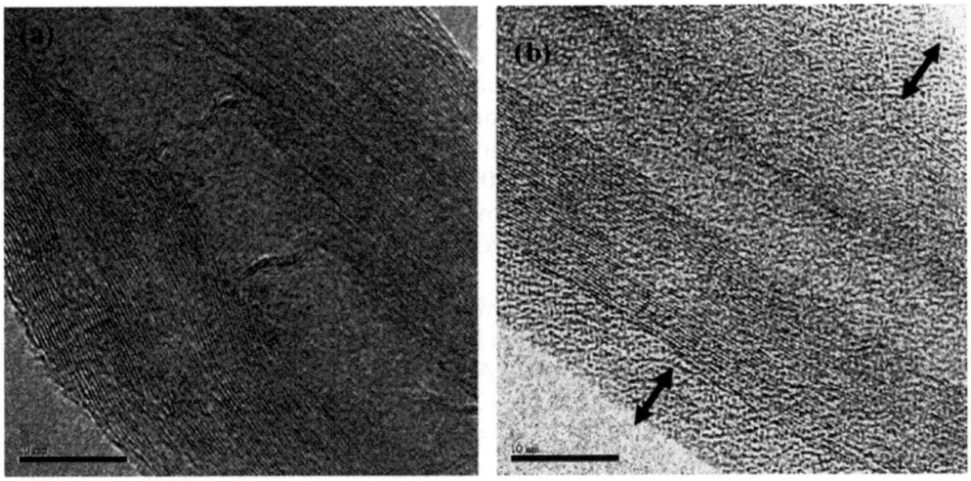

FIGURE 9.5 HRTEM images of CNT1 (a) and CNT2 (b). Arrows show the exterior layers on CNT2.

chemical interactions, according to the results of enthalpy changes and desorption activation energy. The enthalpy changes of nonpolar VOCs, e.g., trichlorethylene, benzene, and n-hexane adsorbed onto CNTs were relatively low, corresponding to physical exothermic adsorption processes. On the contrary, polar acetone had higher enthalpy change and higher desorption activation energy, reflecting chemisorption on the topological defects of CNTs (Shih and Li, 2008).

In another study, highly crystalline multi-walled carbon nanotubes (HC-MWCNTs) were used as the adsorbent for VOCs in contaminated air samples (Sone et al., 2008). The results showed HC-MWCNTs had high selectivity and strong affinity to aromatic VOCs (benzene, toluene, *m*-xylene, *p*-xylene, *o*-xylene, and *p*-dichlorobenzene). The high degree of adsorption is due to the transfer of the π-electrons from the highest occupied molecular orbital of the HC-MWCNTs to the lowest unoccupied molecular orbital of the aromatic VOCs. A CNT sponge was developed as a new type of adsorbent for enriching trace aromatic VOCs, showing adsorption capacities in the order of o-xylene (14.31 mg/g) > ethylbenzene (13.90 mg/g) > toluene (2.45 mg/g) > benzene (0.13 mg/g) (Jang et al., 2020).

Combinations of heat and chemical pretreatment of SWCNTs and MWCNTs were explored to facilitate their adsorption of toluene (Gangupomu et al., 2016). CNTs were purified through heat-treatment at 400°C in steam, followed by nitric acid treatment (3N, 5N, 11N, 16N) for 3–12 h to create openings to facilitate VOC adsorption onto interior sites. The Raman spectroscopy results showed that pretreatment removed impurities and reformed amorphous carbon of SWCNTs. The optimum pre-treatment condition was 3N nitric acid with 3 h, which increased the surface area from 1136 to 1347 m^2/g. More importantly, toluene adsorption capacity of pre-treated SWCNTs reached 383.2 mg/g, comparable to that of commercial GAC (385 mg/g). As for MWCNTs, no substantial difference was detected after treatment, and the highest toluene adsorption capacity of MWCNTs was only 11 mg/g. The different effect of heat and chemical pretreatment on SWCNTs and MWCNTs was attributed to their different structures.

As mentioned above, VOC molecules are mainly adsorbed on the external surface or internal of CNTs, therefore, surface modification affects the adsorption behavior. A study indicated that carboxylation and nitration modification led to the generation of polar functional groups on the MWCNTs surface, and the nanotubes showed stronger adsorption of polar alcohols, but weaker adsorption of nonpolar and aromatic compounds such as hexane, benzene, and toluene (Hussain et al., 2009). Similarly, SWCNTs were chemically modified to develop organophilic analogs by forming amide bonds with oleylamine ($C_{18}H_{37}N$), and the resulting organophilic CNTs exhibited enhanced selectivity to *n*-hexane (Vermisoglou et al., 2007).

Furthermore, a water plasma treatment was applied to modify the surface of MWCNTs, which increased their adsorption capabilities of three chlorinated compounds (trichlorethylene, 1,2-dichlorobenzene and chloroform) (Amade et al., 2014). The adsorption increased with water solubility and the polarity of VOCs, and the presence of aromatic rings resulted in π-stacking interactions to further increase VOC adsorption. Hydroxyl-functionalized CNTs were used to adsorb four VOCs in aqueous solution with the affinity following the order of toluene > ether > acetone > methanol (Li and Mi, 2022). The results also showed that the adsorption affinity decreased with increasing hydroxyl concentration due to the reduction of the originally hydrophobic area of CNTs surface and the additional bonding competition from water molecules. In another study, CNTs were oxidized with nitric acid to obtain CNTs-OX, which showed higher o-xylene adsorption capacity (72.46 mg/g) than that of pristine CNTs (47.17 mg/g) (Srivastava et al., 2018). In order to promote the adsorption for polar VOCs, Hsu and Lu (2009) oxidized SWCNTs with NaOCl to adsorb isopropyl alcohol in air stream, and the adsorption capacity reached 82 mg/g.

The adsorption of VOCs on CNT-based composites has also been widely reported. Yang et al. (2017) evaluated the adsorption behavior of a CNT/ACF material for indoor formaldehyde at low concentration (1.50 mg/m^3). The CNT/ACF was fabricated by in situ growth of CNTs on ACF though a chemical vapor deposition method. The composite removed three times more formaldehyde per weight than the pristine ACF (62.49 mg/g vs. 19.11 mg/g). MWCNTs was chemically bonded with 3-amino-propylsilica (APSG) into a novel adsorbent APSG-MW and its adsorption performance to nine VOCs (benzene, toluene, butyl acetate, ethylbenzene, *o*-xylene, *m*-xylene, styrene, *p*-xylene, *n*-undecane) was measured in form of recovery ($R\%$, determined by dividing the detected amount of target compounds by the total injected amount) (Wang et al., 2010). APSG-MW performed well in adsorbing tested VOCs with recoveries over 80%. The effect of humidity was further investigated considering the real air condition, the recoveries of the nine tested compounds under 50% humidity were close to those of 30% humidity (which is considered as the normal humidity condition), and the recoveries only decreased by 10% at 90% humidity.

The CNT content in composites was proved to an important factor to their VOC adsorption capacities (Liu et al., 2021). In the study, MWCNTs-SiO_2 composites with molar percentages of MWCNTs of 38% (CS2), 52% (CS4), and 66% (CS6) were synthesized using a sol-gel method and tested for toluene adsorption at different temperatures (30–60°C) and water vapor concentrations (1%–5%). Their toluene adsorption capacities followed the order of CS2 < CS4 < CS6 and the adsorption capacity of CS6 was up to 50.28 mg/g. Increasing temperature and water vapor concentration resulted in decreased penetration time and adsorption capacity, respectively. The composites also exhibited good toluene adsorption performance after regeneration, as the surface chemical functional group remained unchanged. The results demonstrate a great potential of the CNT-based composites for adsorbing VOCs.

Aggregation is also one of the main crucial drawbacks limiting the commercial applications of CNT-based materials for VOC removal (Zhang et al., 2017; Zhu et al., 2020). A supporting method may be effective to overcome the aggregation issues. Tulaphol et al. (2016) embedded MWCNTs into SiO_2 particles through the floating-catalyst chemical vapor deposition method, and found the supported CNTs were stable and thus had excellent adsorption performance to gaseous chlorinated phenolics. In addition, surface oxidation and coating with surfactants may also be effective ways to reduce the aggregation of CNTs (Pan and Xing, 2008; Vermisoglou et al., 2007).

9.2.3 Other CNM-based adsorbents

The activated carbon nanofibers (ACnFs) prepared from electrospun polyacrylonitrile/cellulose nanocrystal (PAN-CNC) nanofibers were characterized and tested for their adsorption of cyclohexane (Awad et al., 2021). The addition of CNCs improved fiber uniformity and tensile strength, and decreased fiber diameter. Further, pore size distribution analysis of ACnFs showed large volumes of micropores and mesopores. The specific surface area and total pore volume of ACnFs at PAN to CNC mass ratio of 1:3 were 3497 m^2/g and 2.62 cm^3/g, respectively. The adsorption capacity for cyclohexane reached 1.8 g/g and correlated well with the surface area. In another study, electrospun double-layer polymethyl methacrylate/manganese acetate (core), polyacrylonitrile (shell)-based nanofibers were carbonized and CO_2-activated to produce Mn_3O_4 nanoparticles embedded in hollow activated carbon nanofibers (Mn_3O_4/HACNFs), which demonstrated great ability to adsorb toluene (Kang and Hwang, 2020).

9.3 CNM-based catalysts for VOC removal

In recent years, the research on VOC catalytic oxidation technology has attracted much attention mainly due to the advantages of efficiency and economic feasibility. In this approach, VOCs are oxidized in presence of a suitable catalyst and converted to relatively less harmless products, usually CO_2 and H_2O (Kamal et al., 2016). Catalytic oxidation is also environmentally friendly due to its low temperature operation and the formation of low dioxins and noxious products (Tang et al., 2015; Zhang et al., 2016). Photocatalytic oxidation, plasma catalytic oxidation, and ozone catalytic oxidation have been proved to be efficient for the degradation of VOCs at low concentrations (Yang et al., 2019). Current research in the field of VOC catalytic oxidation mainly focuses on the development of high efficiency and low cost catalysts (Li et al., 2023) because VOC degradation rate highly depends on the catalytic materials during the reaction process (Guo et al., 2021).

Noble metal-supported catalysts and metal oxide-based catalysts are two main types of catalytic nanomaterials applied for VOC degradation. The noble metal supported catalysts have advantages of good catalytic activity and high durability in spite of expensive costs, while metal oxide-based catalysts are widely researched due to comparatively high catalytic performance, low cost and strong resistance to poisoning (Guo et al., 2021). For example, TiO_2 is one of the most widely semiconductor used in photocatalytic VOC degradation, showing the advantages of high photocatalytic activity, low cost, and nontoxic (Guo et al., 2022). However, conventional catalysts often show limited degradation rate of VOCs at low concentrations due to relatively low VOC adsorption ability. On the other hand, the generated intermediate products could aggregate on the catalysts surface to reduce the reaction efficiency, and may also cause secondary pollution after volatilization or dissociation. Therefore, a combination of conventional catalysts with CNMs is encouraged to enhance the adsorption ability and catalytic efficiency. CNMs possess superior surface structures (specific surface area and pore size) and surface chemical properties (function groups, hydrophilicity or hydrophobicity), therefore, presenting multiple benefits of distributing and stabilizing catalyst nanoparticles as well as improving charge transfer (Zou et al., 2019). Furthermore, their high adsorption ability can keep the VOC molecules in the proximity of the reaction active sites, enhance visible light adsorption, promote the formation of radicals, and prevent the generation of by-products and inactivation of photocatalysts (Sunkara et al., 2010). Another property makes CNMs attractive in VOC degradation is their exceptional stability in acidic and basic conditions (Kottappara et al., 2019), leading them ideal for utilization as a catalyst or a catalyst support.

Among various carbon nanostructures, CNTs and graphene nanosheets are most widely researched, especially for the fabrication of composite photocatalysts with TiO_2 (Lee et al., 2008). This is mainly attributed to their enormous specific surface area and high electronic conductivity. This section mainly summarizes the applications of graphene and CNTs in the fabrication of composite catalysts for VOC degradation.

9.3.1 Graphene and its derivatives

Graphene-based composites are excellent catalysts for VOC catalytic oxidation. For example, MnO_2/graphene composites showed effective catalytic ozonation of gaseous toluene (Hu et al., 2014). In the study, MnO_2 content (14.2−94.8 wt%) showed a significant influence on BET surface area and catalytic activity of the composites. At 64.6 wt% content, MnO_2 nanoparticles uniformly distributed throughout the surface of graphene nanosheets, whereas lower or higher contents led to their aggregation or erosion of graphene sheets. As a result, the highest toluene degradation rate (7.89×10^{-6} mol/min/g) was obtained by this composite, which was attributed to the tight connection between the active sites on graphene for adsorption of toluene and decomposition of ozone, and the MnO_2 on graphene for decomposition of ozone to form atomic oxygen species.

GO and rGO have widely used in the fabrication of composite photocatalysts. GO-TiO_2 composites were synthesized using a chemical mixing method, followed by post-treatment under 200−500°C, and utilized for the degradation of 2-ethyl-1-hexanol (2E1H) (Chun and Jo, 2016). The GO-TiO_2 treated at 400°C exhibited the highest adsorption capacity for 2E1H and the best photocatalytic activity. It kept high photocatalytic ability after five cycles with the average photocatalytic efficiency of 99.3%. Another study also confirmed the good reusability of a GO/MnO_x/polymeric carbon nitride (CN) nanosheets for formaldehyde catalytic degradation (Wang et al., 2020b).

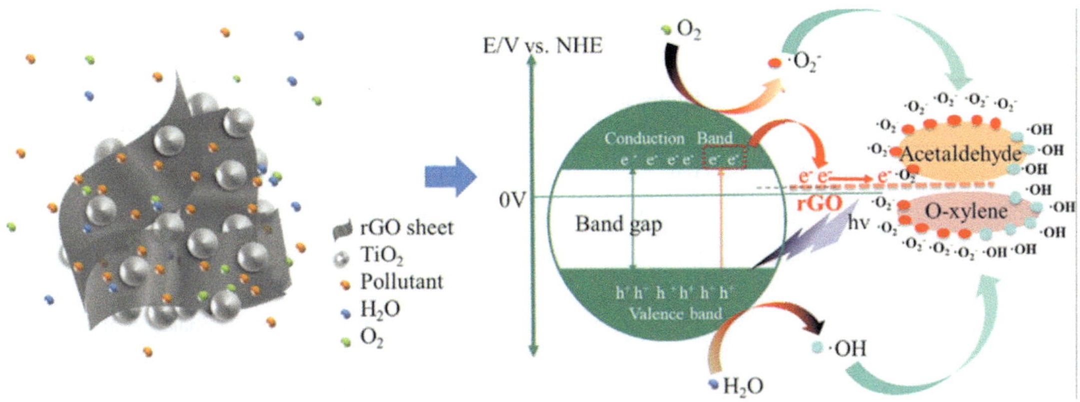

FIGURE 9.6 Photocatalytic degradation mechanism of VOCs rGO-TiO_2.

Lin et al. (2018) used solvothermal process to prepare rGO-TiO_2 nanocomposites for the photocatalytic oxidation of acetaldehyde and o-xylene. The photocatalytic degradation mechanism is shown in Fig. 9.6. Firstly, VOC molecules are adsorbed on the surface of the nanocomposite. Secondly, electrons of TiO_2 are excited to the conduction band, creating e^--h^+ pairs under UV light irradiation, then the charge carriers transfer to O_2 and H_2O to generate different radicals ($\cdot O_2^-$ and •OH). Finally, these radicals react with VOCs on the surfaces of the photocatalyst and degrade them. In the study, the degradation efficiency of o-xylene (54%) was higher than 42% of acetaldehyde, which was attributed to two reasons: (1) o-xylene showed more adsorption affinity than acetaldehyde owing to the π-π conjugation between rGO and aromatic compounds; (2) for the degradation of acetaldehyde, $\cdot O_2^-$ radicals played the predominant role and radical •OH was the secondary (Zeng et al., 2019); while $\cdot O_2^-$ and •OH radicals showed almost the equal role during the degradation of o-xylene. The results also confirm the effect of rGO content in composites on their catalytic efficiencies, which increased with rGO content below 0.5 wt% but decreased when rGO contents increased to 2 wt%.

Graphene and its derivatives have advantages of high transmittance rate of UV-visible lights, fast electrical and thermal conductivity due to special structure of two-dimensional layer of sp^2 hybridized carbon atoms and one-atom thickness (Zou et al., 2019). Further, the delocalization of π network of the layers can efficiently suppress the electron-hole recombination and the photogenerated electrons in graphene layers behave as massless Dirac fermions, leading to the enhanced photocatalytic performance. Roso et al. (2017) compared the adsorption and photocatalytic reactivity of GO, rGO, and few-layer graphene. They found that the rGO-based nanocomposite shown the best methanol degradation efficiency, which mainly due to its enhanced electron mobility through deeper contact with the photocatalyst. Recent studies about photocatalytic efficiency of some graphene-based composites to VOCs are summarized in Table 9.2.

9.3.2 Carbon nanotubes (CNTs)

As indicated previously, CNTs have a specific structure of rolling up the graphene sheets into cylinders, leading to the excellent electrochemical properties and high specific surface

TABLE 9.2 Photocatalytic degradation of VOCs by graphene-based composites (Ahmad et al., 2023; Bilal et al., 2022; Zou et al., 2019).

VOCs	Composites	Photocatalytic efficiency (%)	Main mechanism	References
Toluene	t-TiO_2/rGO	100	Photocatalytic degradation	Zhou et al. (2020)
Toluene	TiO_2/GO	35.4–81.7	Photocatalytic degradation	Azimi et al. (2019)
Toluene	TiO_2/graphene (0.1 wt%)	97.2	Photocatalytic degradation	Ye et al. (2019)
Toluene	Poly ionic liquids@ TiO_2/m-GO	83	Photocatalytic degradation	Shajari et al. (2021)
Toluene	rGO/polypyrrole hybrid aerogel	94.8	Photocatalytic degradation	Yan et al. (2022)
Benzene	Poly ionic liquids@ TiO_2/mGO	97	Photocatalytic degradation	Shajari et al. (2021)
Acetaldehyde	MT@rGO-0.1	70	Photocatalytic degradation	Fang et al. (2020)
Formaldehyde	$BiPO_4$/graphene aerogel	26.4	Photocatalytic degradation	Yang et al. (2021)
Phenol	TiO_2@rGO	97.9	Photocatalytic degradation	Wang et al. (2020a)
Toluene	MnO_2/rGO	98	Photocatalytic degradation	Mallakpour and Khadem (2016)
Toluene	$SrTiO_3$/rGO	100	Photocatalytic degradation	Zhang et al. (2020)
Formaldehyde	rGO/TiO_2	88.3	Photocatalytic degradation	Khusnun et al. (2018)
Formaldehyde	TiO_2/rGO/Ag	90	Photocatalytic degradation	Peng et al. (2020)
Formaldehyde	Amino-TiO_2/rGO	92	Photocatalytic degradation	Kupreenko et al. (2022)
Formaldehyde	Graphene/Fe^{3+}-TiO_2	58	Photocatalytic degradation	Low and Boonamnuayvitaya (2013)
Acetaldehyde	N-doped graphene-Fe_2O_3	55	Photocatalytic degradation	Wang et al. (2015)
Formaldehyde	rGO-TiO_2	88.3	Photocatalytic degradation	Yu et al. (2018a)
Formaldehyde	Ce-GO-TiO_2	83.8	Photocatalytic degradation	Li et al. (2016a)
Benzene	P25/graphene	76.2	Coordination between adsorption and photocatalytic degradation	Zhang et al. (2010)
2-Ethyl-1-hexanol	GO-TiO_2	55.1	Photocatalytic degradation	Chun and Jo (2016)
Bisphenol A	graphene hydrogel-AgBr@rGO	91.4	Coordination between adsorption and photocatalytic degradation	Chen et al. (2017)
Toluene	GO/MIL-125	79.2	Coordination between adsorption and photocatalytic degradation	Jin et al. (2022)

area. Their tubular structure can maintain high shape stability and long residence time of VOC molecules on catalysts (Guo et al., 2022; Zou et al., 2019). For example, in an adsorption-photocatalytic process of styrene by CNTs-TiO_2 nanocomposites, CNTs prolonged adsorption equilibrium, leading to a longer contact time between styrene and the CNTs-TiO_2, and thus enhanced photocatalytic efficiency (An et al., 2012). Another study confirmed that TiO_2 maintained the nanosized dimension on the hollow structure CNTs after the degradation of benzene (Xu et al., 2010). In addition, CNTs can be either metallic, semimetallic, or semiconducting by changing the helicity and diameter of the tube. The special conductivity makes CNTs efficient electron-transfer channel in the adsorption-photocatalytic VOC removal (Li et al., 2011). The photo-generated electrons can move freely toward CNT surface with a low Fermi level, and holes in valence band of a photocatalyst inclines to migrate to the surface and then oxidize VOCs. The electron density calculation (Fig. 9.7a) confirmed that the electron density was higher near CNTs than TiO_2, this difference proved the effective electrons transmission from the conduction band of TiO_2 to CNTs (Khang et al., 2016). Similarly, the reaction mechanism diagram shown in Fig. 9.7b demonstrates the strong interaction between CNTs and TiO_2 to promote the electron transfer process, and the effectiveness under visible light also shows that CNTs have an impact on the energy band structure of TiO_2 (Rahbar and Behpour, 2016).

TiO_2/MWCNTs composites at different CNT concentrations (0.6, 1.2, and 3.0 wt%) improved the degradation efficiency of tetra-chlorophenol (Zouzelka et al., 2016). The composites exhibited strong absorption capacities in both UV and visible range. MWCNTs had a beneficial effect on the photocatalytic performance, a marked increase in photocatalytic degradation rate constant being observed even at very low concentrations of MWCNTs. Compared to the pure catalyst, TiO_2/MWCNTs (0.6 wt%) composites showed about 100% increase in the first order rate reaction constant. The photocatalytic performance tended to be saturated at MWCNT concentration of 1.2 wt%. CNTs promoted the degradation of tetra-chlorophenol by attracting the primarily photoproduced hydroxyl radicals onto tetra-chlorophenol molecules.

CNT-based catalysts are also utilized in low-temperature catalytic oxidation of VOCs. A novel MnO_2/N-doped CNTs (NCNT) composites were prepared with varying MnO_2 content and the one with 40% MnO_2 exhibited the best activity and selectivity for formaldehyde oxidation with a mineralization rate reaching 95% at 30°C (Peng et al., 2020). In another study, MnO_2 was supported on CNTs (MnOx/CNTs) and the ozone catalytic oxidation performances of chlorobenzene were investigated under low temperatures (80–280°C) (Jin et al., 2015). The experimental results showed a synergistic effect between MnOx and CNTs to promote ozone oxidation of chlorobenzene, reaching above 95% with 2300 ppm ozone at temperatures below 120°C. The catalytic oxidation performance of multi-component mixtures of benzene, toluene, ethylbenzene, and o-xylene (BTEX) over Pt/CNTs catalysts was investigated at 40–150°C (Joung et al., 2014). The oxidation activity was presumably promoted due to the higher surface concentration of VOCs afforded by the CNTs. The mixtures were completely oxidized into CO_2 and H_2O without the formation of by-products. The complete oxidation of benzene, toluene, ethylbenzene, and o-xylene were realized at low temperatures of 112, 109, 106, and 104°C, respectively, with 30 wt% Pt/CNT catalyst. Examples of catalytic degradation of VOCs by CNT-based composites are shown in Table 9.3.

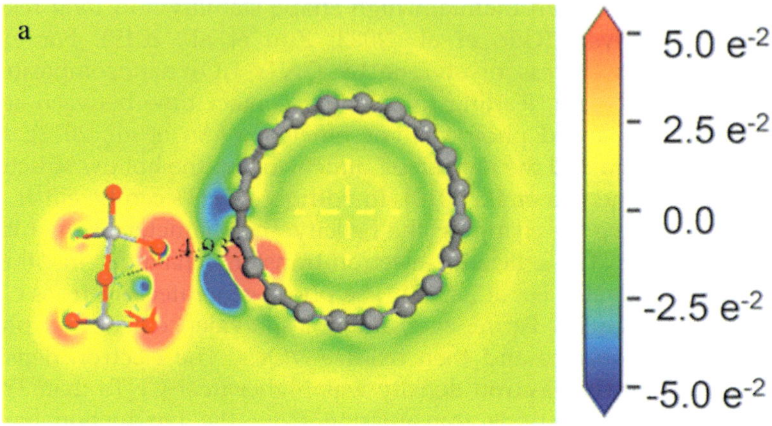

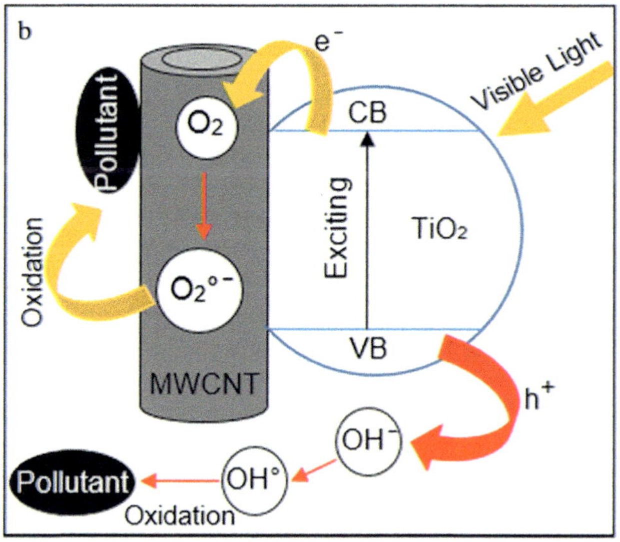

FIGURE 9.7 (a) Steady-state simulation diagram of CNTs-TiO_2, (b) proposed mechanism for photodegradation of pollutant by MWCNT/TiO_2 film.

9.3.3 Other CNM-based catalysts

Carbon quantum dots (CQDs) can also enhance photocatalysis when cooperating with other conventional catalysts. A study decorated TiO_2 nanoparticles with CQDs to examine its potential use in the photocatalytic oxidation of mixed VOCs, e.g., benzene, toluene, and *p*-xylene (Mahmood et al., 2021). The CQDs/TiO_2 demonstrated enhanced photodegradation performance under both UV and visible light illumination. For example, CQDs/TiO_2 containing 0.5 wt% CQDs degraded 64% of mixed VOCs, while pure TiO_2 only exhibited

TABLE 9.3 Photocatalytic degradation of VOCs by CNT-based composites (Zou et al., 2019).

VOCs	Composites	Photocatalytic efficiency (%)	Main mechanism	References
Limonene	CNTs/TiO_2 nanofiber	92	Photocatalytic degradation	Jo and Kang (2015)
Formaldehyde	MnO_2/MWCNT	43	Photocatalytic degradation	Miao et al. (2018)
Benzene	CNTs/TiO_2	64.6	Photocatalytic degradation	Xu et al. (2010)
Styrene	CNTs/TiO_2 sphere	55.4	Coordination between adsorption and photocatalytic degradation	An et al. (2012)
Acetone	MnO_2/TiO_2	50	Photocatalytic degradation	Yu et al. (2011)

44% of photodegradation efficiency. On the other hand, CQDs/TiO_2 demonstrated considerable photocatalytic activity within the visible region, while pure TiO_2 showed no activity. The improved photocatalytic performance was attributed to good light harvesting within the UV-vis region, charge separation, as well as adsorption capability of synthesized catalyst. In another work, CQDs were employed to enhance the photocatalytic performance of TiO_2-based composites for flowing gaseous acetaldehyde removal (Hu et al., 2018). The CQDs/TiO_2 composite (3 wt% of CQDs) maintained a 99% removal efficiency in a 2-h continuous gaseous acetaldehyde flow under fluorescent lamp irradiation ($\lambda > 380$ nm).

The photocatalytic degradation mechanism of toluene and benzene by CQDs based nanocomposites is explained in Fig. 9.8 (Wang et al., 2018). Cadmium sulfide quantum-dots (CdS QDs) and carbon dots (CDs) were used to modify TiO_2 photocatalyst, the prepared ternary CdS/CDs-TiO_2 simultaneously improved the photo-corrosion of CdS and amplified its photocatalytic activity. The introduction of CdS QDs and CDs enhanced light absorbance, prevented the undesirable electron/hole recombination, and promoted charge separation, leading to the continuous formation of radical •OH and radical $\bullet O_2^-$ radicals for converting organic contaminants to CO_2 and H_2O. The study also suggested that the degradation of toluene was more difficult than benzene due to the symmetric structure and conjugation of methyl with benzene ring. For example, the ternary nanocomposites with a CdS QDs to CDs mass ratio of 3:1 achieved 90% degradation efficiencies for benzene within 1 h, but it took 2 h for toluene.

Difference in carbon sources of CQDs may lead to different optical and electronic properties of hybrid materials, affecting photocatalytic degradation activity. CQDs/Bi_2WO_6 hybrid materials were synthesized via a facile in situ hydrothermal method (Liu et al., 2020). The CQDs/Bi_2WO_6 derived from chitosan demonstrated the best photocatalytic activity with 96.9% degradation for toluene and 97.1% for formaldehyde within only 120 min than those derived from other carbon sources including citric acids, glucose, L-ascorbic acid, and sucrose.

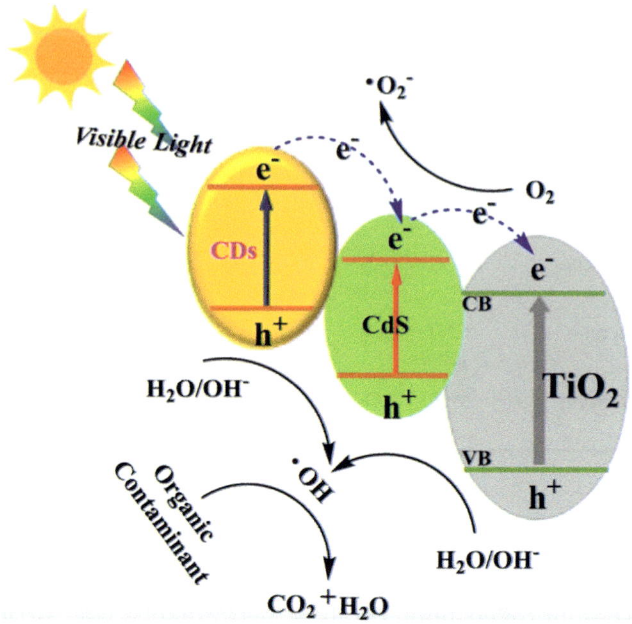

FIGURE 9.8 Photocatalytic degradation mechanism of organic pollutants by CdS/CDs-TiO_2 ternary nanocomposites.

9.4 Summary and perspectives

CNMs are prospective materials in fields of VOC adsorption and catalytic oxidation. Among various CNMs, graphene and its derivatives (e.g., GO and rGO), as well as CNTs and their derivatives showed large adsorption capacity and high degradation efficiency mainly due to their excellent surface characteristics (e.g., specific surface area and pore size), optical and electronic properties. They are widely used to synthesize nanocomposites with conventional materials and exhibited enhanced adsorption capacity and oxidation efficiency. In adsorption process, CNMs usually offer high specific surface area and maintain excellent mechanical strength, while they mainly help adsorb UV-visible light and promote the formation of radicals (•O_2^- and •OH) during photocatalytic oxidation. However, competitive adsorption of water vapor, as well as CO byproduct, water vapor, and coking suppression in catalysis can limit the performance of CNMs in VOC removal. Therefore, further research is still needed to promote the application of CNMs in the adsorption and catalytic oxidation of VOCs. Further research directions include, but not limited to, (1) Syntheses of stable and high-efficient CNMs with tunable adsorption and photocatalytic abilities should be further explored; (2) Studies on CNM adsorption and oxidation of VOC mixtures, or with water vapor, SO_X, or NO_X, should be carried out systematically to mimic the real conditions; (3) CO_2 emissions should be considered in a complete catalytic oxidation of VOCs, while production of toxic byproducts should also be explored if catalytic oxidation is not complete; (4) Combination of adsorption and catalytic degradation seems to be highly effective in VOC removal, thus bifunctional CNM-based adsorbent-catalyst need to be explored and the corresponding adsorption-catalytic mechanisms should be further explained.

References

Agnihotri, S., Zheng, Y., Mota, J.P.B., Ivanov, I., Kim, P., 2007. Practical modeling of heterogeneous bundles of single-walled carbon nanotubes for adsorption applications: estimating the fraction of open-ended nanotubes in samples. Journal of Physical Chemistry C 111, 13747–13755. https://doi.org/10.1021/jp074183o.

Ahmad, A., Ali, M., Al-Sehemi, A.G., Al-Ghamdi, A.A., Park, J.W., Algarni, H., Anwer, H., 2023. Carbon-integrated semiconductor photocatalysts for removal of volatile organic compounds in indoor environments. Chemical Engineering Journal 452, 139436. https://doi.org/10.1016/j.cej.2022.139436.

Amade, R., Hussain, S., Ocaña, I.R., Bertran, E., 2014. Growth and functionalization of carbon nanotubes on quartz filter for environmental applications. Journal of Environmental Engineering and Ecological Science 3, 2. https://doi.org/10.7243/2050-1323-3-2.

An, T., Chen, J., Nie, X., Li, G., Zhang, H., Liu, X., Zhao, H., 2012. Synthesis of carbon nanotube-anatase TiO_2 submicrometer-sized sphere composite photocatalyst for synergistic degradation of gaseous styrene. ACS Applied Materials & Interfaces 4, 5988–5996. https://doi.org/10.1021/am3016476.

Asai, M., Ohba, T., Iwanaga, T., Kanoh, H., Endo, M., Campos-Delgado, J., Terrones, M., Nakai, K., Kaneko, K., 2011. Marked adsorption irreversibility of graphitic nanoribbons for CO_2 and H_2O. Journal of the American Chemical Society 133, 14880–14883.

Awad, R., Haghighat Mamaghani, A., Boluk, Y., Hashisho, Z., 2021. Synthesis and characterization of electrospun PAN-based activated carbon nanofibers reinforced with cellulose nanocrystals for adsorption of VOCs. Chemical Engineering Journal 410, 128412. https://doi.org/10.1016/j.cej.2021.128412.

Azimi, F., Nabizadeh, R., Hassanvand, M.S., Rastkari, N., Nazmara, S., Naddafi, K., 2019. Photochemical degradation of toluene in gas-phase under UV/visible light graphene oxide-TiO_2 nanocomposite: influential operating factors, optimization, and modeling. Journal of Environmental Health Science and Engineering 17, 671–683. https://doi.org/10.1007/s40201-019-00382-x.

Bai, Y., Huang, Z.H., Kang, F., 2013. Synthesis of reduced graphene oxide/phenolic resin-based carbon composite ultrafine fibers and their adsorption performance for volatile organic compounds and water. Journal of Materials Chemistry A 1, 9536–9543. https://doi.org/10.1039/c3ta10545h.

Bao, C., Bi, S., Zhang, H., Zhao, J., Wang, P., Yue, C.Y., Yang, J., 2016. Graphene oxide beads for fast clean-up of hazardous chemicals. Journal of Materials Chemistry A 4, 9437–9446. https://doi.org/10.1039/c6ta01411a.

Bilal, M., Rizwan, K., Rahdar, A., Badran, M.F., Iqbal, H.M.N., 2022. Graphene-based porous nanohybrid architectures for adsorptive and photocatalytic abatement of volatile organic compounds. Environmental Pollution 309, 119805. https://doi.org/10.1016/j.envpol.2022.119805.

Chakrapani, N., Zhang, Y.M., Nayak, S.K., Moore, J.A., Carroll, D.L., Choi, Y.Y., Ajayan, P.M., 2003. Chemisorption of acetone on carbon nanotubes. Journal of Physical Chemistry B 107, 9308–9311. https://doi.org/10.1021/jp034970v.

Chen, F., An, W., Liu, L., Liang, Y., Cui, W., 2017. Highly efficient removal of bisphenol A by a three-dimensional graphene hydrogel-AgBr@rGO exhibiting adsorption/photocatalysis synergy. Applied Catalysis B: Environmental 217, 65–80. https://doi.org/10.1016/j.apcatb.2017.05.078.

Chu, F., Zheng, Y., Wen, B., Zhou, L., Yan, J., Chen, Y., 2018. Adsorption of toluene with water on zeolitic imidazolate framework-8/graphene oxide hybrid nanocomposites in a humid atmosphere. RSC Advances 8, 2426–2432. https://doi.org/10.1039/c7ra12931a.

Chun, H.H., Jo, W.K., 2016. Adsorption and photocatalysis of 2-ethyl-1-hexanol over graphene oxide-TiO_2 hybrids post-treated under various thermal conditions. Applied Catalysis B: Environmental 180, 740–750. https://doi.org/10.1016/j.apcatb.2015.07.021.

Dai, Y., Li, M., Liu, F., Xue, M., Wang, Y., Zhao, C., 2019. Graphene oxide wrapped copper-benzene-1,3,5-tricarboxylate metal organic framework as efficient absorbent for gaseous toluene under ambient conditions. Environmental Science & Pollution Research 26, 2477–2491. https://doi.org/10.1007/s11356-018-3657-8.

Diaz, E., Ordonez, S., Vega, A., 2007. Adsorption of volatile organic compounds onto carbon nanotubes, carbon nanofibers, and high-surface-area graphites. Journal of Colloid and Interface Science 305, 7–16.

Dreyer, D.R., Park, S., Bielawski, C.W., Ruoff, R.S., 2010. The chemistry of graphene oxide. Chemical Society Reviews 39, 228–240.

Esrafili, M.D., Dinparast, L., 2018. The selective adsorption of formaldehyde and methanol over Al- or Si-decorated graphene oxide: a DFT study. Journal of Molecular Graphics and Modelling 80, 25–31. https://doi.org/10.1016/j.jmgm.2017.12.025.

Fang, X., Lu, G., Mahmood, A., Wang, Y., Wang, X., Xie, X., Tang, Z., Sun, J., 2020. A novel ternary mica-titania@rGO composite pearlescent pigment for the photocatalytic degradation of gaseous acetaldehyde. Chemical Engineering Journal 396. https://doi.org/10.1016/j.cej.2020.125312.

Gabal, E., Chatterjee, S., Ahmed, F.K., Abd-Elsalam, K.A., 2019. Carbon nanomaterial applications in air pollution remediation. In: Carbon Nanomaterials for Agri-food and Environmental Applications. Elsevier Inc. https://doi.org/10.1016/B978-0-12-819786-8.00007-4.

Gan, G., Fan, S., Li, X., Zhang, Z., Hao, Z., 2023. Adsorption and membrane separation for removal and recovery of volatile organic compounds. Journal of Environmental Sciences (China) 123, 96–115. https://doi.org/10.1016/j.jes.2022.02.006.

Gangupomu, R.H., Sattler, M.L., Ramirez, D., 2016. Comparative study of carbon nanotubes and granular activated carbon: physicochemical properties and adsorption capacities. Journal of Hazardous Materials 302, 362–374. https://doi.org/10.1016/j.jhazmat.2015.09.002.

Gulati, S., Lingam B, H.N., Kumar, S., Goyal, K., Arora, A., Varma, R.S., 2022. Improving the air quality with functionalized carbon nanotubes: sensing and remediation applications in the real world. Chemosphere 299. https://doi.org/10.1016/j.chemosphere.2022.134468.

Guo, D., Feng, D., Zhang, Y., Zhang, Y., Zhao, Y., Zhou, Z., Sun, J., Quan, C., Chang, G., Sun, S., 2022. Carbon material-TiO_2 for photocatalytic reduction of CO_2 and degradation of VOCs: a critical review. Fuel Processing Technology 231, 107261. https://doi.org/10.1016/j.fuproc.2022.107261.

Guo, Y., Wen, M., Li, G., An, T., 2021. Recent advances in VOC elimination by catalytic oxidation technology onto various nanoparticles catalysts: a critical review. Applied Catalysis B: Environmental 281, 119447. https://doi.org/10.1016/j.apcatb.2020.119447.

Guo, Z., Huang, J., Xue, Z., Wang, X., 2016. Electrospun graphene oxide/carbon composite nanofibers with well-developed mesoporous structure and their adsorption performance for benzene and butanone. Chemical Engineering Journal 306, 99–106. https://doi.org/10.1016/j.cej.2016.07.048.

Gupta, V.K., Kumar, R., Nayak, A., Saleh, T.A., Barakat, M.A., 2013. Adsorptive removal of dyes from aqueous solution onto carbon nanotubes: a review. Advances in Colloid and Interface Science 193–194, 24–34. https://doi.org/10.1016/j.cis.2013.03.003.

Hsu, S.C., Lu, C., 2009. Adsorption kinetic, thermodynamic, and desorption studies of isopropyl alcohol vapor by oxidized single-walled carbon nanotubes. Journal of the Air & Waste Management Association 59, 990–997. https://doi.org/10.3155/1047-3289.59.8.990.

Hu, M., Hui, K.S., Hui, K.N., 2014. Role of graphene in MnO_2/graphene composite for catalytic ozonation of gaseous toluene. Chemical Engineering Journal 254, 237–244. https://doi.org/10.1016/j.cej.2014.05.099.

Hu, Y., Xie, X., Wang, X., Wang, Y., Zeng, Y., Pui, D.Y.H., Sun, J., 2018. Visible-light upconversion carbon quantum dots decorated TiO_2 for the photodegradation of flowing gaseous acetaldehyde. Applied Surface Science 440, 266–274. https://doi.org/10.1016/j.apsusc.2018.01.104.

Hussain, C.M., Saridara, C., Mitra, S., 2009. Modifying the sorption properties of multi-walled carbon nanotubes via covalent functionalization. Analyst 134, 1928–1933. https://doi.org/10.1039/b823316k.

Iijima, S., 1991. Helical microtubules of graphitic carbon. Nature 354, 56–58.

Jang, Y., Bang, J., Seon, Y.S., You, D.W., Oh, J.S., Jung, K.W., 2020. Carbon nanotube sponges as an enrichment material for aromatic volatile organic compounds. Journal of Chromatography A 1617, 460840. https://doi.org/10.1016/j.chroma.2019.460840.

Jin, D., Ren, Z., Ma, Z., Liu, F., Yang, H., 2015. Low temperature chlorobenzene catalytic oxidation over MnOx/CNTs with the assistance of ozone. RSC Advances 5, 15103–15109. https://doi.org/10.1039/c4ra16687f.

Jin, J., Kim, J.P., Wan, S., Kim, K.H., Choi, Y., Li, P., Kang, J., Ma, Z., Lee, J.H., Kwon, O., Kim, D.W., Park, J.H., 2022. Hierarchical pore enhanced adsorption and photocatalytic performance of graphene oxide/Ti-based metal-organic framework hybrid for toluene removal. Applied Catalysis B: Environmental 317, 121751. https://doi.org/10.1016/j.apcatb.2022.121751.

Jo, W.K., Kang, H.J., 2015. Photocatalysis of sub-ppm limonene over multiwalled carbon nanotubes/titania composite nanofiber under visible-light irradiation. Journal of Hazardous Materials 283, 680–688. https://doi.org/10.1016/j.jhazmat.2014.09.067.

Joung, H.J., Kim, J.H., Oh, J.S., You, D.W., Park, H.O., Jung, K.W., 2014. Catalytic oxidation of VOCs over CNT-supported platinum nanoparticles. Applied Surface Science 290, 267–273. https://doi.org/10.1016/j.apsusc.2013.11.066.

Kamal, M.S., Razzak, S.A., Hossain, M.M., 2016. Catalytic oxidation of volatile organic compounds (VOCs) - a review. Atmospheric Environment 140, 117–134. https://doi.org/10.1016/j.atmosenv.2016.05.031.

Kang, S., Hwang, J., 2020. Fabrication of hollow activated carbon nanofibers (HACNFs) containing manganese oxide catalyst for toluene removal via two-step process of electrospinning and thermal treatment. Chemical Engineering Journal 379, 122315. https://doi.org/10.1016/j.cej.2019.122315.

Kauffman, D.R., Star, A., 2008. Carbon nanotube gas and vapor sensors. Angewandte Chemie International Edition 47, 6550–6570. https://doi.org/10.1002/anie.200704488.

Khang, N.C., Van, D.Q., Thuy, N.M., Minh, N.V., Minh, P.N., 2016. Remarkably enhanced photocatalytic activity by sulfur-doped titanium dioxide in nanohybrids with carbon nanotubes. Journal of Physics and Chemistry of Solids 99, 119–123. https://doi.org/10.1016/j.jpcs.2016.06.011.

Khusnun, N.F., Jalil, A.A., Triwahyono, S., Hitam, C.N.C., Hassan, N.S., Jamian, F., Nabgan, W., Abdullah, T.A.T., Kamaruddin, M.J., Hartanto, D., 2018. Directing the amount of CNTs in CuO–CNT catalysts for enhanced adsorption-oriented visible-light-responsive photodegradation of p-chloroaniline. Powder Technology 327, 170–178. https://doi.org/10.1016/j.powtec.2017.12.052.

Kim, J.M., Kim, J.H., Lee, C.Y., Jerng, D.W., Ahn, H.S., 2018. Toluene and acetaldehyde removal from air on to graphene-based adsorbents with microsized pores. Journal of Hazardous Materials 344, 458–465. https://doi.org/10.1016/j.jhazmat.2017.10.038.

Koduru, J.R., Karri, R.R., Mubarak, N.M., 2019. Smart materials, magnetic graphene oxide-based nanocomposites for sustainable water purification. Sustainable Polymer Composites and Nanocomposites.

Kondratyuk, P., Yates, J.T., 2005. Desorption kinetic detection of different adsorption sites on opened carbon single walled nanotubes: the adsorption of n-nonane and CCl_4. Chemical Physics Letters 410, 324–329. https://doi.org/10.1016/j.cplett.2005.05.073.

Kottappara, R., Palantavida, S., Vijayan, B.K., 2019. Enhancing semiconductor photocatalysis with carbon nanostructures for water/air purification and self-cleaning applications. In: Carbon Based Nanomaterials for Advanced Thermal and Electrochemical Energy Storage and Conversion. Elsevier Inc. https://doi.org/10.1016/B978-0-12-814083-3.00007-X.

Kumar, V., Kumar, P., Pournara, A., Vellingiri, K., Kim, K.H., 2018. Nanomaterials for the sensing of narcotics: challenges and opportunities. TrAC, Trends in Analytical Chemistry 106, 84–115. https://doi.org/10.1016/j.trac.2018.07.003.

Kumar, V., Lee, Y.S., Shin, J.W., Kim, K.H., Kukkar, D., Fai Tsang, Y., 2020. Potential applications of graphene-based nanomaterials as adsorbent for removal of volatile organic compounds. Environment International 135, 105356. https://doi.org/10.1016/j.envint.2019.105356.

Kupreenko, S.Y., Strokova, N.E., Il'gova, E.A., Ivanov, A.S., Arkhipova, E.A., Savilov, S.V., 2022. Adsorption of organic solvent vapours on carbon nanotubes, few-layer graphene nanoflakes and their nitrogen-doped counterparts. Adsorption. https://doi.org/10.1007/s10450-021-00349-1.

Lee, C., Wei, X., Kysar, J.W., Hone, J., 2008. Measurement of the elastic properties and intrinsic strength of monolayer graphene. Science (80-) 321, 385–388. https://doi.org/10.1126/science.1157996.

Li, J., Lu, R., Dou, B., Ma, C., Hu, Q., Liang, Y., Wu, F., Qiao, S., Hao, Z., 2012. Porous graphitized carbon for adsorptive removal of benzene and the electrothermal regeneration. Environmental Science & Technology 46, 12648–12654.

Li, B., Mi, C., 2022. On the adsorption of volatile organic compounds on hydroxyl-functionalized carbon nanotubes in aqueous solution. Diamond and Related Materials 125, 108994. https://doi.org/10.1016/j.diamond.2022.108994.

Li, H., Yu, J., Gong, Y., Lin, N., Yang, Q., Zhang, X., Wang, Y., 2023. Perovskite catalysts with different dimensionalities for environmental and energy applications: a review. Separation and Purification Technology 307. https://doi.org/10.1016/j.seppur.2022.122716.

Li, J., Zhang, Q., Lai, A.C.K., Zeng, L., 2016a. Study on photocatalytic performance of cerium–graphene oxide–titanium dioxide composite film for formaldehyde removal. Physica Status Solidi (A) Applications and Materials 213, 3157–3164. https://doi.org/10.1002/pssa.201600261.

Li, M. syue, Wu, S.C., Shih, Y. hsin, 2016b. Characterization of volatile organic compound adsorption on multiwall carbon nanotubes under different levels of relative humidity using linear solvation energy relationship. Journal of Hazardous Materials 315, 35–41. https://doi.org/10.1016/j.jhazmat.2016.04.004.

Li, Q.L., Yuan, D.X., Lin, Q.M., 2004. Evaluation of multi-walled carbon nanotubes as an adsorbent for trapping volatile organic compounds from environmental samples. Journal of Chromatography A 1026, 283–288. https://doi.org/10.1016/j.chroma.2003.10.109.

Li, X., Zhang, L., Yang, Z., Wang, P., Yan, Y., Ran, J., 2020. Adsorption materials for volatile organic compounds (VOCs) and the key factors for VOC adsorption process: a review. Separation and Purification Technology 235. https://doi.org/10.1016/j.seppur.2019.116213.

Li, Y., Miao, J., Sun, X., Xiao, J., Li, Y., Wang, H., Xia, Q., Li, Z., 2016c. Mechanochemical synthesis of Cu-BTC@GO with enhanced water stability and toluene adsorption capacity. Chemical Engineering Journal 298, 191–197. https://doi.org/10.1016/j.cej.2016.03.141.

Li, Z., Gao, B., Chen, G.Z., Mokaya, R., Sotiropoulos, S., Li Puma, G., 2011. Carbon nanotube/titanium dioxide (CNT/TiO_2) core-shell nanocomposites with tailored shell thickness, CNT content and photocatalytic/photoelectrocatalytic properties. Applied Catalysis B: Environmental 110, 50–57. https://doi.org/10.1016/j.apcatb.2011.08.023.

Lin, W., Xie, X., Wang, X., Wang, Y., Segets, D., Sun, J., 2018. Efficient adsorption and sustainable degradation of gaseous acetaldehyde and o-xylene using rGO-TiO_2 photocatalyst. Chemical Engineering Journal 349, 708–718. https://doi.org/10.1016/j.cej.2018.05.107.

Liu, G.Q., Wan, M.X., Huang, Z.H., Kang, F.Y., 2015. Preparation of graphene/metal-organic composites and their adsorption performance for benzene and ethanol. Xinxing Tan Cailiao/New Carbon Materials 30, 566–571. https://doi.org/10.1016/S1872-5805(15)60205-0.

Liu, J.M., Li, L., Fan, H.L., Ning, Z.W., Zhao, P., 2007. Evaluation of single-walled carbon nanotubes as novel adsorbent for volatile organic compounds. Chinese Journal of Analytical Chemistry 35, 830–834. https://doi.org/10.1016/S1872-2040(07)60058-9.

Liu, W., Li, Z., Zhang, S.P., Jian, W.W., Ma, D.Z., 2021. Adsorption performance of multi-walled carbon nanotube-SiO_2 adsorbent for toluene. Ranliao Huaxue Xuebao/Journal of Fuel Chemistry and Technology 49, 861–872. https://doi.org/10.1016/S1872-5813(21)60090-7.

Liu, Y., Zhu, C., Sun, J., Ge, Y., Song, F., Xu, Q., Liu, Y., Zhu, C., Song, F., Zhu, C., Sun, J., Ge, Y., Xu, Q., 2020. In situ assembly of CQDs/Bi2WO6 for highly efficient photocatalytic degradation of VOCs under visible light. New Journal of Chemistry 44, 3455–3462. https://doi.org/10.1039/c9nj04957f.

Lizzio, A.A., DeBarr, J.A., 1996. Effect of surface area and chemisorbed oxygen on the SO_2 adsorption capacity of activated char. Fuel 75, 1515–1522. https://doi.org/10.1016/0016-2361(96)00127-5.

Loh, K.P., Bao, Q., Ang, P.K., Yang, J., 2010. The chemistry of graphene. Journal of Materials Chemistry 2277–2289.

Low, W., Boonamnuayvitaya, V., 2013. Enhancing the photocatalytic activity of TiO_2 co-doping of graphene-Fe^{3+} ions forformaldehyde removal. Journal of Environmental Management 127, 142–149. https://doi.org/10.1016/j.jenvman.2013.04.029.

Mahmood, A., Shi, G., Wang, Z., Rao, Z., Xiao, W., Xie, X., Sun, J., 2021. Carbon quantum dots-TiO_2 nanocomposite as an efficient photocatalyst for the photodegradation of aromatic ring-containing mixed VOCs: an experimental and DFT studies of adsorption and electronic structure of the interface. Journal of Hazardous Materials 401. https://doi.org/10.1016/j.jhazmat.2020.123402.

Mahmoud, A.E.D., 2020. Graphene-based nanomaterials for the removal of organic pollutants: insights into linear versus nonlinear mathematical models. Journal of Environmental Management 270. https://doi.org/10.1016/j.jenvman.2020.110911.

Mallakpour, S., Khadem, E., 2016. Carbon nanotube–metal oxide nanocomposites: fabrication, properties and applications. Chemical Engineering Journal 302, 344–367. https://doi.org/10.1016/j.cej.2016.05.038.

Matsuo, Y., Nishino, Y., Fukutsuka, T., Sugie, Y., 2008. Removal of formaldehyde from gas phase by silylated graphite oxide containing amino groups. Carbon N. Y. 46, 1162–1163. https://doi.org/10.1016/j.carbon.2008.04.015.

Miao, J., Li, C., Liu, H., Zhang, X., 2018. MnO_2/MWCNTs nanocomposites as highly efficient catalyst for indoor formaldehyde removal. Journal of Nanoscience and Nanotechnology 18 (9), 3982–3990.

Na, C.J., Yoo, M.J., Tsang, D.C.W., Kim, H.W., Kim, K.H., 2019. High-performance materials for effective sorptive removal of formaldehyde in air. Journal of Hazardous Materials 366, 452–465. https://doi.org/10.1016/j.jhazmat.2018.12.011.

Novoselov, K.S., Geim, A.K., Morozov, S.V., Jiang, D., Zhang, Y., Dubonos, S.V., Grigorieva, I.V., Firsov, A.A., 2004. Electric field effect in atomically thin carbon films. Science (80-) 306, 666–669.

Pan, B., Xing, B., 2008. Mechanisms of organic chemicals on carbon nanotubes. Environmental Science & Technology 42, 9005–9013.

Pandey, H., Saini, S., Singh, S.P., Gautam, N.K., Singh, S., 2020. Candle soot derived carbon nanoparticles: an assessment of cellular and progressive toxicity using *Drosophila melanogaster* model. Comparative Biochemistry and Physiology, Part C: Toxicology & Pharmacology 228. https://doi.org/10.1016/j.cbpc.2019.108646.

Pei, S., Cheng, H.M., 2012. The reduction of graphene oxide. Carbon 50, 3210–3228. https://doi.org/10.1016/j.carbon.2011.11.010.

Peng, S., Yang, X., Strong, J., Sarkar, B., Jiang, Q., Peng, F., Liu, D., Wang, H., 2020. MnO_2-decorated N-doped carbon nanotube with boosted activity for low-temperature oxidation of formaldehyde. Journal of Hazardous Materials 396. https://doi.org/10.1016/j.jhazmat.2020.122750.

Petit, C., Seredych, M., Bandosz, T.J., 2009. Revisiting the chemistry of graphite oxides and its effect on ammonia adsorption. Journal of Materials Chemistry 19, 9176–9185. https://doi.org/10.1039/b916672f.

Pierson, H.O., 1999. The chemistry of CVD. In: Handbook of Chemical Vapor Deposition. https://doi.org/10.1016/b978-081551432-9.50006-1.

Rahbar, M., Behpour, M., 2016. Multi-walled carbon nanotubes/TiO_2 thin layer for photocatalytic degradation of organic pollutant under visible light irradiation. Journal of Materials Science: Materials in Electronics 27, 8348–8355. https://doi.org/10.1007/s10854-016-4845-2.

Raphey, V.R., Henna, T.K., Nivitha, K.P., Mufeedha, P., Sabu, C., Pramod, K., 2019. Advanced biomedical applications of carbon nanotube. Materials Science and Engineering: C 100, 616–630. https://doi.org/10.1016/j.msec.2019.03.043.

Rasheed, T., Ahmad, N., Nawaz, S., Sher, F., 2020. Photocatalytic and adsorptive remediation of hazardous environmental pollutants by hybrid nanocomposites. Case Studies in Chemical and Environmental Engineering 2. https://doi.org/10.1016/j.cscee.2020.100037.

Ren, X., Chen, C., Nagatsu, M., Wang, X., 2011. Carbon nanotubes as adsorbents in environmental pollution management: a review. Chemical Engineering Journal 170, 395–410. https://doi.org/10.1016/j.cej.2010.08.045.

Rengaraj, S., Yeon, J.W., Kim, Y., Kim, W.H., 2007. Application of Mg-mesoporous alumina prepared by using magnesium stearate as a template for the removal of Nickel: kinetics, isotherm, and error analysis. Industrial & Engineering Chemistry Research 46, 2834–2842. https://doi.org/10.1021/ie060994n.

Roso, M., Boaretti, C., Pelizzo, M.G., Lauria, A., Modesti, M., Lorenzetti, A., 2017. Nanostructured photocatalysts based on different oxidized graphenes for VOCs removal. Industrial & Engineering Chemistry Research 56, 9980–9992. https://doi.org/10.1021/acs.iecr.7b02526.

Shadkam, R., Naderi, M., Ghazitabar, A., Akbari, S., 2021. Adsorption performance of reduced graphene-oxide/cellulose nano-crystal hybrid aerogels reinforced with waste-paper extracted cellulose-fibers for the removal of toluene pollution. Materials Today Communications 28. https://doi.org/10.1016/j.mtcomm.2021.102610.

Shajari, S., Kowsari, E., Seifvand, N., Boorboor Ajdari, F., Chinnappan, A., Ramakrishna, S., Saianand, G., Dashti Najafi, M., Haddadi-asl, V., Abdpour, S., 2021. Efficient photocatalytic degradation of gaseous benzene and toluene over novel hybrid pil@TiO_2/m-go composites. Catalysts 11, 1–24. https://doi.org/10.3390/catal11010126.

Shen, Y., Fang, Q., Chen, B., 2015. Environmental applications of three-dimensional graphene-based macrostructures adsorption, transformation, and detection. Environmental Science & Technology 49, 67–84.

Shih, Y. hsin, Li, M. syue, 2008. Adsorption of selected volatile organic vapors on multiwall carbon nanotubes. Journal of Hazardous Materials 154, 21–28. https://doi.org/10.1016/j.jhazmat.2007.09.095.

Shin, H.J., Kim, K.K., Benayad, A., Yoon, S.M., Park, H.K., Jung, I.S., Jin, M.H., Jeong, H.K., Kim, J.M., Choi, J.Y., Lee, Y.H., 2009. Efficient reduction of graphite oxide by sodium borohydride and its effect on electrical conductance. Advances in Functional Materials 19, 1987–1992. https://doi.org/10.1002/adfm.200900167.

Sone, H., Fugetsu, B., Tsukada, T., Endo, M., 2008. Affinity-based elimination of aromatic VOCs by highly crystalline multi-walled carbon nanotubes. Talanta 74, 1265–1270. https://doi.org/10.1016/j.talanta.2007.08.041.

Srivastava, I., Mishra, S., Singh, P.K., Gupta, T., Sankararamakrishnan, N., 2018. Fast and efficient removal of toluene, ethylbenzene and O-xylene from aqueous phase by functionalized carbon micro/nano composites. Journal of Environmental Chemical Engineering 6, 4917–4926. https://doi.org/10.1016/j.jece.2018.07.008.

Sun, X., Xia, Q., Zhao, Z., Li, Y., Li, Z., 2014. Synthesis and adsorption performance of MIL-101(Cr)/graphite oxide composites with high capacities of n-hexane. Chemical Engineering Journal 239, 226–232. https://doi.org/10.1016/j.cej.2013.11.024.

Sunkara, B., Zhan, J., He, J., McPherson, G.L., Piringer, G., John, V.T., 2010. Nanoscale zerovalent iron supported on uniform carbon microspheres for the in situ remediation of chlorinated hydrocarbons. ACS Applied Materials & Interfaces 2, 2854–2862. https://doi.org/10.1021/am1005282.

Szcześniak, B., Choma, J., Jaroniec, M., 2018a. Effect of graphene oxide on the adsorption properties of ordered mesoporous carbons toward H_2, C_6H_6, CH_4 and CO_2. Microporous and Mesoporous Materials 261, 105–110. https://doi.org/10.1016/j.micromeso.2017.10.054.

Szcześniak, B., Osuchowski, Ł., Choma, J., Jaroniec, M., 2018b. Highly porous carbons obtained by activation of polypyrrole/reduced graphene oxide as effective adsorbents for CO_2, H_2 and C_6H_6. Journal of Porous Materials 25, 621–627. https://doi.org/10.1007/s10934-017-0475-1.

Szulejko, J.E., Kim, K.H., Parise, J., 2019. Seeking the most powerful and practical real-world sorbents for gaseous benzene as a representative volatile organic compound based on performance metrics. Separation and Purification Technology 212, 980–985. https://doi.org/10.1016/j.seppur.2018.11.001.

Tahriri, M., Del Monico, M., Moghanian, A., Tavakkoli Yaraki, M., Torres, R., Yadegari, A., Tayebi, L., 2019. Graphene and its derivatives: opportunities and challenges in dentistry. Materials Science and Engineering: C 102, 171–185. https://doi.org/10.1016/j.msec.2019.04.051.

Tang, W., Liu, G., Li, D., Liu, H., Wu, X., Han, N., Chen, Y., 2015. Design and synthesis of porous non-noble metal oxides for catalytic removal of VOCs. Science China Chemistry 58, 1359–1366. https://doi.org/10.1007/s11426-015-5469-8.

Tulaphol, S., Bunsan, S., Kanchanatip, E., Miao, H.Y., Grisdanurak, N., Den, W., 2016. Influence of chlorine substitution on adsorption of gaseous chlorinated phenolics on multi-walled carbon nanotubes embedded in SiO_2. International journal of Environmental Science and Technology 13, 1465–1474. https://doi.org/10.1007/s13762-016-0984-5.

Vashist, S.K., Zheng, D., Al-Rubeaan, K., Luong, J.H.T., Sheu, F.S., 2011. Advances in carbon nanotube based electrochemical sensors for bioanalytical applications. Biotechnol Adv 29, 169–188. https://doi.org/10.1016/j.biotechadv.2010.10.002.

Vermisoglou, E.C., Georgakilas, V., Kouvelos, E., Pilatos, G., Viras, K., Romanos, G., Kanellopoulos, N.K., 2007. Sorption properties of modified single-walled carbon nanotubes. Microporous and Mesoporous Materials 99, 98–105. https://doi.org/10.1016/j.micromeso.2006.07.035.

Wang, G., Guo, W., Xu, D., Liu, D., Qin, M., 2020a. Graphene oxide hybridised TiO_2 for visible light photocatalytic degradation of phenol. Symmetry (Basel) 12. https://doi.org/10.3390/SYM12091420.

Wang, H., Raziq, F., Qu, Y., Qin, C., Wang, J., Jing, L., 2015. Role of quaternary N in N-doped graphene-Fe_2O_3 nanocomposites as efficient photocatalysts for CO_2 reduction and acetaldehyde degradation. RSC Advances 5, 85061–85064. https://doi.org/10.1039/c5ra17882g.

Wang, L., Liu, J., Zhao, P., Ning, Z., Fan, H., 2010. Novel adsorbent based on multi-walled carbon nanotubes bonding on the external surface of porous silica gel particulates for trapping volatile organic compounds. Journal of Chromatography A 1217, 5741–5745. https://doi.org/10.1016/j.chroma.2010.07.028.

Wang, M., Hua, J., Yang, Y., 2018. Fabrication of CDs/CdS-TiO_2 ternary nano-composites for photocatalytic degradation of benzene and toluene under visible light irradiation. Spectrochimica Acta, Part A: Molecular and Biomolecular Spectroscopy 199, 102–109. https://doi.org/10.1016/j.saa.2018.03.041.

Wang, Z., Yu, H., Xiao, Y., Zhang, L., Guo, L., Zhang, L., Dong, X., 2020b. Free-standing composite films of multiple 2D nanosheets: synergetic photothermocatalysis/photocatalysis for efficient removal of formaldehyde under ambient condition. Chemical Engineering Journal 394. https://doi.org/10.1016/j.cej.2020.125014.

Wu, L., Qin, Z., Zhang, L., Meng, T., Yu, F., Ma, J., 2017. CNT-enhanced amino-functionalized graphene aerogel adsorbent for highly efficient removal of formaldehyde. New Journal of Chemistry 41, 2527–2533. https://doi.org/10.1039/c6nj03643k.

Wu, L., Zhang, L., Meng, T., Yu, F., Chen, J., Ma, J., 2015. Facile synthesis of 3D amino-functional graphene-sponge composites decorated by graphene nanodots with enhanced removal of indoor formaldehyde. Aerosol and Air Quality Research 15, 1028–1034. https://doi.org/10.4209/aaqr.2014.10.0237.

Xu, Y.J., Zhuang, Y., Fu, X., 2010. New insight for enhanced photocatalytic activity of TiO_2 by doping carbon nanotubes: a case study on degradation of benzene and methyl orange. Journal of Physical Chemistry C 114, 2669–2676. https://doi.org/10.1021/jp909855p.

Yan, J., Yu, Y., Xiao, J., Li, Y., Li, Z., 2016. Improved ethanol adsorption capacity and coefficient of performance for adsorption chillers of Cu-btc@GO composite prepared by rapid room temperature synthesis. Industrial & Engineering Chemistry Research 55, 11767–11774. https://doi.org/10.1021/acs.iecr.6b03139.

Yan, S., Song, H., Li, Y., Yang, J., Jia, X., Wang, S., Yang, X., 2022. Integrated reduced graphene oxide/polypyrrole hybrid aerogels for simultaneous photocatalytic decontamination and water evaporation. Applied Catalysis B: Environmental 301. https://doi.org/10.1016/j.apcatb.2021.120820.

Yang, C., Miao, G., Pi, Y., Xia, Q., Wu, J., Li, Z., Xiao, J., 2019. Abatement of various types of VOCs by adsorption/catalytic oxidation: a review. Chemical Engineering Journal 370, 1128–1153. https://doi.org/10.1016/j.cej.2019.03.232.

Yang, F., Wang, M., Zhang, D., Yang, J., Zheng, M., Li, Y., 2020. Chirality pure carbon nanotubes: growth, sorting, and characterization. Chemistry Review 120, 2693–2758. https://doi.org/10.1021/acs.chemrev.9b00835.

Yang, S., Zhu, Z., Wei, F., Yang, X., 2017. Enhancement of formaldehyde removal by activated carbon fiber via in situ growth of carbon nanotubes. Building and Environment 126, 27–33. https://doi.org/10.1016/j.buildenv.2017.09.025.

Yang, W., Tang, S., Wei, Z., Chen, X., Ma, C., Duan, J., Tan, R., 2021. Separate-free $BiPO_4$/graphene aerogel with 3D network structure for efficient photocatalytic mineralization by adsorption enrichment and photocatalytic degradation. Chemical Engineering Journal 421. https://doi.org/10.1016/j.cej.2021.129720.

Ye, H., Wang, Z., Liu, Y., Chen, S., Wang, H., Wu, Z., 2019. Efficient degradation of gas phase toluene by ozone-assisted photocatalytic oxidation on TiO2/graphene composites. Catalysis Letters 149, 2739–2748.

Yu, J., Ma, T., Liu, S., 2011. Enhanced photocatalytic activity of mesoporous TiO_2 aggregates by embedding carbon nanotubes as electron-transfer channel. Physical Chemistry Chemical Physics 13, 3491–3501. https://doi.org/10.1039/c0cp01139h.

Yu, L., Wang, L., Sun, X., Ye, D., 2018a. Enhanced photocatalytic activity of rGO/TiO2 for the decomposition of formaldehyde under visible light irradiation. Journal of Environmental Sciences 73, 138–146. https://doi.org/10.1016/j.jes.2018.01.022.

Yu, L., Wang, L., Xu, W., Chen, L., Fu, M., Wu, J., Ye, D., 2018b. Adsorption of VOCs on reduced graphene oxide. Journal of Environmental Sciences 67, 171–178. https://doi.org/10.1016/j.jes.2017.08.022.

Zare, K., Gupta, V.K., Moradi, O., Makhlouf, A.S.H., Sillanpää, M., Nadagouda, M.N., Sadegh, H., Shahryari-ghoshekandi, R., Pal, A., Wang, Z. jun, Tyagi, I., Kazemi, M., 2015. A comparative study on the basis of adsorption capacity between CNTs and activated carbon as adsorbents for removal of noxious synthetic dyes: a review. J. Nanostructure Chem. 5, 227–236. https://doi.org/10.1007/s40097-015-0158-x.

Zeng, Q., Xie, X., Wang, X., Lu, G., Li, H., Cheng Lee, S., Sun, J., 2019. New insights into the synergistic effect of active radicals and adsorptive ability on the photodegradation of gaseous acetaldehyde over reduced graphene Oxide/P25 composite. Journal of Hazardous Materials 380. https://doi.org/10.1016/j.jhazmat.2019.120814.

Zhang, W., Li, G., Liu, H., Chen, J., Ma, S., Wen, M., Kong, J., An, T., 2020. Photocatalytic degradation mechanism of gaseous styrene over Au/TiO2@CNTs: relevance of superficial state with deactivation mechanism. Applied Catalysis B: Environmental 272. https://doi.org/10.1016/j.apcatb.2020.118969.

Zhang, X., Gao, B., Creamer, A.E., Cao, C., Li, Y., 2017. Adsorption of VOCs onto engineered carbon materials: a review. Journal of Hazardous Materials 338, 102–123. https://doi.org/10.1016/j.jhazmat.2017.05.013.

Zhang, Y., Tang, Z.R., Fu, X., Xu, Y.J., 2010. TiO_2-graphene nanocomposites for gas-phase photocatalytic degradation of volatile aromatic pollutant: is TiO_2-graphene truly different from other TiO_2-carbon composite materials? ACS Nano 4, 7303–7314. https://doi.org/10.1021/nn1024219.

Zhang, Y.N., Niu, Q., Gu, X., Yang, N., Zhao, G., 2019. Recent progress on carbon nanomaterials for the electrochemical detection and removal of environmental pollutants. Nanoscale 11, 11992–12014. https://doi.org/10.1039/c9nr02935d.

Zhang, Z., Jiang, Z., Shangguan, W., 2016. Low-temperature catalysis for VOC removal in technology and application: a state-of-the-art review. Catalysis Today 264, 270–278. https://doi.org/10.1016/j.cattod.2015.10.040.

Zheng, Y., Chu, F., Zhang, B., Yan, J., Chen, Y., 2018. Ultrahigh adsorption capacities of carbon tetrachloride on MIL-101 and MIL-101/graphene oxide composites. Microporous and Mesoporous Materials 263, 71–76. https://doi.org/10.1016/j.micromeso.2017.12.007.

Zhou, X., Huang, W., Shi, J., Zhao, Z., Xia, Q., Li, Y., Wang, H., Li, Z., 2014. A novel MOF/graphene oxide composite GrO@MIL-101 with high adsorption capacity for acetone. Journal of Materials Chemistry A 2, 4722–4730. https://doi.org/10.1039/c3ta15086k.

Zhou, Y., Zhou, L., Zhang, X., Chen, Y., 2016. Preparation of zeolitic imidazolate framework-8/graphene oxide composites with enhanced VOC adsorption capacity. Microporous and Mesoporous Materials 225, 488–493. https://doi.org/10.1016/j.micromeso.2016.01.047.

Zhou, Z., Gao, J., Zhang, G., Dong, Y., Wang, Z., Li, J., Lyu, J., 2020. Optimizing graphene-TiO_2 interface properties via Fermi level modulation for photocatalytic degradation of volatile organic compounds. Ceramics International 46, 5887–5893. https://doi.org/10.1016/j.ceramint.2019.11.040.

Zhu, L., Shen, D., Luo, K.H., 2020. A critical review on VOC adsorption by different porous materials: species, mechanisms and modification methods. Journal of Hazardous Materials 389, 122102. https://doi.org/10.1016/j.jhazmat.2020.122102.

Zou, W., Gao, B., Ok, Y.S., Dong, L., 2019. Integrated adsorption and photocatalytic degradation of volatile organic compounds (VOCs) using carbon-based nanocomposites: a critical review. Chemosphere 218, 845–859. https://doi.org/10.1016/j.chemosphere.2018.11.175.

Zouzelka, R., Kusumawati, Y., Remzova, M., Rathousky, J., Pauporté, T., 2016. Photocatalytic activity of porous multiwalled carbon nanotube-TiO_2 composite layers for pollutant degradation. Journal of Hazardous Materials 317, 52–59. https://doi.org/10.1016/j.jhazmat.2016.05.056.

CHAPTER

10

Metal–organic frameworks (MOFs) for VOC removal

Yuxuan Zhang[1], *Jun Luo*[1], *Xiuyan Wang*[1], *Dunhan Zhang*[1], *Hongxia Xu*[2] *and Xin Hu*[3]

[1]State Key Laboratory of Pollution Control and Resource Reuse, School of the Environment, Nanjing University, Nanjing, China; [2]Key Laboratory of Surficial Geochemistry of Ministry of Education, School of Earth Sciences and Engineering, Hydrosciences Department, Nanjing University, Nanjing, China; [3]State Key Laboratory of Analytical Chemistry for Life Science, Centre of Materials Analysis and School of Chemistry & Chemical Engineering, Nanjing University, Nanjing, China

10.1 Overview

Metal–organic frameworks (MOFs), also known as coordination polymers, are a type of organic–inorganic hybrid crystalline porous material with infinite and uniform framework structures consisting of organic linker units and inorganic metal (or metal cluster) nodes (Yuan et al., 2018). Over the past 2 decades, MOFs have attracted widespread attention worldwide due to their remarkable properties, such as ultra-high specific surface area (up to 3000 m^2/g), excellent thermal stability (>400°C), uniform but tunable pore size, abundant active sites, and ease of functionalization (Zhu et al., 2020). Therefore, potential applications of MOFs in gas storage (Murray et al., 2009; Furukawa and Yaghi, 2009), adsorption (Li et al., 2009a), separation (Li et al., 2012), heterogeneous catalysis (Lee et al., 2009; Yoon et al., 2012), and sensing (Kreno et al., 2012) have been extensively explored. Due to their numerous excellent physicochemical properties, MOFs have been rapidly developed as an alternative of traditional environmental functional materials for the removal of harmful gases (Chaikittisilp et al., 2013). They exhibit extremely high adsorption capacity for various gases at room temperature and have broad application prospects.

At present, the mainstream strategies for treating harmful gases using MOF-based materials can be divided into two categories based on governing mechanisms. One is to use the high specific surface area and pore structure characteristics for adsorption and removal of

Volatile Organic Compound Removal
https://doi.org/10.1016/B978-0-443-15674-8.00003-2

harmful gases. In terms of adsorption, research on MOFs initially focused on their adsorption of H_2 (Murray et al., 2009; Suh et al., 2012; Rowsell and Yaghi, 2006), and gradually expanded to other gases such as CO_2 (Nugent et al., 2013; Farha et al., 2010), CH_4 (Zhou et al., 2007), H_2S (Petit et al., 2010; Sánchez-González et al., 2018), and VOCs (Li et al., 2020b), among which the adsorption of VOCs is currently one of the research hotspots. All studies indicate that MOFs are a class of excellent gas adsorption materials. The other is catalytic degradation of harmful gases into non-toxic or low-toxicity molecules such as CO_2 and H_2O. This method relies on doping MOFs with elements, compounding functional components or further deriving metal oxides for catalytic degradation of VOCs. Functional materials based on MOF-5, MOF-199, MIL-101(Cr), and UiO-66 have been often applied in the adsorption and catalysis of VOCs such as benzene, toluene, xylene, acetone, and formaldehyde (FA).

The goal of this chapter is to systematically summarize the fundamentals of MOFs and their applications as effective adsorbents/catalysts in VOC removal. It not only covers the classification, properties, preparation methods, as well as the performance and mechanisms of MOFs in removing VOCs, but also provides theoretical foundations and technical guidance for their practical application in waste gas treatment. Current challenges and future research directions of MOFs in treating VOCs are outlined at the end.

10.1.1 Classification of MOFs

So far, tens of thousands of MOFs have been synthesized and structurally characterized. MOFs can be classified based on their preparation methods and composition, including IRMOFs (isoreticular metal–organic frameworks), ZIFs (zeolitic imidazolate frameworks), MILs (Materials of Institute Lavoisier frameworks), and UiO (University of Oslo) frameworks, among others.

10.1.1.1 IRMOFs

IRMOF-1 (also known as MOF-5, $Zn_4O(BDC)_3$, where BDC is 1,4-benzenedicarboxylate) was first synthesized by Omar M. Yaghi's group in 1999 (Li et al., 1999). Since then, a series of porous materials, called IRMOFs, have been extended by Yaghi and others based on IRMOF-1 as a prototype, by connecting octahedral Zn–O–C clusters with various organic linkers, which have the same framework structure and similar physicochemical properties. The IRMOF series are pioneers in the field of MOFs and were among the first MOFs to be synthesized and studied, exhibiting high controllability and predictability (Eddaoudi et al., 2002).

Among the numerous IRMOFs, IRMOF-1 is the most representative one, which is a three-dimensional (3D) stereo material formed by using 1,4-benzenedicarboxylic acid as the organic linker to coordinate with Zn ions in an octahedral form, as shown in Fig. 10.1 (Li et al., 1999). It has a high specific surface area and can exist stably even when the solvent molecules in the pore channels are removed or when heated to 300°C.

The pores of IRMOFs exhibit increasing and uniform periodicity, with the highest reaching up to 28.8 Å. Its porosity can increase from 55.8% in IRMOF-5–91.1% in IRMOF-16, which greatly exceeds the porosity of typical zeolites (Eddaoudi et al., 2002). In addition, the design and synthesis of IRMOFs have laid the foundation for the design and synthesis of subsequent MOF materials.

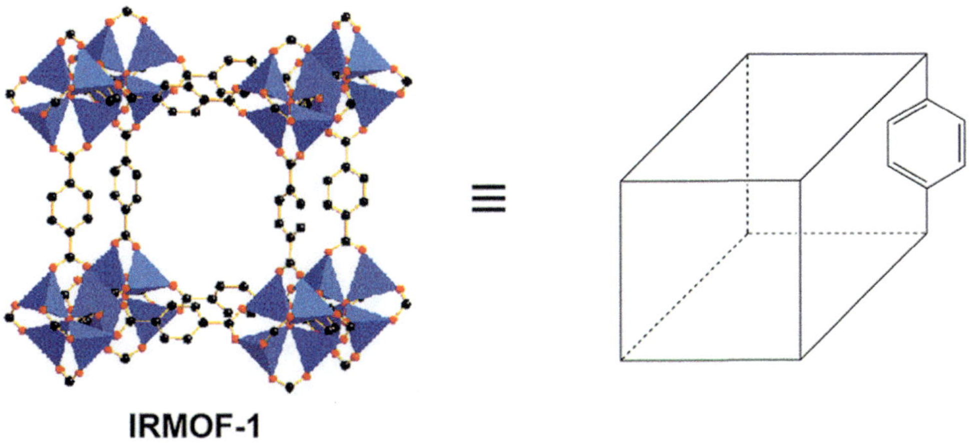

FIGURE 10.1 Structure of IRMOF-1 derived from the nonfunctionalized ligand BDC. Color scheme: For Zn atoms, *blue*; for O atoms, *red*; for C atoms, *black*; for N atoms, *blue* (Zhu et al., 2017). *Reprinted (adapted) with permission from {Metal–organic frameworks for heterogeneous basic catalysis}. Copyright {2017} American Chemical Society.*

10.1.1.2 ZIFs

ZIFs are coordination compounds with zeolite-like structures that are synthesized by reacting zinc or cobalt as the metal source with imidazole or imidazole derivatives as the organic linker in a solvent. They were first proposed by Yaghi's group, and their crystal structure is shown in Fig. 10.2 (Park et al., 2006; Jhinjer et al., 2021). Subsequently, a series of ZIF materials with different pore structures and surface chemical properties have been synthesized by changing the organic linker or functional group, while maintaining the same topology structure. The metal ions have also been extended to Cu and Fe, and their functions have become increasingly diverse (Jhinjer et al., 2021; Xie et al., 2022).

To date, more than 150 structures of ZIFs have been synthesized, including representative materials such as ZIF-8 and ZIF-67. In 1989, Chen's group first synthesized ZIF-8, which was officially named by Yaghi's group after a systematic analysis (Park et al., 2006; Huang et al., 2006). ZIF-8 is composed of zinc atoms and 2-methylimidazole, with a chemical formula of $Zn(Hmim)_2$, and it exhibits excellent stability (Phan et al., 2010; Venna et al., 2010). ZIF-67 is composed of Co(II) ions and 2-methylimidazole, with a cubic crystal symmetry structure (Zhong et al., 2018a,b).

Compared with other MOFs, ZIFs have permanent porosity, high mechanical stability, thermal stability, and chemical stability, making them attractive candidates for many applications such as adsorption, separation, and gas storage (Park et al., 2006; Cheng et al., 2018; Yao and Wang, 2014). The nitrogen-rich organic linkers in ZIF materials can serve as natural nitrogen sources for nitrogen-doped carbon materials. In addition, the compatibility of the metal coordination center with one or more metal ions such as Zn, Co, Ni, Fe, Cu, and Mn is conducive to the construction of multi-functional active sites (Song et al., 2021). ZIFs also have the advantages of simple synthesis, readily available raw materials, and low cost.

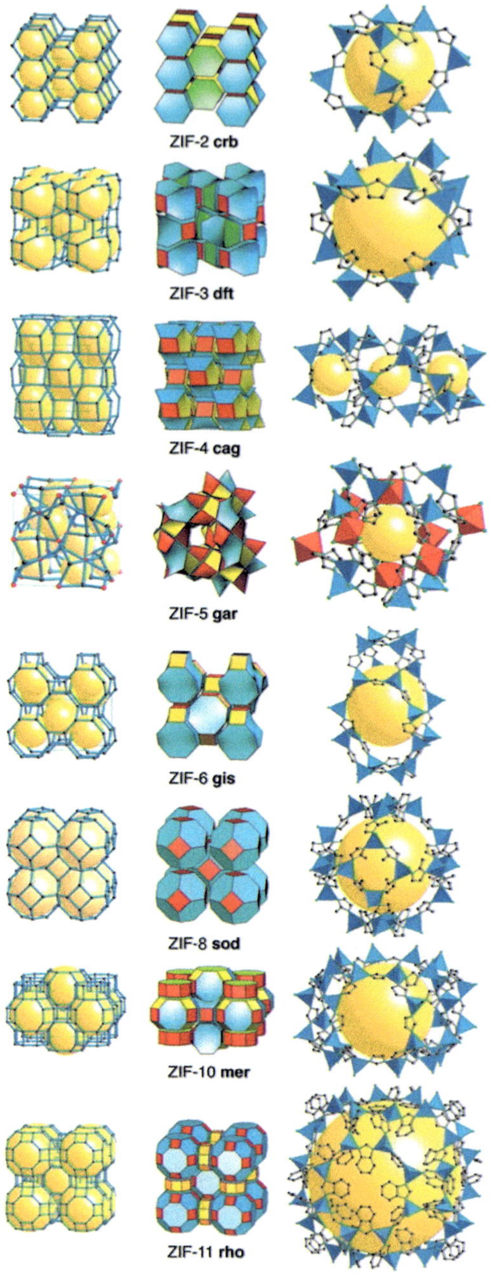

FIGURE 10.2 The ingle crystal X-ray structures of ZIFs. In each row, the net is shown as a stick diagram (left) and as a tiling (center). The largest cage (right) in each ZIF is shown with ZnN_4 tetrahedra in *blue*, and, for ZIF-5, InN6 octahedra in *red*. H atoms are omitted for clarity (Park et al., 2006). *Copyright (2006) National Academy of Sciences, USA.*

10.1.1.3 MILs

MILs were first synthesized by Ferey et al. at the Institute Lavoisier, and were named after the abbreviation of the institute (Férey et al., 2005). Initially, MILs were mainly synthesized by connecting the central metal with organic phosphonic acid or succinic acid (Livage et al., 1998; Riou et al., 1998). Modern MILs are typically composed of trivalent metal ions (such as Fe, Al, Ga, In, V, or Cr) and carboxylate salts, and their main structures are shown in Fig. 10.3 (Janiak and Vieth, 2010).

The most representative MIL series materials developed so far include MIL-53, MIL-100, and MIL-101. MIL-53, with the molecular formula $M^{III}(OH)\cdot[O_2C{-}C_6H_4{-}CO_2]\cdot H_2O$ (M^{III} = Al, Cr, Fe), is a 3D metal–organic framework synthesized from a mixture of soluble metal precursors (aluminum nitrate or chromium nitrate/HF) and a water-soluble solution of 1,4-benzenedicarboxylic acid (1,4-BDC). This structure is composed of infinite intersecting octahedral $Mo_4(OH)_2$ chains interconnected by BDC molecules (Férey et al., 2003). The structural unit of MIL-100 consists of three octahedra interconnected by 1,3,5-benzenetricarboxylic acid (1,3,5-BTC) linkers, and the metal units have been expanded from the initial Cr(III) to Fe(III), Sc(III), and Al(III). MIL-101 is composed of 1,4-BDC combined with a similar trimer, and the metal units include Al(III), Cr(III), V(III), and Sc(III) (Férey et al., 2005; Zhong et al., 2018a,b). MIL materials have permanent porosity, high specific surface area, high stability, and strong flexibility, making them promising adsorbents and catalysts in various fields (Xie et al., 2022; Zhang et al., 2022a).

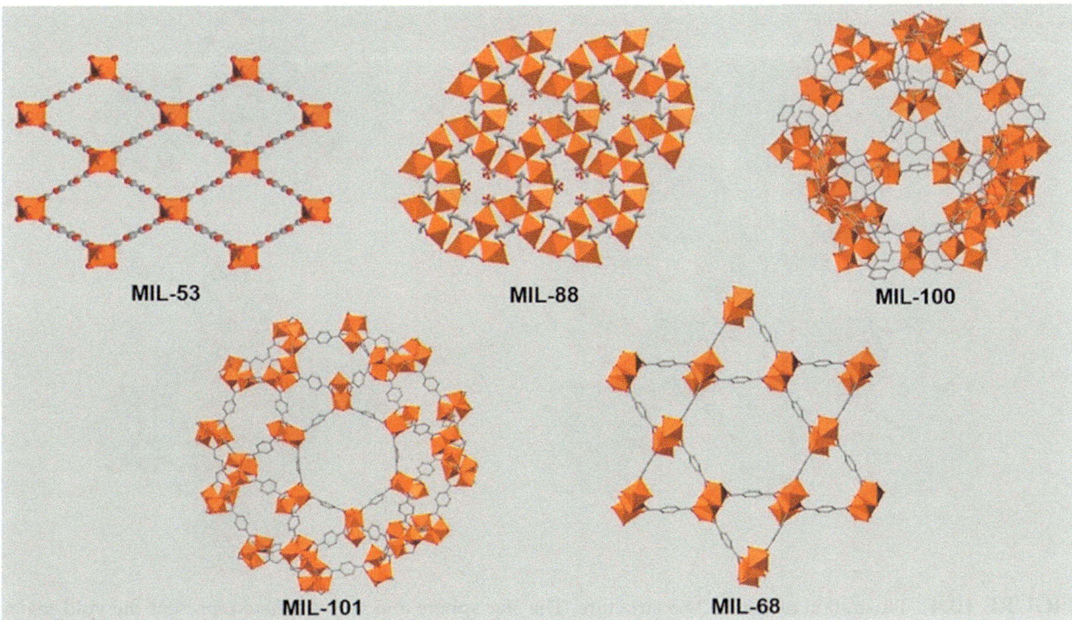

FIGURE 10.3 Polyhedral structures of MIL-53, MIL-88, MIL-68, MIL-100, and MIL-101 (Dapaah et al., 2022). *Reprinted from Dapaah, M.F., Niu, Q., Yu, Y., You, T., Liu, B., Cheng, L., 2022. Efficient persistent organic pollutant removal in water using MIL-metal–organic framework driven Fenton-like reactions: a critical review. Chemical Engineering Journal 431, 134182. Copyright (2022), with permission from Elsevier.*

10.1.1.4 UiOs

The UiO series MOFs were first reported by Lillerud et al., in 2008 (Cavka et al., 2008). They are 3D porous materials constructed with the metal center Zr(IV) and dicarboxylate ligands, and various MOFs with the similar network structure can be obtained by selecting different forms of dicarboxylate (Bai et al., 2016). Since 2008, the UiO series MOFs have developed into a large family with high stability and versatility by combining different geometric and symmetrical organic linkers with Zr nodes (Bai et al., 2016). The representative one is UiO-66 with the formula $[Zr_6O_4(OH)_4(BDC)_6]$, which is composed of zirconium clusters $(Zr_6O_4(OH)_4)$ as secondary building units (SBUs) and 1,4-benzenedicarboxylic acid (BDC) as the linker (Fig. 10.4). UiO-66 has attracted much attention due to its excellent mechanical, chemical, and thermal stability (Feng et al., 2021).

Although many MOFs have been reported, only a few are stable under water, organic solvents, or acidic and alkaline conditions. Among MOFs, the UiO series MOFs exhibit excellent water stability due to the strong Zr–O bonds and specific geometric shape that hinder the inclusion of water molecules and alleviate hydrolysis reactions (Cavka et al., 2008; Lee et al., 2018). In addition, numerous studies have shown that UiO series MOFs are easy to

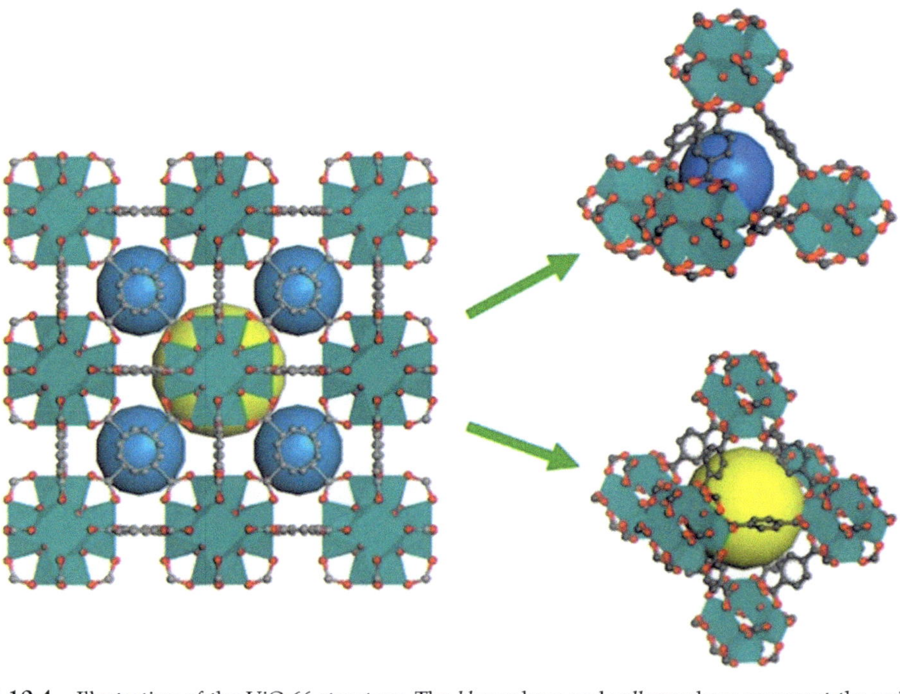

FIGURE 10.4 Illustration of the UiO-66 structure. The *blue* sphere and *yellow* sphere represent the void regions inside the tetrahedral and octahedral cages, respectively. Hydrogen atoms on the organic linkers were omitted for clarity (Ramsahye et al., 2014). *Reprinted (adapted) with permission from Ramsahye, N.A., Gao, J., Jobic, H., Llewellyn, P.L., Yang, Q., Wiersum, A.D., Koza, M.M., Guillerm, V., Serre, C., Zhong, C.L., Maurin, G., 2014. Adsorption and diffusion of light hydrocarbons in UiO-66(Zr): a combination of experimental and modeling tools. The Journal of Physical Chemistry C 118 (47), 27470–27482. Copyright {2014} American Chemical Society.*

functionalize. It can be achieved by using functionalized ligands as starting materials, or by post-synthetic modification or substitution, and the resulting products retain the same topology structure.

10.1.1.5 Other representative MOFs

MOF-199 ($[Cu_3(TMA)_2(H_2O)_3]_n$, where TMA is benzene-1,3,5-tricarboxylate) was synthesized at Hong Kong University of Science and Technology (HKUST) in 1999 and named HKUST-1, also known as Cu-BTC MOF. The MOF-199 framework consists of dimeric metal units connected by TMA. As shown in Fig. 10.5, this polymer forms face-centered cubic crystals containing an intersecting 3D large square pore system (9 Å * 9 Å). The material exhibits excellent thermal stability (up to 240°C) and the channel lining is easily chemically functionalized (Chui et al., 1999).

MOF-177 ($Zn_4O(btb)_2$, where btb = 1,3,5-benzenetribenzoate) is another representative MOF material that contains tetrahedral Zn_4O clusters and benzenetribenzoate ligands. The connection between the tricarboxylic salt unit and the metal cluster is the same as in MOF-5 (Fig. 10.6) (Chae et al., 2004). MOF-177 has been shown to have one of the largest surface areas among known materials, with an apparent Langmuir surface area of up to 4500 m^2/g (Chae et al., 2004; Rowsell and Yaghi, 2005).

MOF-74 ($Zn_2(DHBDC)\ (DMF)_2 \cdot (H_2O)_2$) is synthesized by mixing 2,5-dihydroxy-1,4-benzenedicarboxylic acid (H_2-DHBDC) and $Zn(NO_3)_2 \cdot 4(H_2O)$ in a DMF solution (Rosi et al., 2005). As shown in Fig. 10.7, the remaining coordination sites of the metal are coordinated with DMF, resulting in Lewis acidic open metal sites after DMF removal. Currently, the metal center of MOF-74 has been extended to more than 10 divalent metal ions, including Zn,

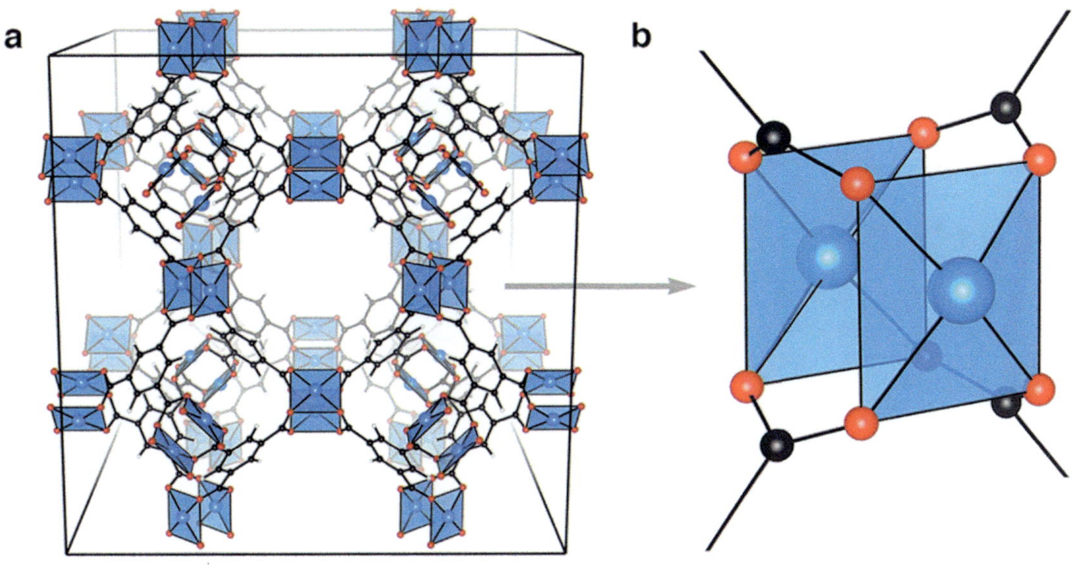

FIGURE 10.5 (a) The coordinatively unsaturated (anhydrous) structure of MOF-199. Each crystallographic unit cell features 32 (b) Cu···Cu paddlewheel motifs, linked by the aromatic tricarboxylate, btc, and 1,3,5-benzene tricarboxylate (Hendon and Walsh, 2015).

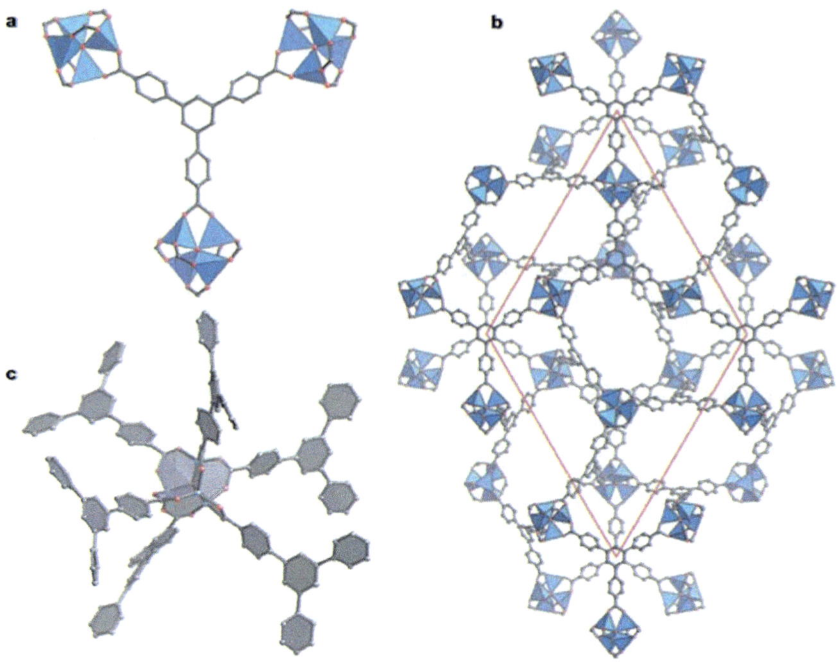

FIGURE 10.6 The structure of MOF-177. (a) A BTB unit linked to three OZn_4 units (H atoms are omitted). ZnO_4 tetrahedra are shown in *blue* and O and C atoms are shown as *red* and *black* spheres, respectively. (b) The structure projected down [001]. Colors as for (a). For clarity only about half the *c*-axis repeat unit is shown. (c) A fragment of the structure radiating from a central OZn_4: Six-membered rings are shown as *gray* hexagons and Zn atoms as *blue* spheres (Chae et al., 2004). *Reproduced with permission from Springer Nature.*

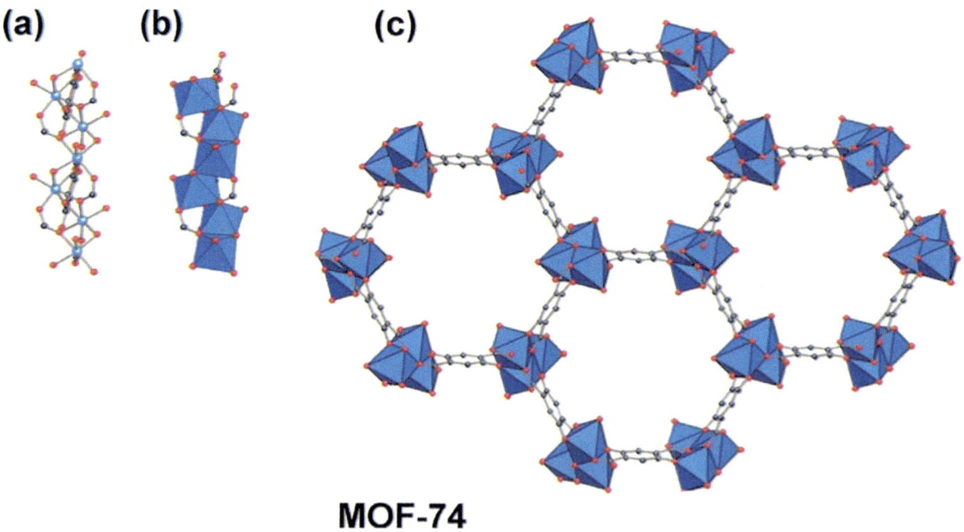

FIGURE 10.7 (a) MOF-74: ball-and-stick representation of SBU; (b) SBU with Zn shown as polyhedra; (c) view of crystalline framework with inorganic SBUs linked together via the benzene ring of 2,5-dihydroxybenzene-1,4-dicarboxylate (DMF and H_2O guest molecules have been omitted for clarity) (Rosi et al., 2005). *Reprinted (adapted) with permission from Rosi, N.L., Kim, J., Eddaoudi, M., Chen, B., O'Keeffe, M., Yaghi, O.M., 2005. Rod packings and metal–organic frameworks constructed from rod-shaped secondary building units. Journal of the American Chemical Society 127 (5), 1504–1518. Copyright {2005} American Chemical Society.*

Cu, Ni, and Co, etc. MOF-74 has high-density open metal sites, high porosity, and surface area, and is considered one of the most promising MOFs (Kim and Hong, 2021).

10.1.2 Synthesis of MOFs

Control of the surface morphology, crystal structure, pore structure, and active sites of MOFs, if achieved, would offer great potential to modify their adsorption properties. The structural features of MOFs depend largely on the synthesis method and conditions. Synthetic methods for MOFs have also made rapid progress in the past 2 decades. The synthesis of MOFs usually takes place in solvents at temperatures ranging from room temperature to 250°C. In addition to room temperature synthesis, traditional methods such as electric heating, microwave heating, mechanochemistry, electrochemistry, and ultrasound have also been used. In some studies, traditional stepwise methods and high-throughput methods have been used (Stock and Biswas, 2012). Different methods result in materials with different morphologies, particle sizes, size distributions and properties. Fig. 10.8 summarizes the various

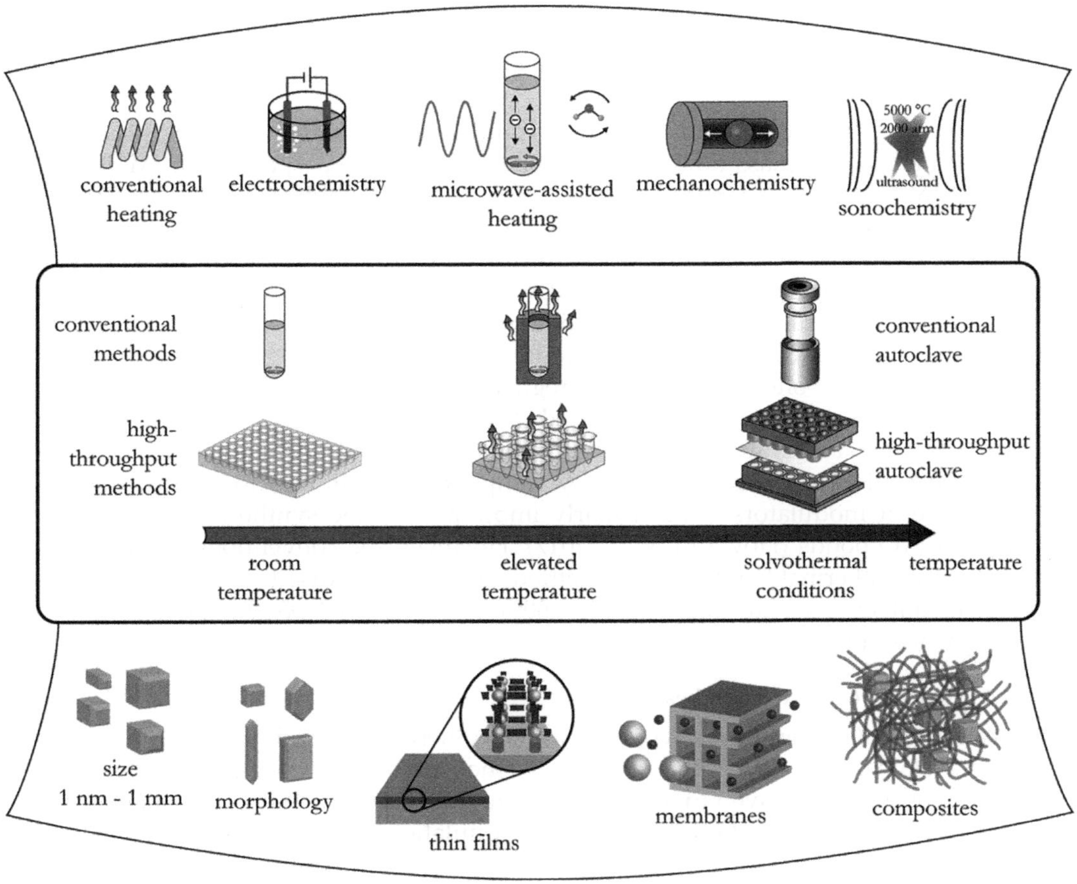

FIGURE 10.8 Overview of synthesis methods, possible reaction temperatures, and final reaction products in MOF synthesis (Stock and Biswas, 2012). *Reprinted (adapted) with permission from Stock, N., Biswas, S. 2012. Synthesis of metal–organic frameworks (MOFs): routes to various MOF topologies, morphologies, and composites. Chemical Reviews 112 (2), 933–969. Copyright {2012} American Chemical Society.*

synthetic methods applied in the past 20 years. This section discusses several synthetic methods commonly used in the preparation of MOFs.

10.1.2.1 *Synthesis of MOFs*

10.1.2.1.1 Hydrothermal/solvothermal synthesis

Hydrothermal/solvothermal method is a traditional and classical method for preparing MOFs. This method generally refers to directly mixing metal salts and organic bridging ligands in a specific solvent (such as water or organic solvent) in a sealed high-pressure reactor. The reactants undergo a reaction under the self-generated pressure of the system through heating. The reaction and crystallization temperature of MOF materials are usually between 80 and 200°C. Through this method, MOF crystals can be stably grown from hot solutions over a period of hours to days (Bhattarai et al., 2021). Due to the relatively high pressure and temperature, this method solves the problem of reactants being insoluble under ambient conditions, which is beneficial for the single crystal growth of MOF products. It can synthesize MOF single crystals that can be used for single crystal X-ray diffraction experiments (usually larger than 0.1 mm). In addition, the advantages of this method include the uniform size of the synthesized MOFs, good crystallinity, large specific surface area, adjustable topology and morphology, etc. Therefore, it has been widely used since the early stages of MOF development and is currently the most commonly used synthesis method (Zhang et al., 2022a).

For hydrothermal/solvothermal methods, conditions such as solvent type, reaction temperature, time, reactant mass ratio, and pH value all affect the preparation of MOFs. Water, ethanol, N,N-dimethylformamide (DMF), and N,N-dimethylacetamide are the most commonly used solvents. The removal of these solvent molecules from the pores is a complicated process, generally achieved by vacuum drying or washing with solvents such as ethanol or methanol (Remya and Kurian, 2019). Solvothermal/hydrothermal methods typically choose a temperature higher than the boiling point of the solvent as the reaction temperature (Stock and Biswas, 2012). As for modulators, formic acid, trifluoroacetic acid, and hydrochloric acid are often used as acid regulators, and sometimes other inorganic salts such as sodium acetate crystals are also used. The main function of modulators is to utilize nonstructural unit linkers to form dynamic bonds with metal precursors and to help slow down the formation of structural bonds by competing with linkers for metal coordination sites. The use of modulators is particularly important for the synthesis of Zr-MOFs with strong Zr(IV)–O bonds (Howarth et al., 2017). However, the conventional heating method used in the hydrothermal/solvothermal method also has some disadvantages, such as harmful reagents (Bitzer et al., 2020), low yield ($<50\%$) (Férey et al., 2005), high energy consumption, long reaction time, etc.

10.1.2.1.2 Microwave-assisted synthesis

Synthesizing MOFs through solvothermal/hydrothermal methods generally takes several days for crystallization. Since shorter reaction times are more attractive for industrial-scale production, using microwave radiation to shorten the reaction time becomes a better option. The microwave-assisted method is similar in formulation and procedure to the solvothermal/hydrothermal method, except that microwave heating replaces traditional oven heating

(Zhang et al., 2022a). Compared to the solvothermal/hydrothermal method (several days), the time to synthesize MOFs using microwave-assisted method is significantly reduced (typically less than 1 h). Using the hotspots and overheating effects of microwaves can greatly accelerate the dissolution of precursors and the condensation of metal–oxygen networks, making the acceleration effect of crystallization obvious. Therefore, using microwave radiation to rapidly and reproducibly synthesize MOFs is suitable for industrial production (Bhattarai et al., 2021; Jhung et al., 2007). Moreover, because radiation directly interacts with the solution/reactants, microwave-assisted heating is an energy-efficient heating method (Stock and Biswas, 2012).

Microwave-assisted methods have been successfully used to prepare various MOFs based on metal (III)/(II)-carboxylate salts. Among them, MIL-100(Cr) was first studied and reported by Jhung et al., in 2005 (Jhung et al., 2005). The X-ray diffraction spectra of MIL-100(Cr) obtained by microwave irradiation at 220°C for 4 h were in good agreement with those of samples synthesized by conventional hydrothermal methods at 220°C for 4 days, and the yield (44%) was similar to that of the hydrothermal method (45%). They also successfully prepared MIL-101(Cr) using microwave radiation for less than 60 min. In another study, the purified MIL-101 (Cr) was stable for several months in air (Jhung et al., 2007). Similarly, MIL-53(Fe) was successfully prepared by microwave irradiation for 30 min at 220°C (Horcajada et al., 2010). Several studies have systematically compared MOF-199 synthesized by microwave-assisted methods with other methods (hydrothermal, solvothermal, etc.) and concluded that microwave-assisted heating is the preferred method for rapidly synthesizing MOF-199 crystals in a short time (Seo et al., 2009; Schlesinger et al., 2010). Although microwave radiation has potential advantages such as fast crystallization time, phase selectivity (Laybourn et al., 2017), high product purity, easy morphological control (Lee et al., 2013; Park et al., 2004), and high energy efficiency (Jhung et al., 2007; Ni and Masel, 2006; Klinowski et al., 2011), this method is still not popular in synthesizing inorganic–organic hybrid materials. At the same time, this method is also considered to have disadvantages such as small and uneven crystal size of synthesized MOFs (Jhung et al., 2007).

10.1.2.1.3 Mechanochemical synthesis

Mechanochemical synthesis mainly utilizes mechanical force to break intramolecular bonds, thereby leading to chemical reactions (Stock and Biswas, 2012). According to whether solvents are involved in the reaction process, mechanochemical synthesis can generally be divided into three types: neat grinding (NG) which does not use any solvents; liquid-assisted grinding (LAG), which only adds trace amounts of solvents, and it can accelerate the mechanochemical reaction and exhibit structure-directing properties (Friščić and Fábián, 2009); and ionic and liquid-assisted grinding (ILAG), which adds a small amount of solvent and trace salt additives, and studies have shown that it can also exhibit structure-directing properties (Beldon et al., 2010; Friščić et al., 2010). Mechanochemical synthesis can effectively avoid the use of organic solvents. In addition, some studies have found that metal oxides can replace metal salts as starting materials under certain conditions, avoiding the environmental problems that may be caused by byproducts other than water. Shorter reaction times (generally around 10–60 min) and higher yields are also the main advantages of mechanochemical synthesis (Stock and Biswas, 2012; Chen et al., 2019b; Crawford and Casaban, 2016).

Mechanochemical synthesis was first applied to the preparation of MOFs in 2006 (Pichon et al., 2006), in which a microporous metal–organic framework $Cu(INA)_2$ (INA = isonicotinic acid) was successfully obtained by solvent-free grinding of $Cu(OAc)_2 \cdot H_2O$ and isonicotinic acid without any heating methods. In mechanochemical synthesis, factors such as the type of metal salt and solvent, presence/absence of solvent, and grinding conditions (time and frequency) have been shown to affect the preparation and properties of the resulting products. For example, several studies have investigated the effects of these factors on the resulting products of MOF-199 (Schlesinger et al., 2010; Pichon and James, 2008; Yuan et al., 2010; Klimakow et al., 2010; Yang et al., 2011a). The results showed that copper acetate is the most effective starting material for reacting with carboxylic acid, which may be related to the alkalinity of its anions and the solvation effect of acetate byproducts (Pichon and James, 2008; Yang et al., 2011a). Residual reactants and byproducts such as acetic acid in the product can be removed by washing and thermal activation, and washing is necessary to increase the surface area of the product (Yuan et al., 2010). The addition of solvents such as DMF and ethanol before grinding can lead to better crystallinity of MOF-199 compared to using water of hydration alone, which results in an enhanced BET surface area (Schlesinger et al., 2010; Yang et al., 2011a). In addition, products obtained by shorter grinding times exhibit mesoporous and microporous characteristics, while the mesopore volume of the product decreases with longer grinding times (Yuan et al., 2010). However, this method cannot control the particle size well and products may contain ultrafine particles due to mechanical grinding (Szcześniak et al., 2020).

10.1.2.1.4 Electrochemical synthesis

The electrochemical synthesis of MOFs was first reported by researchers from BASF in 2005 (Mueller et al., 2006). Electrochemical synthesis mainly uses electrolysis principles to continuously introduce metal ions into a reaction medium containing dissolved organic bridging ligands and conductive salts through anodic dissolution. It excludes the influence of anions, such as nitrates, perchlorates, or chlorides, which are mainly related to large-scale production processes. In addition, the continuous operation of electrochemical synthesis has a higher solid-phase MOF yield than ordinary batch reactions (Stock and Biswas, 2012).

Researchers from BASF have made groundbreaking work in using electrochemical methods for MOF synthesis. They successfully established electrochemical synthesis procedures for Zn- and Cu-carboxylate salts and obtained four high-porosity Zn-MOF and Cu-MOFs (Muller et al., 2005). To date, research on electrochemical synthesis of MOFs has mainly focused on MOF-199. The electrochemical synthesis of MOF-199 was first reported by Mueller et al. (2006), followed by more detailed discussion by Hartmann et al. (2008). Mueller et al. found significant differences in the local fine structure of copper between conventionally synthesized and electrochemically prepared MOF-199. Similarly, Hartmann et al. proposed that MOF-199 obtained by solvothermal conditions, under atmospheric pressure and reflux, and through electrochemical reactions had significant differences in specific surface area and total pore volume. In another study, the yield of MOF-199 obtained by electrochemical synthesis was relatively low compared to other synthesis methods, and the product had a small surface area and low crystallinity, mainly due to the conductive salt added blocking the pores in the MOF framework and being difficult to remove (Schlesinger et al., 2010). In addition, the disadvantages of this method, such as hazardous reagents and high equipment and maintenance costs, also limit its widespread application (Joaristi et al., 2012).

10.1.2.1.5 Ultrasonic synthesis

Ultrasonic synthesis mainly uses the extremely high temperature ($\sim$5000 K) and pressure ($\sim$1000 bar) or shear force generated by the instantaneously bursting bubbles to promote the formation of excited-state molecules, bond breaking, and further chemical reactions (Stock and Biswas, 2012). At present, only a few studies have used ultrasonic synthesis to prepare MOFs, and it is mostly used for known structure MOFs that have been obtained by other methods, such as MOF-199 (Schlesinger et al., 2010), MOF-5 (Tranchemontagne et al., 2008), and MOF-177 (Tranchemontagne et al., 2008).

Ultrasonic waves were first applied to the synthesis of MOFs in 2008. The method successfully prepared mesoporous $Zn_3(BTC)_2 \cdot 12H_2O$ by ultrasonically treating a mixture of zinc acetate and H_3BTC in ethanol and water (Qiu et al., 2008). The yield and size of obtained MOFs increased with the reaction time. Son et al. successfully synthesized MOF-5 crystals using ultrasonic synthesis with N-methyl-2-pyrrolidone (NMP) as the solvent and investigated the effects of synthesis time, temperature, and power level (Son et al., 2008). Compared with crystals synthesized under conventional convective heating (24 h, about 900 μm), the synthesis time (about 30 min) and size (5–25 μm) of the product were significantly reduced. Sabouni et al. further combined ultrasonic and microwave radiation energy for the synthesis of MOF-5 and successfully activated the sample by solvent exchange, obtaining a sample with a higher surface area (2473 m^2/g) (Sabouni et al., 2010). In the process of using ultrasound to synthesize MOF-199, the presence of DMF can enable MOF-199 to be synthesized within 1 min, while it cannot be obtained at room temperature without DMF (Khan and Jhung, 2009). Moreover, adding different concentrations of DMF can also affect the surface area, pore volume, morphology, and phase of the product. Compared with solvothermal synthesis, ultrasonic synthesis has been proven to be an efficient, high-yield, and low-cost method (Qiu et al., 2008). Moreover, the equipment is cheaper than microwave devices and has very low power consumption, since there is no induction period required before the ultrasound instrument becomes stable and the synthesis time is very short (Son et al., 2008). In addition to its short synthesis time and low cost, the advantages of this method also include simple operation (Li et al., 2009b; Vaitsis et al., 2019), but there are also limitations such as its unsuitability for industrial production.

10.1.2.2 Modification of MOFs

Improving the structural properties of pristine MOFs through modification to achieve higher performance has been the focus of pioneering research in recent years. Based on existing research, the modification of MOF materials used for VOC removal can be mainly divided into three categories: functionalization, compounding and derivatization.

10.1.2.2.1 Functionalization

Functionalization mainly refers to the modification of MOF by modifying organic ligands or introducing unsaturated metal sites.

Modification of organic ligands is most commonly achieved by using amino-functionalized organic ligands to obtain modified MOFs containing $-NH_2$ groups as functional components (Garibay and Cohen, 2010). MOFs-NH_2 and pristine MOFs can be regarded as similar MOFs with different connectors (Jhung et al., 2012). Through this process, some additional advantages can be gained from MOF materials, and the ability for this type of modification is a common feature of most MOFs. As a result, amino-functionalized MOFs can form hydrogen bonds with guest molecules through $-NH_2$ groups, thereby enhancing the adsorption capacity of VOCs. Besides, N orbitals also play an important role in reducing the band gap, making it easier to absorb hydrogen and improve the electron transfer efficiency. For example, $-NH_2$ modification can increase the visible light absorption range and acidity of MOFs, which can effectively catalyze the degradation of VOCs even in the presence of water (Zhang et al., 2022b).

Another modification method is to enhance the VOC removal ability of MOFs by introducing unsaturated metal sites in MOFs. There are mainly two methods to achieve this. One is to remove weakly bound guest water and residual solvent molecules from metal coordination sites through heating or vacuum, thereby forming new unsaturated metal sites. The other is to introduce metal ions into MOFs through post-synthesis modification.

10.1.2.2.2 Compounding

Compounding mainly refers to the use of platforms provided by MOFs to integrate functional materials such as organic molecules, polymers, carbon-based materials, metal nanoparticles (MNPs), and metal oxides to form MOF-based composite materials with new physical and chemical properties (Li et al., 2022). It has been demonstrated that the combination of the tunable pore size of MOFs with the multiple functions of other materials exhibits a synergistic effect of different functional units. The improved performance of composite materials is significantly better than that of individual components alone. Currently, MOF-based composite adsorbents have shown good performance in VOC removal.

Currently, the functional components used in the development and application of MOF-based adsorbents mainly include organic molecules, polymers, and carbon-based materials. Many studies have shown that the addition of organic molecules and polymers can not only significantly improve the hydrophobic properties of MOF materials, but also effectively construct the three-dimensional structure of MOFs. By introducing carbon-based materials, the adsorption performance, water stability, and renewable performance of MOF materials can be effectively improved.

However, MOF-based catalysts use more functional components, mainly noble metals and metal oxides. This is mainly because this system integrates MOFs with large surface area and high porosity to provide VOC adsorption sites and highly dispersed metal species to provide catalytic active sites, which can effectively improve the catalytic performance of MOF-based materials.

10.1.2.2.3 Derivatization

Derivatization mainly refers to obtaining corresponding pyrolyzed carbon/metal oxides through the pyrolysis of MOFs as precursors. Due to the extremely high surface area and abundant and ordered channels, MOFs are considered as ideal precursors for the preparation of porous metal oxides (PMOs). In recent years, these derivatives not only inherit the properties of MOFs such as large specific surface area and high porosity, but also have excellent chemical stability, showing outstanding adsorption capacity and high stability in VOC removal (Wang et al., 2023; Jiang et al., 2019).

Currently, in addition to the direct derivatization of pristine monometallic and bimetallic MOFs, a large number of studies have explored the derivatization of modified MOFs as precursors to obtain metal oxides with better activity, such as the pyrolysis derivatization of MOFs immersed in another metal salt solution and the pyrolysis derivatization of MOFs doped with noble metal nanoparticles. In addition, some studies have also combined the derivatized products of MOFs with other functional components to obtain composite materials based on MOF derivatives.

10.2 MOF-based adsorbents for VOC removal

Adsorption can separate VOCs from the atmosphere based on the adsorption selectivity of the adsorbent, by utilizing the van der Waals interaction or chemical bond interaction between the adsorbent surface and VOC molecules. Compared with other methods, adsorption is regarded as one of the most promising VOC control technologies due to its high removal efficiency, low energy consumption, low cost, and mature process (Xie et al., 2022). MOFs have a large specific surface area and abundant surface chemical functional groups. In addition, the combination of metal cations and ligands can produce variable geometric shapes and electronic structures in a wide range of pore sizes and specific surface areas (Yin et al., 2015). Moreover, the open metal sites on the pore surface of MOFs can also be used to enhance the adsorption of various VOCs and can maintain their permanent structure and crystalline order after regeneration, making MOFs have great potential for application in VOC adsorption (Mishra et al., 2014).

Activated carbon (AC), zeolite, and hypercrosslinked polymeric resin (HPR) were once listed by the US EPA as the three main adsorbents for controlling VOCs, among which activated carbon is the most widely used adsorbent. Compared with these adsorbents, MOFs have the largest specific surface area (Fig. 10.9). The average specific surface area of MOFs is 2653 m^2/g, with a maximum of 4293 m^2/g, which is 1.89 times, 3.32 times, and 2.16 times that of AC, zeolite, and HPR, respectively. The average pore volume of MOFs is 1.37 cm^3/g, with a maximum of 2.42 cm^3/g, which is higher than the other materials. In particular, MOFs exhibit significantly higher adsorption capacity for VOCs than AC, zeolite, and HPR, with an average adsorption amount of 796.2 mg/g and a maximum of 1375.0 mg/g, which are 1.7 times, 5.8 times, and 2.1 times that of AC, zeolite, and HPR, respectively (Li et al., 2020b).

Currently, IRMOFs series (MOF-5, MOF-177), MILs series (MIL-101, MIL-53), UiO series (UiO-66), and ZIF series (ZIF-67) are often used for VOC adsorption. This section summarizes the research on the removal of VOCs by different types of MOF adsorbents, including the mechanisms and influencing factors involved in the adsorption.

10.2.1 Adsorption mechanisms

The adsorption of gas molecules on the surface of adsorbent can be classified into physical adsorption (physisorption) and chemical adsorption (chemisorption). Physical adsorption mainly relies on intermolecular forces and is a common phenomenon that exists on the surface of solids (adsorbents) brought into contact with adsorbed gases (adsorbates). There are intermolecular forces between any molecules. Physical adsorption is nonselective, has a low activation energy, and is easy to adsorb and desorb. It can occur as monolayer adsorption or

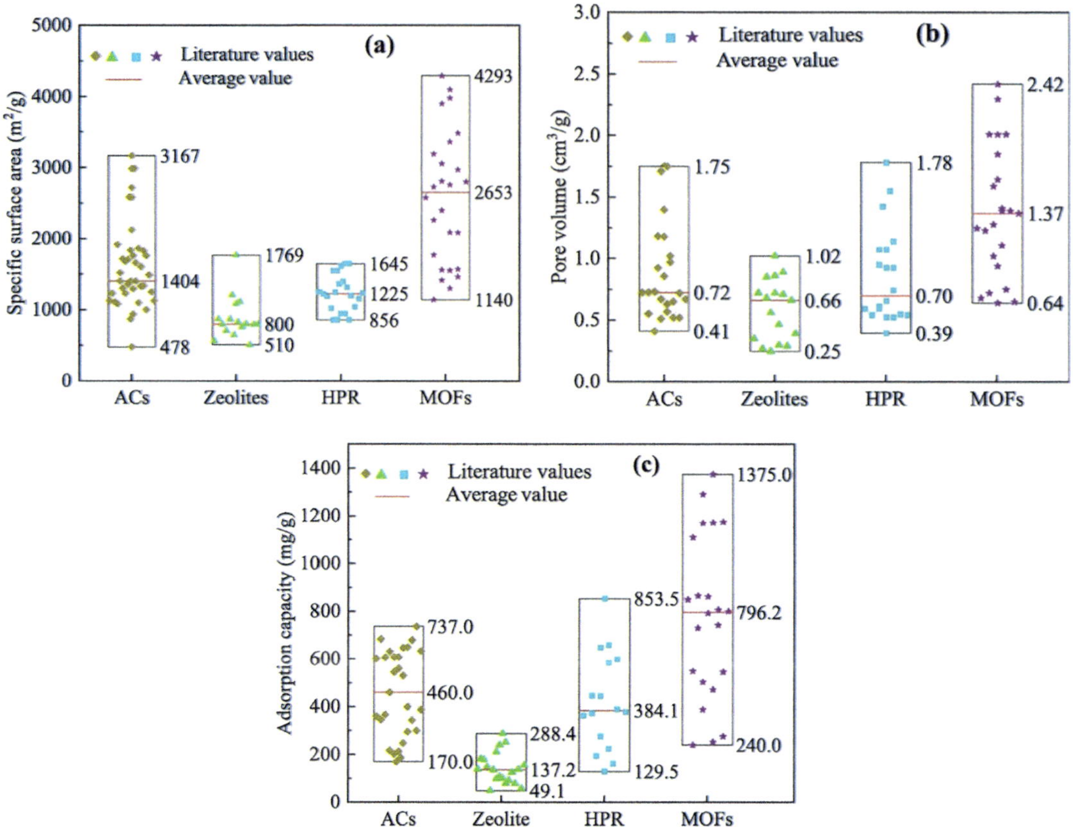

FIGURE 10.9 Comparison of different adsorption materials (ACs; zeolites; HPR; MOFs). (a) Specific surface area; (b) pore volume; (c) adsorption capacity (Li et al., 2020b). *Reprinted from Li, X., Zhang, L., Yang, Z., Wang, P., Yan, Y., Ran, J., 2020. Adsorption materials for volatile organic compounds (VOCs) and the key factors for VOCs adsorption process: a review. Separation and Purification Technology 235, 116213, Copyright (2020), with permission from Elsevier.*

multilayer adsorption, and is greatly influenced by the specific surface area and pore characteristics of the adsorbent. Adsorption that occurs through the formation of chemical bonds between molecules is called chemisorption. Chemical adsorption is selective, has a high activation energy, is difficult to adsorb, and is not easily desorbed, often requiring higher temperatures. Chemical adsorption always occurs as monolayer adsorption.

Currently, various studies have investigated the adsorption mechanisms of VOC molecules on the surface of MOF-based adsorbents, which involve multiple interactions but mainly physical adsorption. This section mainly discusses several commonly discussed mechanisms, including hydrogen bonds, pore filling, $\pi-\pi$ interactions, and covalent interactions.

10.2.1.1 Hydrogen bond

The adsorption of some VOC molecules on the surface of MOF-based adsorbents is mainly due to the hydrogen bonding interaction between the polarized hydrogen of the VOC molecules and the chemical functional groups (such as hydroxyl, carboxyl, amino, etc.) on the adsorbent surface, which binds the VOC molecules to the adsorbent surface. The impact of hydrogen bonding mechanism is generally greater than that of van der Waals forces or $\pi-\pi$ interactions (Vellingiri et al., 2017).

For example, ZIF series materials and amine-modified MOFs (such as UiO-66-NH_2) have N− containing functional groups in their frameworks, and it is known that these groups can form hydrogen bonds with VOC molecules. In the study of Vellingiri et al., UiO-66-NH_2 showed a higher adsorption capacity for toluene (252 mg/g) compared to UiO-66 (166 mg/g) and MOF-199 (159 mg/g), which can be attributed to the hydrogen bonds between toluene and the micropores of the MOF framework (Vellingiri et al., 2017). Vo et al. found that while the $-NH_2$ groups in UiO-66-NH_2 samples reduced the porosity of the adsorbent, they enhanced the adsorption capacity for toluene (180 mg/g), which was much higher than that of pristine UiO-66 (139 mg/g). They believed that the hydrogen bonding interaction between UiO-66-NH_2 and toluene was the main factor determining the degree of toluene capture (The Ky et al., 2020). Similarly, MOF-based materials with O-containing groups carried by organic linkers can also effectively remove VOCs through hydrogen bond interactions. Ahmed et al. evaluated the adsorption capacity of MOF-199 for 19 target VOCs and found that the adsorption capacity of phenolic and indole compounds in MOF-199 was significantly higher than that of other VOCs. This may be due to the strong hydrogen bonding between the −OH and −NH functional groups and the carboxyl (−COOH) groups on the adsorbent surface (Ahmed et al., 2016).

10.2.1.2 Pore filling

The pore filling mechanism involves gas molecules being adsorbed onto the walls of pores in an adsorbent through van der Waals forces. As the pressure increases, gas molecules can form capillary condensation in the mesopores. This is an adsorption mechanism widely reported in previous studies for the adsorption of various vapors into porous materials, including zeolites, resins, activated carbons, and MOFs. In the study of Vellingiri et al., UiO-66 exhibited a toluene adsorption capacity of 166 mg/g. They attributed the adsorption mechanism of UiO-66 to the size of its pores, which lead to the accumulation of toluene molecules inside the cage. For instance, the average pore size of UiO-66 was determined to be 8.5 Å, which is larger than the kinetic diameter of toluene molecules (5.85 Å). Therefore, during the host−guest interaction process, the pores of UiO-66 may accommodate toluene molecules (Vellingiri et al., 2017). Yang et al. investigated the adsorption capacity of MIL-101 for polar acetone and nonpolar benzene, toluene, ethylbenzene, and xylene. They found that the volume adsorption capacity (V_a) of several target VOCs showed a negative linear relationship with their molecular cross-sectional areas. Based on this phenomenon, it was inferred that the selected VOCs were adsorbed into MIL-101 due to continuous micropore filling (Yang et al., 2011b). Additionally, another evidence of the micropore filling mechanism of VOC adsorption in MIL-101 was the relationship between the first derivative (dQ) of the amount adsorbed and the relative pressure (P/P_0) of Benzene and Toluene, which showed two peaks. These two peaks were attributed to the two micropore windows present in the MIL-101 framework, and thus corresponded to the filling of VOC molecules into the 12 and 18 Å micropores of MIL-101.

10.2.1.3 $\pi-\pi$ interactions

MOFs containing benzene ring structures can form $\pi-\pi$ interactions with aromatic VOCs during the adsorption process, which is stronger than the C−π and O−π interactions between aromatic VOCs and carbon materials or zeolites. Therefore, compared with AC and zeolite, benzene-ring-containing VOCs are more easily adsorbed by benzene-ring-containing MOFs. In addition, there is another type of π–π interaction, mainly seen between open metal sites in MOF frameworks and VOC molecules, but it has been relatively less discussed. Ma et al. synthesized a series of iron-based MOFs and applied them to the adsorption and removal of toluene, among which MIL-100(Fe) exhibited a maximum equilibrium adsorption capacity of 663 mg/g, which was higher than MIL-101(Fe) and MIL-53(Fe) (Ma et al., 2021). Based on the high-resolution C 1s XPS spectroscopy results, the area ratio of the O–C=O peak of pristine MIL-100(Fe) increased from 2.85% to 6.35% after the adsorption of toluene, which may be due to the generation of π–π components. Therefore, the π–π stacking interaction formed between the H_3BTC ligand and the benzene ring in MIL-100(Fe) is the main adsorption mechanism of MIL-100(Fe) for toluene. Duan et al. reported a simulation of the toluene adsorption density on MIL-100(Fe) (Duan et al., 2020). It is found that most of the toluene molecules are distributed near the organic ligands, while only a few molecules are concentrated near the Fe sites, which can prove the π–π stacking interaction between the toluene molecules and H_3BTC ligands. In addition to the benzene ring structure contained in the original MOF, the benzene ring structure can also be introduced through the modification process, thereby enhancing the π–π interaction to improve the adsorption effect of aromatic VOCs. For example, in Li's work, as the content of H_2NDC increased, the toluene adsorption capacity of UiO-66-NDC increased to 209 mg/g, which can be explained by the increased efficiency of the π–π interaction between the toluene molecule and the naphthalene ring (Li et al., 2021). MOF-199 is known for its open metal sites and high surface area. Therefore, another π–π interaction may occur between toluene and Cu ions (open metal sites) during the adsorption of toluene by MOF-199 (Vellingiri et al., 2017).

10.2.1.4 Covalent interactions

Up to now, covalent interaction also plays a certain role in the adsorption of VOC molecules by MOF-based materials. However, compared with the above several mechanisms of action, the chemical adsorption mechanism is relatively rare in the process of VOC removal. Due to the tendency of amine groups to form covalent bonds with FA, amine-functionalized adsorbents have attracted great interest in FA adsorption studies. In the work of Wang et al., the amine functional groups were connected to the open metal sites of Cr(III) to enhance the adsorption capacity of FA by MIL-101(Cr) (Wang et al., 2016). Compared with original MIL-101 (100.3 mg/g), the FA adsorption capacity of ethylenediamine-functionalized MOF (ED-MIL-101) increased to 164.86 mg/g. The adsorption mechanism between FA molecules and ED-MIL-101 involves proton exchange through the lone pair of amine nitrogen attacking the carbonyl group of FA, leading to the formation of an unstable hemiacetal intermediate. Finally, the hemiacetal amine intermediate is dehydrated to form an imine. Analysis of the used ED-MIL-101 by FTIR further confirmed the formation of imine with a new band at around 1640–1690 cm^{-1}.

10.2.2 Key adsorption factors

The properties of adsorbent materials, such as specific surface area, pore volume, pore size distribution, and surface functional groups, affect the adsorption of VOCs. However, because the current VOC adsorption by MOFs is mainly based on physical adsorption, it is greatly influenced by the specific surface area and pore structure characteristics. In addition, VOC properties such as molecular weight, molecular diameter, boiling point, polarity, and saturation vapor pressure can also affect the adsorption process. The adsorption process is also influenced by adsorption conditions such as temperature, humidity, and VOC concentration. The combined effects of various factors lead to different affinities between VOC molecules and adsorbents, which in turn affect the adsorption rate and capacity (Li et al., 2020b).

10.2.2.1 Characteristics of MOF-based adsorbents

10.2.2.1.1 Specific surface area and pore volume

For any adsorbent, a large specific surface area means excellent adsorption performance, because the surface area provides a site for the adsorption process, which increases the interaction probability between the adsorbent and VOCs. The effect of pore volume on the adsorption of VOC by the adsorbent is related to the specific surface area. Sometimes in order to increase the adsorption capacity of the adsorbent for VOCs, closed pores may be opened or new pores may be created to increase the specific surface area of the adsorbent. The specific surface area of the adsorbent is usually calculated by the Brunauer–Emmett–Teller (BET) method and the specific surface area and pore volume of the micropores are calculated by the t-Plot method (Li et al., 2020b). When comparing the toluene adsorption performance of MIL-101(Fe), MIL-100(Fe), and MIL-53(Fe), Ma et al. found that MIL-100 exhibited the highest BET surface area and pore volume (1398.7 m^2/g, 1.085 cm^3/g), which was much higher than MIL-101 (873.3 m^2/g, 0.737 cm^3/g) and MIL-53 (268.6 m^2/g, 0.338 cm^3/g) (Ma et al., 2021). At the same time, MIL-100(Fe) showed the highest toluene adsorption capacity, with the equilibrium adsorption capacity in the order of MIL-100(Fe) (663 mg/g) > MIL-101(Fe) (180 mg/g) > MIL-53(Fe) (114 mg/g). It can be concluded that the toluene adsorption capacity of the three Fe-based MILs follows the same order as their surface area and pore volume. In the work of Jangodaz et al., despite the pore volume of MIL-101(Cr) (0.9438 cm^3/g) being slightly smaller than that of MIL-53(Fe) (1.34 cm^3/g), MIL-101(Cr) showed higher adsorption capacity for ethylbenzene than MIL-53(Fe) due to its much higher BET surface area (2052 m^2/g) compared to MIL-53(Fe) (1170 m^2/g) (Jangodaz et al., 2018). These results indicate that increasing the surface area is more effective than increasing the pore volume for VOC adsorption.

10.2.2.1.2 Pore size distribution

According to pore size, the pores of materials can be divided into macropores (>50 nm), mesopores (2–50 nm), micropores (<2 nm) and narrow micropores (<1 nm). The pore volume and pore size distribution of adsorbent materials are usually measured by Barrett–Joymer–Halenda (BJH) method, while adsorption isotherms are used for support. MOFs have a small pore size range, mainly concentrated in micropores, which is beneficial for the adsorption of small molecule VOCs. In addition, the pore size can be adjusted by modifying the preparation methods to adapt to different VOCs. Ma et al. proposed that micropores of MIL-100(Fe) provide significant surface area, thus offering abundant adsorption sites for VOCs. When the microporous structure is filled with adsorbate, diffusion and mass transfer of the adsorbate within the pores are limited. Mesopores, on the other hand, provide high VOC diffusion coefficients, facilitating the mass diffusion and transfer of the adsorbate. When the main pore size of a porous adsorbent is one to three times of the kinetic diameter of the target VOC molecule, the adsorption capacity can be significantly enhanced. MIL-100(Fe) is composed of two types of mesoporous cages (cage diameter: 2.5 and 2.9 nm). The dynamic diameter of toluene is 0.59 nm, which is smaller than the pore size of MIL-100(Fe). Therefore, the high adsorption capacity of MIL-100(Fe) for toluene can be attributed to its abundant micropores and mesopores, as well as its appropriate graded porous structure (Ma et al., 2021).

10.2.2.2 Characteristics of VOCs

10.2.2.2.1 Molecular size

According to the size exclusion theory, only when the VOC molecule diameter is smaller than the adsorbent pore diameter can the VOC molecules enter the pores of the adsorbent. Therefore, the molecular size of VOC determines whether it can be adsorbed by a specific adsorbent. When exploring the adsorption of various VOCs by MIL-101, Yang et al. found that acetone, benzene, and toluene rapidly diffused into the pores of MIL-101 at $P/P_0 < 0.1$. While at this stage, MIL-101 hardly adsorbed ethylbenzene and xylene, which can be attributed to their larges molecular size than acetone, benzene, and toluene (Yang et al., 2011b).

In addition to molecular diameter, Yang et al. quantified the size of VOC molecules using molecular cross-sectional area, which is the orthographic projection of a molecule where the molecular geometry is accepted as a sphere. The molecular size of the selected VOCs and their possible and determined volumetric adsorption on MIL-101 are shown in Table 10.1. The results showed that the volumetric adsorption capacity (V_a) of the selected VOCs exhibited a negative linear relationship with their molecular cross-sectional area

TABLE 10.1 Molecule dimension and the possible and determined volume adsorption capacities of selected VOCs into MIL-101 (Yang et al., 2011b).

Compounds	X	Y	Z	D_{XY} (nm)	$V_{0.55}-V_{XY}$ (mL/g)	D_{YZ} (nm)	$V_{0.55}-V_{YZ}$ (mL/g)	D_{XZ} (nm)	$V_{0.55}-V_{XZ}$ (mL/g)	V_a (mL/g)
Acetone	6.6	4.129	5.233	0.778	1.392	0.666	1.51	0.842	1.228	1.645 ± 0.090
Benzene	6.628	3.277	7.337	0.739	1.393	0.804	1.291	0.989	1.078	1.477 ± 0.081
Toluene	6.625	4.012	8.252	0.775	1.291	0.918	1.138	1.058	0.918	1.270 ± 0.164
Ethylbenzene	6.625	5.285	9.361	0.847	1.288	1.075	0.997	1.147	0.928	1.228 ± 0.128
m-Xylene	8.994	3.949	7.315	0.982	1.084	0.831	1.228	1.159	0.918	0.846 ± 0.100
o-Xylene	7.269	3.834	7.826	0.822	1.26	0.871	1.228	1.068	1.003	0.866 ± 0.205
p-Xylene	6.618	3.81	9.146	0.746	1.357	0.991	1.075	1.129	0.945	1.246 ± 0.096

X, Y and Z are the molecular width; thickness and length, respectively; $V_{0.55}$ is the adsorbed volume of N_2 into MIL-101 at $P/P_0 = 0.55$, D_{XY}, D_{YZ}, D_{XZ} are diameters of circumcircles of the XY, YZ and XZ plane of molecules, calculated by the following equations, $D_{XY} = (X^2 + Y^2)^{0.5}$, $D_{YZ} = (Y^2 + Z^2)^{0.5}$, $D_{XZ} = (X^2 + Z^2)^{0.5}$, where X, Y and Z are the molecular width, thickness and length (Table 10.1), respectively; V_{XY}, V_{YX} and V_{XZ} are the adsorbed volumes of N_2 into MIL-101 pores whose sizes are smaller than D_{XY}, D_{YZ} and D_{XZ}, respectively; V_{XY}, V_{YX} and V_{XZ} are obtained directly from the relationship between pore diameter and accumulative pore volume of N_2 on MIL-101(please see the detailed illustration in Supporting Information); V_a is the adsorbed capacity in volume.

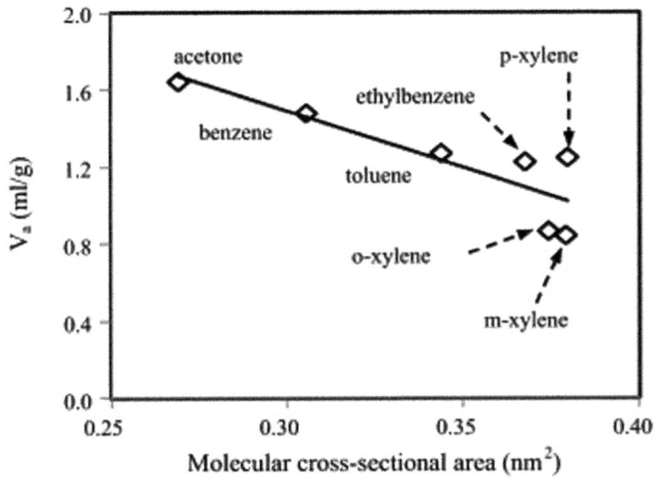

FIGURE 10.10 Relationship between molecular cross-sectional area and volume adsorption capacity (V_a) of seven selected VOCs (Yang et al., 2011b). *Reprinted from Yang, K., Sun, Q., Xue, F., Lin, D., 2011. Adsorption of volatile organic compounds by metal–organic frameworks MIL-101: influence of molecular size and shape. Journal of Hazardous Materials 195, 124–131, Copyright (2011), with permission from Elsevier.*

(Fig. 10.10), which can be attributed to the pore type (i.e., cylindrical pore) of MIL-101. As the molecular cross-sectional area of VOCs increased, the decrease in volumetric adsorption capacity of VOCs was due to the fact that some VOC molecules could not enter a small portion of the cylindrical micropores of MIL-101 because their size was smaller than that of the VOC molecules.

10.2.2.2.2 Molecular shape

Besides molecular size, the shape of VOC molecules may also affect their adsorption onto adsorbents. For VOC molecules, their shape is commonly accepted as rectangular solid, having three important dimensions, i.e., width (X), thickness (Y) and length (Z). A rectangular VOC molecule is composed of six rectangle planes. The six rectangle planes are three pairs of congruent pairs, i.e., XY, YZ, and XZ. Therefore, they could possibly enter pores of adsorbents with one of the three congruent rectangle pairs. Thus adsorption capacity of VOCs by filling into pores of adsorbents is dependent on the length of diagonal line of the molecular rectangle plane (i.e., the diameter of circumcircle of the plane including D_{XY}, D_{YZ} and D_{XZ}) by which VOC molecules entered into MIL-101 pores (Yang et al., 2011b). For example, in the work of Yang et al., m-xylene and o-xylene have molecular cross-sectional areas almost equal to those of p-xylene, but their adsorption capacity is lower than that of p-xylene. This can be attributed to the difference in molecular shape caused by the different positions of the methyl side chains in the xylene molecules (Fig. 10.11). In most cases, VOC molecules (such as acetone, benzene, toluene, ethylbenzene, and p-xylene) enter the MIL-101 pores with the plane having the minimum diameter, showing that $V_{0.55} - V_{YZ}$ values of acetone and $V_{0.55} - V_{XY}$ values of toluene, benzene, ethylbenzene, and p-xylene are equal to their V_a values (Table 10.1). However, the V_a values of o-xylene and m-xylene are equal to their $V_{0.55} - V_{XZ}$ values, which means that they fill the MIL-101 pores with the XZ plane (i.e., the plane with the largest diameter). One possible reason is that the methyl groups in the xylene molecules interact with the active sites of MIL-101 before their benzene rings.

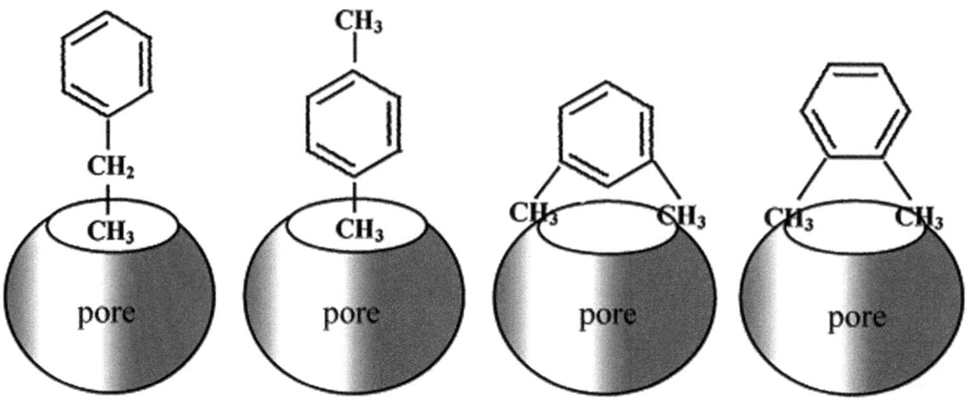

FIGURE 10.11 Scheme of ethylbenzene, p-xylene, o-xylene and m-xylene entering into MIL-101 pores (Yang et al., 2011b). *Reprinted from Yang, K., Sun, Q., Xue, F., Lin, D., 2011. Adsorption of volatile organic compounds by metal–organic frameworks MIL-101: influence of molecular size and shape. Journal of Hazardous Materials 195, 124–131, Copyright (2011), with permission from Elsevier.*

Therefore, o-xylene and m-xylene cannot fill the pores in MOFs with diameters smaller than D_{XZ}. However, the two methyl groups in p-xylene are connected to the benzene ring at an angle of approximately 180 degrees, forming a linear structure, which allows p-xylene molecules to enter MIL-101 pores with diameters larger than D_{XY}, as shown in Fig. 10.11. Therefore, the volumetric adsorption capacity of p-xylene in MIL-101 is higher than that of o-xylene and m-xylene (Yang et al., 2011b).

10.2.2.2.3 Molecular polarity

It has been reported that polar VOCs such as acetone and isopropanol tend to adsorb onto adsorbents with polar functional groups such as hydroxyl and amino groups, while nonpolar or weakly polar VOCs such as benzene, toluene, and xylene tend to adsorb onto adsorbents without polar functional groups. For example, by introducing amino groups through post-synthesis modification, MIL-125-NH_2 showed excellent removal effects on various VOCs, especially for polar VOCs. Kim et al. found that the adsorption capacity of VOCs followed the order of their polarity at each temperature (p-xylene < toluene < benzene < acetone < isopropanol). For instance, at 293 K and $P/P_0 = 0.95$, the highest adsorption capacity of MIL-125-NH_2 for isopropyl alcohol reached up to 452 mg/g, while the adsorption capacity for p-xylene was only 337 mg/g (Kim et al., 2018).

10.2.2.3 Adsorption conditions

10.2.2.3.1 Temperature

For most porous adsorbents, the adsorption of VOCs is mainly attributed to physical adsorption, where the adsorption temperature plays an important role. Generally, the adsorption capacity of VOCs on MOF-based adsorbents decreases with increasing temperature. Kim et al. investigated the equilibrium adsorption capacity of MIL-125-NH_2 for VOCs at 293, 298, and 303 K (Kim et al., 2018). As the adsorption temperature increased from 293 to 303 K, the adsorption capacity of MIL-125-NH_2 for benzene, toluene, p-xylene, acetone, and isopropanol decreased by 58%, 48%, 17%, 39%, and 47%, respectively. Most of previous studies have focused on VOC adsorption at temperatures ranging from 25 to 40°C. Vellingiri et al. innovatively studied the adsorption characteristics of MOFs for toluene at high temperatures (Vellingiri et al., 2017). As the adsorption temperature increased from 293 to 373 K, the adsorption capacities of UiO-66-NH_2, MOF-199, UiO-66, and ZIF-67 decreased by 81.3%, 79.9%, 86.0%, and 91.3%, respectively. The effect of temperature on VOC adsorption capacity confirms the physical adsorption mechanism.

10.2.2.3.2 Humidity

Water vapor is an omnipresent component that directly affects the adsorption behavior of gas on the surface of adsorbents. Multiple studies have shown that the presence of water molecules can affect the adsorption of VOCs on MOF-based materials, as they may compete for adsorption sites. According to the Dubinin–Serpinsky theory, water molecules may competitively occupy adsorption sites in the pores through surface oxygen functional group reactions, hydrogen bonds, and capillary condensation. Generally, water molecules first adsorb on the surface oxygen functional groups of the adsorbent, and then adsorb more water molecules through hydrogen bonds. Subsequently, with the increase of water vapor pressure, water clusters are formed, followed by capillary condensation in the pores of the MOF-based material (Zhu et al., 2020; Zhang et al., 2017). Therefore, from a practical perspective, it is necessary to study the adsorption process in the presence of water. Vellingiri et al. investigated the impact of two different RH conditions (25% and 50%) at 298 K on the breakthrough curves of toluene for four selected MOFs (UiO-66, UiO-66-NH_2, MOF-199, ZIF-67) (Vellingiri et al., 2017). In all cases, the increase in RH led to a shorter breakthrough time. For example, the adsorption capacity values of UiO-66, UiO-66-NH_2, MOF-199, and ZIF-67 (at 25 RH%) were 157, 233, 151, and 185 mg/g, respectively. In comparison, the values measured at 50 RH% were significantly reduced to 66.1, 120, 31.9, and 63.1 mg/g, respectively. In the work of Zhang et al., the adsorption performance of UiO-66 for toluene significantly decreased with increasing relative humidity (Zhang et al., 2019a,b). The adsorption amounts of toluene vapor at 0%, 30%, 50%, and 70% RH were 151, 82, 50, and 27 mg/g, respectively. This could mainly be attributed to the adsorption of water molecules, which not only occupy adsorption sites but also cause blockages, preventing toluene molecules from entering the pores and thus inhibiting the adsorption of toluene.

10.2.2.3.3 Relative pressure

Relative pressure (P/P_0) is an important factor that affects the adsorption of gas molecules on the surface of solid adsorbents, and its influence on adsorption is generally studied through adsorption isotherm research. Static adsorption isotherms usually use pure pollutant vapors for adsorption, allowing the material to fully contact and reach equilibrium with the pollutant over a long time, which can reflect the effect of relative pressure on adsorption performance as well as the adsorption capacity of the material for specific pollutants.

In Li et al.'s study, they investigated the adsorption isotherm of toluene on UiO-66-NDC (Li et al., 2021). The results showed that the adsorption isotherm of toluene could be divided into three stages based on the change of adsorption capacity. When P/P_0 was less than 0.16, the adsorption capacity of the sample to toluene rapidly increased with the increase of P/P_0. It has been reported that low pressure is beneficial for the adsorption of VOCs, so this phenomenon may be attributed to the continuous micropore filling effect under low pressure. As P/P_0 further increased to 0.75, the adsorption capacity only slightly increased, entering the plateau region, indicating that surface multilayer adsorption occurred. When P/P_0 exceeded 0.75, the adsorption capacity again rapidly increased with the increase of P/P_0, which is generally attributed to the accumulation of a large number of gas molecules in the pores, leading to capillary condensation.

10.2.3 MOF-based adsorbents

10.2.3.1 Pristine MOFs

Pristine MOFs mainly refer to MOF materials without any modification. Currently, various types of pristine MOFs have been applied for adsorption removal of VOCs, including MIL series, UiO series, ZIF series, MOF-5, and MOF-199, etc. Table 10.2 summarizes some representative pristine MOFs used for adsorption removal of VOCs.

TABLE 10.2 Representative MOFs for VOC adsorption.

Adsorbent	Synthesis method	Adsorbate	Adsorption capacity (mg/g)	Adsorption condition	References
MIL-53(Fe)	Hydrothermal method	Toluene	114^{S}	$P/P_0 \sim 0.8$, 25°C	Ma et al. (2021)
MIL-100(Fe)	Hydrothermal method	Toluene	663^{S}	$P/P_0 \sim 0.8$, 25°C	Ma et al. (2021)
MIL-101(Fe)	Hydrothermal method	Toluene	180^{S}	$P/P_0 \sim 0.8$, 25°C	Ma et al. (2021)
MIL-101(Fe)	Solvothermal method	Toluene	98.3^{C}	$P/P_0 = 0.0026$, 298 K	Vellingiri et al. (2017)
MIL-101(Cr)	Hydrothermal method	Benzene	103.4^{C}	Adsorbent dosage: 0.02 g/L, $C_0 = 0 \sim 2000$ mg/m^3, room temperature	Wang et al. (2018)
MIL-53(Fe)	Solvothermal method	Ethylbenzene	115.68^{C}	1 atm, 25°C, 250 ppm	Jangodaz et al. (2018)
MIL-101(Cr)	Hydrothermal method	Ethylbenzene	2320.51^{C}	1 atm, 25°C, 250 ppm	Jangodaz et al. (2018)
MIL-101(Cr)	Microwave irradiation method	Benzene	1304.44^{S}	$P/P_0 = 0.5$, 30°C	Jhung et al. (2007)
MIL-101(Cr)	Microwave irradiation method	p-Xylene	1275.10^{C}	6 mbar, 288 K	Zhao et al. (2011)
MIL-101(Cr)	Hydrothermal method	Benzene	1291 ± 77^{S}	$P/P_0 = 0.55 \pm 0.05$, 25°C	Yang et al. (2011b)
		Toluene	1096 ± 142^{S}	$P/P_0 = 0.55 \pm 0.05$, 25°C	

TABLE 10.2 Representative MOFs for VOC adsorption.—cont'd

Adsorbent	Synthesis method	Adsorbate	Adsorption capacity (mg/g)	Adsorption condition	References
		Ethylbenzene	1105 ± 116^S	$P/P_0 = 0.55 \pm 0.05$, 25°C	
		m-Xylene	727 ± 88^S	$P/P_0 = 0.55 \pm 0.05$, 25°C	
		o-Xylene	758 ± 176^S	$P/P_0 = 0.55 \pm 0.05$, 25°C	
		p-Xylene	1067 ± 83^S	$P/P_0 = 0.55 \pm 0.05$, 25°C	
		Acetone	1291 ± 71^S	$P/P_0 = 0.55 \pm 0.05$, 25°C	
MIL-101(Cr)	Hydrothermal method	Formaldehyde	100.30^S	150 ppm	Wang et al. (2016)
UiO-66	Solvothermal method	Benzene	166^C	$P/P_0 = 0.0026$, 298 K	Vellingiri et al. (2017)
UiO-66	Solvothermal method	Benzene	27.1^D	50 ppm, 298 K	Vikrant et al. (2019)
UiO-66	Solvothermal method	Toluene	167^S	34,000 ppmv, 3.4 kPa of toluene, 296 K	Ramos et al. (2022)
UiO-66	Microwave irradiation method	Toluene	130^D	1000 ppm, 0.3 L/min, 25°C	The Ky et al. (2019)
UiO-66	Solvothermal method	Styrene	569.4^C	200 ± 1 ppmv, 25 mL/min	Liu et al. (2021a)
ZIF-67	Solvothermal method	Benzene	224^C	$P/P_0 = 0.0026$, 298 K	Vellingiri et al. (2017)
MOF-5	/	Toluene	32.9^C	$P/P_0 = 0.0026$, 298 K	Vellingiri et al. (2017)
MOF-199	Solvothermal method	Toluene	159^C	$P/P_0 = 0.0026$, 298 K	Vellingiri et al. (2017)
MOF-199	Solvothermal method	Benzene	793.64^C	$P/P_0 \sim 0.08$, 288 K	Zhao et al. (2015)
MOF-199	Mechanochemical synthesis method	Toluene	571.27^S	298 K	Li et al. (2016)
MOF-199	/	Benzene	250.75^S	298 K	Ma et al. (2011)
		Toluene	234.04^S	298 K	
		Methanol	164.69^S	298 K	
		Ethanol	163.09^S	298 K	

(Continued)

TABLE 10.2 Representative MOFs for VOC adsorption.—cont'd

Adsorbent	Synthesis method	Adsorbate	Adsorption capacity (mg/g)	Adsorption condition	References
MOF-5	Solvothermal method	Formaldehyde	1.66[D]	298 K, 12% RH, 25 ppm	Vikrant et al. (2019)
MOF-177	Solvothermal method	Acetone	589 ± 22[C]	25 ± 0.5°C, 1.0 atm, 600 mL/min	Yang et al. (2013)
		Benzene	800 ± 19[C]	25 ± 0.5°C, 1.0 atm, 600 mL/min	
		Toluene	585 ± 40[C]	25 ± 0.5°C, 1.0 atm, 600 mL/min	
		Ethylbenzene	271 ± 11[C]	25 ± 0.5°C, 1.0 atm, 600 mL/min	
		o-Xylene	257 ± 9[C]	25 ± 0.5°C, 1.0 atm, 600 mL/min	
		m-Xylene	271 ± 15[C]	25 ± 0.5 °C, 1.0 atm, 600 mL/min	
		p-Xylene	213 ± 12[C]	25 ± 0.5°C, 1.0 atm, 600 mL/min	
		Ethenylbenzene	234 ± 25[C]	25 ± 0.5°C, 1.0 atm, 600 mL/min	
MOF-74	Solvothermal method	Benzene	96[D]	25°C, 25 mL/min	Britt et al. (2008)
IRMOF-3	Solvothermal method	Benzene	56[D]	25°C, 25 mL/min	Britt et al. (2008)
IRMOF-62	Solvothermal method	Benzene	109[D]	25°C, 25 mL/min	Britt et al. (2008)

[S]represents the maximum adsorption capacity obtained through static adsorption, [D]represents the maximum adsorption capacity obtained through dynamic breakthrough experiment, and [C]represents the maximum adsorption capacity obtained through calculated models.

MILs have high stability, significant pore volume, and uniform pore size, which are favorable for the adsorption of VOC molecules. Yang et al. applied MIL-101(Cr) prepared by a hydrothermal method for the removal of various typical VOCs, including nonpolar benzene, toluene, ethylbenzene, xylene, and polar acetone (Yang et al., 2011b). The adsorption capacity of MIL-101(Cr) for all selected VOCs was higher than that of reported adsorbents such as zeolite and activated carbon, among which the saturated adsorption capacity for benzene was estimated to be 1291 ± 77 mg/g at $P/P_0 = 0.55 \pm 0.05$, which is close to the previously reported value. There was a negative linear relationship between the volume adsorption capacity (V_a) of selected VOCs and their molecular cross-sectional areas, which indicated that the adsorption of selected VOCs into MIL-101 could be attributed to continuous micropore filling. Jhung et al. synthesized nanoporous MIL-101(Cr) under microwave irradiation, which greatly shortened the synthesis time (Jhung et al., 2007). Through the gas phase adsorption experiment, it is estimated that the adsorption capacity of MIL-101(Cr) to benzene at $P/P_0 = 0.5$ and 30°C is 1304.44 mg/g, and the complete adsorption time only required 253 s. In addition, MIL-101(Cr) was also shown to have excellent adsorption performance for benzene in liquid phase in this study. Unlike the toxic Cr element, Fe, with its nontoxic, environmentally friendly, and low-cost advantages, makes Fe-based MOFs popular materials in many applications such as degradation, Fenton process, and catalysis. Ma et al. systematically synthesized three Fe-based MIL series MOFs, MIL-100(Fe), MIL-53(Fe), and MIL-101(Fe), by a hydrothermal method and compared their toluene adsorption performance. Through static adsorption experiments, MIL-100(Fe) with the highest specific surface area (1398.7 m^2/g) showed the highest toluene adsorption capacity (663 mg/g), mainly relying on the $\pi-\pi$ stacking interaction between the organic ligand and the benzene ring (Ma et al., 2021).

Among various MOFs, the UiO series based on the tetravalent metal Zr are a class of highly stable MOFs. Their stability is attributed to the high oxygen affinity of Zr(IV), which makes UiO series MOFs widely studied in the fields of adsorption separation and heterogeneous catalysis. Vikrant et al. tested the adsorption performance of UiO-66 for gaseous benzene under different partial pressures prevailing in the practical application of air quality management (Vikrant et al., 2019a,b). Under the tested concentration of 50 ppm, the maximum adsorption capacity of UiO-66 for benzene is about 27.1 mg/g. Compared with AC and MOF-199, the lower performance of UiO-66 for benzene adsorption (0.1–50 ppm) might be explained by the formation of a monolayer formed on U6 (Langmuir fit) while a multilayer (Freundlich fit) formed on AC and MOF-199. Typically, UiO-66 is prepared by solvothermal synthesis in an autoclave with long synthesis times (typically 24 h) and low yields. To address this issue, Vo et al. synthesized UiO-66 using microwave irradiation in a continuous tube reactor. The experimental results showed that UiO-66 was successfully prepared in a short reaction time of 10 min, and possessed high yield, high porosity, and high crystallinity. At 25°C, the prepared UiO-66 showed a toluene adsorption capacity of 130 mg/g, which is comparable to that of UiO-66 obtained by solvothermal synthesis but much higher than those of MOF-5 (32.9 mg/g), MIL-101(Fe) (98.3 mg/g), and Zeolite (30.7 mg/g) (The Ky et al., 2019).

For other representative pristine MOFs, Vellingiri et al. evaluated the ability of different types of pristine MOFs, including UiO-66, ZIF-67, MOF-199, and MIL-101(Fe), to remove toluene under environmental conditions using MOF-5 as a reference (Vellingiri et al., 2017). The results showed that ZIF-67 exhibited the highest toluene saturated adsorption capacity (224 mg/g), mainly due to the single-layer coverage on the uniform surface of ZIF-67. UiO-66 (166 mg/g) and MOF-199 (159 mg/g) showed similar adsorption capacities, relying on the accumulation of toluene molecules and the π-complexation between the open Cu sites and toluene molecules, respectively. MIL-101(Fe) (98.3 mg/g) and MOF-5 (32.9 mg/g) had the lowest toluene uptake, mainly due to their lower surface area and the absence of functional groups in the framework structure. Yang et al. prepared MOF-177 using a solvothermal method and observed its large adsorption capacity (>200 mg/g) for multiple VOCs, with the highest removal efficiency observed for benzene (~800 mg/g), acetone (~589 mg/g), and toluene (585 mg/g). This is mainly attributed to the relatively large surface area and pore volume of MOF-177 (Yang et al., 2013). Britt et al. measured the adsorption capacity of various

MOFs for harmful gases by kinetic breakthrough experiments and found that MOF-74 had a dynamic adsorption capacity of 96 mg/g for benzene, mainly due to its coordination unsaturated metal sites (highly reactive 5-coordinate zinc species) and potentially reactive oxo group (Britt et al., 2008). In addition, they found that IRMOF-3 (56 mg/g) and IRMOF-62 (109 mg/g) also had a certain ability to remove benzene, with the former mainly relying on the amino functional groups.

10.2.3.2 Modified MOFs

10.2.3.2.1 Functionalized MOFs

In recent years, improving the structural properties of pristine MOFs to achieve higher adsorption of VOCs and other pollutants has been the focus of pioneering research. This section mainly introduces modified MOFs, which are obtained by modifying pristine MOFs with organic ligands (such as introducing $-NH_2$) or introducing unsaturated metal sites to obtain improved properties, rather than introducing another independent functional material.

10.2.3.2.1.1 Organic ligand modification At present, several studies have demonstrated that amino-functionalized MOFs can form hydrogen bonds with guest molecules through $-NH_2$ groups, thereby enhancing the adsorption capacity of VOCs. In Vellingiri's study, UiO-66-NH_2 exhibited a maximum adsorption capacity of 252 mg/g for gaseous toluene, which was much higher than that of UiO-66 (166 mg/g). This is mainly attributed to the hydrogen bonds between toluene molecules and the amino functional group, while UiO-66 mainly relied on the accumulation of toluene molecules in the pores (Vellingiri et al., 2017). By amine functionalizing MIL-125, Kim et al. found that MIL-125-NH_2 had significantly better adsorption capacity for various VOCs than other MOFs, which may be due to the synergistic effect provided by the amino groups and the micropores in the framework (Kim et al., 2018). In addition, the adsorption capacity of MIL-125-NH_2 increased in the order of increasing polarity (o-xylene < toluene < benzene < acetone < isopropanol). This may be attributed to the fact that relatively polar VOCs (acetone and isopropanol) prefer to adsorb onto adsorbents with polar functional groups (such as hydroxyl and amino groups), while nonpolar VOCs (BTX) tend to adsorb onto adsorbents without polar functional groups. Due to the tendency of N− containing groups to form covalent bonds with FA, amine-functionalized adsorbents have also been used for FA adsorption studies. In the study of Vikrant et al., they found that UiO-66-NH_2 exhibited excellent adsorption capacity (69.7 mg/g) for formaldehyde under near-environmental conditions (25 ppm FA, 298 K, 12% RH), which is much better than other test adsorbents (Vikrant et al., 2019a,b). The superiority of UiO-66-NH_2 was mainly attributed to the interactions between the amino group and FA coupled with interactions between its open metal sites and the FA carbonyl groups.

In addition to amino modification, the researchers also explored other possible pathways for organic ligand modification. Fluorination is considered an effective method for generating modified MOFs with hydrophobic frameworks. Wang et al. synthesized a novel fluorinated MOF, F-MIL-53(Al), by using monocarboxylic acid modulator to regulate the competitive coordination between tetrafluoroterephthalic acid and 1,4-dicarboxybenzene, significantly improving its hydrophobicity (Wang et al., 2022a). The toluene absorption capacity of F-MIL-53(Al) reached 254.85 mg/g under humid conditions, which was 33.56% higher than that of MIL-53(Al), and its performance did not decrease after 10 cycles of recovery at 50% RH. In the work of Li et al., nonpolar functional groups naphthyl groups were introduced into UiO-66 to increase hydrophobicity (Li et al., 2021). UiO-66-NDC was synthesized by using 1,4-naphthalenedicarboxlic acid to substitute terephthalic acid in a solvothermal process. The modified adsorbents showed excellent adsorption performance for toluene in humid conditions (50% RH), with a capacity of 76 mg/g, which was 69% higher than that of pristine UiO-66 (45 mg/g).

10.2.3.2.1.2 Metal ion doping In addition to organic ligand modification, many researchers have also attempted to enhance the VOC adsorption performance of MOFs by introducing unsaturated metal sites into MOFs. Wang et al. prepared MIL-101(Cr) and Cu@MIL-101(Cr) by a hydrothermal method for the removal of various VOCs (Wang et al., 2018). Although the specific surface area of Cu@MIL-101(Cr) decreased from 3367 to 2518 m^2/g, it exhibited a higher benzene adsorption capacity (114.4 mg/g) than MIL-101(Cr) (103.4 mg/g), which may be attributed to the introduction of Cu(II) ions providing more adsorption sites for MIL-101(Cr). Similarly, in Yang's study, NU-1000(Zr) and Fe(III)-doped NU-1000(Zr) were prepared via a hydrothermal method and used for the removal of toluene (Yang et al., 2022). When Fe/Zr = 0.1, the surface of the material had a high adsorption capacity for toluene, up to 231 mg/g, significantly better than NU-1000(Zr) (162 mg/g), which is due to the less doping of Fe elements and more defective sites in the structure. However, as the amount of introduced Fe increased, the adsorption performance of Fe2-NU-1000 (Fe/Zr = 0.3) and Fe3-NU-1000 (Fe/Zr = 0.5) decreased, with the adsorption capacities for toluene being 134 and 142 mg/g, respectively, and the specific surface area and pore volume showing similar trends. The results of SEM and EDS analysis revealed that this could be due to excessive Fe doping, which destroyed the original structure of the framework and reduced the number of Fe loading sites.

10.2.3.2.2 MOF-based composite adsorbent

At present, the functional components used to prepare MOF-based composite adsorbents mainly include organic molecules, polymers, and carbon-based materials.

10.2.3.2.2.1 Organic molecules/polymers Compared to many attractive features of MOFs, one of their drawbacks is their low adsorption performance in humid environments. The preferred method for preparing hydrophobic MOFs is to use organic molecules/polymers as hydrophobic coatings to modify MOFs, which has the advantages of low cost, simple and controllable operation, and so on. Wang et al. prepared PEG@MIL-101 by incorporating hydrophobic short-chain polyethylene glycol (PEG) as a hydrophobic barrier into MIL-101(Cr) (Wang et al., 2021). Characterization and simulation results showed that the terminal −OH group of the PEG chain coordinated with Cr sites in MIL-101(Cr), forming hydrophobic pores with a "cage-in-cage" geometry (11.2 Å). The PEG chains significantly hindered the diffusion of water into the MOF, while allowing the diffusion of toluene molecules. As a result, PEG@MIL-101 exhibited 30.8% higher adsorption capacity for toluene (431.22 mg/g at 1 mbar) and 31.5% higher diffusion coefficient (1.88×10^{-2}) than pristine MIL-101. Breakthrough curve showed 3.3 times of toluene working capacity on PEG5@MIL-101 at 80% RH compared to pristine MIL-101(Cr). In a study of Azmi et al., MIL-101(Cr)@PDMS composites with enhanced hydrophobicity were successfully obtained by vapor deposition of polydimethylsiloxane (PDMS) on MIL-101(Cr), and its surface area and porosity were comparable to those of unmodified MIL-101(Cr) (Azmi et al., 2022). At 40% RH and an initial concentration of 188 ppm, MIL-101(Cr)@PDMS showed 60% higher toluene uptake (170 mg/g) than MIL-101 (107 mg/g), which could be directly attributed to the shift in the mesopore condensation/pore filling processes in its now hydrophobic mesopores. However, at 40% RH and an initial concentration of 3750 ppm, MIL-101(Cr) exhibited higher uptake (507 mg/g) than the composite material (426 mg/g), highlighting the importance of the strong chemical interactions between the natural MOF framework and toluene at these higher toluene concentrations.

In addition to enhancing hydrophobicity, MOFs are often combined with organic molecules/polymers to construct MOF-based materials with tangible structures. Cui et al. synthesized foam MOF-199 in-situ on carboxymethyl cellulose (CMC) (Cui et al., 2019). The competitive covalent bonding of $Cu(OAc)_2$ to cellulose and H_3BTC had an important influence on the crystal structure and interfacial properties of foam MOF-199. As a result, the irregular particles of MOF-199 transformed into regular cubes. The newly formed interface between cellulose and MOF-199 enhanced the adsorption capacity for toluene. The high surface area (490–545 m^2/g), more exposed Cu(II) sites, and enhanced pore dispersion force enabled MOF-199@CMC to exhibit excellent adsorption capacity (193.49–285.63 mg/g) for toluene at ultra-low relative pressures. Similarly, Pinto et al. prepared a composite MOF foam by directly synthesizing UiO-66 on a polyurethane foam template, which not only retained the large pore structure and flexibility of the polyurethane foam, but also exhibited the microporosity, high surface area, and adsorption performance of UiO-66 (Pinto et al., 2013).

10.2.3.2.2.2 Carbon-based materials In addition to organic molecules/polymers, another class of materials widely used in the preparation of MOF-based composite adsorbents is carbon-based materials. In recent years, carbon-based materials such as graphene, graphene oxide (GO), graphene quantum dots (GQDs), activated carbon (AC), and biochar (BC) have received extensive attention from researchers, with GO being the most studied.

Recent studies have demonstrated that the introduction of GO can enhance the adsorption performance and water stability of MOFs. Additionally, GO has an aromatic structure and can coexist with the organic groups at the edges of MOFs, which lays the foundation for the covalent attachment between GO and MOFs. Li et al. synthesized MOF-199@GO composite material via a solvent-free mechanochemical method within 30 min (Li et al., 2016). The results showed that the maximum toluene uptake capacity of MOF-199@GO at 298 K could reach 838.47 mg/g, which was 47% higher than that of MOF-199 and far exceeded that of conventional AC and zeolite. More interestingly, the water stability of MOF-199@GO was significantly enhanced, maintaining its structure and porosity after soaking in water for 10 h, while MOF-199 completely lost its structure and porosity. This may be attributed to the coordination between the O-containing groups in GO and the Cu(II) metal sites in MOF-199. Zhao et al. successfully prepared a series of UiO-67@GO composites via a solvothermal synthesis method, with a maximum toluene adsorption capacity of up to 876 mg/g (Zhao et al., 2022). The improved adsorption performance is attributed to GO, which enhances the $\pi-\pi$ interaction, promotes the generation of defects, and provides more adsorption sites. With the combined effect of defects, $\pi-\pi$ interaction, and hydrogen bonding, UiO-67@GO exhibits efficient toluene removal performance.

Besides being susceptible to coexisting water, the low regeneration efficiency of used MOF sorbents by traditional regeneration methods also limits the widespread application of MOF-based adsorbents. To address this issue, Liu et al. attempted to desorb VOCs from a photodynamic GO@MIL-101 nanocomposite adsorbent under near-infrared light irradiation (Liu et al., 2022). The results showed that MIL-101 not only had inherent VOC adsorption ability in the composite material, but also had strong light-capturing ability throughout the UV—vis—NIR region. More importantly, the GO nanosheets could convert incident light into heat energy when exposed to UV—vis or UV—vis—NIR light. Then the localized thermal energy would rapidly transfer and distribute to the MIL-101 nanosheets located on the GO, resulting in GO@MIL-101 nanocomposite material achieving a higher surface temperature than pristine MIL-101 and GO. Finally, under the action of locally induced heat, the adsorbed ethyl acetate was completely released from the GO@MIL-101 nanocomposite, realizing efficient regeneration of the MOF-based adsorbents.

The atomic density of MOF is low, which cannot generate strong dispersion forces to effectively adsorb small gas molecules. Additionally, due to its low atomic density, the thermal conductivity of porous MOF is unusually low, further limiting the performance of MOF adsorption systems. To address this issue, Sun et al. combined MOF with GO having a high-density atomic array to enhance the effective adsorption of small gas molecules by MOF and improve its thermal performance (Sun et al., 2017). As a result, the MIL-101@GO composite material exhibited an enhanced thermal conductivity of up to 0.369 W/mK at 303 K, which is 60% higher than that of MIL-101 alone, demonstrating that the introduction of GO effectively improves the thermal performance of MIL-101.

10.2.3.2.2.3 Other functional materials Although organic/polymeric molecules and carbon-based materials are most commonly used to prepare MOF composites, researchers have been exploring alternatives to broaden their potential applications. Zhang et al. used another type of MOF (MOF-5) as a structure-modifying agent to successfully synthesize improved defective UiO-66 (Zhang et al., 2019a,b). They found that the addition of MOF-5 affects the self-assembly of metal particles and organic ligands. Compared with the original UiO-66, UiO-66@MOF-5 has more defects in its structure units. By adjusting the ratio between MOF-5 and Zr(IV) ions, a modified defective UiO-66 with high p-xylene adsorption performance of 257 mg/g was obtained, which is 1.7 times that of the original UiO-66, indicating that the missing junction defects caused by the additional MOFs contribute to improving the adsorption performance of UiO-66. Saini et al. prepared composite structure of zeolite and metal—organic frameworks, referred as ZMF, by immobilization of fine MOF-199 powder on foam shaped Zeolite Socony Mobil-5 (ZSM-5) Zeolitic structure, referred as ZF, achieving the construction of MOFs from microscopic to macroscopic levels (Saini and Pires, 2017). MOF-199 formed a uniform and well-dispersed coating on the porous framework of ZF. Compared with the ZF, MOF-199@ZSM-5 showed higher VOC adsorption capacity, which was attributed to the increase in surface area, total pore volume, and micropore volume. In addition, MOF-199@ZSM-5 exhibited enhanced adsorption selectivity, particularly for benzene/cyclohexane and benzene/cyclohexane mixtures, which was mainly attributed to the easy entry of VOC molecules into the MOF-199 surface fixed within the molecular sieve structure, demonstrating the synergistic effect of MOF-199 and ZSM-5. Romas et al. used [BMIM][CH_3COO] ionic liquid to modify UiO-66 and conducted toluene adsorption tests

(Ramos et al., 2022). It was found that 1 wt% loading of [BMIM][CH_3COO] on MIL-101, UiO-66, and UiO-66-NH_2 produced a synergistic effect, increasing the toluene adsorption capacity of the IL/MOF by 16%, 14%, and 5%, respectively. However, due to the strong interaction between the IL and MOF, an antagonistic effect of IL loading was observed as the IL loading increased to 10 wt%, leading to a decrease in toluene adsorption performance on UiO-66 and UiO-66-NH_2.

10.2.3.2.3 MOF derivatives

In addition to functionalized MOFs and MOF-based composite materials, MOF-derived materials provide a platform for functionalized adsorbents in VOC adsorption by virtue of hierarchical porous structures and active heteroatom doping sites. In terms of VOC removal by adsorption, MOF derivatives can be mainly divided into direct derivatization of pristine MOFs and derivatization of modified MOFs.

10.2.3.2.3.1 Thermal decomposition of pristine MOFs As a representative of Fe-MOFs, MIL-100(Fe) is constructed by the connection of iron octahedral trimers and carboxylic acid salts. However, in the synthesis process, the iron sites are coordinated by H_2O and anions (–OH, –F), limiting the function of the metal center. It is shown that more iron-coordinated unsaturated metal sites (Fe CUSs) can be exposed through thermal activation, that is, pre-coordinated molecules and impurities in pores can be evacuated under high-temperature vacuum. These Fe CUSs can act as Lewis acid sites and active centers, while the oxygen-containing functional groups of oxygenated VOCs (OVOCs) are highly polar and act as Lewis bases. Therefore, the Lewis acid–base interaction between Fe sites and OVOCs can enhance the adsorption ability of the adsorbent. Based on this concept, Pei et al. used MIL-100(Fe) as a precursor to obtain a series of iron-based MOF derivatives via high-temperature calcination, which were used to adsorb and remove OVOCs, such as methanol, formaldehyde, and acetone, under dynamic conditions (Pei et al., 2021). The results showed that the breakthrough adsorption capacity of MIL-100(Fe) derivatives for methanol was 61.5% higher than that of pristine MIL-100(Fe). In addition to the abundant Fe CUSs exposed by high-temperature calcination, the layered porous structure and high specific surface area also played a promoting role.

However, porous carbons prepared by direct carbonization of MOFs often exhibit a low specific surface area. The addition of extra carbon sources such as glucose and sucrose can effectively enhance the specific surface area and adsorption potential of MOF-derived materials. In the work of Wang et al., a new MOF-199-derived porous carbon was synthesized using MOF-199 as a precursor, glucose as an extra carbon source, and KOH as an activating agent (Wang et al., 2020). The results showed that the specific surface area and pore volume of the MOF-199-derived porous carbon were 2320 m^2/g and 1.05 m^3/g, respectively. At 25°C, it exhibited a high adsorption capacity for benzene up to 999.81 mg/g, which was superior to the original MOF-199 and some conventional adsorbents. Furthermore, the MOF-199-derived porous carbon also showed excellent water resistance and a high adsorption selectivity for C_6H_6/H_2O of up to 16.3.

Compared with traditional carbon-based materials, nitrogen-doped carbons and carbon nitrides have significant advantages in gas adsorption applications. Nitrogen doping into the carbon framework makes the carbon surface more polar and produces more active sites, resulting in a significant increase in the adsorption heat of the interaction between gas molecules and the adsorbent. Based on this concept, Ma et al. used nitrogen-rich ZIF-8 as a template, and urea as a secondary nitrogen source, to synthesize a highly microporous carbon material with a high nitrogen content (25.7%) and chemical stoichiometry close to C_3N type through carbonization (Ma et al., 2019). The content and type of nitrogen functional groups can be adjusted by the pyrolysis temperature. The results showed that the methanol adsorption capacity was determined by the electrostatic interaction at low relative pressure, since the nitrogen functional groups enhanced the electrostatic potential of the carbon surface. The dispersion interaction became dominant at high relative pressure, so the methanol adsorption mainly depended on the pore structure.

10.2.3.2.3.2 Thermal decomposition of MOF-based composites In addition to traditional porous carbons, composite adsorbents based on MOF derivatives with different morphologies can be obtained by compositing MOFs with functional materials and carbonizing, such as carbon aerogels and carbon nanofibers. Zhang et al. synthesized ZIF-8/agarose composite porous carbon aerogels using ZIF-8 nanoparticles and agarose (AG) as raw materials through a hydrothermal synthesis method and subsequent carbonization, and studied their toluene adsorption performance (Zhang et al., 2021). The results showed that the carbon aerogels had a large specific surface area and uniform microporous structure, and the adsorption capacity for toluene at 50% RH could reach 511.2 mg/g. The adsorption mechanism was investigated by DFT analysis and attributed to pore filling effect, $\pi-\pi$ interaction, and electron donor/acceptor interaction. Yang et al. synthesized ZIF-8 and polyacrylonitrile (ZIF-8/PAN) composite fibers by electrospinning, and obtained hierarchical porous N-doped carbon nanofibers through pyrolysis under N_2 atmosphere (Yang et al., 2021). With the synergistic effect of multi-level porous structure and abundant nitrogen active sites, the N-doped carbon nanofibers exhibited a high adsorption capacity for benzene up to 694 mg/g.

10.3 MOF-based catalysts for VOC removal

Although adsorption for VOC removal has advantages such as high removal efficiency, simple operation, and low cost, it does not decompose the VOC molecules. When external environmental conditions change, such as an increase in temperature or a decrease in pressure, the VOC molecules stored in the adsorbent may be released, leading to secondary pollution. In addition, when the adsorbent becomes saturated, it is disposed of as solid waste and becomes a potential environmental threat. To address this issue, the degradation and transformation of VOCs into nontoxic or low-toxicity substances is a promising method (Li et al., 2020c).

At present, many studies have demonstrated the excellent adsorption performance of MOF-based materials toward VOCs, which also provides favorable conditions for the catalytic degradation of VOCs. MOFs contain a significant amount of metal ($\sim$20–40 wt%) in the form of metal nodes, which possess free and exchangeable active sites. The presence of these metal sites endows MOFs with heterogeneous catalytic properties (Bahri et al., 2017).

The pathways for the degradation of VOCs using MOF-based catalysts mainly include catalytic oxidation and photocatalytic degradation. As photocatalysts, previous studies have shown that appropriate organic ligands in MOFs can serve as light-harvesting centers, and subsequently activate the metal cluster nodes upon light excitation, thus making MOFs behave like inorganic semiconductors, which have potential for use in photocatalysis (Yao et al., 2018). However, pristine MOFs exhibit low photosensitivity, insufficient electron-hole separation ability under visible light irradiation, and photodegradation problems (Gao et al., 2020). For the catalytic oxidation process of VOCs, MOF-based materials mainly use adsorbed O_2 (O_{ad}) molecules or lattice O_2 (O_{Latt}) in the structure to achieve catalytic oxidation of VOC molecules. To further improve the catalytic efficiency of MOF-based catalysts, noble metal/metal oxide loading or derivatization or metal oxide incorporation is generally adopted.

This section provides an overview of studies investigating the degradation of VOCs using various types of MOF materials, as well as the underlying mechanisms and factors influencing the degradation process.

10.3.1 Degradation mechanisms

10.3.1.1 Catalytic oxidation

Catalytic oxidation is a conventional VOC reduction technique, where both noble metal-based catalysts and transition metal-based catalysts are used for the degradation of VOCs. Compared to thermal incineration, catalytic oxidation can be performed at lower temperatures, as it provides a different reaction pathway. In the catalytic oxidation process, the oxidation activation energy can be reduced by changing the reaction mechanism and adsorbing active species such as VOCs and O_2. To date, several studies have evaluated the VOC oxidation ability of noble metal-based catalysts (such as Pt, Pd, Au, and Ag) and transition metal-based catalysts (such as Mn, Co, Ti, and Ce) coated on MOF supports.

Although extensive research has been conducted to explore catalytic elimination of VOCs from different sources and types, there are only few comprehensive summaries of the VOC oxidation mechanisms involved in the degradation process. The catalytic oxidation of VOCs primarily occurs between VOC molecules and oxygen species, which can be derived from various sources, including gas streams, catalyst surfaces, and catalyst lattices, depending on the free energy and diffusion rate. Therefore, the VOC oxidation reaction pathways and kinetics differ among different VOCs and catalysts. Three kinetic models have been proposed to describe the surface reaction mechanisms of catalytic VOC oxidation based on the source of oxygen, namely Langmuir–Hinshelwood (L–H) model, Eley–Rideal (E–R) model, and Mars–van Krevelen (MVK) model (Fig. 10.12) (Chung et al., 2019).

The reaction mechanism represented by the L–H model can mainly be described as follows: VOC and O_2 are first adsorbed on the catalyst surface, then they react with each other to form products, and finally, the products are desorbed from the active site. For VOC oxidation, VOC and O_2 can adsorb on adjacent active sites (single-site L–H model) or different active sites (dual-site L–H model). For the dual-site L–H model, the diffusion coefficients of VOC and O_2 play an important role. The E–R model can be described as oxygen first adsorbing on the catalyst surface, followed by the reaction with VOC in the gas phase. In the E–R mechanism, the rate of VOC oxidation is the limiting step. The MVK model can be described as VOC first adsorbing on the catalyst surface, then reacting with O_{Latt}. Subsequently, the oxygen vacancies generated in the previous step can be reoxidized by free

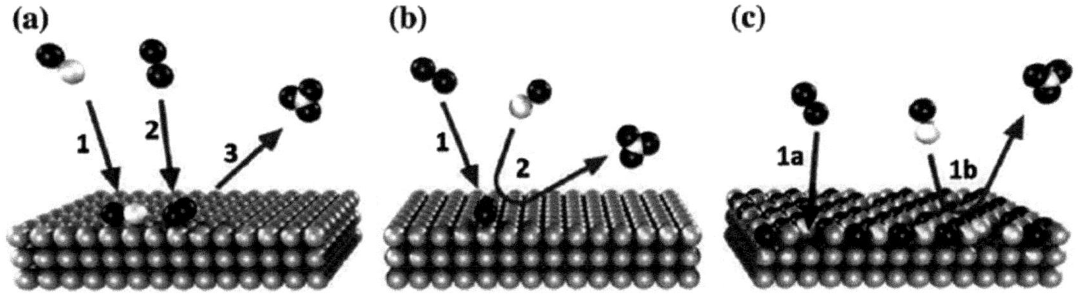

FIGURE 10.12 Possible reaction routes for catalysis: (a) Langmuir–Hinshelwood, (b) Eley–Rideal, and (c) Mars–van Krevelen (Chung et al., 2019). *Reprinted from Chung, W.-C., Mei, D.-H., Tu, X., Chang, M.-B., 2019. Removal of VOCs from gas streams via plasma and catalysis. Catalysis Reviews 61, 270–331, Copyright (2019), with permission from the publisher (Taylor & Francis Ltd, http://www.tandfonline.com*

oxygen. Unlike the L–H and E–R models, the oxygen involved in the oxidation is mainly attributed to O_{Latt} rather than surface oxygen (O_{sur}). The L–H and E–R models are typically used to elucidate the oxidation mechanism of VOCs on noble metal catalysts, whereas the MVK model is often used to explain the oxidation mechanism of hydrocarbon VOCs on metal oxides (Chung et al., 2019).

Hou et al. proposed that toluene oxidation on UiO-66-Ce involves both the MVK and L–H mechanisms, and the MVK mechanism may prevail at lower temperatures (Hou et al., 2022). They suggested that during the reaction, toluene molecules can be adsorbed on metal sites. The adsorbed toluene molecules then react with oxygen in adjacent lattices to form various intermediate products. During the reaction, the consumption of O_{Latt} leads to the formation of oxygen vacancies, and then gaseous oxygen molecules in the reaction gas are adsorbed and re-filled into adjacent oxygen vacancies, forming new O_{Latt}. The O_{Latt} then interacts with the intermediate species adsorbed on the catalyst to generate CO_2 and H_2O. A schematic diagram is shown in Fig. 10.13.

Chen et al. investigated the possible reaction pathways for toluene oxidation on CeO_2-MOF (Chen et al., 2018a,b). To study the influence of O_{Latt} on catalytic activity, experiments were conducted without gas-phase oxygen. The activation of toluene occurred only via surface O_{Latt} at high temperatures, in the absence of gas-phase oxygen. However, the introduction of O_2 under the same conditions significantly accelerated the transformation of toluene to intermediate species. Based on in-situ DRIFTS analysis, the reaction pathway for toluene combustion could be performed in continuous steps (Fig. 10.14), consistent with the L–H model. Toluene and oxygen molecules initially adsorb onto the catalyst surface and quickly react to generate aldehyde-like species, followed by benzoic acid-like species, and ultimately, CO_2 and H_2O are formed.

10.3.1.2 Photocatalytic oxidation

The heterogeneous catalytic reaction involves several consecutive steps, including adsorption and photocatalysis: (1) Mass transfer of VOC pollutants from the gas phase to the surface of the hybrid material; VOC molecules are adsorbed on catalysts through surface interactions, such as hydrophobic interactions, hydrogen bonds, π–π bonding, and van der Waals

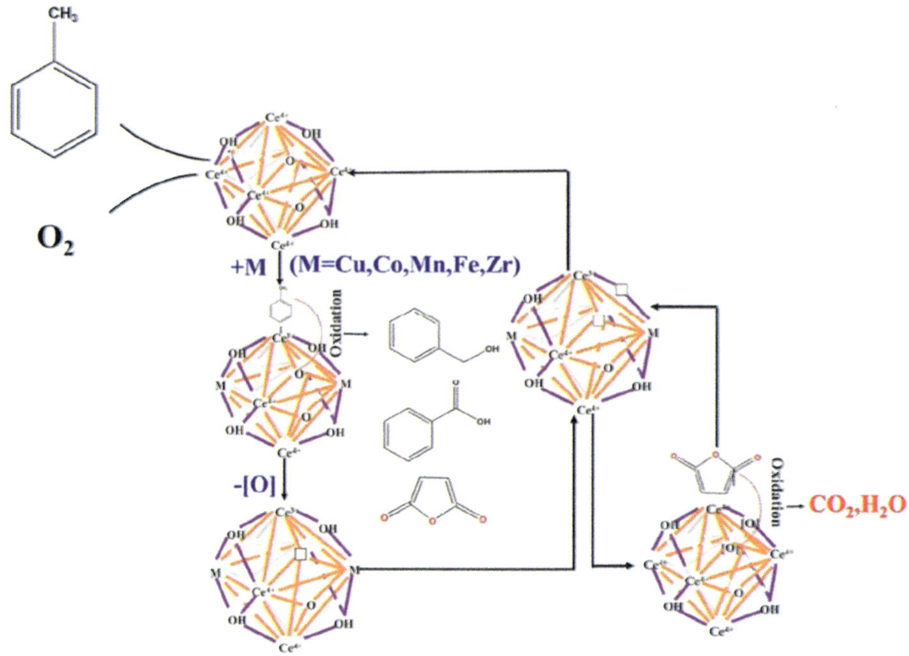

FIGURE 10.13 A possible reaction mechanism and pathway of toluene over the doped UiO-66-Ce catalysts (Hou et al., 2022). *Reprinted from Hou, X., Bian, Y., Yang, L., 2022. One-step synthesis of transition metal modified UiO-66-Ce metal–organic framework: catalytic oxidation of toluene and investigation of the mechanism. Microporous and Mesoporous Materials 345, 112214, Copyright (2022), with permission from Elsevier.*

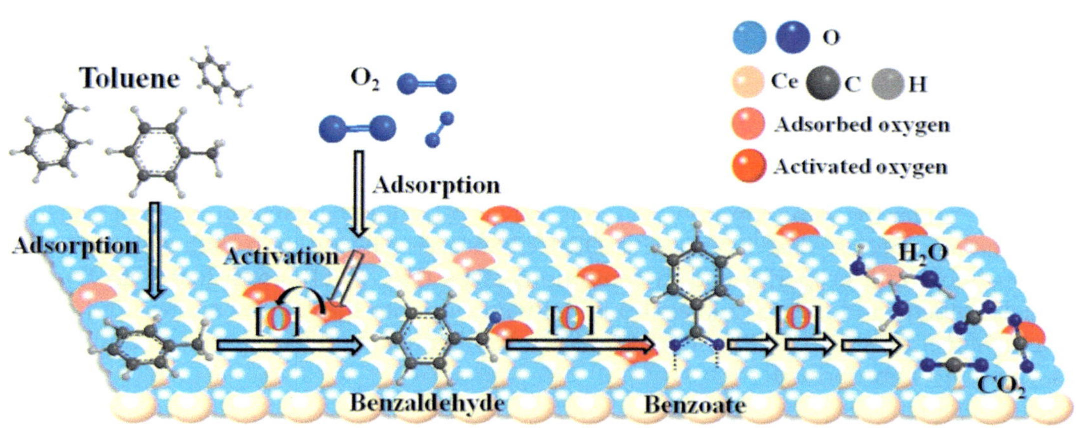

FIGURE 10.14 Mechanism of toluene degradation over CeO_2-MOF (Chen et al., 2018a,b). *Reprinted from Chen, X., Chen, X., Yu, E., Cai, S., Jia, H., Chen, J., Liang, P., 2018. In-situ pyrolysis of Ce-MOF to prepare CeO_2 catalyst with obviously improved catalytic performance for toluene combustion. Chemical Engineering Journal 344, 469–479, Copyright (2018), with permission from Elsevier.*

interactions; concentration gradients facilitate the diffusion of adsorbed VOCs to the photocatalyst. (2) When the incident light energy is greater than the bandgap of the photocatalyst, electrons in the valence band (VB) transfer to the conduction band (CB), creating electron–hole pairs; holes and electrons generated on the surface have redox abilities, forming hydroxyl radicals (•OH) and superoxide radicals ($\bullet O_2^-$); these radicals oxidize the VOC pollutants to CO_2, H_2O, or other intermediate products. (3) The product is released and the photocatalyst is regenerated.

The photocatalytic oxidation process is essentially a reaction between reducible reactants on the catalyst surface and oxygen-containing molecules, typically described by Langmuir–Hinshelwood (L–H), first-order, second-order, and power-law kinetic models (Zou et al., 2019). The L–H model is used to determine the rate of multi-phase catalytic mechanisms based on certain assumptions derived from the Langmuir adsorption isotherm. For example, reactant molecules exist in a monolayer and reversibly chemisorb onto the surface; the adsorption process involves only one adsorbate species per surface site; the energy of the adsorbate is the same at any surface site and is not influenced by adjacent adsorbate species. However, the L–H model does not consider intermediate species in the reaction, limiting its development. Furthermore, the L–H model is influenced by factors such as catalyst properties, light source, mass transfer parameters, operating conditions, and kinetic parameters. At low initial concentrations of VOC pollutants, the L–H model can be simplified to a first-order reaction rate equation, while at high concentrations, a semi-empirical power-law kinetic models can be well expressed. In addition, the L–H model can be developed to describe experimental results as multiple compounds compete for reaction on the catalyst.

To photodegrade organic pollutants, h^+, $\cdot O_2^-$, and $\cdot OH$ are the most common active agents. Li et al. investigated the mechanism for the excellent toluene removal efficiency of Fe-MOF (Li et al., 2020c). Since the reduction potential of $O_2/\cdot O_2^-$ is about −0.33 V versus NHE, electrons on the LUMO position of Fe-MOF (−0.36 V vs. NHE) can produce $\cdot O_2^-$, but $\cdot O_2^-$ is not considered the main free radical in the system due to the relatively weak reaction driving force. Direct oxidation of H_2O to produce $\cdot OH$ ($H_2O/\cdot OH$, 2.34 V vs. NHE) also seems impractical because the HOMO is located at 1.94 V. However, H_2O can be oxidized to H_2O_2, which only requires 0.69 V versus NHE. As is well known, Fe(II) can promote the production of $\cdot OH$ from H_2O_2 via the Fenton reaction ($H_2O_2/\cdot OH$, 1.07 V vs. NHE). Finally, VOCs are mainly degraded by $\cdot OH$ or h^+.

Zhang et al. evaluated the photocatalytic performance of P–NH_2-MIL-125 toward toluene and explored the mechanism (Zhang et al., 2022b). To determine the main active species in photocatalytic degradation, scavenger experiments were performed, where 2,2,6,6-tetramethylpiperidine-1-oxyl (TEMPO) was used as $\cdot OH$ scavenger and p-benzoquinone (PBQ) was used as $\cdot O_2^-$ scavenger. The addition of TEMPO and PBQ significantly reduced the photocatalytic degradation, indicating that $\cdot O_2^-$ and $\cdot OH$ radicals played a dominant role in the photocatalytic degradation of toluene and chlorobenzene. The EPR spectrum of P–NH_2-MIL-125 was obtained using DMPO as a spin trap reagent under visible light excitation, which also supported this conclusion.

Gao et al. investigated the degradation mechanism of synchronous N/Zn-modified MIL-125(Ti) toward acetaldehyde (Gao et al., 2020). During the degradation process, photocatalysis was triggered by incident visible light, which generated electrons and holes in the system. These electrons were excited to the CB of MIL-125(N–Ti_9Zn_1) and transferred to the ligands of MOF, leaving behind holes in the VB of MOF (Eq. 10.1). In this process, the dopants (Zn and N) in MOF acted as shallow traps, effectively avoiding the recombination of electron–hole pairs. Then, O_2 was attracted by these electrons and produced $\cdot O_2^-$ (Eq. 10.2), while a small amount of H_2O was attracted by the active sites in MOF and produced $\cdot OH$ with higher oxidation ability (Eqs. 10.3 and 10.4). The gaseous CH_3CHO was first adsorbed and enriched on the surface of MIL-125(N–Ti_9Zn_1), and then attacked by holes or produced $\cdot OH/\cdot O_2^-$, ultimately being oxidized into intermediates CH_3COOH and CO_2/H_2O (Eqs. 10.5–10.7). The water produced under air conditions can promote the degradation of CH_3CHO in the current catalytic system.

$$\mathrm{MIL-125(N{-}Ti_9Zn_1)} + h\upsilon \rightarrow \mathrm{MIL-125(N{-}Ti_9Zn_1)}(e^-/h^+) \rightarrow (e^- + h^+) \tag{10.1}$$

$$e^- + O_2 \rightarrow \cdot O_2^- \tag{10.2}$$

$$h^+ + H_2O \rightarrow H^+ + OH^- \tag{10.3}$$

$$h^+ + OH^- \rightarrow \cdot OH \tag{10.4}$$

$$h^+/\cdot OH/\cdot O_2^- + CH_3CHO \rightarrow CH_3COOH \tag{10.5}$$

$$h^+/\cdot OH/\cdot O_2^- + CH_3COOH \rightarrow H_2O + CO_2 \tag{10.6}$$

$$h^+/\cdot OH/\cdot O_2^- + CH_3CHO \rightarrow H_2O + CO_2 \tag{10.7}$$

10.3.2 Key catalysis factors

Due to the fact that heterogeneous catalytic processes are controlled by both gas adsorption onto the surface of solid adsorbents and subsequent gas catalytic degradation, the characteristics of adsorbent materials, VOC properties, and environmental conditions discussed in Section 10.2.2 inevitably affect the catalytic degradation of VOCs, with similar underlying mechanisms. In order to avoid redundancy, this section focuses on other influencing factors beyond those discussed in Section 10.2.2.

10.3.2.1 Characteristics of MOF-based catalysts

10.3.2.1.1 Specific surface area

Multiple studies have shown that a larger surface area can enhance the activity of catalysts by increasing the dispersion of active species and the availability of active sites. This can improve the degradation of VOCs by promoting adsorption. Therefore, surface area plays an important role in achieving high VOC degradation efficiency. In the study of Hou et al., UiO-66(CeCu) exhibited the highest catalytic efficiency among various bimetallic catalysts (UiO-66(Ce/Zr), UiO-66(Ce/Cu), UiO-66(Ce/Co), UiO-66(Ce/Mn), UiO-66(Ce/Fe)), achieving almost 100% catalytic efficiency at 250°C. This may be due to the largest specific surface area of UiO-66(CeCu), which resulted in its strongest toluene adsorption ability and promoted its oxidative activity toward toluene (Hou et al., 2022). In the work of Zhang et al., it was found from the characterization results that the specific surface area of the materials was in the order of P–NH_2-MIL-125 > NH_2-MIL-125 > P–NO_2-MIL-125. Interestingly, the order of catalytic performance was consistent with the BET specific surface area, and P–NH_2-MIL-125 showed the highest photocatalytic activity toward toluene. After irradiation for 20 min, the degradation rate of toluene in the presence of P–NH_2-MIL-125 reached 96%. This may be due to the enlargement of the pore size and volume of the catalyst, which improves the catalytic efficiency of the active sites located in the carrier channels. In addition, the larger BET surface area exposes more Lewis active sites, which is favorable for the adsorption of toluene on the surface (Zhang et al., 2022b).

10.3.2.1.2 Morphology of catalysts

Generally, the various morphologies of metal oxides derived from MOFs, resulting from different preparation methods, have distinct physicochemical properties that can impact their catalytic efficiency for VOC degradation. Dong et al. successfully synthesized various 3D Co-MOFs including 3D nanocube Co-MOFs (Co-MOFs-NC), 3D mesh nanosheet Co-MOFs (Co-MOFs-MNS), and 3D prismatic nanotube Co-MOFs (Co-MOFs-PNT), from which $CeCoO_x$-NC, $CeCoO_x$-MNS, and $CeCoO_x$-PNT were derived (Dong et al., 2021). In the catalytic oxidation process of toluene, the activity increased in the order of CoO_x < $CeCoO_x$-PNT < $CeCoO_x$-NC < $CeCoO_x$-MNS, indicating that the morphology of $CeCoO_x$ samples is crucial for regulating their catalytic performance. In Chen's work, compared with commercial Cr_2O_3 (Cr_2O_3–C), Cr_2O_3-M derived from MIL-101(Cr) is composed of small-sized Cr_2O_3 nanocrystals with 3D penetrating channels, irregular mesopores, and enhanced surface area, which are conducive to the diffusion of VOC molecules and contact with active sites (Chen et al., 2019a). The formation of more defects and oxygen vacancies during pyrolysis is beneficial for catalytic activity. The introduction of MnO_x weakens the strength of Cr–O bonds, promotes the dissociation of O_{Latt} from crystals, and provides more active oxygen and oxygen vacancies to participate in catalytic reactions.

10.3.2.1.3 Type of active metals

In multiple studies, the substitution of different active metals in MOF-based catalysts has shown varying effects on catalytic performance due to differences in preparation methods, substitution sites, and amounts. It may even lead to different reaction mechanisms. Li et al. extended their study from Pt/UiO-66 NPs to Pd/UiO-66 NPs and Ru/UiO-66 NPs to investigate the influence of active metal species on the catalytic degradation performance of ethyl acetate and CO_2 yield (Li et al., 2022). Under 260°C, Pt/UiO-66 NPs (0.50%) achieved complete catalytic degradation of ethyl acetate, while the degradation rates of Pd/UiO-66 NPs (0.63%) and Ru/UiO-66 NPs (0.98%) were about 89.5% and 87.6%, respectively. Meanwhile, the CO_2 yield of Pt/UiO-66 NPs was much lower than that of Pt/UiO-66 NPs, which may be due to the generation of some intermediates. The results indicate that the type of metal plays an important role in the effective catalytic degradation of VOCs on metal/UiO-66 NPs catalysts. Ma et al. evaluated the influence of Cu and Ni substitution on the catalytic performance of Co_3O_4 in the oxidation of o-xylene (Ma et al., 2022). The results showed that Cu substitution significantly promoted the catalytic oxidation of o-xylene, while Ni greatly inhibited it. Co_3O_4 catalysts with different molar ratios of Cu and Ni substitution exhibited similar improved and reduced performance, which may be attributed to different substitution sites and amounts. Firstly, both $CuCo_2O_4$ and $NiCo_2O_4$ with spinel structure are derived from bimetallic MOFs: Cu substituted for both Co_{Oh} and Co_{Td} sites, while Ni only substituted for Co_{Oh} sites. Different substitution sites resulted in higher and lower Co_{Oh}/Co_{Td} ratios for $CuCo_2O_4$ and $NiCo_2O_4$, respectively. The electron transfer between Cu/Ni and Co cations through adjacent O_{Latt} also increased and decreased the Co(III)/Co(II) ratio of $CuCo_2O_4$ and $NiCo_2O_4$. High Co_{Oh}/Co_{Td} and Co(III)/Co(II) ratios are usually associated with the reducibility and oxygen mobility of Co_3O_4 spinel, which contribute to excellent catalytic performance for VOC combustion. Therefore, different substitution sites and electron interactions between Cu- and Ni-substituted Co_3O_4 changed the geometric distribution and oxidation state of CoOh and Co(III), which further affected their redox properties for the catalytic oxidation of o-xylene.

10.3.2.1.4 Valence state of active metals

By changing the ratio of metal ions with different valence states in the precursor, or by doping with other metals, it is possible to change the valence distribution of active metal ions in metal oxides derived from MOFs. By adjusting the valence state distribution, oxygen vacancies and unsaturated chemical bonds can be formed on the surface of the catalyst, thereby enhancing the catalytic oxidation activity of the catalyst. Liu et al. prepared a series of Fe-MOFs for toluene degradation by varying the ratio of Fe ions with different valence states in the precursor (Li et al., 2020c). Characterization results showed that FeMOF-0.5 displayed the lowest Fe(II)/Fe(III) content ratio of 0.98, which may be due to the predominance of Fe(III) in the precursor, while FeMOF-0.25 had the highest Fe(II)/Fe(III) ratio of 1.53. Since ·OH has been demonstrated to be the main radical species driving VOC oxidation, FeMOF-0.25, which had a rich content of Fe(II), could generate more ·OH formation via the following reaction: $H_2O_2 + Fe(II) \rightarrow Fe(III) + OH^- + \cdot OH$. Therefore, toluene could be efficiently and completely degraded to CO_2 and H_2O by the enriched ·OH. Houet al. found that bimetallic catalysts (UiO-66(Ce/Zr), UiO-66(Ce/Cu), UiO-66(Ce/Co), UiO-66(Ce/Mn), and UiO-66(Ce/Fe)) exhibited improved catalytic efficiency for toluene compared to the monometallic catalyst UiO-66(Ce) (Hou et al., 2022). It is well known that doping other metals into Ce-based catalysts can greatly enhance their charge transfer ability ($Ce(IV) + M^{n+} \rightarrow Ce(III) + M^{(n+1)+}$). The escape of O_{Latt} in CeO_2 fluorite phase can lead to the formation of oxygen vacancies and Ce(III) species ($Ce(VI) + e^- \rightarrow Ce(III) + OVs$). Therefore, it is also important to calculate the contents of Ce(III) and Ce(IV) species. The surface Ce(III) content of the UiO-66-CeCu species sample was higher than that of other catalysts, and the presence of Ce(III) may cause charge imbalance, leading to the formation of oxygen vacancies and unsaturated chemical bonds on the catalyst surface, indicating its rich oxygen vacancies. By changing the ratio of Zn and Co, Gao et al. prepared a series of $ZnCoO_x$ catalysts for the catalytic degradation of VOCs and compared them with Co_3O_4 (Guo et al., 2022). The results showed that under different conditions, $Zn_{0.05}CoO_x$ exhibited the best mineralization activity for the elimination of different VOCs (including benzene, toluene, and cyclohexane), followed by Co_3O_4 and $Zn_{0.4}CoO_x$. This may be attributed to the fact that the Zn-doped catalysts have higher

Co(III)/(Co(III) + Co(II)) values, indicating that the doping of Zn can change the oxidation state of Co species. This result further supports the high VOC degradation ability of the catalysts, as Co(III) species are believed to be the active sites for the catalytic degradation of VOCs.

10.3.2.1.5 Heavy metal/metal oxide loading

As the main active components for catalytic degradation of VOCs, the loading of active metals and metal oxides directly affect the efficiency of VOC degradation. An appropriate increase in the loading of active metals or metal oxides can promote the interaction of VOC molecules on the catalyst, but an excessive loading of active components may also have adverse effects on VOC degradation, such as covering adsorption sites and affecting the crystal structure of the catalyst. Li et al. investigated the conversion rate of ethyl acetate and CO_2 production on Pt/UiO-66 NPs with different Pt loading amounts (Li et al., 2022). The experimental results showed that pristine UiO-66 NPs exhibited poor catalytic activity for the oxidation of ethyl acetate, and the production of CO_2 was almost negligible. In comparison, with the doping of Pt and the increase of Pt loading amount from 0.32% to 1.87%, the conversion rate of ethyl acetate and CO_2 production on Pt/UiO-66 catalyst gradually increased at the same temperature.

However, sometimes increasing the loading amount of active metal can adversely affect the performance of the catalyst. Tang et al. found that, as the Ag loading amount increased, Ag@m-MIL-100 (Fe) exhibited stronger PL intensity, representing a decrease in the electron—hole separation effect of catalyst (Tang et al., 2022). This may be attributed to the excess Ag particles on the m-MIL-100 (Fe) surface being partially oxidized to Ag_2O, which can act as recombination centers for photogenerated electrons and holes. In addition, photocurrent results also revealed the efficiency of electron—hole separation in photocatalysts, which is consistent with the PL trend. Further increasing the Ag content in Ag@m-MIL-100(Fe) led to the formation of Ag_2O particles in the catalyst, which hindered charge transfer at the catalyst-electrolyte interface and resulted in a large radius of the EIS curve, leading to higher interface charge transfer resistance of the catalyst.

Chen et al. tested the VOC degradation performance of a series of Mn/Cr_2O_3-M materials (Chen et al., 2019a). Among them, 15Mn/Cr_2O_3-M showed the best catalytic activity for toluene oxidation, which may be due to the improved reducibility and oxygen migration rate resulting from the appropriate doping of MnO_x. However, as the amount of MnO_x further increased, the activity no longer increased additionally and was even actually slightly affected, especially at high temperatures (above 270°C). Considering the characterization such as TEM, EDS, TPR, and TPD analyses, the existence of more isolated MnO_x in 25Mn/Cr_2O_3-M should not be helpful for catalytic activity, as there is a lack of strong interaction between excess MnO_x and Cr_2O_3.

10.3.2.1.6 Type of ligands

In the MOF-based catalysts for VOC catalytic degradation, the most widely used modification ligands are N-containing groups. It has been demonstrated that different N-containing groups and N-doping at different sites have different effects on the catalytic activity and stability of MOFs. To investigate the effect of N-containing groups on the photocatalytic activity of MIL-125, Gao et al. synthesized NH_2-MIL-125(Ti) using NH_2-BDC and compared it with pyrrole N-coordinated MOFs (Gao et al., 2020). The results showed that NH_2 groups could extend the photoresponse to the visible region, effectively reducing the bandgap (Eg = 2.25 eV), which is similar to pyrrole N-doped MIL-125(Ti). However, the catalytic activity of NH_2-MIL-125(Ti) sharply declined when the concentration of CH_3CHO decreased to a certain degree, indicating that the degradation of CH_3CHO led to the deactivation of the reaction catalyst. Compared with NH_2-MIL-125(Ti), pyrrole N-coordinated MOFs (MIL-125(N–Ti_9Zn_1)) on metal positions exhibited sustainable catalytic activity throughout the reaction process and showed faster adsorption rates. It is apparent that the position and properties of alkali nitrogen doping have different effects on the catalytic activity and stability of MOFs.

10.3.2.1.7 Surface acidity/alkalinity

The number of Lewis and Brønsted acid sites in the catalyst is the main factor affecting the catalytic activity for VOC degradation. NH_3-TPD is often used to study the acidity of catalysts, where the total amount of desorbed NH_3 and the maximum desorption temperature correspond to the number and strength of acid sites on the catalyst surface. In the work of Zhang et al., through NH_3-TPD and pyridine adsorption infrared (Pyridine IR) spectroscopy analysis, the peak intensity of P–NH_2-MIL-125 was the highest among all samples, indicating that the total acid content of this material was the highest and the acid activity was the strongest. Therefore, P–NH_2-MIL-125 has a better photocatalytic effect on the degradation of chlorobenzene. In addition, they also distinguished the Lewis and Brønsted acid sites on P–NH_2-MIL-125 and P–NO_2-MIL-125 through further pyridine IR studies. The Lewis acid content in P–NH_2-MIL-125 and P–NO_2-MIL-125 was 25.34 and 22.21, respectively, consistent with the order of catalytic performance (Zhang et al., 2022b). Dong et al. conducted NH_3-TPD analysis to evaluate the acidity differences of different forms of $CeCoO_x$-NC, $CeCoO_x$-MNS, and $CeCoO_x$-PNT (Dong et al., 2021). Their results showed that $CeCoO_x$-MNS exhibited more abundant acid sites, while $CeCoO_x$-PNT catalyst had the lowest surface acidity, consistent with the order of activity for catalytic VOC degradation. According to previous reports, weak acid sites are favorable for the adsorption of toluene molecules, especially Lewis acid sites and lattice oxygen contributions can greatly weaken and break the C–H bands from the methyl (–CH_3) of toluene. Gao et al. found that the surface basicity of MIL-125(Ti) increased after Zn and N doping (Gao et al., 2020). According to the hard and soft acid–base theory (HSAB), the hardness of Zn(II) is much lower than that of Ti(IV), so the substitution of Zn in Ti-based MOF will reduce the surface acidity sites and promote the adsorption of CH_3CHO acidic vapor on MOF. In addition, the pyrrole N-coordination can also enhance the surface basicity, thus further enhancing the adsorption capacity of MOF for CH_3CHO. Therefore, both factors lead to an increase in the surface basicity of MIL-125(Ti), resulting in excellent catalytic activity for the degradation of CH_3CHO.

10.3.2.1.8 Surface oxygen vacancies

Oxygen vacancies have been widely recognized as a key factor for catalytic oxidation of VOCs. Several studies have shown that surface oxygen vacancies can act as electron donors to enhance the activation performance of oxygen. While exploring the catalytic degradation ability of different forms of $CeCoO_x$ for VOCs, Dong et al. observed an increase in the surface content of O_{Latt} in the following order: $CeCoO_x$-NC < $CeCoO_x$-PNT < $CeCoO_x$-MNS, indicating that $CeCoO_x$-MNS catalyst had more O_{Latt} species, and their removal resulted in the generation of oxygen vacancies. Therefore, it was confirmed that $CeCoO_x$-MNS catalyst had more oxygen vacancies. Combining with the activity, it was hypothesized that the more oxygen vacancies on the surface of $CeCoO_x$-MNS catalyst, the more favorable it is to increase the adsorption of toluene and oxygen molecules, thereby accelerating the reaction rate (Dong et al., 2021).

10.3.2.2 Characteristics of VOCs

10.3.2.2.1 Space velocity

In heterogeneous catalysis, space velocity is one of the main factors affecting catalytic efficiency. Space velocity generally refers to the mass (or volume) of reactants passing through unit mass (or volume) of catalyst per unit time. It is usually adjusted by changing the gas flow rate, and its effect on catalytic efficiency is similar to that of initial concentration. Chen et al. invetigated the effect of weight hourly space velocity (WHSV) on the catalytic activity of CeO_2-MOF (Chen et al., 2018a). Their results showed that as the WHSV increased from 20,000 to 120,000 mL/(g h), the T10%, T50%, T90%, and T100% of the catalyst all increased to higher temperatures. In addition, compared with CeO_2—P, CeO_2-MOF catalyst showed better WHSV tolerance, which may be attributed to structural differences. CeO_2-MOF with many irregular mesoporous channels provides more active sites and allows reactant and product molecules to diffuse rapidly. Chen et al. found that at 1000 ppm concentration, the catalytic oxidation of toluene on 15Mn/Cr_2O_3-M showed a worse trend as the WHSV increased from 20,000 to 120,000 mL/(g h). For low-concentration toluene (150 ppm), a similar behavior was observed. The apparent E_a calculated by the Arrhenius plot showed a positive correlation with the space velocity, consistent with the catalytic activity of the catalyst (Chen et al., 2019a).

10.3.2.2.2 Type of VOCs

In the process of heterogeneous catalysis, different types of VOC molecules have different bond energies, resulting in different activation energies required for dissociation. In addition, there are differences in the reaction sites of different VOC molecules on the catalyst surface. Wang et al. found that the $CeCuO_x$ catalyst exhibited high activity for the oxidation of toluene, methanol, acetone, and o-xylene (Wang et al., 2019). Methanol (80°C) had the lowest T50% for catalytic combustion, followed by acetone (87°C), o-xylene (142°C) and toluene (150°C). It is generally accepted that the oxidation of organic molecules on CeO_2 catalysts follows the MVK mechanism. Oxygenated aliphatic hydrocarbons react faster than aromatic molecules as the latter requires the extraction of hydrogen from the benzene and methyl groups on four to five adjacent oxygen sites and reacts to produce water. Therefore, methanol and acetone are fully oxidized at temperatures below 125°C, while toluene and o-xylene require higher oxidation temperatures (200°C). Dong et al. investigated the catalytic degradation ability of $CeCoO_x$-MNS for various aromatic VOCs (Dong et al., 2021). Their results showed that the T10%, T50%, and T90% of benzene oxidation were 122°C, 172°C, and 204°C, respectively, which were lower than those of toluene and δ-xylene. In addition, the T100% of δ-xylene was 320°C, higher than that of benzene (282°C) and toluene (300°C). Thus, the activity of aromatic compounds decreased in the following order:

benzene > toluene > δ-xylene, which may be attributed to the higher dissociation energy of the C–H bond on the benzene ring than that on the methyl group. Previous work reported that the dissociation energies of C–H bonds on the benzene ring and methyl group were 110 and 75 kcal/mol, respectively, indicating that the C–H bond on the methyl group is relatively easy to oxidize and react with oxygen. In addition, the steric hindrance of the methyl group is also important in determining its activity, and the steric hindrance of the dimethyl group in δ-xylene is stronger than that of the methyl group in toluene. Therefore, it is believed that the number of methyl groups on the benzene ring is the key factor determining the activity when the number of benzene rings in the reactant molecules is the same. Guo et al. found that on the same $ZnCoO_x$ catalyst, cyclohexane molecules were more easily eliminated, followed by benzene and toluene (Guo et al., 2022). The aromatic molecules with benzene ring characteristics can be regarded as Lewis basic hydrocarbons, which can be adsorbed on the derived metal oxide catalyst through the acid–base interaction between Lewis base aromatics and Lewis acid sites of the catalyst. In contrast, it is speculated that cyclohexane molecules with C–C bonds exhibit two reaction pathways with acidic sites on the derived catalyst surface. Cyclohexane molecules can be transformed into nonclassical five-coordinated carbonium ions through protonation effect of the Brønsted acid sites, whereas classical carbonium ions would be generated due to the loss of H^- from cyclohexane molecules through the capture of Lewis acid sites. Then, these two types of carbonium ions will be transformed into intermediate substances and final products over the catalyst (Bai et al., 2016). Based on this analysis, it suggests that both Brønsted and Lewis acid sites are beneficial to the catalytic elimination of cyclohexane. However, only the Lewis acid sites are responsible for the oxidation of benzene and toluene. Therefore, cyclohexane can be more effectively degraded under the same conditions compared to the catalytic elimination of aromatic hydrocarbons (benzene and toluene).

10.3.2.3 Catalysis conditions

10.3.2.3.1 Temperature

Generally, with increasing temperature, the physical adsorption of gas-phase molecules on the catalyst surface may be inhibited, which is not conducive to further catalytic degradation. Temperature may also have certain impacts on the degradation pathway and product selectivity of VOC molecules. Li et al. evaluated the adsorption and degradation of ethyl acetate on UiO-66 NPs at different temperatures, such as 30°C, 140°C, and 260°C (Fig. 10.15) (Li et al., 2022). At 30°C, the outlet concentration of ethyl acetate sharply decreased to the complete adsorption point and began to steadily increase to 90 min, and the complete breakthrough time was 360 min, indicating that UiO-66 NPs exhibited excellent adsorption capacity for ethyl acetate at 30°C. However, at 140°C, the complete breakthrough of UiO-66 NPs occurred within 120 min. When the temperature was raised to 260°C, the outlet ethyl acetate appeared rapidly and reached adsorption equilibrium at 70% of the initial concentration within 60 min, which may be due to partial catalytic oxidation of ethyl acetate on UiO-66 NPs.

Long et al. prepared nanosized Co-ZIF catalysts for the selective oxidation of toluene to benzaldehyde under mild conditions (Long et al., 2022). As shown in Fig. 10.16, the experimental results show that the reaction temperature has a great influence on the catalytic performance of Co-ZIF, which is manifested in the different selectivity of products at different

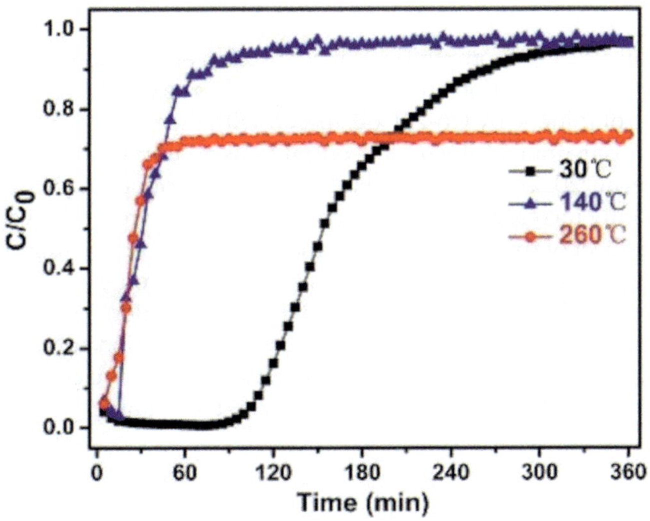

FIGURE 10.15 Catalytic degradation of ethyl acetate on UiO-66 NPs at 30°C, 140°C, 260°C (Li et al., 2022). *Reprinted from Li, J., Xu, Z., Wang, T., Xie, X., Li, D., Wang, J., Huang, H., Ao, Z., 2022. A versatile route to fabricate metal/UiO-66 (metal = Pt, Pd, Ru) with high activity and stability for the catalytic oxidation of various volatile organic compounds. Chemical Engineering Journal, 448, 136900, Copyright (2022), with permission from Elsevier.*

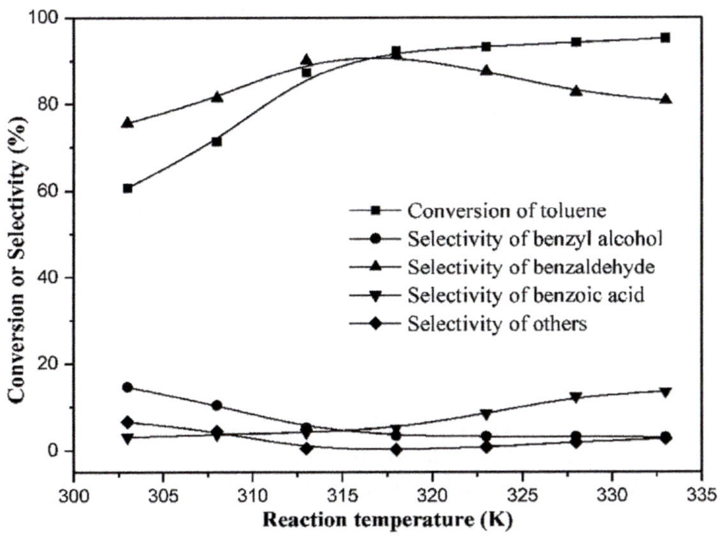

FIGURE 10.16 Effect of reaction temperature on selective oxidation of toluene (Long et al., 2022).

temperatures. As the reaction temperature increased from 303 to 318K, the conversion of toluene increased from 60.72% to 92.30%, and the selectivity of benzaldehyde increased from 75.64% to 91.31%. However, as the reaction temperature further increased, the increasing trend of the conversion of toluene slowed down, and the selectivity of benzaldehyde decreased significantly. At the same time, the selectivity of benzoic acid in the product increases rapidly, indicating that higher reaction temperatures may promote the conversion of benzaldehyde to benzoic acid.

10.3.2.3.2 VOC concentration

The initial concentration of VOCs may affect the competition for adsorption and catalytic sites on the adsorbent surface, thus affecting the degradation efficiency and reaction selectivity. In the study by Long et al., the conversion rate of toluene by Co-ZIF decreased with increasing toluene concentration, indicating that the appropriate initial toluene concentration is an important factor affecting the main product yield (Long et al., 2022). As the initial toluene concentration increased, the selectivity of benzyl alcohol significantly increased, indicating that the oxidation reaction was incomplete. Chen et al. investigated the effect of the initial concentration of toluene on the degradation effect when exploring the catalytic oxidation activity of $15Mn/Cr_2O_3$-M to toluene (Chen et al., 2019a). The experimental results showed that the catalytic degradation of low-concentration toluene on $15Mn/Cr_2O_3$-M was easier than that of high-concentration toluene. The apparent activation energy (E_a) calculated by the Arrhenius plot showed a positive correlation with the toluene concentration, consistent with the catalytic activity of the catalyst.

10.3.3 MOF-based catalysts

10.3.3.1 Pristine MOF

The application of pristine MOFs in the degradation of gaseous VOCs is still extremely limited, and there are only few studies in the literature. Li et al. synthesized a series of isomorphic Fe-MOFs by controlling the composition of metal salts, which can achieve repeatable and efficient VOC removal through alternating adsorption removal and photodegradation regeneration steps (Li et al., 2020c). In the work, FeMOF-0.25 exhibited the best toluene removal ability, and 100% of toluene could be removed at 460 ppm under simulated sunlight irradiation of 100 mW/cm^2, with 85% of adsorbed toluene being converted into harmless CO_2. This can be attributed to the highest Fe(II) content in the FeMOF-0.25 sample. The exploration of the degradation mechanism showed that $\cdot OH$ is the main active oxidant, and samples with abundant Fe(II) content can generate more $\cdot OH$ through the reaction of $H_2O_2 + Fe(II) \rightarrow Fe(III) + OH^- + \cdot OH$. Moreover, Fe(III) can be easily reduced to Fe(II) by absorbing electrons, ensuring the continuous formation of $\cdot OH$ while inhibiting hole–electron recombination. Therefore, toluene can be efficiently and completely degraded into CO_2 and H_2O by abundant $\cdot OH$. Long et al. prepared a nanosized Co-ZIF catalyst for the catalytic oxidation of toluene under mild conditions (Long et al., 2022). As toluene can be oxidized to benzyl alcohol, and is also easily oxidized to benzaldehyde, which is ultimately converted to the final oxidation product, benzoic acid. Considering the control of selective oxidation of reactants, toluene was selectively oxidized to benzaldehyde by adding NHPI as a reaction initiator to the solution. After exploring the catalytic mechanism, it was found that Co-ZIF can achieve selective catalytic oxidation of toluene through the deprotonation and protonation reactions of NHPI. Under the optimized reaction conditions, the Co-ZIF catalyst achieved a toluene conversion rate of 92.30% and a benzaldehyde selectivity of 91.31%.

10.3.3.2 Modified MOFs

10.3.3.2.1 Functionalized MOFs

Similar to MOF-based adsorbents, the currently modified MOFs for VOC degradation are mainly based on organic ligand modification and metal ion doping. The former is a typical method to enhance the photocatalytic activity of MOF materials at the molecular level, which can increase the light-responsive range of the ligand and change the bandgap energy of the MOF material.

10.3.3.2.1.1 Organic ligand modification Amino modification, as a method of organic ligand modification, has received wide attention in current research. Zhang et al. used plastic-based terephthalic acid to synthesize MIL-125, which was further modified by N-based groups ($-NH_2$ and $-NO_2$) (Zhang et al., 2022b). DFT calculations showed that the N orbitals make an important contribution to reducing the bandgap, making it easier to absorb hydrogen and improve electron transfer efficiency. Compared with $-NO_2$ modification, $-NH_2$ modification increased the visible light absorption range and acidity of MIL-125, resulting in effective catalytic degradation of chlorobenzene and toluene, even in the presence of water. The mechanism exploration results showed that $\cdot O_2^-$ and $\cdot OH$ radicals played a dominant role in the photocatalytic degradation of toluene and chlorobenzene. Gao et al. used a one-pot method to synthesize MIL-125(Ti) with simultaneous N/Zn in-situ doping to enhance the surface charge separation and cycling stability of MIL-125(Ti) for the degradation of gaseous acetaldehyde under high humidity conditions (Gao et al., 2020). The experimental results showed that the introduction of N and Zn narrowed the bandgap of MIL-125(Ti) (from 3.40 to 1.85 eV) and increased the electron conductivity. In addition, it was also shown to enhance the surface alkalinity and hydrophobicity of MIL-125(Ti). The synergistic effect of N/Zn co-doping enhanced the ability to activate O_2 and H_2O molecules into $\cdot O_2^-$ and $\cdot OH$, and the reaction kinetic rate of formaldehyde was increased by 10 times compared with MIL-125(Ti).

10.3.3.2.1.2 Metal ion doping Cerium oxide (CeOy) is a rare earth element with high oxygen storage capacity, good redox behavior, and ease of generating oxygen vacancies. To increase the deoxygenation rate of metal oxides on catalysts and thus improve the total oxidation rate of VOCs, a second or third metal component can be added. Inspired by this, Hou et al. prepared UiO-66-Ce and a series of UiO-66-CeM (M = Cu, Co, Mn, Fe, Zr) through a fast and simple one-step synthesis method to study the catalytic oxidation of toluene (Hou et al., 2022). Toluene molecules first adsorb on the metal sites of the MOF and then react with the oxygen in the adjacent lattice to form various intermediate products. Finally, oxygen in the gas phase is adsorbed on oxygen vacancies, interacts with the intermediate adsorbed on the catalyst, and ultimately is converted into CO_2 and H_2O. Due to the large number of oxygen vacancies and more exposed metal sites, toluene can be converted at relatively low reaction temperatures.

10.3.3.2.2 MOF-based composite catalysts

Currently, MOF-based catalysts for VOC degradation mainly include the following categories: (1) supported noble metal@MOFs composite catalysts; (2) supported metal oxide@-MOFs composite catalysts; (3) supported noble metal@metal oxide@MOFs composite catalysts.

10.3.3.2.2.1 Supported noble metal@MOFs composite catalysts Several studies have prepared MOF composite materials (MNPs@MOFs) with highly dispersed metal nanoparticles (Pt, Pd, Ru, Ag) supported on them via impregnation reduction, $NaBH_4$ reduction, ethylene glycol reduction, and other methods for catalytic oxidation and photocatalytic degradation of VOCs. Bi et al. prepared different Pd-UiO-66 catalysts using three different reduction methods, including pre-impregnation, $NaBH_4$-reduction, and ethylene glycol-reduction, and investigated the effects of different synthesis methods on the structure of the catalysts and the catalytic activity for toluene oxidation (Bi et al., 2020). For the Pd–U–H catalyst prepared by pre-impregnation method, when H_2 was used to reduce the Pd–U catalyst, the Pd was first reduced to sub-nanometric palladium clusters. And then, the Pd NPs were formed, after reacting with reaction atmosphere. However, when the stronger reducing agents $NaBH_4$ and EG were used to reduce the Pd–U catalyst, Pd could be reduced in one step, forming highly dispersed Pd NPs. Among them, Pd–U-EG prepared by ethylene glycol reduction showed the highest catalytic efficiency for toluene, which may be attributed to its higher active Pd species and O_{Latt} content. The experimental results showed that the reaction mechanism of toluene combustion on Pd–U-EG may be: toluene is rapidly converted to benzaldehyde, which is then oxidized to benzoic acid and maleic acid, and finally decays to CO_2 and H_2O. Li et al. prepared a series of metal (Pt, Pd, Ru)/MOFs catalysts by impregnation reduction for the catalytic oxidation of various VOCs, including ethyl acetate, n-hexane, and toluene (Li et al., 2022). Among them, Pt/UiO-66 NPs showed the best VOC degradation performance, indicating that the metal species played an important role in the effective catalytic degradation of VOCs on metal/UiO-66 NPs catalysts. Based on the experimental results, the possible reaction mechanism is that VOCs (such as ethyl acetate) molecules are first adsorbed on the surface of Pt/UiO-66 NPs, then oxidized into intermediates (such as ethanol, acetaldehyde, and acetic acid), which are further oxidized into CO_2 and H_2O by the O_{Latt} of PtO. With the generation of oxygen vacancies, the Pt sites are reduced to Pt^0. Finally, Pt^0 is activated by molecular oxygen and re-oxidized to PtO to eliminate oxygen vacancies, forming a Pt^0/PtO redox cycle.

There have been reports on both stepwise and one-pot synthesis methods for constructing MNPs@MOFs composite materials. Solution impregnation is a typical stepwise method, but the particle size of the MNPs obtained by this method is mainly determined by the pore structure of the MOF host, and sometimes the pore structure of the MOF may limit the growth of the MNPs or cause their aggregation. In-situ one-pot synthesis methods also have drawbacks such as difficult control of reaction parameters and interference of metal precursors with MOF nucleation and growth. Therefore, Tang et al. proposed a simple method for the controllable synthesis of MNPs@MOFs composite materials with the desired microstructure (Tang et al., 2022). Unlike the typical stepwise method, they dispersed MIL-100(Fe) gel (m-MIL-100(Fe)) instead of crystalline MIL-100(Fe) particles (c-MIL-100(Fe)) in water to form MIL-100(Fe) colloid. Ag precursor was then added to the colloid under constant stirring, and then reduced by a photochemical method to obtain uniformly sized (2–3 nm) Ag NPs. Further removal of solvents yielded Ag NPs@MIL-100(Fe) monoliths (Ag@m-MIL-100(Fe)), with Ag NPs evenly distributed in the MIL-100(Fe) monoliths. The formation of Schottky junctions between Ag NPs and m-MIL-100(Fe) made it possible for effective charge transfer at the interface and promoted the separation of electron-hole pairs in m-MIL-100(Fe). Compared with the original m-MIL-100(Fe) and Ag/c-MIL-100(Fe), the Ag@m-MIL-100(Fe) composite material showed significantly enhanced photocatalytic degradation efficiency for o-xylene, effectively oxidizing it into CO_2 and H_2O under illumination.

10.3.3.2.2.2 Supported metal oxide@MOFs composite catalysts To prepare metal oxide@MOFs composite catalysts for VOCs degradation, TiO_2 has been studied extensively due to its high activity, stability, nontoxicity, and low cost, which can effectively combine with the large surface area, interconnected 3D structure, photoresponsive characteristics, and semiconductor properties of MOFs to achieve a synergistic effect of "$1 + 1 > 2$". Zhang et al. successfully prepared TiO2@UiO-66-NH_2 nanocomposites with different mass ratios by a simple solvent evaporation method for photocatalytic oxidation of toluene and acetaldehyde under UV light (Zhang et al., 2020). Compared with pure TiO_2 and UiO-66-NH_2, the prepared TiO2@UiO-66-NH_2 exhibited the highest photocatalytic activity, with conversion rates of toluene and acetaldehyde of 72.7% and 70.74%, respectively, which were about 9.7 and 10.5 times that of UiO-66-NH_2. The photoelectrochemical properties indicated that the photogenerated carriers could be effectively separated and transferred through the interface between TiO_2 and UiO-66-NH_2. In addition, the possible mechanism of VOCs oxidation was proposed based on free radical scavenger experiments, demonstrating that the free radicals $\cdot O_2^-$ and h^+ were the main active species. To further improve the lifetime of photogenerated electron-hole pairs in MOFs, Chen et al. chose Fe_2O_3 as the functional component to synthesized MIL-100(Fe)/α-Fe_2O_3 composite materials with artificial Z-type structure by a simple one-step hydrothermal method, and selected o-xylene as the target pollutant (Chen et al., 2021). The experimental results confirmed the formation of the Z-type structure, which significantly prolonged the lifetime of active electrons and holes with strong reduction and oxidation ability by the Z-type channel for electron transfer, and promoted the production of reactive $\cdot O_2^-$ and $\cdot OH$ radicals. In addition, the reversible conversion of Fe(III) and Fe(II) under light irradiation also played a key role in the effective production of oxidized o-xylene and active radicals. The MIL-100(Fe)/α-Fe_2O_3 composite material showed 100% removal efficiency for o-xylene under 250 W xenon lamp irradiation and 90% high removal efficiency for o-xylene under visible light ($\lambda \geq 420$ nm).

10.3.3.2.2.3 Supported noble metal@metal oxide@MOFs composite catalysts Supported noble metal@metal oxide@MOFs composite catalysts have also been demonstrated to have excellent VOC degradation ability, as they may induce synergistic interactions between metal nanoparticles (NPs) and metal oxides (especially TiO_2). Liu et al. simultaneously encapsulated Au and TiO_2 into visible light-responsive NH_2-UiO-66, using Au nanoclusters as electron mediators, to synthesize a charge-separated Z-type Au$-TiO_2$@NH_2-UiO-66 photocatalyst for efficient VOC degradation (Liu et al., 2021a,b). Due to the enhanced interface electron transfer by Au electron mediators and the high redox ability of photogenerated carriers in the Z-type heterojunction, the Au$-TiO_2$@NH_2-UiO-66 nanocomposite material greatly improved the separation efficiency of photogenerated electron-hole pairs, thereby producing abundant strong oxidizing species, including photogenerated holes, $\cdot O_2^-$, and $\cdot OH$ radicals, for VOC mineralization. Within 360 min of xenon lamp irradiation, the mineralization efficiency of 1 wt% Au$-TiO_2$@NH_2-UiO-66 photocatalyst reached 85% for the degradation of 94.6% ethyl acetate.

10.3.3.2.3 MOFs derivatives

Carbon/metal oxides synthesized by MOFs pyrolysis have been shown to possess enhanced physicochemical properties and improved catalytic performance for VOC degradation, which can increase the catalytic efficiency of VOC oxidation (Chen et al., 2019a). The catalysts based on MOF derivatives for VOC degradation mainly include the following categories: (1) direct derivatization of MOFs (including traditional monometallic and bimetallic MOFs); (2) derivatization of monometallic MOFs with mixed metal salts; (3) derivatization of MOFs with supported metal nanoparticles; and (4) loading other functional materials after MOFs derivatization.

10.3.3.2.3.1 Thermal decomposition of pristine monometallic/bimetallic MOFs For the pathway of preparing catalysts directly by MOF oxidation, monometallic MOFs such as Ce-MOF, Ti-MOF, and Co-MOF are the most used MOF types. This is mainly attributed to the excellent properties of corresponding metal oxides (such as CeO_2, TiO_2, and Co_3O_4) in catalytic oxidation and photocatalysis of organic compounds. CeO_2 has good oxygen storage capacity and abundant acid sites, which are very helpful for the catalytic combustion process. Particularly, the Ce(IV) to Ce(III) redox transition caused by a large number of oxygen vacancies or defects in the lattice makes CeO_2 highly reducible. To develop CeO_2 catalysts with improved catalytic performance, Chen et al. synthesized mesoporous CeO_2 catalysts (CeO_2-MOF) by the pyrolysis of Ce-MOF precursor (Ce-(1,3,5-BTC) $(H_2O)_6$) and evaluated the catalytic performance of the samples by toluene combustion (Chen et al., 2018a,b). The results showed that CeO_2-MOF/350 catalyst obtained by pyrolysis at 350°C exhibited enhanced toluene oxidation activity, with T10%, T50%, and T90% conversion rates at 180°C, 211°C, and 223°C, respectively, under the conditions of SV = 20,000 mL/(g h) and toluene concentration of 1000 ppm. The high catalytic performance of CeO_2-MOF/350 can be reasonably attributed to a series of superior properties, including 3D penetrating mesoporous channels, larger specific surface area, smaller average crystal size, higher Ce(III)/Ce(IV) and O_{Sur}/O_{Latt} ratios, higher oxygen storage capacity, higher oxygen vacancy concentration, better low-temperature reducibility, more active oxygen, and more acidic sites.

In recent years, phase-junction TiO_2 with the same chemical composition but different crystal structures has been highly studied to improve photocatalytic performance of materials. Research has shown that phase-junction TiO_2 composed of an appropriate arrangement of rutile and anatase can effectively enhance photocatalytic activity. Inspired by this, Liu et al. successfully synthesized N-doped mesoporous carbon confined mixed-phase TiO_2 nanocomposites (TiO_2@C–N(x)) by pyrolyzing NH_2-MIL-125 in Ar and CO_2 atmospheres (Liu et al., 2019). The combination of TiO_2 and graphite carbon can further improve the photocatalytic activity of TiO_2, and N doping can extend the light absorption range from UV to visible light, expanding the application of the material. In addition, using CO_2 as a phase-transition promoter and mild oxidant for rutile-anatase phase transition, TiO_2@C–N(x) obtained showed significantly improved photocatalytic activity and mineralization efficiency compared to TiO_2@C–N annealed without CO_2 atmosphere. By using the optimal TiO_2@C–N(30) nanocomposite, the mineralization efficiency can reach 51.9% when the styrene is degraded by 62.4% under visible light irradiation for 240 min.

In fact, the pyrolysis of bimetallic MOFs can lead to the formation of multi-component metal oxide nanomaterials (MMONs), which exhibit stronger activity in catalytic degradation of VOCs than single metal oxides (Chen et al., 2020). Currently, the most commonly used bimetallic MOFs in this field are based on Co, Cu, Ce, Zn, Ni, and Mn, among which Co, Ce, and Cu are the most studied and used. Generally, two metal salts and an organic bridging ligand are mixed in a specific solvent (such as water or organic solvent) to prepare bimetallic MOFs, which are then further calcined to obtain the corresponding derivatives.

Due to the relatively low Co—O bond energy, cobalt oxide exhibits excellent reduction ability and high concentration of surface electrophilic oxygen species. In addition, Co(III) is suitable for being isovalently substituted into CeO_2 by forming a stable half-filled orbital. Meanwhile, CeO_2 possesses good oxygen storage capacity and abundant acidic sites, which are helpful for catalytic combustion processes. Particularly, a large number of oxygen vacancies or defects in the CeO_2 lattice can cause the Ce(IV) to Ce(III) redox transition, leading to its good reducibility (Chen et al., 2018a). Bimetallic Co—Ce metal oxide catalysts have been extensively studied for oxidizing various VOCs. Li et al. successfully obtained $CoCeO_x$ by preparing a bimetallic CoCe MOF and achieving the isovalent substitution of cobalt in the CeO_2 lattice (Li et al., 2020a). Characterization results confirmed the successful insertion of cobalt atoms and the concomitant increase in defects and oxygen vacancies, which were closely related to the catalytic activity for the oxidation of methanol, acetone, toluene, and o-xylene. The prepared $CoCeO_x$ achieved T50 and T90 in toluene oxidation at 212°C and 227°C, respectively, significantly lower than the corresponding temperatures of Co_3O_4/CeO_2 nanocubes. Furthermore, the $CoCeO_x$ catalyst could completely oxidize organic compounds into CO_2 and H_2O, while the reaction on the control material produced a large amount of CO.

Inspired by this, Wang et al. synthesized CeCuBDC bimetallic MOFs via a solvothermal method by precisely controlling the solvent/water ratio (Wang et al., 2019). Subsequently, they further prepared mixed transition metal oxide $CeCuO_x$ by thermal decomposition of CeCu-MOF to generate crystal strain and abundant defects at grain boundaries for high catalytic performance. Experimental results showed that the prepared $CeCuO_x$ exhibited good catalytic oxidation performance for common VOCs in air, such as methanol, acetone, toluene, and o-xylene, all of which were superior to CeO_2 and CuO, mainly due to the Cu(I)—OV—Ce(III) structure.

In addition to the active species, the catalyst support also influences the catalytic oxidation of VOCs. Metal foam has received widespread attention as a catalyst support due to its three-dimensional channels and good heat transfer properties. In the work of Zheng et al., $Cu(OH)_2$ nanorods were rapidly synthesized on the surface of Cu foam via an in-situ oxidation—reduction method, serving as a template and precursor for synthesizing CoCu-MOF (Zheng et al., 2022). $Cu(OH)_2$ nanorods overcome the self-aggregation of MOF derivatives, and the interfacial interaction of Co and Cu favors the formation of oxygen vacancies, which makes the optimized catalyst exhibit excellent acetone oxidation performance.

10.3.3.2.3.2 Thermal decomposition of monometallic MOFs mixed with metal salts In addition to preparing bimetallic MOFs by one-step method followed by pyrolysis to obtain bimetallic oxides, there are many studies that use a process of immersing a prepared single-metal MOF in a solution of another metal salt, followed by thermal treatment of the resulting solid to obtain mixed metal oxides. This has also been proven to be an effective method, which can not only improve the physicochemical properties of the catalyst, but also enhance the interaction between the two different oxides. Currently, the most extensively studied combinations are based on the pairwise combination of MOFs and metal salts based on Mn, Co, Ce, Cr and Cu.

MnO_x exhibits activity comparable to noble metal catalysts in the removal of gaseous pollutants due to its polymorphism and multivalence, as well as its rich oxygen species and good reducibility. Meanwhile, Cr_2O_3 has good thermal stability and resistance to organic oxidation poisoning. Based on these theories, Chen et al. prepared a series of MnO_x/Cr_2O_3 composite materials based on MIL-101(Cr) using a combination of impregnation of potassium permanganate solution, in-situ redox precipitation, and thermal decomposition (Chen et al., 2019a). Compared with commercial Cr_2O_3 or pure Cr_2O_3 obtained by thermal decomposition of MIL-101(Cr), MnO_x/Cr_2O_3 showed significantly enhanced catalytic activity for the oxidation of toluene. Its good durability and tolerance may be related to better crystal structure stability, oxygen vacancies, and reducibility. Further addition of MnO_x did not further improve the catalytic performance, possibly because excess MnO_x existed as isolated phases, lacking strong interaction between the two metal oxides. In-situ DRIFTS results confirmed that O_{Latt} played a crucial role in toluene combustion, and the pathway of toluene combustion was through rapid conversion to aldehyde substances, followed by benzoic acid substances, and finally forming CO_2 and H_2O.

In addition to impregnation with metal salts, transition metal acetylacetonate complexes ($M(acac)_2$) have also been reported as good precursors for doping with foreign elements. Based on previous studies, it is very promising to use $M(acac)_2$ as a precursor to improve the dispersibility for the preparation of MMON. In the study by Chen et al., they prepared a series of MMON by calcining Mn-BTC nanorods with various $M(acac)_2$ (M = Cu, Ce, Zn, Co), and evaluated their catalytic performance in the oxidation removal of VOCs (Chen et al., 2020). This strategy overcame the problems of poor repeatability and low dispersibility in traditional precipitation methods for preparing MMON. Among them, the Mn–Ce-based MMON showed the highest activity in the benzene oxidation reaction, with a T90% lower than 245°C. This may be attributed to the highest oxygen vacancy density in Mn–Ce-based MMON, especially Ce(III) being conducive to the formation of surface oxygen vacancies. In addition to oxygen vacancy density, the number of surface Mn(II) is another factor determining catalytic performance, and a lower amount of Mn(II) is favorable for the catalytic oxidation of benzene.

10.3.3.2.3.3 Thermal decomposition of MOFs with supported metal nanoparticles So far, the preparation of supported noble metal catalysts by calcining MOF integrated precursors with noble metal nanoparticles has received great attention and is often adopted in catalytic applications. MOFs, as common porous materials, have high surface area and small pores, which allows them to encapsulate small noble metal NPs in the pores. Therefore, noble metal NPs supported by metal oxides can be formed during the pyrolysis process, which has two advantages: (1) reducing the contact between NPs and inhibiting the aggregation of NPs to some extent; (2) strengthening the interaction between active noble metal NPs and metal oxide supports. This is believed to be beneficial for improving the catalytic efficiency of VOC oxidation (Chen et al., 2018a,b).

Chen et al. synthesized three-dimensional mesoporous Cr_2O_3 catalysts with penetrating Pt loading (Pt@M-Cr_2O_3) by pyrolysis of MIL-101-Cr containing pre-impregnated Pt NPs (Chen et al., 2018a,b). Pre-impregnation of Pt in MIL-101-Cr not only partially inhibited the aggregation of Pt NPs during the pyrolysis of MOF into M-Cr_2O_3, but also enhanced the interaction between them. The obtained Pt@M-Cr_2O_3 exhibited excellent catalytic performance for toluene combustion, with T10%, T50%, and T90% of 120, 140, and 144°C, respectively, at 1000 ppm of toluene and 20,000 mL/(g h) of space velocity. In addition, in-situ DRIFTS studies showed that the toluene degradation on Pt@M-Cr_2O_3 proceeded through the rapid conversion of benzoic acid substances to benzyl and aldehyde substances, which were then oxidized to maleic anhydride in the aromatics ring-opening reaction, and finally decomposed into CO_2 and H_2O. Similarly, Xiao et al. prepared Pt@Co_3O_4 to promote toluene degradation. The experimental results showed that the electron metal-support interaction (EMSI) effect induced partial electron transfer from Co_3O_4 to Pt species to produce electron-rich Pt and Co(III) species (Xiao et al., 2022). Electrophilic O_2 molecules can be activated by obtaining electrons from the electron-rich Pt species, forming nucleophilic oxygen species, which is conducive to attacking the C—H bond in toluene.

10.3.3.2.3.4 Loading other functional materials after pyrolysis In addition to the two categories of catalysts based on MOF derivatization mentioned above, there is another type that involves loading functional components (such as MNPs or metal oxides) onto MOF-derived materials. Qin et al. obtained nitrogen-doped graphene carbon (GC-N) using ZIF-8 as a precursor, and then loaded TiO_2 particles onto the surface of GC-N to obtain GC-N-TiO_2 through a hydrothermal method, which was applied to the photocatalytic degradation of toluene (Qin et al., 2020). The experimental results showed that the catalytic efficiency of GC-N-TiO_2 for toluene degradation was higher than that of the control materials (such as N-doped TiO_2, pure TiO_2, ZIF-8/TiO_2, and rGO-TiO_2). TiO_2 generates photogenerated holes and electrons under UV light irradiation. The holes can directly oxidize toluene or react with H_2O to generate ·OH radicals that can oxidize toluene. Due to its high conductivity and redox ability, the electrons can be transferred to GC-N containing abundant pyridine N, and the adjacent carbon atoms can adsorb O_2 and reduce it to H_2O or H_2O_2. The generated H_2O and H_2O_2 can react with the holes and $\cdot O_2^-$ to produce more ·OH, which can attract toluene molecules. In addition, GC-N obtained by carbonizing ZIF-8 exhibits good chemical stability, which can prevent the structure from being destroyed during the photocatalytic reaction. Wang et al. prepared Mn_2O_3 with different morphologies derived from MOFs using a simple method, and uniformly loaded ultra-small Pt NPs onto the Mn_2O_3 substrate to explore their catalytic performance in VOCs degradation (Wang et al., 2022b). MOF-derived oxides inherit the high adsorption capacity of MOFs, which increases the instantaneous reactant concentration around the catalyst and promotes the forward oxidation reaction. With excellent conductivity, Pt indirectly enhances the migration rate of ROS by promoting electron transfer. Weak Mn ions in Mn_2O_3 combine with lattice oxygen to help form oxygen vacancies. Through the EMSI effect, partial charges are transferred from Mn_2O_3 to Pt species, producing electron-rich Pt. Gaseous oxygen promotes C—H attack by accepting electrons to form nucleophilic oxygen species. The strong EMSI effect significantly enhances the redox properties of the catalyst and the activity of oxygen species.

10.4 Summary and perspectives

Due to their unique properties, MOFs have become highly attractive materials because of their large surface area, ultra-high porosity, wide pore size distribution, low density, good thermal stability, and the presence of adsorption functional sites composed of a series of ligands. They have been used in various applications, with significant progress made in recent years in the removal of harmful gases including VOCs. To date, researchers have developed many MOF-based materials for the effective adsorption and catalytic degradation of VOCs, mainly including adding various metals and functional groups ($-NH_2$, etc.) to MOFs, incorporating MOFs with other materials (organic molecules, polymers, carbon-based materials, etc.), and obtaining various derivatives by using MOFs as precursors. Despite significant progress in recent years, further research is needed to promote the application of MOF-based materials in the removal of VOCs. Potential future research directions include:

10.4.1 Preparation of MOF-based materials with bulk structures

Most of the MOF-based materials for VOC adsorption and catalytic degradation in existing literature are presented in the form of nanoparticles. However, adsorbents/catalysts for industrial reactors should be prepared in the form of bulk or honeycomb structures. Therefore, self-assembling nanosized MOFs into 3D materials or placing them on appropriate substrates while maintaining the activity, selectivity, and stability of the nano-materials to prepare bulk or honeycomb materials is still a huge challenge.

10.4.2 Further clarification of the removal mechanism of VOCs

It is necessary to use characterization techniques and computational simulation methods to better understand the interaction between VOCs and MOF-based adsorbents/catalysts. Molecular simulation can be applied to study the adsorption of VOCs, obtain quantitative adsorption data that are difficult to obtain experimentally, and further explain the adsorption phenomenon at the molecular level.

In addition, it is crucial to examine detailed reaction mechanisms between the catalyst and VOCs. Further investigation of the relationship between the MOF-based material structure and activity through characterization techniques and computational simulations may provide theoretical guidance for developing new catalysts that can prevent the formation of harmful by-products.

10.4.3 Adjusting the VOC degradation pathway

In many studies, the degradation of VOCs is often accompanied by the production of various harmful by-products (such as aldehydes, carboxylic acids, and other aromatic/aliphatic compounds), some of which are even more dangerous than the parent pollutants themselves. In addition, the most commonly used method for VOC elimination is to completely catalytically oxidize VOCs into H_2O and CO_2. Although this method can efficiently oxidize VOCs, it also causes resource waste and CO_2 emissions. From the perspective of resource utilization and economic benefits, selectively catalyzing VOCs to useful products is considered a new emerging alternative solution for VOC control.

10.4.4 Explore practical application conditions

Currently, most VOC removal experimental studies are carried out on a laboratory scale, and the experimental conditions are idealized, such as high adsorbent/catalyst mass, low flow rate, low humidity, and single pollutant systems. However, in real exhaust gas, VOCs exist in the form of mixtures and are easily affected by water vapor, which can affect the selectivity and adsorption and degradation capabilities of MOF-based materials. Therefore, more work needs to be done to demonstrate and validate the selective adsorption of MOFs in real environments, such as the presence of multiple pollutants and different humidity conditions.

10.4.5 Improving MOF stability

For highly polluted VOCs containing oxygen, chlorine, nitrogen, and sulfur, several studies have found that MOF-based materials undergo significant deactivation when removing such VOCs. Therefore, reasonable modification of the adsorbent/catalyst to achieve anti-deactivation effects is an important goal of such research. Improving the strength and heat resistance of MOFs to optimize the regeneration and recycling performance can achieve the goal of reducing costs.

References

Ahmed, E., Deep, A., Kwon, E.E., Brown, R.J.C., Kim, K.-H., 2016. Performance comparison of MOF and other sorbent materials in removing key odorants emitted from pigpen slurry. Scientific Reports 6. https://doi.org/10.1038/srep31283.

Azmi, L.H.M., Cherukupally, P., Hunter-Sellars, E., Ladewig, P.B., Williams, R.D., 2022. Fabrication of MIL-101-polydimethylsiloxane composites for environmental toluene abatement from humid air. Chemical Engineering Journal 429. https://doi.org/10.1016/j.cej.2021.132304.

Bahri, M., Haghighat, F., Rohani, S., Kazemian, H., 2017. Metal organic frameworks for gas-phase VOCs removal in a NTP-catalytic reactor. Chemical Engineering Journal 320, 308–318. https://doi.org/10.1016/j.cej.2017.02.087.

Bai, Y., Dou, Y., Xie, L.-H., Rutledge, W., Li, J.-R., Zhou, H.-C., 2016. Zr-based metal–organic frameworks: design, synthesis, structure, and applications. Chemical Society Reviews 45 (8), 2327–2367. https://doi.org/10.1039/C5CS00837A.

Beldon, P.J., Fábián, L., Stein, R.S., Thirumurugan, A., Cheetham, A.K., Friščić, T., 2010. Rapid room-temperature synthesis of zeolitic imidazolate frameworks by using mechanochemistry. Angewandte Chemie International Edition 49 (50), 9640–9643. https://doi.org/10.1002/anie.201005547.

Bhattarai, D.P., Pant, B., Acharya, J., Park, M., Ojha, G.P., 2021. Recent progress in metal–organic framework-derived nanostructures in the removal of volatile organic compounds. Molecules 26 (16), 4948. https://doi.org/10.3390/molecules26164948.

Bi, F., Zhang, X., Chen, J., Yang, Y., Wang, Y., 2020. Excellent catalytic activity and water resistance of UiO-66-supported highly dispersed Pd nanoparticles for toluene catalytic oxidation. Applied Catalysis B: Environmental 269, 118767. https://doi.org/10.1016/j.apcatb.2020.118767.

Bitzer, J., Heck, S.-L., Kleist, W., 2020. Tailoring the breathing behavior of functionalized MIL-53(Al,M)-NH_2 materials by using the mixed-metal concept. Microporous and Mesoporous Materials 308. https://doi.org/10.1016/j.micromeso.2020.110329.

Britt, D., Tranchemontagne, D., Yaghi, O.M., 2008. Metal–organic frameworks with high capacity and selectivity for harmful gases. Proceedings of the National Academy of Sciences of the United States of America 105 (33), 11623–11627. https://doi.org/10.1073/pnas.0804900105.

Cavka, J.H., Jakobsen, S., Olsbye, U., Guillou, N., Lamberti, C., Bordiga, S., Lillerud, K.P., 2008. A new zirconium inorganic building brick forming metal organic frameworks with exceptional stability. Journal of the American Chemical Society 130 (42), 13850–13851. https://doi.org/10.1021/ja8057953.

Chae, H.K., Siberio-Pérez, D.Y., Kim, J., Go, Y., Eddaoudi, M., Matzger, A.J., O'Keeffe, M., Yaghi, O.M., Materials, D., Discovery, G., 2004. A route to high surface area, porosity and inclusion of large molecules in crystals. Nature 427 (6974), 523–527. https://doi.org/10.1038/nature02311.

Chaikittisilp, W., Ariga, K., Yamauchi, Y., 2013. A new family of carbon materials: synthesis of MOF-derived nanoporous carbons and their promising applications. Journal of Materials Chemistry A 1 (1), 14–19. https://doi.org/10.1039/c2ta00278g.

Chen, X., Chen, X., Yu, E., Cai, S., Jia, H., Chen, J., Liang, P., 2018a. In situ pyrolysis of Ce-MOF to prepare CeO_2 catalyst with obviously improved catalytic performance for toluene combustion. Chemical Engineering Journal 344, 469–479. https://doi.org/10.1016/j.cej.2018.03.091.

Chen, X., Chen, X., Cai, S., Chen, J., Xu, W., Jia, H., Chen, J., 2018b. Catalytic combustion of toluene over mesoporous Cr_2O_3-supported platinum catalysts prepared by in situ pyrolysis of MOFs. Chemical Engineering Journal 334, 768–779. https://doi.org/10.1016/j.cej.2017.10.091.

Chen, X., Cai, S., Yu, E., Chen, J., Jia, H., 2019a. MnO_x/Cr_2O_3 composites prepared by pyrolysis of Cr-MOF precursors containing in situ assembly of MnO_x as high stable catalyst for toluene oxidation. Applied Surface Science 475, 312–324. https://doi.org/10.1016/j.apsusc.2018.12.277.

Chen, D., Zhao, J., Zhang, P., Dai, S., 2019b. Mechanochemical synthesis of metal–organic frameworks. Polyhedron 162, 59–64.

Chen, B., Yang, X., Zeng, X., Huang, Z., Xiao, J., Wang, J., Zhan, G., 2020. Multicomponent metal oxides derived from Mn-BTC anchoring with metal acetylacetonate complexes as excellent catalysts for VOCs and CO oxidation. Chemical Engineering Journal 397. https://doi.org/10.1016/j.cej.2020.125424.

Chen, L., Wang, X., Rao, Z., Tang, Z., Wang, Y., Shi, G., Lu, G., Xie, X., Chen, D., Sun, J., 2021. In-situ synthesis of Z-scheme MIL-100(Fe)/alpha-Fe_2O_3 heterojunction for enhanced adsorption and visible-light photocatalytic oxidation of O-xylene. Chemical Engineering Journal 416. https://doi.org/10.1016/j.cej.2021.129112.

Cheng, N., Ren, L., Xu, X., Du, Y., Dou, S.X., 2018. Recent development of zeolitic imidazolate frameworks (ZIFs) derived porous carbon based materials as electrocatalysts. Advanced Energy Materials 8 (25). https://doi.org/10.1002/aenm.201801257, 1801257.

Chui, S.S.Y., Lo, S.M.F., Charmant, J.P.H., Orpen, A.G., Williams, I.D., 1999. A chemically functionalizable nanoporous material $[Cu_3(TMA)_2(H_2O)_3]_n$. Science 283 (5405), 1148–1150. https://doi.org/10.1126/science.283.5405.1148.

Chung, W.-C., Mei, D.-H., Tu, X., Chang, M.-B., 2019. Removal of VOCs from gas streams via plasma and catalysis. Catalysis Reviews 61 (2), 270–331. https://doi.org/10.1080/01614940.2018.1541814.

Crawford, D.E., Casaban, J., 2016. Recent developments in mechanochemical materials synthesis by extrusion. Advanced Materials 28 (27), 5747–5754. https://doi.org/10.1002/adma.201505352.

Cui, X., Sun, X., Liu, L., Huang, Q., Yang, H., Chen, C., Nie, S., Zhao, Z., Zhao, Z., 2019. In-situ fabrication of cellulose foam HKUST-1 and surface modification with polysaccharides for enhanced selective adsorption of toluene and acidic dipeptides. Chemical Engineering Journal 369, 898–907. https://doi.org/10.1016/j.cej.2019.03.129.

Dapaah, M.F., Niu, Q., Yu, Y., You, T., Liu, B., Cheng, L., 2022. Efficient persistent organic pollutant removal in water using MIL-metal–organic framework driven Fenton-like reactions: a critical review. Chemical Engineering Journal 431, 134182. https://doi.org/10.1016/j.cej.2021.134182.

Dong, F., Han, W., Guo, Y., Han, W., Tang, Z., 2021. $CeCoO_x$-MNS catalyst derived from three-dimensional mesh nanosheet Co-based metal–organic frameworks for highly efficient catalytic combustion of VOCs. Chemical Engineering Journal 405. https://doi.org/10.1016/j.cej.2020.126948.

Duan, C., Yu, Y., Yang, P., Zhang, X., Li, F., Li, L., Xi, H., 2020. Engineering new defects in MIL-100(Fe) via a mixed-ligand approach to effect enhanced volatile organic compound adsorption capacity. Industrial & Engineering Chemistry Research 59 (2), 774–782. https://doi.org/10.1021/acs.iecr.9b05751.

Eddaoudi, M., Kim, J., Rosi, N., Vodak, D., Wachter, J., O'Keeffe, M., Yaghi, O.M., 2002. Systematic design of pore size and functionality in isoreticular MOFs and their application in methane storage. Science 295 (5554), 469–472. https://doi.org/10.1126/science.1067208.

Farha, O.K., Özgür Yazaydın, A., Eryazici, I., Malliakas, C.D., Hauser, B.G., Kanatzidis, M.G., Nguyen, S.T., Snurr, R.Q., Hupp, J.T., 2010. De novo synthesis of a metal–organic framework material featuring ultrahigh surface area and gas storage capacities. Nature Chemistry 2 (11), 944–948. https://doi.org/10.1038/nchem.834.

Feng, X., Jena, H.S., Krishnaraj, C., Leus, K., Wang, G., Chen, H., Jia, C., Van Der Voort, P., 2021. Generating catalytic sites in UiO-66 through defect engineering. ACS Applied Materials & Interfaces 13 (51), 60715–60735. https://doi.org/10.1021/acsami.1c13525.

Férey, G., Latroche, M., Serre, C., Millange, F., Loiseau, T., Percheron-Guégan, A., 2003. Hydrogen adsorption in the nanoporous metal-benzenedicarboxylate $M(OH)(O_2C-C_6H_4-CO_2)$ ($M = Al^{3+}, Cr^{3+}$), MIL-53. Chemical Communications (24), 2976–2977. https://doi.org/10.1039/B308903G.

Férey, G., Mellot-Draznieks, C., Serre, C., Millange, F., Dutour, J., Surblé, S., Margiolaki, I., 2005. A chromium terephthalate-based solid with unusually large pore volumes and surface area. Science 309 (5743), 2040–2042. https://doi.org/10.1126/science.1116275.

Friščić, T., Fábián, L., 2009. Mechanochemical conversion of a metal oxide into coordination polymers and porous frameworks using liquid-assisted grinding (LAG). CrystEngCommun 11 (5), 743–745. https://doi.org/10.1039/B822934C.

Friščić, T., Reid, D.G., Halasz, I., Stein, R.S., Dinnebier, R.E., Duer, M.J., 2010. Ion- and liquid-assisted grinding: improved mechanochemical synthesis of metal–organic frameworks reveals salt inclusion and anion templating. Angewandte Chemie International Edition 49 (4), 712–715. https://doi.org/10.1002/anie.200906583.

Furukawa, H., Yaghi, O.M., 2009. Storage of hydrogen, methane, and carbon dioxide in highly porous covalent organic frameworks for clean energy applications. Journal of the American Chemical Society 131 (25), 8875–8883. https://doi.org/10.1021/ja9015765.

Gao, Z., Wang, J., Muhammad, Y., Zhang, Y., Shah, S.J., Hu, Y., Chu, Z., Zhao, Z., Zhao, Z., 2020. Enhanced moisture-resistance and excellent photocatalytic performance of synchronous N/Zn-decorated MIL-125(Ti) for vaporous acetaldehyde degradation. Chemical Engineering Journal 388. https://doi.org/10.1016/j.cej.2020.124389.

Garibay, S.J., Cohen, S.M., 2010. Isoreticular synthesis and modification of frameworks with the UiO-66 topology. Chemical Communications 46 (41), 7700–7702. https://doi.org/10.1039/C0CC02990D.

Guo, Y., Wen, M., Song, S., Liu, Q., Li, G., An, T., 2022. Enhanced catalytic elimination of typical VOCs over $ZnCoO_x$ catalyst derived from in situ pyrolysis of ZnCo bimetallic zeolitic imidazolate frameworks. Applied Catalysis B-Environmental 308. https://doi.org/10.1016/j.apcatb.2022.121212.

Hartmann, M., Kunz, S., Himsl, D., Tangermann, O., Ernst, S., Wagener, A., 2008. Adsorptive separation of isobutene and isobutane on $Cu_3(BTC)_2$. Langmuir 24 (16), 8634–8642. https://doi.org/10.1021/la8008656.

Hendon, C.H., Walsh, A., 2015. Chemical principles underpinning the performance of the metal–organic framework HKUST-1. Chemical Science 6 (7), 3674–3683. https://doi.org/10.1039/C5SC01489A.

Horcajada, P., Chalati, T., Serre, C., Gillet, B., Sebrie, C., Baati, T., Eubank, J.F., Heurtaux, D., Clayette, P., Kreuz, C., Chang, J.-S., Hwang, Y.K., Marsaud, V., Bories, P.-N., Cynober, L., Gil, S., Férey, G., Couvreur, P., Gref, R., 2010. Porous metal–organic-framework nanoscale carriers as a potential platform for drug delivery and imaging. Nature Materials 9 (2), 172–178. https://doi.org/10.1038/nmat2608.

Hou, X., Bian, Y., Yang, L., 2022. One-step synthesis of transition metal modified UiO-66-Ce metal–organic framework: catalytic oxidation of toluene and investigation of the mechanism. Microporous and Mesoporous Materials 345. https://doi.org/10.1016/j.micromeso.2022.112214.

Howarth, A.J., Peters, A.W., Vermeulen, N.A., Wang, T.C., Hupp, J.T., Farha, O.K., 2017. Best practices for the synthesis, activation, and characterization of metal–organic frameworks. Chemistry of Materials 29 (1), 26–39. https://doi.org/10.1021/acs.chemmater.6b02626.

Huang, X.-C., Lin, Y.-Y., Zhang, J.-P., Chen, X.-M., 2006. Ligand-directed strategy for zeolite-type metal–organic frameworks: zinc(II) imidazolates with unusual zeolitic topologies. Angewandte Chemie International Edition 45 (10), 1557–1559. https://doi.org/10.1002/anie.200503778.

Jangodaz, E., Alaie, E., Safekordi, A.A., Tasharrofi, S., 2018. Adsorption of ethylbenzene from air on metal–organic frameworks MIL-101(Cr) and MIL-53(Fe) at room temperature. Journal of Inorganic and Organometallic Polymers and Materials 28 (5), 2090–2099. https://doi.org/10.1007/s10904-018-0896-6.

Janiak, C., Vieth, J.K., 2010. MOFs, MILs and more: concepts, properties and applications for porous coordination networks (PCNs). New Journal of Chemistry 34 (11), 2366–2388. https://doi.org/10.1039/C0NJ00275E.

Jhinjer, H.S., Singh, A., Bhattacharya, S., Jassal, M., Agrawal, A.K., 2021. Metal–organic frameworks functionalized smart textiles for adsorptive removal of hazardous aromatic pollutants from ambient air. Journal of Hazardous Materials 411. https://doi.org/10.1016/j.jhazmat.2021.125056.

Jhung, S.-H., Lee, J.-H., Chang, J.-S., 2005. Microwave synthesis of a nanoporous hybrid material, chromium trimesate. Bulletin of the Korean Chemical Society 26 (6), 880–881.

Jhung, S.H., Lee, J.-H., Yoon, J.W., Serre, C., Ferey, G., Chang, J.-S., 2007. Microwave synthesis of chromium terephthalate MIL-101 and its benzene sorption ability. Advanced Materials 19 (1), 121. https://doi.org/10.1002/adma.200601604.

Jhung, S.H., Khan, N.A., Hasan, Z., 2012. Analogous porous metal–organic frameworks: synthesis, stability and application in adsorption. CrystEngCommun 14 (21), 7099–7109. https://doi.org/10.1039/C2CE25760B.

Jiang, Y., Gao, J., Zhang, Q., Liu, Z., Fu, M., Wu, J., Hu, Y., Ye, D., 2019. Enhanced oxygen vacancies to improve ethyl acetate oxidation over MnO_x-CeO_2 catalyst derived from MOF template. Chemical Engineering Journal 371, 78–87. https://doi.org/10.1016/j.cej.2019.03.233.

Joaristi, A.M., Juan-Alcaniz, J., Serra-Crespo, P., Kapteijn, F., Gascon, J., 2012. Electrochemical synthesis of some archetypical Zn^{2+}, Cu^{2+}, and Al^{3+} metal organic frameworks. Crystal Growth & Design 12 (7), 3489–3498. https://doi.org/10.1021/cg300552w.

Khan, N.A., Jhung, S.-H., 2009. Facile syntheses of metal–organic framework Cu3 (BTC) 2 (H2O) 3 under ultrasound. Bulletin of the Korean Chemical Society 30 (12), 2921–2926.

Kim, H., Hong, C.S., 2021. MOF-74-type frameworks: tunable pore environment and functionality through metal and ligand modification. CrystEngCommun 23 (6), 1377–1387. https://doi.org/10.1039/D0CE01870H.

Kim, B., Lee, Y.-R., Kim, H.-Y., Ahn, W.-S., 2018. Adsorption of volatile organic compounds over MIL-125-NH2. Polyhedron 154, 343–349. https://doi.org/10.1016/j.poly.2018.08.010.

Klimakow, M., Klobes, P., Thünemann, A.F., Rademann, K., Emmerling, F., 2010. Mechanochemical synthesis of metal–organic frameworks: a fast and facile approach toward quantitative yields and high specific surface areas. Chemistry of Materials 22 (18), 5216–5221. https://doi.org/10.1021/cm1012119.

Klinowski, J., Paz, F.A.A., Silva, P., Rocha, J., 2011. Microwave-assisted synthesis of metal–organic frameworks. Dalton Transactions 40 (2), 321–330. https://doi.org/10.1039/C0DT00708K.

Kreno, L.E., Leong, K., Farha, O.K., Allendorf, M., Van Duyne, R.P., Hupp, J.T., 2012. Metal–organic framework materials as chemical sensors. Chemical Reviews 112 (2), 1105–1125. https://doi.org/10.1021/cr200324t.

Laybourn, A., Katrib, J., Ferrari-John, R.S., Morris, C.G., Yang, S., Udoudo, O., Easun, T.L., Dodds, C., Champness, N.R., Kingman, S.W., Schröder, M., 2017. Metal–organic frameworks in seconds via selective microwave heating. Journal of Materials Chemistry A 5 (16), 7333–7338. https://doi.org/10.1039/C7TA01493G.

Lee, J., Farha, O.K., Roberts, J., Scheidt, K.A., Nguyen, S.T., Hupp, J.T., 2009. Metal–organic framework materials as catalysts. Chemical Society Reviews 38 (5), 1450–1459. https://doi.org/10.1039/b807080f.

Lee, Y.-R., Kim, J., Ahn, W.-S., 2013. Synthesis of metal–organic frameworks: a mini review. Korean Journal of Chemical Engineering 30 (9), 1667–1680. https://doi.org/10.1007/s11814-013-0140-6.

Lee, Y., Chang, Y., Lee, D., Hsu, J., 2018. Water stable metal–organic framework as adsorbent from aqueous solution: a mini-review. Journal of the Taiwan Institute of Chemical Engineers 93, 176–183. https://doi.org/10.1016/j.jtice.2018.06.035.

Li, H., Eddaoudi, M., O'Keeffe, M., Yaghi, O.M., 1999. Design and synthesis of an exceptionally stable and highly porous metal–organic framework. Nature 402 (6759), 276–279. https://doi.org/10.1038/46248.

Li, J., Kuppler, R.J., Zhou, H., 2009a. Selective gas adsorption and separation in metal–organic frameworks. Chemical Society Reviews 38 (5), 1477–1504. https://doi.org/10.1039/B802426J.

Li, Z., Qiu, L., Xu, T., Wu, Y., Wang, W., Wu, Z., Jiang, X., 2009b. Ultrasonic synthesis of the microporous metal–organic framework Cu3 (BTC) 2 at ambient temperature and pressure: an efficient and environmentally friendly method. Materials Letters 63 (1), 78–80. https://doi.org/10.1016/j.matlet.2008.09.010.

Li, J., Sculley, J., Zhou, H., 2012. Metal–organic frameworks for separations. Chemical Reviews 112 (2), 869–932. https://doi.org/10.1021/cr200190s.

Li, Y., Miao, J., Sun, X., Xiao, J., Li, Y., Wang, H., Xia, Q., Li, Z., 2016. Mechanochemical synthesis of Cu-BTC@GO with enhanced water stability and toluene adsorption capacity. Chemical Engineering Journal 298, 191–197. https://doi.org/10.1016/j.cej.2016.03.141.

Li, Y., Han, W., Wang, R., Weng, L., Serrano-Lotina, A., Banares, M.A., Wang, Q., Yeung, K.L., 2020a. Performance of an aliovalent-substituted CoCeOx catalyst from bimetallic MOF for VOC oxidation in air. Applied Catalysis B: Environmental 275, 119121. https://doi.org/10.1016/j.apcatb.2020.119121.

Li, X., Zhang, L., Yang, Z., Wang, P., Yan, Y., Ran, J., 2020b. Adsorption materials for volatile organic compounds (VOCs) and the key factors for VOCs adsorption process: a review. Separation and Purification Technology 235. https://doi.org/10.1016/j.seppur.2019.116213.

Li, P., Kim, S., Jin, J., Chun, D.H., Park, J.H., 2020c. Efficient photodegradation of volatile organic compounds by iron-based metal–organic frameworks with high adsorption capacity. Applied Catalysis B-Environmental 263. https://doi.org/10.1016/j.apcatb.2019.118284.

Li, W., Wang, W., Sun, J., Ma, X., Dong, Y., 2021. Hydrophobic modification of UiO-66 by naphthyl ligand substitution for efficient toluene adsorption in a humid environment. Microporous and Mesoporous Materials 326. https://doi.org/10.1016/j.micromeso.2021.111357.

Li, J., Xu, Z., Wang, T., Xie, X., Li, D., Wang, J., Huang, H., Ao, Z., 2022. A versatile route to fabricate Metal/UiO-66 (Metal = Pt, Pd, Ru) with high activity and stability for the catalytic oxidation of various volatile organic compounds. Chemical Engineering Journal 448. https://doi.org/10.1016/j.cej.2022.136900.

Liu, H., Ma, Y., Chen, J., Wen, M., Li, G., An, T., 2019. Highly efficient visible-light-driven photocatalytic degradation of VOCs by CO_2-assisted synthesized mesoporous carbon confined mixed-phase TiO_2 nanocomposites derived from MOFs. Applied Catalysis B-Environmental 250, 337–346. https://doi.org/10.1016/j.apcatb.2019.03.054.

Liu, H., Xu, M., Li, G., Zhang, W., An, T., 2021a. Solar-light-triggered regenerative adsorption removal of styrene by silver nanoparticles incorporated in metal–organic frameworks. Environmental Science: Nano 8 (2), 543–553. https://doi.org/10.1039/d0en01011a.

Liu, H., Chang, X., Liu, X., Li, G., Zhang, W., An, T., 2021b. Boosting the photocatalytic degradation of ethyl acetate by a Z-scheme Au-TiO_2@NH_2-UiO-66 heterojunction with ultrafine Au as an electron mediator. Environmental Science: Nano 8 (9), 2542–2553. https://doi.org/10.1039/d1en00225b.

Liu, H., Li, N., Feng, M., Li, G., Zhang, W., An, T., 2022. Near-infrared light induced adsorption-desorption cycle for VOC recovery by integration of metal–organic frameworks with graphene oxide nanosheets. Environmental Science: Nano 9 (5), 1858–1868. https://doi.org/10.1039/d2en00103a.

Livage, C., Egger, C., Nogues, M., Férey, G., 1998. Hybrid open frameworks (MIL-n). Part 5 Synthesis and crystal structure of MIL-9: a new three-dimensional ferrimagnetic cobalt (II) carboxylate with a two-dimensional array of edge-sharing Co octahedra with 12-membered rings. Journal of Materials Chemistry 8 (12), 2743–2747. https://doi.org/10.1039/A802714E.

Long, W., Chen, Z., Huang, Y., Kang, X., 2022. Selective oxidation of toluene to benzaldehyde using Co-ZIF nanocatalyst. International Journal of Molecular Sciences 23 (21). https://doi.org/10.3390/ijms232112881.

Ma, F.-J., Liu, S.-X., Liang, D.-D., Ren, G.-J., Wei, F., Chen, Y.-G., Su, Z.-M., 2011. Adsorption of volatile organic compounds in porous metal–organic frameworks functionalized by polyoxometalates. Journal of Solid State Chemistry 184 (11), 3034–3039. https://doi.org/10.1016/j.jssc.2011.09.002.

Ma, X., Li, L., Zeng, Z., Chen, R., Wang, C., Zhou, K., Su, C., Li, H., 2019. Synthesis of nitrogen-rich nanoporous carbon materials with C_3N-type from ZIF-8 for methanol adsorption. Chemical Engineering Journal 363, 49–56. https://doi.org/10.1016/j.cej.2019.01.132.

Ma, X., Wang, W., Sun, C., Li, H., Sun, J., Liu, X., 2021. Adsorption performance and kinetic study of hierarchical porous Fe-based MOFs for toluene removal. Science of the Total Environment 793. https://doi.org/10.1016/j.scitotenv.2021.148622.

Ma, Y., Wang, L., Ma, J., Liu, F., Einaga, H., He, H., 2022. Improved and reduced performance of Cu- and Ni-substituted Co_3O_4 catalysts with varying Co-oh/Co-Td and Co3+/Co2+ ratios for the complete catalytic oxidation of VOCs. Environmental Science & Technology. https://doi.org/10.1021/acs.est.2c02450.

Mishra, P., Edubilli, S., Mandal, B., Gumma, S., 2014. Adsorption characteristics of metal–organic frameworks containing coordinatively unsaturated metal sites: effect of metal cations and adsorbate properties. Journal of Physical Chemistry C 118 (13), 6847–6855. https://doi.org/10.1021/jp412622b.

Mueller, U., Schubert, M., Teich, F., Puetter, H., Schierle-Arndt, K., Pastré, J., 2006. Metal–organic frameworks—prospective industrial applications. Journal of Materials Chemistry 16 (7), 626–636. https://doi.org/10.1039/B511962F.

Muller, U., Puetter, H., Hesse, M., Wessel, H., 2005. International Patent, WO 49892 (2005).

Murray, L.J., Dincă, M., Long, J.R., 2009. Hydrogen storage in metal–organic frameworks. Chemical Society Reviews 38 (5), 1294–1314. https://doi.org/10.1039/B802256A.

Ni, Z., Masel, R.I., 2006. Rapid production of metal–organic frameworks via microwave-assisted solvothermal synthesis. Journal of the American Chemical Society 128 (38), 12394–12395. https://doi.org/10.1021/ja0635231.

Nugent, P., Belmabkhout, Y., Burd, S.D., Cairns, A.J., Luebke, R., Forrest, K., Pham, T., Ma, S., Space, B., Wojtas, L., Eddaoudi, M., Zaworotko, M.J., 2013. Porous materials with optimal adsorption thermodynamics and kinetics for CO_2 separation. Nature 495 (7439), 80–84. https://doi.org/10.1038/nature11893.

Park, S.-E., Chang, J.-S., Hwang, Y.K., Kim, D.S., Jhung, S.H., Hwang, J.S., 2004. Supramolecular interactions and morphology control in microwave synthesis of nanoporous materials. Catalysis Surveys from Asia 8 (2), 91–110. https://doi.org/10.1023/B:CATS.0000026990.25778.a8.

Park, K.S., Ni, Z., Côté, A.P., Choi, J.Y., Huang, R., Uribe-Romo, F.J., Chae, H.K., O'Keeffe, M., Yaghi, O.M., 2006. Exceptional chemical and thermal stability of zeolitic imidazolate frameworks. Proceedings of the National Academy of Sciences 103 (27), 10186–10191. https://doi.org/10.1073/pnas.0602439103.

Pei, Y., Qin, J., Wang, J., Hu, Y., 2021. Fe-based metal organic framework derivative with enhanced Lewis acidity and hierarchical pores for excellent adsorption of oxygenated volatile organic compounds. The Science of the Total Environment 790. https://doi.org/10.1016/j.scitotenv.2021.148132, 148132-148132.

Petit, C., Mendoza, B., Bandosz, T.J., 2010. Hydrogen sulfide adsorption on MOFs and MOF/graphite oxide composites. ChemPhysChem 11 (17), 3678–3684. https://doi.org/10.1002/cphc.201000689.

Phan, A., Doonan, C.J., Uribe-Romo, F.J., Knobler, C.B., O'Keeffe, M., Yaghi, O.M., 2010. Synthesis, structure, and carbon dioxide capture properties of zeolitic imidazolate frameworks. Accounts of Chemical Research 43 (1), 58–67. https://doi.org/10.1021/ar900116g.

Pichon, A., James, S.L., 2008. An array-based study of reactivity under solvent-free mechanochemical conditions—insights and trends. CrystEngCommun 10 (12), 1839–1847. https://doi.org/10.1039/B810857A.

Pichon, A., Lazuen-Garay, A., James, S.L., 2006. Solvent-free synthesis of a microporous metal–organic framework. CrystEngCommun 8 (3), 211–214. https://doi.org/10.1039/B513750K.

Pinto, M.L., Dias, S., Pires, J., 2013. Composite MOF foams: the example of UiO-66/polyurethane. ACS Applied Materials & Interfaces 5 (7), 2360–2363. https://doi.org/10.1021/am303089g.

Qin, J., Wang, J., Yang, J., Hu, Y., Fu, M., Ye, D., 2020. Metal organic framework derivative-TiO_2 composite as efficient and durable photocatalyst for the degradation of toluene. Applied Catalysis B-Environmental 267. https://doi.org/10.1016/j.apcatb.2020.118667.

Qiu, L.-G., Li, Z.-Q., Wu, Y., Wang, W., Xu, T., Jiang, X., 2008. Facile synthesis of nanocrystals of a microporous metal–organic framework by an ultrasonic method and selective sensing of organoamines. Chemical Communications (31), 3642–3644. https://doi.org/10.1039/B804126A.

Ramos, V.C., Han, W., Yeung, K.L., 2022. Functionalization of metal organic frameworks by BMIM CH_3COO ionic liquid for toluene capture. Environmental Research 215. https://doi.org/10.1016/j.envres.2022.114341.

Ramsahye, N.A., Gao, J., Jobic, H., Llewellyn, P.L., Yang, Q., Wiersum, A.D., Koza, M.M., Guillerm, V., Serre, C., Zhong, C.L., Maurin, G., 2014. Adsorption and diffusion of light hydrocarbons in UiO-66(Zr): a combination of experimental and modeling tools. Journal of Physical Chemistry C 118 (47), 27470–27482. https://doi.org/10.1021/jp509672c.

Remya, V., Kurian, M., 2019. Synthesis and catalytic applications of metal–organic frameworks: a review on recent literature. International Nano Letters 9, 17–29. https://doi.org/10.1007/s40089-018-0255-1.

Riou, D., Roubeau, O., Férey, G., 1998. Composite microporous compounds. Part I: Synthesis and structure determination of two new vanadium alkyldiphosphonates (MIL-2 and MIL-3) with three-dimensional open frameworks. Microporous and Mesoporous Materials 23 (1–2), 23–31. https://doi.org/10.1016/S1387-1811(98)00042-0.

Rosi, N.L., Kim, J., Eddaoudi, M., Chen, B., O'Keeffe, M., Yaghi, O.M., 2005. Rod packings and metal–organic frameworks constructed from rod-shaped secondary building units. Journal of the American Chemical Society 127 (5), 1504–1518. https://doi.org/10.1021/ja045123o.

Rowsell, J.L.C., Yaghi, O.M., 2005. Strategies for hydrogen storage in metal–organic frameworks. Angewandte Chemie International Edition 44 (30), 4670–4679. https://doi.org/10.1002/anie.200462786.

Rowsell, J.L.C., Yaghi, O.M., 2006. Effects of functionalization, catenation, and variation of the metal oxide and organic linking units on the low-pressure hydrogen adsorption properties of metal–organic frameworks. Journal of the American Chemical Society 128 (4), 1304–1315. https://doi.org/10.1021/ja056639q.

Sabouni, R., Kazemian, H., Rohani, S., 2010. A novel combined manufacturing technique for rapid production of IRMOF-1 using ultrasound and microwave energies. Chemical Engineering Journal 165 (3), 966–973. https://doi.org/10.1016/j.cej.2010.09.036.

Saini, V.K., Pires, J., 2017. Development of metal organic from work-199 immobilized zeolite foam for adsorption of common indoor VOCs. Journal of Environmental Sciences 55, 321–330. https://doi.org/10.1016/j.jes.2016.09.017.

Sánchez-González, E., Mileo, P.G.M., Sagastuy-Breña, M., Álvarez, J.R., Reynolds, J.E., Villarreal, A., Gutiérrez-Alejandre, A., Ramírez, J., Balmaseda, J., González-Zamora, E., Maurin, G., Humphrey, S.M., Ibarra, I.A., 2018. Highly reversible sorption of H_2S and CO_2 by an environmentally friendly Mg-based MOF. Journal of Materials Chemistry A 6 (35), 16900–16909. https://doi.org/10.1039/C8TA05400B.

Schlesinger, M., Schulze, S., Hietschold, M., Mehring, M., 2010. Evaluation of synthetic methods for microporous metal–organic frameworks exemplified by the competitive formation of [Cu2 (btc) 3 (H2O) 3] and [Cu2 (btc)(OH)(H2O)]. Microporous and Mesoporous Materials 132 (1–2), 121–127. https://doi.org/10.1016/j.micromeso.2010.02.008.

Seo, Y.-K., Hundal, G., Jang, I.T., Hwang, Y.K., Jun, C.-H., Chang, J.-S., 2009. Microwave synthesis of hybrid inorganic–organic materials including porous Cu3 (BTC) 2 from Cu(II)-trimesate mixture. Microporous and Mesoporous Materials 119 (1–3), 331–337. https://doi.org/10.1016/j.micromeso.2008.10.035.

Son, W.-J., Kim, J., Kim, J., Ahn, W.-S., 2008. Sonochemical synthesis of MOF-5. Chemical Communications (47), 6336–6338. https://doi.org/10.1039/B814740J.

Song, X., Jiang, Y., Cheng, F., Earnshaw, J., Na, J., Li, X., Yamauchi, Y., 2021. Hollow carbon-based nanoarchitectures based on ZIF: inward/outward contraction mechanism and beyond. Small 17 (2). https://doi.org/10.1002/smll.202004142, 2004142.

Stock, N., Biswas, S., 2012. Synthesis of metal–organic frameworks (MOFs): routes to various MOF topologies, morphologies, and composites. Chemical Reviews 112 (2), 933–969. https://doi.org/10.1021/cr200304e.

Suh, M.P., Park, H.J., Prasad, T.K., Lim, D.-W., 2012. Hydrogen storage in metal–organic frameworks. Chemical Reviews 112 (2), 782–835. https://doi.org/10.1021/cr200274s.

Sun, X., Lv, D., Chen, Y., Wu, Y., Wu, Q., Xia, Q., Li, Z., 2017. Enhanced adsorption performance of aromatics on a novel chromium-based MIL-101@graphite oxide composite. Energy & Fuels 31 (12), 13985–13990. https://doi.org/10.1021/acs.energyfuels.7b02665.

Szcześniak, B., Borysiuk, S., Choma, J., Jaroniec, M., 2020. Mechanochemical synthesis of highly porous materials. Materials Horizons 7 (6), 1457–1473. https://doi.org/10.1039/D0MH00081G.

Tang, Z., Tao, H., Wang, X., Chen, L., Song, C., Lu, G., Xie, X., Sun, J., 2022. Quasi-in situ synthesis of Ag NPs@m-MIL-100(Fe) for the enhanced photocatalytic elimination of flowing xylenes. ACS Applied Materials & Interfaces. https://doi.org/10.1021/acsami.2c15811.

The Ky, V., Van Nhieu, L., Yoo, K.S., Song, M., Kim, D., Kim, J., 2019. Facile synthesis of UiO-66(Zr) using a microwave-assisted continuous tubular reactor and its application for toluene adsorption. Crystal Growth & Design 19 (9), 4949–4956. https://doi.org/10.1021/acs.cgd.9b00170.

The Ky, V., Van Nhieu, L., Van Cuong, N., Song, M., Kim, D., Yoo, K.S., Park, B.J., Kim, J., 2020. Microwave-assisted continuous-flow synthesis of mixed-ligand UiO-66(Zr) frameworks and their application to toluene adsorption. Journal of Industrial and Engineering Chemistry 86, 178–185. https://doi.org/10.1016/j.jiec.2020.03.001.

Tranchemontagne, D.J., Hunt, J.R., Yaghi, O.M., 2008. Room temperature synthesis of metal–organic frameworks: MOF-5, MOF-74, MOF-177, MOF-199, and IRMOF-0. Tetrahedron 64 (36), 8553–8557. https://doi.org/10.1016/j.tet.2008.06.036.

Vaitsis, C., Sourkouni, G., Argirusis, C., 2019. Metal organic frameworks (MOFs) and ultrasound: a review. Ultrasonics Sonochemistry 52, 106–119. https://doi.org/10.1016/j.ultsonch.2018.11.004.

Vellingiri, K., Kumar, P., Deep, A., Kim, K.-H., 2017. Metal–organic frameworks for the adsorption of gaseous toluene under ambient temperature and pressure. Chemical Engineering Journal 307, 1116–1126. https://doi.org/10.1016/j.cej.2016.09.012.

Venna, S.R., Jasinski, J.B., Carreon, M.A., 2010. Structural evolution of zeolitic imidazolate framework-8. Journal of the American Chemical Society 132 (51), 18030–18033. https://doi.org/10.1021/ja109268m.

Vikrant, K., Na, C.-J., Younis, S.A., Kim, K.-H., Kumar, S., 2019a. Evidence for superiority of conventional adsorbents in the sorptive removal of gaseous benzene under real-world conditions: test of activated carbon against novel metal–organic frameworks. Journal of Cleaner Production 235, 1090–1102. https://doi.org/10.1016/j.jclepro.2019.07.038.

Vikrant, K., Cho, M., Khan, A., Kim, K.-H., Ahn, W.-S., Kwon, E.E., 2019b. Adsorption properties of advanced functional materials against gaseous formaldehyde. Environmental Research 178. https://doi.org/10.1016/j.envres.2019.108672.

Wang, Z., Wang, W., Jiang, D., Zhang, L., Zheng, Y., 2016. Diamine-appended metal–organic frameworks: enhanced formaldehyde-vapor adsorption capacity, superior recyclability and water resistibility. Dalton Transactions 45 (28), 11306–11311. https://doi.org/10.1039/c6dt01696k.

Wang, D., Wu, G., Zhao, Y., Cui, L., Shin, C.-H., Ryu, M.-H., Cai, J., 2018. Study on the copper(II)-doped MIL-101(Cr) and its performance in VOCs adsorption. Environmental Science and Pollution Research 25 (28), 28109–28119. https://doi.org/10.1007/s11356-018-2849-6.

Wang, Q., Li, Z., Bañares, M.A., Weng, L., Gu, Q., Price, J., Han, W., Yeung, K., 2019. A novel approach to high-performance aliovalent-substituted catalysts—2D bimetallic MOF-derived CeCuOx microsheets. Small 15 (42). https://doi.org/10.1002/smll.201903525, 1903525.

Wang, C., Yin, H., Tian, P., Sun, X., Pan, X., Chen, K., Chen, W.-J., Wu, Q.-H., Luo, S., 2020. Remarkable adsorption performance of MOF-199 derived porous carbons for benzene vapor. Environmental Research 184. https://doi.org/10.1016/j.envres.2020.109323.

Wang, J., Muhammad, Y., Gao, Z., Shah, S.J., Nie, S., Kuang, L., Zhao, Z., Qiao, Z., Zhao, Z., 2021. Implanting polyethylene glycol into MIL-101(Cr) as hydrophobic barrier for enhancing toluene adsorption under highly humid environment. Chemical Engineering Journal 404. https://doi.org/10.1016/j.cej.2020.126562.

Wang, H., Lu, S., Liu, Q., Han, R., Lu, X., Song, C., Ji, N., Ma, D., 2022a. Synthesis of hierarchical-porous fluorinated metal–organic frameworks with superior toluene adsorption properties. ChemSusChem 15 (16). https://doi.org/10.1002/cssc.202200702.

Wang, Y., Li, X., Xiao, J., Chen, D., Li, N., Xu, Q., Li, H., He, J., Lu, J., 2022b. Metal–organic frameworks-derived manganese trioxide with uniformly loaded ultrasmall platinum nanoparticles boosting benzene combustion. Science of the Total Environment 839. https://doi.org/10.1016/j.scitotenv.2022.156345.

Wang, X., Wu, L., Wang, Z., Feng, Y., Liu, Y., Dai, H., Wang, Z., Deng, J., 2023. Photothermal synergistic catalytic oxidation of ethyl acetate over MOFs-derived mesoporous N-TiO_2 supported Pd catalysts. Applied Catalysis B-Environmental 322. https://doi.org/10.1016/j.apcatb.2022.122075.

Xiao, M., Yu, X., Guo, Y., Ge, M., 2022. Boosting toluene combustion by tuning electronic metal-support interactions in in situ grown Pt@Co_3O_4 catalysts. Environmental Science & Technology 56 (2), 1376–1385. https://doi.org/10.1021/acs.est.1c07016.

Xie, Y., Lyu, S., Zhang, Y., Cai, C., 2022. Adsorption and degradation of volatile organic compounds by metal–organic frameworks (MOFs): a review. Materials 15 (21). https://doi.org/10.3390/ma15217727.

Yang, H., Orefuwa, S., Goudy, A., 2011a. Study of mechanochemical synthesis in the formation of the metal–organic framework Cu3 (BTC) 2 for hydrogen storage. Microporous and Mesoporous Materials 143 (1), 37–45. https://doi.org/10.1016/j.micromeso.2011.02.003.

Yang, K., Sun, Q., Xue, F., Lin, D., 2011b. Adsorption of volatile organic compounds by metal–organic frameworks MIL-101: influence of molecular size and shape. Journal of Hazardous Materials 195, 124–131. https://doi.org/10.1016/j.jhazmat.2011.08.020.

Yang, K., Xue, F., Sun, Q., Yue, R., Lin, D., 2013. Adsorption of volatile organic compounds by metal–organic frameworks MOF-177. Journal of Environmental Chemical Engineering 1 (4), 713–718. https://doi.org/10.1016/j.jece.2013.07.005.

Yang, X., Wu, X., Chen, Z., Li, W., Sun, Q.-J., Guo, Z., Liang, X., He, Y., 2021. Hierarchically porous N-doped carbon nanofibers derived from ZIF-8/PAN composites for benzene adsorption. Journal of Applied Polymer Science 138 (20). https://doi.org/10.1002/app.50431.

Yang, J., Gao, M., Wang, S., Zhang, M., Chen, L., Su, J., Huang, Y., Zhang, Y., Wang, X., Shen, B., 2022. Experimental and simulation studies of the adsorption of methylbenzene by Fe(III)-doped NU-1000 (Zr). ACS Applied Materials & Interfaces. https://doi.org/10.1021/acsami.2c11700.

Yao, J., Wang, H., 2014. Zeolitic imidazolate framework composite membranes and thin films: synthesis and applications. Chemical Society Reviews 43 (13), 4470–4493. https://doi.org/10.1039/C3CS60480B.

Yao, P., Liu, H., Wang, D., Chen, J., Li, G., An, T., 2018. Enhanced visible-light photocatalytic activity to volatile organic compounds degradation and deactivation resistance mechanism of titania confined inside a metal–organic framework. Journal of Colloid and Interface Science 522, 174–182. https://doi.org/10.1016/j.jcis.2018.03.075.

Yin, Z., Zhou, Y., Zeng, M., Kurmoo, M., 2015. The concept of mixed organic ligands in metal–organic frameworks: design, tuning and functions. Dalton Transactions 44 (12), 5258–5275. https://doi.org/10.1039/C4DT04030A.

Yoon, M., Srirambalaji, R., Kim, K., 2012. Homochiral metal–organic frameworks for asymmetric heterogeneous catalysis. Chemical Reviews 112 (2), 1196–1231. https://doi.org/10.1021/cr2003147.

Yuan, W., Garay, A.L., Pichon, A., Clowes, R., Wood, C.D., Cooper, A.I., James, S.L., 2010. Study of the mechanochemical formation and resulting properties of an archetypal MOF: Cu3 (BTC) 2 (BTC= 1,3,5-benzenetricarboxylate). CrystEngCommun 12 (12), 4063–4065. https://doi.org/10.1039/C0CE00486C.

Yuan, S., Feng, L., Wang, K., Pang, J., Bosch, M., Lollar, C., Sun, Y., Qin, J., Yang, X., Zhang, P., Wang, Q., Zou, L., Zhang, Y., Zhang, L., Fang, Y., Li, J., Zhou, H.-C., 2018. Stable metal–organic frameworks: design, synthesis, and applications. Advanced Materials 30 (37). https://doi.org/10.1002/adma.201704303.

Zhang, X., Gao, B., Creamer, A.E., Cao, C., Li, Y., 2017. Adsorption of VOCs onto engineered carbon materials: a review. Journal of Hazardous Materials 338, 102–123. https://doi.org/10.1016/j.jhazmat.2017.05.013.

Zhang, X., Lv, X., Shi, X., Yang, Y., Yang, Y., 2019a. Enhanced hydrophobic UiO-66 (University of Oslo 66) metal–organic framework with high capacity and selectivity for toluene capture from high humid air. Journal of Colloid and Interface Science 539, 152–160. https://doi.org/10.1016/j.jcis.2018.12.056.

Zhang, X., Shi, X., Chen, J., Yang, Y., Lu, G., 2019b. The preparation of defective UiO-66 metal organic framework using MOF-5 as structural modifier with high sorption capacity for gaseous toluene. Journal of Environmental Chemical Engineering 7 (5). https://doi.org/10.1016/j.jece.2019.103405.

Zhang, J., Hu, Y., Qin, J., Yang, Z., Fu, M., 2020. TiO_2-UiO-66-NH_2 nanocomposites as efficient photocatalysts for the oxidation of VOCs. Chemical Engineering Journal 385. https://doi.org/10.1016/j.cej.2019.123814.

Zhang, L., Sun, Q., Su, T., Wang, C., Yang, Y., 2021. MOF-derived 3D porous carbon aerogels as an efficient adsorbent for toluene in humid air. Journal of Environmental Chemical Engineering 9 (6). https://doi.org/10.1016/j.jece.2021.106722.

Zhang, H., Hu, X., Li, T., Zhang, Y., Xu, H., Sun, Y., Gu, X., Gu, C., Luo, J., Gao, B., 2022a. MIL series of metal organic frameworks (MOFs) as novel adsorbents for heavy metals in water: a review. Journal of Hazardous Materials 429. https://doi.org/10.1016/j.jhazmat.2022.128271.

Zhang, X., Yue, K., Rao, R., Chen, J., Liu, Q., Yang, Y., Bi, F., Wang, Y., Xu, J., Liu, N., 2022b. Synthesis of acidic MIL-125 from plastic waste: significant contribution of N orbital for efficient photocatalytic degradation of chlorobenzene and toluene. Applied Catalysis B-Environmental 310. https://doi.org/10.1016/j.apcatb.2022.121300.

Zhao, Z., Li, X., Li, Z., 2011. Adsorption equilibrium and kinetics of p-xylene on chromium-based metal organic framework MIL-101. Chemical Engineering Journal 173 (1), 150–157. https://doi.org/10.1016/j.cej.2011.07.051.

Zhao, Z., Wang, S., Yang, Y., Li, X., Li, J., Li, Z., 2015. Competitive adsorption and selectivity of benzene and water vapor on the microporous metal organic frameworks (HKUST-1). Chemical Engineering Journal 259, 79–89. https://doi.org/10.1016/j.cej.2014.08.012.

Zhao, Q., Zhao, Z., Rao, R., Yang, Y., Ling, S., Bi, F., Shi, X., Xu, J., Lu, G., Zhang, X., 2022. Universitetet i Oslo-67 (UiO-67)/graphite oxide composites with high capacities of toluene: synthesis strategy and adsorption mechanism insight. Journal of Colloid and Interface Science 627. https://doi.org/10.1016/j.jcis.2022.07.059.

Zheng, Y., Su, Y., Pang, C., Yang, L., Song, C., Ji, N., Ma, D., Lu, X., Han, R., Liu, Q., 2022. Interface-enhanced oxygen vacancies of $CoCuO_x$ catalysts in situ grown on monolithic Cu foam for VOC catalytic oxidation. Environmental Science & Technology 56 (3), 1905–1916. https://doi.org/10.1021/acs.est.1c05855.

Zhong, G., Liu, D., Zhang, J., 2018a. The application of ZIF-67 and its derivatives: adsorption, separation, electrochemistry and catalysts. Journal of Materials Chemistry A 6 (5), 1887–1899. https://doi.org/10.1039/C7TA08268A.

Zhong, G., Liu, D., Zhang, J., 2018b. Applications of porous metal–organic framework MIL-100(M) (M = Cr, Fe, Sc, Al, V). Crystal Growth & Design 18 (12), 7730–7744. https://doi.org/10.1021/acs.cgd.8b01353.

Zhou, W., Wu, H., Hartman, M.R., Yildirim, T., 2007. Hydrogen and methane adsorption in metal–organic frameworks: a high-pressure volumetric study. Journal of Physical Chemistry C 111 (44), 16131–16137. https://doi.org/10.1021/jp074889i.

Zhu, L., Liu, X., Jiang, H., Sun, L., 2017. Metal–organic frameworks for heterogeneous basic catalysis. Chemical Reviews 117 (12), 8129–8176. https://doi.org/10.1021/acs.chemrev.7b00091.

Zhu, L., Shen, D., Luo, K.H., 2020. A critical review on VOCs adsorption by different porous materials: species, mechanisms and modification methods. Journal of Hazardous Materials 389. https://doi.org/10.1016/j.jhazmat.2020.122102.

Zou, W., Gao, B., Ok, Y.S., Dong, L., 2019. Integrated adsorption and photocatalytic degradation of volatile organic compounds (VOCs) using carbon-based nanocomposites: a critical review. Chemosphere 218, 845–859. https://doi.org/10.1016/j.chemosphere.2018.11.175.

CHAPTER

11

Metal nanoparticles (MNPs) for VOC removal

Xing Liu and Jun Luo
Nanjing University, Nanjing, China

11.1 Overview of MNPs for VOCs removal

Volatile organic compounds (VOCs) are among the most crucial air contaminants. They mainly originate from diverse chemical process industries, including petrochemicals, polymer syntheses, paint, and coatings (He et al., 2019). The World Health Organization (WHO) reported that air pollution caused around seven million premature deaths in 2012, of which 4.3 million were related to indoor air pollution (Luengas et al., 2015). Among the variety of indoor pollutants, Gallego et al. (2009) identified 130 different VOCs in private homes, with alkanes being the largest group (32%), followed by aromatic hydrocarbons (17%) and esters and alkenes (7%).

Due to the severe biological toxicity and photochemical behaviors of VOCs even at very low concentrations (Ji et al., 2020; Mellouki et al., 2015), the development of environmentally benign methods for efficient removal of gaseous VOCs from polluted air has received considerable attention (Nahm et al., 2012; He et al., 2019; Ding et al., 2020). The treatment methods for VOCs are generally classified into two categories: recovery and destruction (Kask et al., 2020). The former technique includes absorption, condensation, membrane separation, etc. (Zhang et al., 2017). The latter contains thermal incineration, catalytic oxidation, low-temperature plasma, photocatalytic oxidation, etc. (Konsolakis et al., 2013; Młotek et al., 2015).

The adsorption technique has been intensively investigated because it is economic, highly efficient, and also holds great potential to recover both valuable VOCs and adsorbents for further reuse (Liu et al., 2021). However, common adsorbents (e.g., activated carbon, polymeric resins, zeolite) possess limited adsorption capability and difficulties in regeneration, which have restricted their applications in the removal of VOCs (Vellingiri et al., 2016). Therefore, the development of advanced adsorbents which can offer high adsorption capability and excellent regenerability, is extremely imperative.

Volatile Organic Compound Removal
https://doi.org/10.1016/B978-0-443-15674-8.00002-0

Catalysis as another important means to remove VOCs has been studied by a large number of researchers, due to the advantages of high efficiency, complete degradation, and so on. Nevertheless, traditional catalysts such as TiO_2, MnO, and ZnO, etc., are difficult to adapt to most complex working environments due to their limited catalytic capacity and harsh reaction conditions.

Aiming at solving these challenges, metal nanoparticles (MNPs) came into the scene because of the electrical conductivity of metal particles, chemical bond relationship with various functional groups, and synergistic effect with inorganic materials (Fu et al., 2021). They have inherent advantages for the removal of VOCs, attracting increasing attention from scholars. Many studies have been conducted to explore the applications of MNPs in VOC removal, especially with respect to adsorption and catalysis (including photocatalysis and thermocatalysis) technologies.

Remarkably, MNPs have shown even better performance in catalysis than adsorption technology due to their unique properties such as high electrical conductivity and thermal conductivity, as well as the characteristics of nanomaterials such as surface, quantum size, and quantum tunneling effects (Zhu et al., 2022). This is reflected by the number of research papers on MNPs in VOCs removal in the past 20 years (Fig. 11.1).

11.2 MNP-based adsorbents for VOC removal

During the adsorption/desorption of VOCs on MNP-based materials, MNPs would exert a crucial effect on the process (Troppova et al., 2017). Usually MNPs play a role of strengthening the adsorption performance of the adsorbents (Barsotti et al., 2016). The adsorption

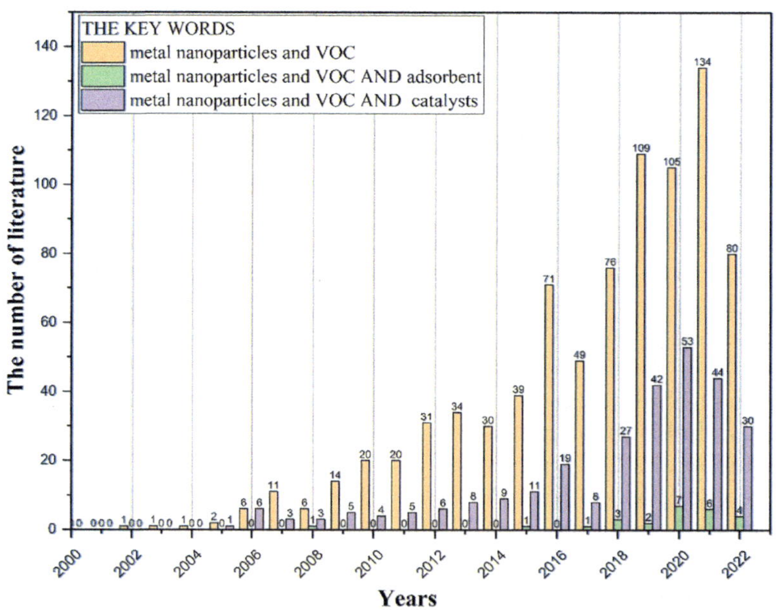

FIGURE 11.1 Number of research papers on the application of MNPs in VOC removal between 2000 and 2022. *Based on the* Web of Science.

capacity and mechanism of MNP-based adsorbents for VOCs are closely related to the type of MNPs, supporting materials, loading methods, etc. (Zhou et al., 2014; Choi et al., 2020; Yin et al., 2022)

11.2.1 Synthesis of MNP-based adsorbents

Most of the MNPs are loaded in a support material such as silica, alumina, zeolite, or porous carbonaceous material to exert their effect on VOC removal. The loading method plays a decisive role in the distribution of MNPs on the supports to affect their adsorption performance. Numerous MNP loading methods have been developed including colloidal deposition method (Liu et al., 2021), solute mixed sedimentation (Adebayo et al., 2020), heteroatoms (Zhou et al., 2022) and so on.

Ojha et al. (2019) synthesized a g-CN/ATO hybrid nanocomposite for VOC adsorption. They mixed the ATO nanoparticle (synthesized from $SnCl_4 \cdot 5H_2O$ and $SbCl_3$) and graphitic carbon nitride(g-CN)at 1:1 weight ratio. The mixture was ultrasonicated in a 50 mL ethanol for 2 h, dried slowly at 90°C for 24 h, and subsequently heat-treated at 150°C for 3 h. After that, the nanocomposite g-CN/ATO was successfully synthesized. Baur et al. (2015) loaded a series of MNPs on activated carbon fibers (ACF) for VOC removal. They first immerged ACF in metal precursor solutions, and then used thermal decomposition to obtain the targeted MNP-loaded ACF adsorbent. Pei et al. (2013) used a similar method to load silver (Ag) nanoparticle on activated carbon (AC) surface. They used trisodium citrate (Na_3Ct) and dimethylethanolamine (DMEA) to reduce Ag^+ to obtain the Ag nanoparticle-coated AC nanocomposites (Ag/ACs). All these MNP-based materials have shown an excellent adsorption performance to VOCs.

11.2.2 Advantages of MNP-based adsorbents

Numerous studies have shown that the addition of MNPs can greatly improve the adsorption performance of conventional adsorbents to various VOCs (Baur et al., 2015; Yang et al., 2019; Fu et al., 2021; Zhou et al., 2022). The advantages of MPN-loaded adsorbents over the pristine ones are summarized as follows:

(1) High specific surface area
High specific surface area (SSA) and porous structure are crucial parameters to the performance of VOC adsorbents at ambient temperature and pressure (Barsotti et al., 2016). There are many reports of VOC removal by porous adsorbents like zeolite (Zhou et al., 2014), AC (Tefera et al., 2014; Lin et al., 2015), graphene (Guo et al., 2016), metal-organic frameworks (MOFs) (Kumar et al., 2016b), and metal-doped metal oxides (Ghosh et al., 2017). g-CN is a soft polymeric semiconductor that can easily be applied to the surface of other materials, and MNP-loaded g-CN has been applied for the adsorption of VOCs (Muñoz-Batista et al., 2014; Hou et al., 2016). In the research of Ojha et al. (2019), g-CN/ATO showed much higher Brunauer-Emmett-Teller (BET) surface area than ATO and g-CN (Fig. 11.2). The large SSA of the hybrid nanocomposite was attributed to the exfoliation of stalked g-CN sheets and less aggregate of ATO nanoparticles in between the sheets,

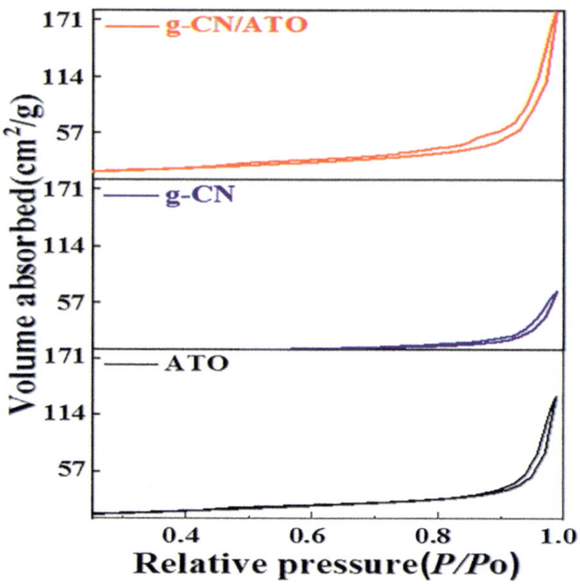

FIGURE 11.2 N_2 adsorption–desorption isotherms (type IV) of the adsorbents (Ojha et al., 2019).

increasing its the porosity significantly (Wang et al., 2014). The similar results have been reported in several studies (Barsotti et al., 2016; Ojha et al., 2019; Zhou et al., 2022).

(2) Strong intermolecular interactions
MNPs in the composites can strengthen molecular interactions between the adsorbent and VOCs. Ojha et al. (2019) summarized the principles of chloroform, benzene, toluene and xylene on g-CN/ATO as follows: (1) Compared with the pristine adsorbent, the surface polarity of the nanocomposite is higher, increasing the π–π attraction of VOC molecules; (2) Exposed metal cation sites of ATO in pores can interact with nonpolar VOC molecules through cation–π interaction; (3) Presence of surface functional groups on g-CN can also adsorb VOCs via π–π interaction. It is recognized that MNP surface can form a π-complex with aromatic part of VOC compounds. This π-complex is a noncovalent weak π–π interaction between the benzene ring of VOCs such as toluene and benzene and MNPs (Zhong et al., 2012; Ojha et al., 2019).

11.2.3 Regeneration of MNP-based adsorbents

The regeneration performance of a adsorbent is a pivotal factor, directly related to whether the adsorbent can be promoted and applied (Liu et al., 2021). However, conventional adsorbents have common drawbacks of poor thermal stability and flammability (Yin et al., 2022). In this case, a feasible method is to optimize the properties of the adsorbents through the modification of MNPs. In the study of Zhou et al. (2022), 20%Mg@MIL-101/PC showed a high acetone adsorption capacity of 1309 mg/g, and the desorption process from 1 to 10 cycles was carried out at 423 K. It was found that its saturated adsorption capacity decreased by only 12% over 10 adsorption-desorption consecutive cycles due to the presence of MNPs. In another study, UiO-66 showed a very slow styrene desorption rate after long irradiation and adsorption capacities decreased dramatically from 561.7 to 290.2 mg/g after five runs. After loading silver nanoparticles, however, the modified UiO-66 maintained high adsorption of styrene (Liu et al., 2021).

11.2.4 Applications of MNP-based adsorbents

11.2.4.1 Removal of indoor VOCs

Improved indoor air quality is essential for human health since people spend almost 90% of their life indoors (Klepeis et al., 2001). Among numerous indoor air pollutants, VOCs such as formaldehyde, toluene, xylene, and styrene are among the most concerning ones. Usually, the concentrations of VOCs indoors are higher than outdoors. Thus, it is significant to develop technologies for efficient removal of VOCs in indoor environments, which, however, has been a challenge. The adsorbent usually needs to have a high VOC adsorption capacity at room temperature and pressure, requiring extremely high removal rate in the relatively closed environment.

Exposure to typical indoor VOCs such as formaldehyde and acetaldehyde may cause headache, irritation, diseases, and even cancer (Mundt et al., 2017). Wang et al. (2018a) used the coprecipitation method to synthesize a multiple structured adsorbent composited by Au, ZnO, and ZIF-8 for formaldehyde removal. In the composite adsorbent, ZnO maintains the material conductivity, ZIF-8 improves the selective gas adsorption, and the plasmonic effect of Au nanorods enhances the visible-light-driven photocatalysis of ZnO at room temperature. Therefore, the absorbent of Au@ZnO@ZIF-8 demonstrated excellent ability of simultaneous adsorption and degradation of formaldehyde at room temperature. Choi et al., 2020 modified AC with Mn, Ag, Mn/Ag nanoparticles via radio-frequency thermal plasma. The resulting MNP-ACs showed high adsorption of formaldehyde and acetaldehyde at room temperature. Among them, Mn/Ag-AC displayed the highest adsorption of the two indoor VOCs, much higher than the parent AC and Mn-AC.

Benzene series VOCs in room have also received much attention. As already stated, the g-CN/ATO hybrid nanocomposite adsorbent developed by Ojha et al. (2019) had an excellent performance for removing indoor chloroform, benzene, toluene, xylene and styrene.

11.2.4.2 Removal of industrial VOCs

Due to their toxicity and strong impact on the environment even at low concentrations (Lillo-Rodenas et al., 2005), VOC concentrations in industrial gas emissions are subjected to legal limits (Ghoshal and Manjare, 2002). Reducing VOC concentrations in emissions have been a challenge due to a large number of impurity molecules (possibly overwhelming in number) competing with VOCs for adsorption sites. This reduces the adsorbent activity for VOCs. For the removal of industrial VOCs, selective adsorption is thus particularly important. Various MNP-based adsorbents have thus been developed to enhance the selectivity toward specific VOCs in industrial gas emissions. For example, Bhardwaj et al. (2020) found that noble MNP (Au, Pd, and Pt) functionalized MoS_2 showed excellent selectivity of low concentrations of VOCs such as acetone and benzene.

11.3 MNP-based catalysts for VOC removal

Catalytic oxidation of VOCs is highly desirable because of its energy savings, low cost, and operation safety. High activity, selectivity, and stability are the main evaluation indicators for catalysts for VOC oxidation. Catalytic activity of MNP-based materials in VOC oxidation depends on multiple aspects, including electronic and coordination structures, metal dispersion, support properties, interactions between active metals and supports, and so on. Due to the great activity and excellent regeneration ability, noble metals (Pt, Pd, Ru, etc.) catalysts have attracted the attention of many researchers (Kumar et al., 2016a). They are not only supported on commonly used oxides including SiO_2 and γ-Al_2O_3 (Darif et al., 2016), but also loaded on porous materials to improve the dispersion of active sites and opportunity of contact (Bian et al., 2014).

Notably, porous structure generally prolongs the duration time of reactants and reaction intermediates over the surfaces of the catalysts, improving catalytic performance (Adebayo et al., 2020). Meanwhile, the VOC selectivity of the catalysts, a key indicator second only to catalytic efficiency, can also be effectively modulated, because an increasing in the local VOC concentrations can shift the reaction equilibrium (Zhang et al., 2013). Regarding the catalysts that contain confined active MNPs, the introduction of a confined space as a second coordination sphere around the active metal species has been demonstrated to be a viable strategy for enhancing selectivity and activity in heterogeneous catalytic reactions (Leenders et al., 2015). This conclusion provides theoretical support for MOF-MNPs catalysts for VOC removal.

The electronic interactions between the MNPs and the host nanostructure can subtly change the electronic state, affecting the performance of the catalysts (Miners et al., 2016). Moreover, the spatial restriction of active MNPs in confined spaces prevents sintering and aggregation, meanwhile blocks the molecules that are likely poisoning the active sites, resulting in a prolonged activity (Luo et al., 2019).

There are many types of catalysis processes for VOC removal. Based on the triggering factors of the reactions, this chapter mainly focuses on photocatalysis and thermocatalysis of VOCs.

11.3.1 Synthesis of composite catalysts

11.3.1.1 MNPs and supports

Highly efficient and stable catalysts are crucial to the catalytic degradation of VOCs. Therefore, the choice of materials and synthesis methods are particularly important. On the one hand, nanostructured noble metals (e.g., Pt, Pd, Ru, Mn, Co, and Ag) are equipped with an excellent catalytic activity for VOC oxidation, but their practical application is still limited by some weakness such as easy agglomerate, high cost, and so on (Li et al., 2016; Mishra et al., 2018; Jakubek et al., 2019; Mo et al., 2019). Thus, there is increasing interest in developing synthetic methods to reduce the total metal loading on the supporting substrates (Huang et al., 2015). On the other hand, a good support always has an excellent adhesion property for MNPs including physical adhesion, chemical bond adsorption, etc. In addition, the supports usually have a porous structure to hold as many MNPs as possible to increase the contact opportunity between VOCs and catalysts. Common MNP supports include diatomaceous (Tomatis et al., 2018), graphitic carbon nitride (Zhao et al., 2015), zeolite (Zhu et al., 2022), TiO_2 (Li et al., 2016), etc.

It is worth noting that MOFs have a natural advantage as a support for MNPs. As described in Chapter 10, MOF structures are crystalline materials with one-dimensional, two-dimensional, or three-dimensional porous network structures formed by self-assembly of metal ions or groups and organic ligands through coordination bonds. Compared with traditional supports, this special structure enables MOFs to have high porosity, uniform and adjustable pore size, and controllable functional groups (Becheca et al., 2015; Zhang et al., 2015a). Therefore, according to the synergistic effect between the optical and dynamic properties of MOFs and MNPs, many researchers have explored the application of MNP-MOF composites for VOC removal, showing promising results (Qi et al., 2015; Xu et al., 2015).

11.3.1.2 Synthesis methods

A significant number of studies have been devoted to the development of MNP-based catalysts using various method aiming to dispersed the nanoparticles on the supports uniformly (Xu et al., 2015; Zhao et al., 2015; Sun et al., 2020). The most commonly used methods include (1) sol-gel method, (2) thermal decomposition technique, and (3) hydrothermal method.

Sol-gel method: This method refers to producing solid materials from small molecules. This process is a combination of two steps, hydrolysis and polycondensation of suitable molecular precursors, leading to the gradual formation of a solid-phase network (Filipovic et al., 2019). After this process, the monomers are converted into a colloidal solution that acts as the precursor for an integrated network of either discrete particles or network polymers. Then to obtain the final sample, the gel is dried and calcinated to remove the organic components.

Thermal decomposition technique: In this method, metal salts are first deposited on appropriate substrates by impregnation or deposition-precipitation (Li et al., 2015), and then they are calcinated at high temperature to obtain the final MNP-based catalysts. Liu et al. (2012) reported the preparation of Pt/SiO_2 catalysts by flame spray pyrolysis using tetraethoxysilane (TEOS) as a precursor of SiO_2. Platinum acetylacetonate and TEOS were dissolved in toluene with a certain ratio to prepare the precursor solution, which was then dispersed to a flame to initiate combustion. Finally, the as-formed Pt/SiO_2 particles were collected by a glass fiber filter with the aid of a vacuum pump. In another study, Anandkumar et al. (2015) synthesized Au/CeO_2 nanocomposite catalysts using similar method.

Hydrothermal method: Hydrothermal synthesis is a technique to crystallize substances from high-temperature aqueous solutions under high vapor pressure in which generally insoluble materials can be dissolved (Li et al., 2016). In an autoclave, a temperature gradient is maintained between the opposite ends of the growth chamber. The nutrient solute dissolves at the hotter end, while deposits on a seed crystal at the cooler end, resulting in the formation of desired products. This method is particularly suitable for the growth of good-quality MNP crystals accompanied by maintaining good control over their composition.

11.3.2 Characterization methods

Within the last decades, the development of characterization techniques and testing methods has led to the blossom of research in various synthesis. Herein, the main representation methods of characterization are introduced briefly.

SSA and pore volume are important physical properties of both catalytic and adsorption materials. They are indicators directly related to the catalytic efficiency of catalysts in contact with VOCs. SSAs are often determined with the BET methods using N_2 adsorption and desorption isotherms. Other information such as micropore/mesoporous pore volume and pore size distribution can be also calculated based the isotherms.

When characterizing the physical properties of MNP-based catalyst surfaces, scanning electron microscopy (SEM) and transmission electron microscopy (TEM) would be the first choice. SEM scans the sample with a narrow focused high-energy electron beam, generates various physical information through the interaction between the beam and the material. The resolution of emerging SEM can reach 1 nm. In addition, the combination of SEM and other analytical instruments such as energy dispersive X-ray spectroscopy can reveal the microscopic morphology and simultaneously carry out the analysis of the composition of the material micro-area. TEM uses an electron beam as the light source to identify the fine structure of less than 200 nm that cannot be recognize clearly by the optical microscope. The resolution of TEM can reach 0.2 nm. With the help of SEM and TEM, the loading of MNPs on the surface of the supports can be identified. In the research of Zhang et al. (2015b), they prepared Ag-based catalysts loaded on different supports (TiO_2, Al_2O_3, and CeO_2) by the impregnation method for the catalytic oxidation of formaldehyde at low temperature. The characterization results revealed that the type of supports had a dramatic influence on the Ag particle size and dispersion. This conclusion benefited from TEM and SEM, which provided the direct visualizations of different MNP-based catalysts.

At present, there are some methods that can qualitatively or even quantitatively analyze chemical properties of MNP-based catalysts surfaces. For example, X-ray diffraction (XRD) can analyze the composition of ENPs on the surface of supports, their particle size, crystal form, crystallinity, etc. X-ray photoelectron spectroscopy (XPS) can provide the information about the content and morphology of various elements on the surface of the MNP-based catalysts. Meanwhile, it can also be used to determine the existence of MNPs and whether they are involved in the whole process of adsorption/catalysis. Furthermore, Fourier transform infrared spectrometer (FT-IR) is a commonly used tool to detect functional groups on the surface of catalysts.

Characterization of MNP-based catalysts is very important work. It is not only used as direct evidence for the success of synthesis, but also provides a reliable basis for scholars to study the controlling factors of the catalytic oxidation of VOCs.

11.3.3 Photocatalysis of VOCs

11.3.3.1 Advantages of MNP-based photocatalysis

Up to now, several experimental methods have been applied to synthesize MNP-composite photocatalysts (Liu et al., 2010; Yan et al., 2010b). In these methods, photocatalysts are formed through self-assembly or precipitation of MNPs on the porous supports (Zhao et al., 2015). MNP-based photocatalysis can be suitably pertained for the removal of VOC pollutants because it can mineralize them into CO_2 and H_2O without producing secondary pollutants. The properties of the MNPs in the composites change radically when the size reaches to nanoscale range. Changing in MNP size affects the electrical and optical properties of the MNP-based photocatalysts, offering tunable photocatalytic activities for their applications in removing VOCs (Theerthagiri et al., 2018).

11.3.3.2 Mechanism of MNP-based photocatalysis

Firstly, the random growth process may lead to serious consequences, such as overgrowth or aggregation of MNPs, as well as their inhomogeneous distribution on supports (Marikutsa et al., 2022). A feasible strategy is to utilize the molecular properties and forces between the supports and MNPs to achieve uniform loading. For example, in the supramolecular network, neighboring tetranuclear Cd–S clusters are connected through π-π interactions with high order, so, after calcination, the CdS nanoparticles are embedded in the g-C_3N_4 matrix homogeneously (Wang et al., 2018b).

Secondly, the process of photo-generated electron-hole pairs generation and recombination is one of the most important factors controlling the photocatalytic performance of MNP-based catalysts. It is worth noting that MNPs can significantly affect this process. For example, pure g-C_3N_4 suffers from rapid recombination of photo-generated electron-hole pairs resulting in low photocatalytic activity. The surface plasmon resonance (SPR) endowed in noble metals of MNPs can increase the visible light utilization and probably show special synergistic effects with g-C_3N_4 (Miranda et al., 2013). The work of Yan et al. (2010a) demonstrated that synthesized nanoscopic Au-semiconductor heterojunctions effectively accelerated the electron transfer on the intimate interface between g-C_3N_4 and Au nanoparticles. Yang et al. (2013) found that, compared to that of pristine g-C_3N_4, the photocatalytic activity of Ag/g-C_3N_4 was sharply enhanced toward the degradation of p-nitrophenol, attributed to the enhanced visible light utilization efficiency due to the SPR absorption of silver nanoparticles.

The enhanced photocatalytic activities of MNP-g-C_3N_4 composites are mainly achieved by the following processes. (1) The SPR effect of noble MNPs causes intense local electromagnetic fields, which can speed up the formation rate of holes and electrons within g-C_3N_4 (Bai et al., 2014); (2) the favorable Fermi level of noble MNPs facilitates the separation of electrons and holes, which in turn enhances the quantum efficiency of g-C_3N_4, due to the intimate combination of noble metal/g-C_3N_4 nanohybrids (Ingram et al., 2011); (3) the transfer of electrons shifts the Fermi level to more negative potential, thereby improving the reducibility of electrons in the Fermi level close to the conduction band of g-C_3N_4 (Cao et al., 2012) (4) the efficient utilization of sunlight can be realized due to SPR absorption in the visible-light region as well as UV light response of the interband transition of noble MNPs (Zhao et al., 2015).

In addition, it has been confirmed that sulfur vacancies in MNP-based composites can play an important role in the enhancement of photocatalytic activity (Qiang and Xia, 2020; Li et al., 2023). Wang et al. (2018b) synthesized sulfur vacancy rich CdS-based photocatalysts using CdS MNP and g-C_3N_4 through calcination (CdS@g-C_3N_4). The result of electron paramagnetic resonance spectrum (EPR) illustrated that CdS@g-C_3N_4 (A) (in a vacuum oven at 70°C for 10 h) possessed more sulfur vacancies than CdS@g-C_3N_4 (B) (same condition for 4h) and CdS@g-C_3N_4 (C) (same condition for 6h). As a result, CdS@g-C_3N_4 (A) showed the highest photodegradation of acetaldehyde. The authors speculate that the degradation of acetaldehyde completes in three steps: Firstly, oxygen atoms from acetaldehyde and H_2O are captured by sulfur vacancies of CdS@g-C_3N_4 and fastened on its surface. Secondly, positive holes on the valence band of g-C_3N_4 react with surface bound H_2O to produce a hydroxyl radical (·OH). The negative electrons on the conduction band of CdS are taken by O_2 and generate a super oxide radical (O_2^-). Finally, the hydroxyl radical (·OH) and super oxide radical (O_2^-) react with acetaldehyde, leading to its oxidation and degradation.

11.3.3.3 Applications of MNP-based photocatalysts

In the research of Wang et al. (2018b), CdS@g-C_3N_4 exhibited stronger photo adsorption capacity than pure CdS MNPs and g-C_3N_4. The results showed that, in the absence of CdS@g-C_3N_4, the concentration of acetaldehyde remained unchanged. In sharp contrast, the addition of CdS@g-C_3N_4 led to an obvious decrease of acetaldehyde. After visible light illumination for 120 min, about 95.2% of acetaldehyde were degraded by CdS@g-C_3N_4, and accompanied by the increase of CO_2 with the conversion rates of 92.1%. To study the universality of CdS@g-C_3N_4 for VOC removal, another typical VOC pollutant, benzene, was chosen as a model compound. In 120 min, about 85.2% benzene was photodegraded by CdS@g-C_3N_4 and the CO_2 conversion rate was as high as 90.6%.

11.3.4 Thermocatalysis of VOCs

11.3.4.1 Advantages of MNP-based thermocatalysis

In recent years, aiming to reduce the temperature for the complete oxidation of VOCs, great efforts have been devoted to investigating the applications of MNPs in thermocatalytic removal of VOCs (Tomatis et al., 2018; Mo et al., 2019; Zhu et al., 2022). Various metal elements, metal composites, and supports have been examined for photodegradation of VOCs (Tomatis et al., 2016). Noble metals such as Pd, Pt, and Au have been widely studied for their high activity and selectivity even at relative low temperatures (Wang et al., 2015; Fu et al., 2016). Researchers have investigated different MNPs and support materials, aiming to improve activity and selectivity of the prepared catalysts toward the complete thermal degradation of VOCs (Liu et al., 2013; Zeng et al., 2015). The relationship between the thermocatalysis performance and physical or chemical features of nanomaterials has also been determined to develop efficient and stable catalysts.

11.3.4.2 Mechanism of MNP-based thermocatalysis

The utilization of strong metal-support interactions (SMSIs) between precious metals and metal oxide supports is a promising approach to improve the thermo-catalytic property and durability of catalysts. The formation of SMSIs can significantly improve these properties of the catalysts. The SMSIs can hinder the growth of metal nanoparticle crystals, promoting the bonding structures and electron transfers between the metals and the supports (Fujiwara et al., 2017). For example, a series of characterization results indicated that the SMSIs between the Pt NPs and MnO_2 improved low-temperature reducibility, optimized surface chemical states, and the weakening of surface lattice oxygen species (Mo et al., 2019). Moreover, the special role of the SMSIs in the Pt/MnO_2 catalysts greatly improved the surface oxygen vacancies and the mobility of the surface lattice oxygen to enhance their thermo-catalytic activities.

Supported Pt NP catalysts have been widely studied in the field of heterogeneous catalysis due to their outstanding activity for VOC degradation (Yang et al., 2018; Mo et al., 2019). Size of Pt NPs in the composite catalysts can influence their structure-property relationships. The biggest contribution of Pt/MnO_2 SMSIs is the weakening of surface Mn-O bonds and the improved mobility of the surface lattice oxygen. Mo et al. (2019) speculated that the toluene oxidation simultaneously follows the Langmuir–Hinshelwood (L-H) mechanism and the Mars-van Krevelen (MvK) mechanism on the surface of the Pt/MnO_2-based catalysts. In their studies, the concentrations of the surface Mn^{4+} and active oxygen species (O_{ads}, O_2^- or O_2^{2-}) on Pt/MnO_2 were generally higher than those of the single MnO_2 supports. It has been reported that the O_{ads} species plays significant roles in VOC oxidation (Wang et al., 2012; Tang et al., 2014).

In the research of Li et al. (2022), the catalytic mechanism and reaction pathways over MNP/MOFs were revealed by the study of active sites and reaction intermediates via XPS, in situ diffuse reflectance infrared Fourier transform spectroscopy (DRIFTS), and proton transfer reaction-time of flight mass spectrometer (PTR-TOF-MS). Fig. 11.3 shows that VOC (e.g., ethyl acetate) molecules were firstly adsorbed on the surface of Pt/UiO-66 nanocomposites and then oxidized into intermediates (e.g., ethanol, acetaldehyde, and acetic acid), furtherly oxidized into CO_2 and H_2O by the lattice oxygen of the PtO (Li et al., 2022). The Pt sites were reduced to Pt^0 with the production of oxygen vacancies. Finally, Pt^0 was activated by molecular oxygen and re-oxidized into the PtO to eliminate the oxygen vacancies, forming a Pt^0/PtO redox cycle.

11.3.4.3 Applications of MNP-based thermocatalysis

As mentioned earlier, catalyst selection is crucial for the removal of targeted VOCs. By comparing the catalytic oxidation activity of pure MnO_2 and Pt/MnO_2 for toluene, Mo et al. (2019) found that the temperature required for the thermo-catalytic oxidation of 90% of toluene (T90) by single MnO_2 was 263°C, while that by Pt/MnO_2 (Pt loading 0.39wt%) rapidly reduced to 170°C. In addition, they also studied the effect of different loadings of Pt NPs on the activity of Pt/MnO_2 to toluene. The experimental results showed that, with an increase in the content (from 0.4 to 0.8wt%) of the Pt NPs on the surface of MnO_2, the thermo-catalytic oxidation of toluene increased simultaneously. When 0.8wt% Pt NPs was loaded on MnO_2, the temperature needed for 90% toluene degradation was only 158°C.

Mn and Co in either single or binary metal catalysts have been widely studied for their thermal decomposition activity toward benzene, toluene, ethylbenzene, and xylene. They are often loaded on diatomite for thermocatalysis of VOCs because of the excellent physical properties of diatomite, including peculiar porous structure, thermal stability, and hydrophobicity. In the research of Tomatis et al. (2018), diatomite was used as a support to prepare Mn/Co binary metal catalysts at different metal loadings (5–10wt% Mn and 5 to 15wt% Co) by incipient wetness deposition. Their results showed that Mn/Co@diatomite equipped with a great activity toward the complete oxidation of benzene to CO_2 and H_2O.

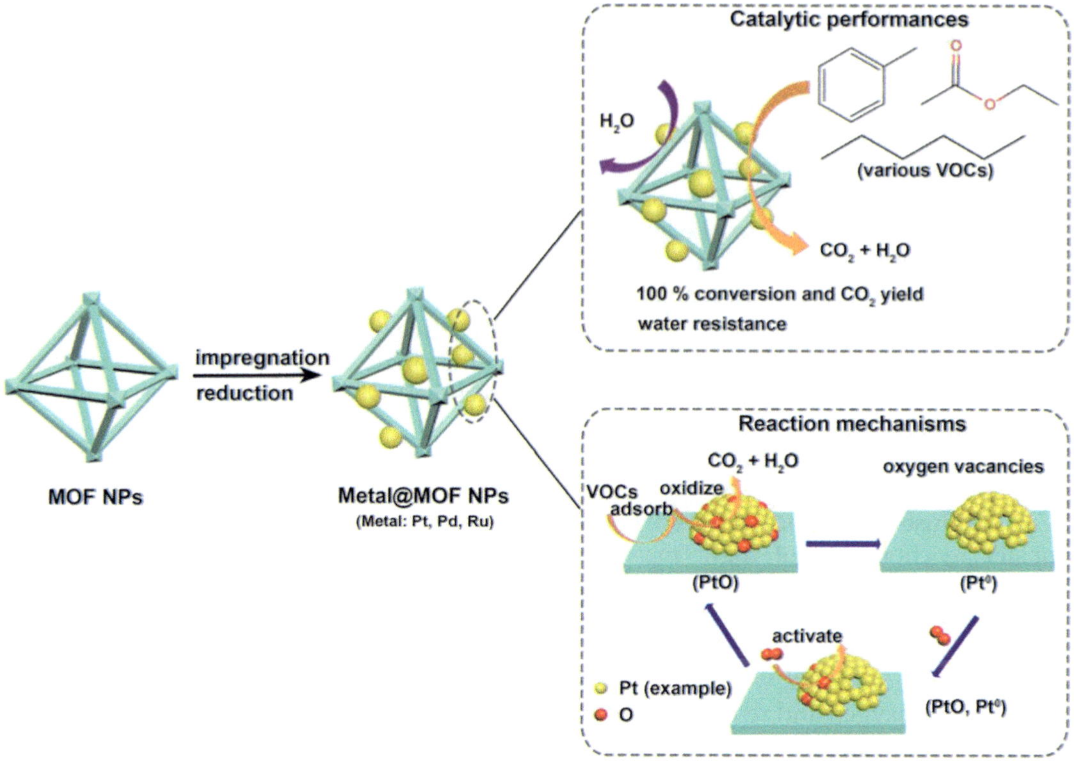

FIGURE 11.3 Thermo-catalytic degradation of VOCs by MNP/MOF catalysts (Li et al., 2022).

In a recent study, Li et al. (2022) proposed a flexible and universal strategy to prepare the highly dispersed MNPs (Pt, Pd, and Ru) supported on MOFs through the impregnation-reduction method for the catalytic oxidation of VOCs (Fig. 11.3). The resultant Pt/UiO-66 catalyst with 0.50% Pt loading exhibited nearly 100% conversion efficiency and CO_2 yield for the catalytic oxidation of various VOCs including ethyl acetate (260°C), n-hexane (260°C) and toluene (180°C). In addition, there was no obvious difference in the morphology, size, distribution, crystallinity, and composition of the fresh and used Pt/UiO-66 NPs from the TEM images, XRD spectra, and XPS analysis. Furthermore, catalysts exhibited high reusability, good water resistance as well as excellent stability over 150 h on the on-stream reactions for the degradation of ethyl acetate.

Zeolite-supported noble-metal catalysts are one of the most promising catalysts to achieve complete thermal oxidation of VOCs. Recently, Zhu et al. (2022) synthesized Pt/Z-En catalysts which Pt nanoparticles (-2 nm) encapsulated in zeolite materials by a one-pot synthesis strategy. Their results showed that the optimum catalytic performance for toluene reached 98% at a temperature as low as 214°C. More importantly, the Pt/Z-En catalysts demonstrated a long-term stability for continuous applications.

11.3.5 Resistance to poisoning

H_2O, SO_2, alkali metals (e.g., K, Na, and Ca), and heavy metals can cause the catalysts to show different degrees of poisoning symptoms during VOC degradation (Ren et al., 2021; Xiang et al., 2023). F or the exhaust gas to be treated, these factors are unavoidable in most cases. Therefore, the resistance of the MNP-based catalysts to poisoning is also directly related to their promotion and applications in VOC removal.

SO_2 can react with NH_3 and O_2 to generate stable sulfated species, which are deposited onto the surface of catalysts to cause irreversible loss of the surface-active centers, leading to the deactivation of selective catalytic reduction. Alkali metals can react with the Bronsted acid sites in catalysts to reduce their catalytic activities (Gao et al., 2014). Heavy metals (e.g., Zn and Pb) can decrease the reducibility, surface acidity, and NO adsorption ability of catalysts (Guo et al., 2015). Moreover, the metal poisoning of MNP-based catalysts caused by the adsorption of halogen compounds also occurs during VOC removal, leading to the deactivation of catalysts (Zhu et al., 2022).

Scientists have developed various techniques to solve these problems. Wang et al. (2020) prepared Cu/ZSM-5-PAD and Cu/ZSM-5-CIM catalysts using a polyacrylic acid assisted deposition strategy (PAD) and a conventional impregnation method (CIM), respectively. They found that unlike Cu/ZSM-5-CIM, most of the metal oxide particles in Cu/ZSM-5-PAD were stably confined within the zeolite channels. As a result, Cu/ZSM-5-PAD showed excellent catalytic performance and presented satisfactory hydrothermal stability to resist H_2O/SO_2 poisoning. This result indicates that different loading methods of MNPs can affect catalysts' ability to resist poisoning (Fei et al., 2021). Based on this idea, Zhu et al. (2022) successful encapsulated Pt NPs in ZSM-5 zeolites (Pt/Z-En) using tetraammineplatinum nitrate ($Pt(NH_3)_4(NO_3)_2$) as the metal precursor. They found that the resultant MPN-zeolite catalysts with a hydrophobic surface had excellent selectivity to preferably adsorb VOCs, while restricting the adsorption of poisons such as halogen compounds.

Several studies have demonstrated that MNP-based VOC catalysis decreases with increasing moisture content in exhaust gas, due to the competitive adsorption of H_2O and VOC molecules on catalyst surfaces (Gao et al., 2014; Guo et al., 2015; Ren et al., 2021). However, hydrophobic modification of MNP-based catalysts can increase their resistance to moisture poisoning. For example, the unfavorable effect of H_2O on VOC catalytic oxidation was mitigated by increasing the Si/Al molar ratios of the catalysts (Zhu et al., 2022).

11.4 Summary and perspectives

Adsorption and catalysis are the most effective and feasible technologies for VOC removal. Many studies have shown that the presence of MNPs can significantly enhance the performance of adsorbents and catalysts in removing VOCs.

MNPs have been loaded on surface of tradition adsorbents via the methods of colloidal deposition, solute mixed sedimentation, heteroatoms, and so on. In addition, the mechanism of enhanced adsorption of VOCs by MNP-based adsorbents has been determined, which include increasing of surface area, increasing active sites, strengthening the intermolecular interactions, and so on. It is worth noting that the loaded MNPs can also significantly increase the regeneration and reusability of adsorbents, promoting their popularization and applications in VOC removal.

Compared with adsorption, MNPs are more popular in VOC catalytic oxidation. They have been loaded on supporting catalysts by various means. The resultant composite catalysts have been applied in photocatalytic oxidation or thermo-catalytic oxidation of VOCs. On the one hand, MNP-based photocatalysis can achieve the complete degradation of VOCs, relying on photo-generated electron-hole pairs, SPR effect, etc. to produce radicals for the degradation. On the other hand, MNP-based thermocatalysis can significantly reduce the decomposition temperature of VOCs, attributing to SMSIs and strong interaction between MNPs and VOCs. In addition, studies have shown that the addition of MNPs can significantly enhance the ability of the catalysts to resist poisoning by SO_2, water, halogen compounds, heavy metals, etc.

In a nutshell, previous studies have demonstrated that the addition of MNPs into adsorbents and catalysts can significantly enhance their ability to remove VOCs from the environment. The opportunities and challenges for future research and development on the applications of MNPs in VOC removal are summarized below:

(1) The synthesis process of MNP-based materials has a tremendous effect on their properties. For example, the loading of MNPs on supports may require a calcinate stage, in which the time of calcination directly affects the distribution and morphology of MNPs on the supports to indirectly affect their VOC removal performances. Therefore, in-depth research on different synthesis methods and conditions should be carried out in the future to optimize MNP-based materials for VOC removal.

(2) Studies have shown that MNPs can effectively promote the regeneration and reusability of adsorbents for VOC removal. Further research in this direct to improve stability and reusability of MNP-based adsorbent is necessary.

(3) Deactivation of catalysts is one of the biggest challenges to their practical applications in VOC removal. While the presence of MNPs can increase the resistance of catalysts to poisons, they may also become the targets. Further investigations thus should be conducted to develop strategies for protecting the activity sites of MNP-based materials, especially the sites of multifunctional composites, during VOC removal under realistic conditions.

(4) Selective catalysis of VOCs is important in practical applications. MNP-based catalysts have demonstrated good selectivity for catalytic degradation of VOCs. However, further studies are still needed to explore the fundamental mechanisms and theories to promote selective catalysis of VOCs by MNP-based catalysts.

(5) Previous studies have mainly focused on the development of MNPs for VOC removal, while less attention has been paid to the supports, which are equally important. Additional studies are needed to explore and optimize the selection of supports for MNPs to enhance their properties and VOC removal performance.

References

Adebayo, B.O., Lawson, S., Rownaghi, A.A., Rezaei, F., 2020. Analysis of equilibrium and dynamic adsorption of benzene vapor over unimodal and bimodal silica-based mixed-metal oxides. Chemical Engineering Journal 396, 125273.

Anandkumar, M., Ramamurthy, C.H., Suresh Babu, K., Thirunavukkarasu, C., 2015. Influence of age on the free-radical scavenging ability of CeO_2 and Au/CeO_2 nanoparticles. Journal of Materials Science 50, 2522–2531.

Bai, X., Zong, R., Li, C., 2014. Enhancement of visible photocatalytic activity via $Ag@C_3N_4$ core-shell plasmonic composite. Applied Catalysis B. Environmental 147 (5), 82–91.

Barsotti, E., Tan, S.P., Saraji, S., Piri, M., Chen, J.H., 2016. A review on capillary condensation in nanoporous media: implications for hydrocarbon recovery from tight reservoirs. Fuel 184, 344–361.

Baur, G.B., Yuranov, I., Kiwi-Minsker, L., 2015. Activated carbon fibers modified by metal oxide as effective structured adsorbents for acetaldehyde. Catalysis Today 249, 252–258.

Becheca, M.L., Kinadjian, N., Ollis, D., Backov, R., Lacombe, S., 2015. Comparison of kinetics of acetone, heptane and toluene photocatalytic mineralization over TiO_2 microfibers and Quartzel (R) mats. Applied Catalysis B: Environmental 179, 78–87.

Bhardwaj, R., Selamneni, V., Thakur, U.N., Sahatiya, P., Hazra, A., 2020. Detection and discrimination of volatile organic compounds by noble metal nanoparticle functionalized MoS_2 coated biodegradable paper sensors. New Journal of Chemistry 44, 16613–16625.

Bian, X.J., Qian, K., Liao, L., Zhou, X.F., Liu, B.H., 2014. Platinum nanoparticles supported on hierarchical carbon foams for electrocatalytic oxidation of methanol. Revista De Pesquisa Cuidado E Fundamental 3, F11–F14.

Cao, S.W., Yin, Z., Barber, J., Boey, F.Y.C., Loo, S.C.J., Xue, C., 2012. Preparation of Au-$BiVO_4$ heterogeneous nanostructures as highly efficient visible-light photocatalysts. ACS Applied Materials & Interfaces 4, 418–423.

Choi, W.S., Wang, M.J., Kwon, Y.J., Jung, S.H., Youn, Y.J., 2020. Radio-frequency thermal plasma treated activated carbon impregnated with Mn and Ag for volatile organic compounds adsorption. Journal of Nanoscience and Nanotechnology 20, 4423–4427.

Darif, B., Ojala, S., Pirault-Roy, L., Bensitel, M., Bouchra, Pirault-Roy, Laurence, 2016. Study on the catalytic oxidation of DMDS over Pt-Cu catalysts supported on Al_2O_3, $AlSi_20$ and SiO_2. Applied Catalysis B. Environmental: An International Journal Devoted to Catalytic Science and Its Applications 181, 24–33.

Ding, X., Liu, H., Chen, J., Wen, M., Li, G., An, T., Zhao, H., 2020. In situ growth of well-aligned Ni-MOF nanosheets on nickel foam for enhanced photocatalytic degradation of typical volatile organic compounds. Nanoscale 12, 9462–9470.

Fei, X., Wang, P., Zhang, D., Wang, H., Wu, Z., 2021. Confined catalysts application in environmental catalysis: current research progress and future prospects. ChemCatChem 13, 2313–2336.

Filipovic, S., Pavlovic, V.P., Mitric, M., Levic, S., Mitrovic, N.C., Maricic, A., Vlahovic, B., Pavlovic, V.B., 2019. Synthesis and characterization of $BaTiO_3/\alpha$-Fe_2O_3 core/shell structure. Journal of Advanced Ceramics 8, 133–147.

Fu, X., Liu, Y., Yao, W., Wu, Z., 2016. One-step synthesis of bimetallic Pt-Pd/MCM-41 mesoporous materials with superior catalytic performance for toluene oxidation. Catalysis Communications 83, 22–26.

Fu, Y., Hu, F., Li, H., Cui, L., Qian, G., Zhang, D., Xu, Y., 2021. Application and mechanisms of microalgae harvesting by magnetic nanoparticles (MNPs). Separation and Purification Technology 265, 118519.

Fujiwara, K., Okuyama, K., Pratsinis, S.E., 2017. Metal–support interactions in catalysts for environmental remediation. Environmental Science: Nano 4, 2076–2092.

Gallego, E., Roca, X., Perales, J.F., Guardino, X., 2009. Determining indoor air quality and identifying the origin of odour episodes in indoor environments. Journal of Environmental Sciences 21, 333–339.

Gao, S., Wang, P., Chen, X., Wang, H., Wu, Z., Liu, Y., Weng, X., 2014. Enhanced alkali resistance of $CeO_2/SO_4^{2-}-ZrO_2$ catalyst in selective catalytic reduction of NO_x by ammonia. Catalysis Communications 43, 223–226.

Ghosh, A., Maity, A., Banerjee, R., Majumder, S.B., 2017. Volatile organic compound sensing using copper oxide thin films: addressing the cross sensitivity issue. Journal of Alloys and Compounds 692, 108–118.

Ghoshal, A.K., Manjare, S.D., 2002. Selection of appropriate adsorption technique for recovery of VOCs: an analysis. Journal of Loss Prevention in the Process Industries 15, 413–421.

Guo, R., Wang, Q., Pan, W., Chen, Q., Ding, H., Yin, X., Yang, N., Lu, C., Wang, S., Yuan, Y., 2015. The poisoning effect of heavy metals doping on Mn/TiO_2 catalyst for selective catalytic reduction of NO with NH_3. Journal of Molecular Catalysis A: Chemical 407, 1–7.

Guo, Z., Huang, J., Xue, Z., Wang, X., 2016. Electrospun graphene oxide/carbon composite nanofibers with well-developed mesoporous structure and their adsorption performance for benzene and butanone. Chemical Engineering Journal 306, 99–106.

He, C., Cheng, J., Zhang, X., Douthwaite, M., Pattisson, S., Hao, Z.P., 2019. Recent advances in the catalytic oxidation of volatile organic compounds: a review based on pollutant sorts and sources. Chemical Reviews 119, 4471–4568.

Hou, H., Gao, F., Wang, L., Shang, M., Yang, Z., Zheng, J., Yang, W., 2016. Superior thoroughly mesoporous ternary hybrid photocatalysts of $TiO_2/WO_3/g\text{-}C_3N_4$ nanofibers for visible-light-driven hydrogen evolution. Journal of Materials Chemistry A 4, 6276–6281.

Huang, H., Xu, Y., Feng, Q., Leung, D.Y.C., 2015. Low temperature catalytic oxidation of volatile organic compounds: a review. Catalysis Science and Technology 5, 2649–2669.

Ingram, D.B., Christopher, P., Bauer, J.L., Linic, S., 2011. Predictive model for the design of plasmonic metal/semiconductor composite photocatalysts. ACS Catalysis 1, 1441–1447.

Jakubek, T., Hudy, C., Indyka, P., Nowicka, E., Golunski, S., Kotarba, A., 2019. Effect of noble metal addition to alkali-exchanged cryptomelane on the simultaneous soot and VOC combustion activity. Catalysis Communications 132, 105807.

Ji, Y.M., Shi, Q.J., Li, Y.X., An, T.C., Zheng, J., Peng, J.F., Gao, Y.P., Chen, J.Y., Li, G.Y., Wang, Y., Zhang, F., Zhang, A.L., Zhao, J.Y., Molina, M.J., Zhang, R.Y., 2020. Carbenium ion-mediated oligomerization of methylglyoxal for secondary organic aerosol formation. Proceedings of the National Academy of Sciences of the United States of America 117, 13294–13299.

Kask, M., Bolobajev, J., Krichevskaya, M., 2020. Gas-phase photocatalytic degradation of acetone and toluene, and their mixture in the presence of ozone in continuous multi-section reactor as possible air post-treatment for exhaust from pulsed corona discharge. Chemical Engineering Journal 399, 125815.

Klepeis, N.E., Nelson, W.C., Ott, W.R., Robinson, J.P., Tsang, A.M., Switzer, P., Behar, J.V., Hern, S.C., Engelmann, W.H., 2001. The National Human Activity Pattern Survey (NHAPS): a resource for assessing exposure to environmental pollutants. Journal of Exposure Analysis and Environmental Epidemiology 11, 231–252.

Konsolakis, M., Carabineiro, S.A., Tavares, P.B., Figueiredo, J.L., 2013. Redox properties and VOC oxidation activity of Cu catalysts supported on Ce(1)-xSmxOdelta mixed oxides. Journal of Hazardous Material 261, 512–521.

Kumar, J., Deo, G., Kunzru, D., 2016a. Preferential oxidation of carbon monoxide on $Pt/\gamma\text{-}Al_2O_3$ catalyst: effect of adding ceria and nickel. International Journal of Hydrogen Energy 41, 18494–18501.

Kumar, P., Kim, K.-H., Kwon, E.E., Szulejko, J.E., 2016b. Metal-organic frameworks for the control and management of air quality: advances and future direction. Journal of Materials Chemistry A 4, 345–361.

Leenders, L.H.A.M., Gramage-Doria, R., Bruina, B.d., Reek, J.N.H., 2015. Transition metal catalysis in confined spaces. Chemical Society Reviews 44, 433–448.

Li, J., Liu, H., Deng, Y., Liu, G., Chen, Y., Yang, J., 2016. Emerging nanostructured materials for the catalytic removal of volatile organic compounds. Nanotechnology Reviews 5, 147–181.

Li, J., Xu, Z., Wang, T., Xie, X., Li, D., Wang, J., Huang, H., Ao, Z., 2022. A versatile route to fabricate Metal/UiO-66 (Metal = Pt, Pd, Ru) with high activity and stability for the catalytic oxidation of various volatile organic compounds. Chemical Engineering Journal 448, 136900.

Li, L., Chai, Z., Jin, W., Sun, H., He, J., Wu, G., Xia, W., 2023. Sulfur vacancy in SnS2 nanoflake adjusted by precursor and improved photocatalytic performance. Journal of Alloys and Compounds 932, 167658.

Li, S., Jia, M., Gao, J., Wu, P., Yang, M., Yang, Y., Zhang, W., 2015. Infrared studies of the promoting role of water on the reactivity of Pt/FeO_x catalyst in low-temperature oxidation of carbon monoxide. Journal of Physical Chemistry C 119.

Lillo-Rodenas, M.A., Cazorla-Amoros, D., Linares-Solano, A., 2005. Behaviour of activated carbons with different pore size distributions and surface oxygen groups for benzene and toluene adsorption at low concentrations. Carbon 43, 1758–1767.

Liu, G., Niu, P., Sun, C., Smith, S.C., Chen, Z., Lu, G.Q.M., Cheng, H.-M., 2010. Unique electronic structure induced high photoreactivity of sulfur-doped graphitic C3N4. Journal of the American Chemical Society 132, 11642–11648.

Liu, G., Yang, K., Li, J., Tang, W., Xu, J., Liu, H., Yue, R., Chen, Y., 2012. Surface diffusion of Pt clusters in/on SiO2 matrix at elevated temperatures and their improved catalytic activities in benzene oxidation. Journal of Physical Chemistry C 118, 22719–22729.

Liu, H.L., Xu, M.J., Li, G.Y., Zhang, W.P., An, T.C., 2021. Solar-light-triggered regenerative adsorption removal of styrene by silver nanoparticles incorporated in metal-organic frameworks. Environmental Science: Nano 8, 543–553.

Lin, S.-Y., Zhang, R.-J., Jiang, X., Yang, T.-T., Lao, J.-C., Zhu, H.-W., 2015. Gas adsorption properties of carbon materials and their applications in air purification. New Carbon Materials 30, 502–510.

Liu, Z.S., Chen, J.Y., Peng, Y.H., 2013. Activated carbon fibers impregnated with Pd and Pt catalysts for toluene removal. Journal of Hazardous Materials 256–257, 49–55.

Luengas, A., Barona, A., Hort, C., Platel, V., Gallastegui, G., Elias, A., 2015. A review of indoor air treatment technologies. Reviews in Environmental Science and Biotechnology 499–522.

Luo, W., Gan, J., Huang, Z., Chen, W., Duan, X., 2019. Boosting HER performance of Pt-based catalysts immobilized on functionalized vulcan carbon by atomic layer deposition. Frontiers in Materials 6, 251.

Marikutsa, A., Khmelevsky, N., Rumyantseva, M., 2022. Synergistic effect of surface acidity and PtO_x catalyst on the sensitivity of nanosized metal-oxide semiconductors to benzene. Sensors 22, 6520.

Mellouki, A., Wallington, T.J., Chen, J., 2015. Atmospheric chemistry of oxygenated volatile organic compounds: impacts on air quality and climate. Chemical Reviews 115, 3984–4014.

Miners, S.A., Rance, G.A., Khlobystov, A.N., 2016. Chemical reactions confined within carbon nanotubes. Chemical Society Reviews 45, 4727–4746.

Miranda, C., Mansilla, H., Yáñez, J., Obregón, S., Colón, G., 2013. Improved photocatalytic activity of $g\text{-}C_3N_4/TiO_2$ composites prepared by a simple impregnation method. Journal of Photochemistry and Photobiology A: Chemistry 253, 16–21.

Mishra, A., Mehta, A., Kainth, S., Basu, S., 2018. Effect of different plasmonic metals on photocatalytic degradation of volatile organic compounds (VOCs) by bentonite/M-TiO_2 nanocomposites under UV/visible light. Applied Clay Science 153, 144–153.

Młotek, M., Reda, E., Jóźwik, P., Krawczyk, K., Bojar, Z., 2015. Plasma-catalytic decomposition of cyclohexane in gliding discharge reactor. Applied Catalysis A: General 505, 150–158.

Mo, S., Zhang, Q., Zhang, M., Zhang, Q., Li, J., Fu, M., Wu, J., Chen, P., Ye, D., 2019. Elucidating the special role of strong metal–support interactions in Pt/MnO_2 catalysts for total toluene oxidation. Nanoscale Horizons 4, 1425–1433.

Mundt, K.A., Gallagher, A.E., Dell, L.D., Natelson, E.A., Boffetta, P., Gentry, P.R., 2017. Does occupational exposure to formaldehyde cause hematotoxicity and leukemia-specific chromosome changes in cultured myeloid progenitor cells? Critical Reviews in Toxicology 47, 592–602.

Muñoz-Batista, M.J., Kubacka, A., Fernández-García, M., 2014. Effect of $g\text{-}C_3N_4$ loading on TiO_2-based photocatalysts: UV and visible degradation of toluene. Catalysis Science and Technology 7, 2006–2015.

Nahm, S.W., Shim, W.G., Park, Y.-K., Kim, S.C., 2012. Thermal and chemical regeneration of spent activated carbon and its adsorption property for toluene. Chemical Engineering Journal 210, 500–509.

Ojha, D.P., Song, J.H., Kim, H.J., 2019. Facile synthesis of graphitic carbon-nitride supported antimony-doped tin oxide nanocomposite and its application for the adsorption of volatile organic compounds. Journal of Environmental Sciences 79, 35–42.

Pei, L.X., Zhou, J.R., Zhang, L.Z., 2013. Preparation and properties of Ag-coated activated carbon nanocomposites for indoor air quality control. Building and Environment 63, 108–113.

Qi, L., Cheng, B., Ho, W., Liu, G., Yu, J., 2015. Hierarchical Pt/NiO hollow microspheres with enhanced catalytic performance. Chemnanomat 1, 58–67.

Qiang, T., Xia, Y., 2020. Facile-controllable synthesis of SnS2 with sulfur vacancy for boosted Cr(VI) photoreduction performance. Journal of Alloys and Compounds 845, 156155.

Ren, S., Liang, W., Fang, H., Zhu, Y., 2021. Performance and poisoning analysis of organic sulfur resistance of Pd-Ce catalyst in catalytic oxidation of VOCs. Journal of Environmental Chemical Engineering 9, 106640.

Sun, W., Yang, Z., Xu, Y., Shi, Y., Shen, Y., Liu, G., 2020. Fabrication of Pd3@Beta for catalytic combustion of VOCs by efficient Pd3 cluster and seed-directed hydrothermal syntheses. RSC Advances 10, 12772–12779.

Tang, W., Wu, X., Li, D., Wang, Z., Liu, G., Liu, H., Chen, Y., 2014. Oxalate route for promoting activity of manganese oxide catalysts in total VOCs' oxidation: effect of calcination temperature and preparation method. Journal of Materials Chemistry A 2, 2544–2554.

Tefera, D.T., Hashisho, Z., Philips, J.H., Anderson, J.E., Nichols, M., 2014. Modeling competitive adsorption of mixtures of volatile organic compounds in a fixed-bed of beaded activated carbon. Environmental Science and Technology 48, 5108–5117.

Theerthagiri, J., Chandrasekaran, S., Salla, S., Elakkiya, V., Senthil, R.A., Nithyadharseni, P., Maiyalagan, T., Micheal, K., Ayeshamariam, A., Arasu, M.V., Al-Dhabi, N.A., Kim, H.-S., 2018. Recent developments of metal oxide based heterostructures for photocatalytic applications towards environmental remediation. Journal of Solid State Chemistry 267, 35–52.

Tomatis, M., Xu, H.-H., He, J., Zhang, X.-D., 2016. Recent development of catalysts for removal of volatile organic compounds in flue gas by combustion: a review. Journal of Chemistry 2016, 1–15.

Tomatis, M., Xu, H., Wei, C., Bishop, M., He, J., Wang, C., Zhao, M., Xiao, H., Yu, H., Behera, S., Tang, B., 2018. A comparative study of Mn/Co binary metal catalysts supported on two commercial diatomaceous earths for oxidation of benzene. Catalysts 8, 111–139.

Troppova, I., Matejova, L., Kubonova, L., Strasak, T., Studentova, S., Kustrowski, P., Obalova, L., 2017. Molecular dimensions and porous structure of activated carbons for sorption of xylene and isooctane. Chemical Engineering & Technology 40, 6–17.

Vellingiri, K., Kumar, P., Kim, K.H., 2016. Coordination polymers: challenges and future scenarios for capture and degradation of volatile organic compounds. Nano Research 9, 3181–3208.

Wang, D., Li, Z., Zhou, J., Fang, H., He, X., Jena, P., Zeng, J.-B., Wang, W.-N., 2018a. Simultaneous detection and removal of formaldehyde at room temperature: Janus Au@ZnO@ZIF-8 nanoparticles. Nano-Micro Letters 10, 1–11.

Wang, F., Dai, H., Deng, J., Bai, G., Ji, K., Liu, Y., 2012. Manganese oxides with rod-, wire-, tube-, and flower-like morphologies: highly effective catalysts for the removal of toluene. Environmental Science & Technology 46, 4034–4041.

Wang, J., Fan, C., Ren, Z., Fu, X., Qian, G., Wang, Z., 2014. N-doped TiO_2/C nanocomposites and N-doped TiO_2 synthesised at different thermal treatment temperatures with the same hydrothermal precursor. Dalton Transactions 43, 13783–13791.

Wang, Y., Dai, C., Chen, B., Wang, Y., Shi, C., Guo, X., 2015. Nanoscale HZSM-5 supported Pt Ag bimetallic catalysts for simultaneous removal of formaldehyde and benzene. Catalysis Today 258, 616–626.

Wang, Y., Ji, X., Meng, H., Qu, L., Wu, X., 2020. Fabrication of high-silica Cu/ZSM-5 with confinement encapsulated Cu-based active species for NH_3-SCR. Catalysis Communications 138, 105969.

Wang, Y., Xu, X., Lu, W., Huo, Y., Bian, L., 2018b. A sulfur vacancy rich CdS based composite photocatalyst with g-C_3N_4 as a matrix derived from a Cd-S cluster assembled supramolecular network for H2 production and VOC removal. Dalton Transactions 47, 4219–4227.

Xiang, L., Zhang, L., Shao, J., Lin, F., Wang, Z., Yan, B., Chen, G., 2023. Tentative exploration on water-promoted catalytic ozonation of Cl-VOCs and feasibility of application in complex flue gas over monolithic cobalt catalyst. Journal of Hazardous Materials 441, 129997.

Xu, Z., Yu, J., Jaroniec, M., 2015. Efficient catalytic removal of formaldehyde at room temperature using AlOOH nanoflakes with deposited Pt. Applied Catalysis B: Environmental 163, 306–312.

Yan, D., Wang, X., Thomas, A., Antonietti, M., 2010a. Making metal-carbon nitride heterojunctions for improved photocatalytic hydrogen evolution with visible light. ChemCatChem 2, 834–838.

Yan, S.C., Lv, S.B., Li, Z.S., Zou, Z.G., 2010b. Organic-inorganic composite photocatalyst of g-C(3)N(4) and TaON with improved visible light photocatalytic activities. Dalton Transactions 39, 1488–1491.

Yang, C.T., Miao, G., Pi, Y.H., Xia, Q.B., Wu, J.L., Li, Z., Xiao, J., 2019. Abatement of various types of VOCs by adsorption/catalytic oxidation: a review. Chemical Engineering Journal 370, 1128–1153.

Yang, E., Jang, E.J., Lee, J.G., Yoon, S., Lee, J., Musselwhite, N., Somorjai, G.A., Kwak, J.H., An, K., 2018. Acidic effect of porous alumina as supports for Pt nanoparticle catalysts in n-hexane reforming. Catalysis Science and Technology 8, 3295–3303.

Yang, Y., Guo, Y., Liu, F., Yuan, X., Guo, Y., Zhang, S., Guo, W., Huo, M., 2013. Preparation and enhanced visible-light photocatalytic activity of silver deposited graphitic carbon nitride plasmonic photocatalyst. Applied Catalysis B: Environmental 142–143, 828–837.

Yin, T., Meng, X., Wang, S., Yao, X., Liu, N., Shi, L., 2022. Study on the adsorption of low-concentration VOCs on zeolite composites based on chemisorption of metal-oxides under dry and wet conditions. Separation and Purification Technology 280, 119634.

Zeng, J., Liu, X., Wang, J., Lv, H., Zhu, T., 2015. Catalytic oxidation of benzene over MnOx/TiO2 catalysts and the mechanism study. Journal of Molecular Catalysis A: Chemical 408, 221–227.

Zhang, F., Li, X., Zhao, Q., Zhang, Q., Tadé, M., 2015a. Fabrication of α-Fe_2O_3/In_2O_3 composite hollow microspheres: a novel hybrid photocatalyst for toluene degradation under visible light. Journal of Colloid and Interface Science 457, 18–26.

Zhang, H., Pan, X., Han, X., Liu, X., Wang, X., Shen, W., Bao, X., 2013. Enhancing chemical reactions in a confined hydrophobic environment: an NMR study of benzene hydroxylation in carbon nanotubes. Chemical Science 4, 1075–1078.

Zhang, J., Li, Y., Zhang, Y., Chen, M., Wang, L., Zhang, C., He, H., 2015b. Effect of support on the activity of Ag-based catalysts for formaldehyde oxidation. Scientific Reports 5, 12950.

Zhang, X., Gao, B., Creamer, A.E., Cao, C., Li, Y., 2017. Adsorption of VOCs onto engineered carbon materials: a review. Journal of Hazardous Materials 338, 102–123.

Zhao, Z., Sun, Y., Dong, F., 2015. Graphitic carbon nitride based nanocomposites: a review. Nanoscale 7, 15–37.

Zhong, L., Lee, C.-S., Haghighat, F., 2012. Adsorption performance of titanium dioxide (TiO_2) coated air filters for volatile organic compounds. Journal of Hazardous Materials 243, 340–349.

Zhou, K., Li, D., Zhou, C., Chen, H., Guo, Y., Zeng, Z., Ma, W., Li, L., 2022. Metal heteroatom (Mg, Cu and Co) and porous carbon co-doped MIL-101 composites with superior acetone capture capacity. Chemical Engineering Journal 430, 132656.

Zhou, L., Chen, Y.-L., Zhang, X.-H., Tian, F.-M., Zu, Z.-N., 2014. Zeolites developed from mixed alkali modified coal fly ash for adsorption of volatile organic compounds. Materials Letters 119, 140–142.

Zhu, X., He, X., Guo, L., Shi, Y., Zhao, N., Qiao, C., Dai, L., Tian, Y., 2022. Hydrophobic modification of ZSM-5-encapsulated uniform Pt nanoparticles for catalytic oxidation of volatile organic compounds. ACS Applied Nano Materials 5.

CHAPTER

12

Nano metal oxides (NMOs) for VOC removal

Hao Li[1,2], *Shuping Wu*[2], *Guoqiang Chen*[1,3], *Bin Fu*[1], *Yiran Zhao*[1,2], *Feiyue Li*[3] *and Ping Qu*[1]

[1]Institute of Agricultural Resources and Environment, Jiangsu Academy of Agricultural Sciences, Key Laboratory of Agricultural Environment in downstream of Yangtze Plain, Ministry of Agriculture and Rural Affairs, Jiangsu Collaborative Innovation Center for Solid Organic Waste Resource Utilization, Nanjing, China; [2]Research School of Polymeric Materials, School of Materials Science & Engineering, Jiangsu University, Zhenjiang, China; [3]College of Resources and Environment Science, Anhui Science and Technology University, Fengyang, China

12.1 Introduction

Organic compounds with the boiling point below 250°C and saturated vapor pressure equivalent to the atmospheric pressure (101.325 kPa) are identified as volatile organic compounds (VOCs), which include alkanes, olefins, alcohols, ketones, aldehydes, aromatics, paraffins, halogenated hydrocarbons, etc. (Enesca and Cazan, 2020).

They are major contributors to air pollution due to their toxic nature or as a precursor of ozone and photochemical smog. Common VOCs such as formaldehyde, benzene, toluene, propene, phenol, and acetone are harmful to human beings. They are emitted from industrial, transportation (outdoor sources), and household (indoor sources) activities, such as power generation, vehicle emission, solvent use, and from office supplies and restaurant and domestic cooking. Fig. 12.1 shows VOC emission from different sources in China in 2015 (Yang et al., 2019).

Nanoparticles refer to particles with particle size ranging from 1 to 100 nm. Due to the particularity of their size, nanoparticles have a series of unique properties related to quantum size effect, surface and interface effect, macroscopic quantum tunneling effect, etc. Metal oxides are oxides composed of metal elements and oxygen elements, such as Zn and O to form ZnO, which are widely used as catalysts in the field of catalysis as well as adsorbents in the

Volatile Organic Compound Removal
https://doi.org/10.1016/B978-0-443-15674-8.00009-3

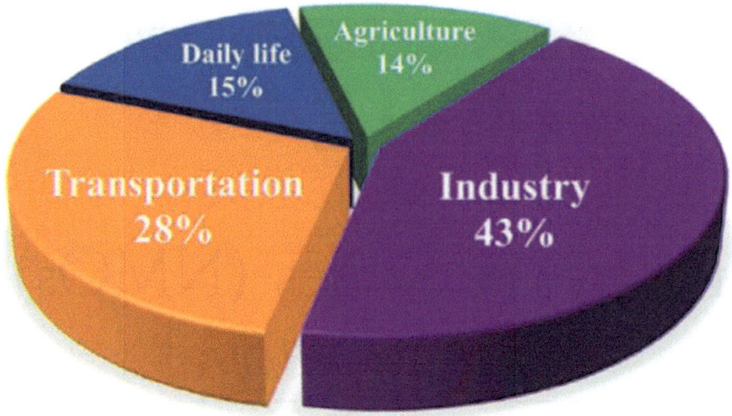

FIGURE 12.1 VOC emission from different sources in China in 2015 (Yang et al., 2019).

field of adsorption. Nano metal oxides (NMOs) refer to nanoscale metal oxides with the combined characteristics of nanoparticles and metal oxides. As a result, NMOs have demonstrated strong adsorption and catalytic oxidation ability to various VOCs.

The main emission reduction technologies of VOCs include absorption, adsorption, biological method, plasma technology, thermal incineration, and catalytic oxidation technology. Among them, NMO-based catalytic oxidation technology is considered as one of the most promising ones, which have the characteristics of high treatment efficiency, low energy consumption, and low secondary pollution. In comparison to other VOC removal technologies including adsorption, NMO-based catalytic oxidation technology has the advantages of clean process, less by-products and simple operation process. In the process of VOC oxidation, NMOs have demonstrated promising catalytic performance (Shah and Li, 2019). This chapter thus mainly focuses on catalytic removal of VOCs by NMOs (Section 12.2–12.4). It also provides a brief overview of NMOs as VOC adsorbent (Section 12.5). Perspectives and future development of NMOs in VOC removal are outlined at the end of the chapter.

12.2 Overview of NMOs as catalysts for VOC removal

In recent years, various metal-based catalysts including transition metal oxides, mesoporous metal oxides, macroporous metal oxides, macroporous composite metal oxides and their supported transition metal oxides, and precious metals have been applied for catalysis (Liu et al., 2016, 2019, 2022). Supported noble metal catalysts, despite their relatively high cost, are preferred because of their high specific activities at low temperatures. Loading of noble metals like Pt, Pd, and Rh has been used to enhance the performance of traditional catalysts such as TiO_2 for photocatalytic oxidation of VOCs (e.g., benzene) in gas phase. This technology, however, may produce polychlorinated by-products in the form of chlorine-containing

VOCs, which are more toxic. Regarding metal oxide catalysts, pure transition metal oxides, mixed transition metal oxides, and perovskites are recognized as efficient low-cost catalysts for VOC oxidation. The d orbital of the transition metal is not saturated, which is easy to interact with the reactant molecules and has good catalytic capacity. Many studies have shown that transition metals, such as Mn, Co, Cu, Ce, V and their oxides, have a good performance in the catalytic oxidation of VOCs with the complete conversion temperature lower than 300°C (Shah and Li, 2019). Composite metal oxides and their supported precious metals are also excellent catalytic materials for eliminating VOCs. For example, perovskite like oxides (A_2BO_4), spinel like oxides (AB_2O_4) and hexaaluminate ($AAl_{12}O_{19}$) have excellent catalytic activity for the oxidation of VOCs (Almaie et al., 2022). It is accepted that the catalytic activity of a metal oxide is associated with its surface area, pore structure, oxygen nonstoichiometry, and reducibility. Making composite metal oxide into porous or nano structure can significantly improve its specific surface area and dispersion of precious metals, promote the diffusion and activation adsorption of reactant molecules, and thus improve its catalytic performance for VOC oxidation (Yang et al., 2019).

Current efforts for NMO-based VOC removal technology mainly concentrate on developing efficient and low-cost catalysts which can catalyze the complete oxidation of VOCs into CO_2 and H_2O at low temperatures (He et al., 2019). Three models have been developed to describe the key processes controlling the catalytic oxidation of VOCs.

(1) MVK (Marse-van Krevelen) model, also known as redox model. According to this model, the reaction occurs between the adsorbed VOCs and the lattice oxygen on the catalyst rather than the gas phase oxygen. The reaction is divided into two steps: in the first step, VOCs are adsorbed on the surface of the catalyst and react with lattice oxygen on the surface of the catalyst to generate CO_2 and H_2O, and metal oxides in the catalyst are reduced at the same time; in the second step, the reduced metal oxide is oxidized by O_2 in the gas phase.

(2) L-H (Langmuir-Hinshelwooch) model. According to this model, the reaction occurs between adsorbed VOCs and adsorbed oxygen, that is, after the catalyst adsorbs VOCs and O_2, the redox reaction occurs. The adsorption of VOCs and O on the same active site is called the single-point L-H model, while the adsorption on different active sites is called the two-point L-H model.

(3) E-R (Eley-Rideal) model. The model suggests that the adsorbed oxygen species and VOCs in the gas phase react on the surface of the catalyst. After the reaction of reactive oxygen species and VOCs, the reaction products desorbed from the catalyst surface.

The applicability of each mechanism model varies with the nature of NMO catalysts and VOCs. Different types of catalysts follow different mechanisms for the catalytic oxidation of VOCs, which are mainly affected by the nature of loading and active center and catalytic reverse conditions. MVK is a good model to explain the catalytic oxidation of VOCs by metal oxide catalysts. In addition, it has also been used to explain the mechanism of hydrocarbon oxidation reaction on Pt based catalysts. L-H and E-R models are often applied to describe VOC oxidation catalyzed by noble metal oxide catalysts (Liu et al., 2016).

12.3 NMOs for catalytic oxidation of VOCs

12.3.1 NMO catalysts for VOC oxidation

Catalytic oxidation is the most important method for the removal of VOCs, especially at low concentrations. In recent years, great efforts have been devoted to the research on applications of NMO-based materials for catalytic removal of VOCs. In order to develop efficient and stable catalysts to reduce the temperature for the complete oxidation of VOCs, many studies have explored the relationship between the catalytic performance and physicochemical features of NMOs. It has been demonstrated that NMOs such as MnO_x, CeO_2, Co_3O_4, FeO_x, CuO_x, and NiO_x have excellent catalytic ability for the deep oxidation of VOCs (Liu et al., 2022). Summary of NMO catalysts for VOC oxidation is listed in Table 12.1.

MnO_x-based catalyst is favored in the catalytic system for VOC (e.g., hydrocarbon, ketone, aldehydes, esters, and alcohol) abatement especially in the post catalysis system, due to its advantages such as polymorphism, multiple valences, high oxygen mobility, and adsorption capacity (Chang et al., 2021; Lee et al., 2020; Li et al., 2014; Shang et al., 2021; Xiong et al., 2021; Yang et al., 2020). CeO_2-based catalyst has been widely used in VOC (e.g., hydrocarbon, aldehydes, and alcohol) decomposition process due to its good thermal stability, high oxygen storage capacity, oxygen vacancies, bulk oxygen mobility, and the presence of redox couple (Ce^{3+}/Ce^{4+}) (Wang et al., 2017; Wang et al., 2019; Zhu et al., 2015). Cobalt oxides are popular in catalytic removal of VOCs (e.g., hydrocarbon, aldehydes, and esters) due to their high concentration of electrophilic oxide species, large content of oxygen vacancies, weak oxygen bond strength, high turnover frequency for oxidation-reduction reactions, and excellent CO oxidation at room temperature (Kamal et al., 2016; Yao et al., 2019; Zhu et al., 2017; Jiang et al., 2019). Iron oxide-based catalysts are used to eliminate VOCs (e.g., hydrocarbon, aldehydes, esters, and chlorinated VOCs) in the catalytic system due to its relatively low cost, high activity, environmental friendliness, and the O_2 transportation via facile conversion between Fe^{2+} and Fe^{3+} oxidation states (Chang et al., 2021; Lu et al., 2015; Sultana et al., 2018; Xiong et al., 2021). CuO_x-based catalyst has the merits of sufficient catalytic activity and relatively low cost, which is applied in the catalytic removal of VOCs (e.g., ketone, aldehydes, and chlorinated VOCs) (Veerapandian et al., 2019; Zeng et al., 2020; Zhu et al., 2015). Nickel oxides are favored in VOC (e.g., hydrocarbon and oxygenated VOCs) decomposition attributing to their superior activity of CC and CH bonds on Ni particles, high capability of electron transfer, and low price (Liu et al., 2019; Liu et al., 2017; Muroyama et al., 2010; Sun et al., 2019; Xu et al., 2020).

In addition, metal oxides are considered as "active" supports for noble metal catalysts, by both allowing the reduction of the noble metal loaded and increasing their life time (Yang et al., 2019). Dispersed Pt/CeO_2 catalyst was decorated on conductive nickel foam substrate for toluene degradation (Du et al., 2022). CO and acetaldehyde were efficiently degraded using nano-sized Pt catalysts supported on CeO_2 and CeO_2/ZSM-5 composite (Fuku et al., 2013). Ceria promoted the reducibility of Pt and Al_2O_3 (Abbasi et al., 2011). Cu was doped nanostructures to confine CeO_2 growth in boron-rich carbon frameworks for acetone oxidation with enhanced redox properties (Liu et al., 2021). Ce-modified mesoporous gamma-Al_2O_3 supported Pd−Pt nanoparticle catalysts achieved the complete oxidation of benzene (Zhu et al., 2019). Cu doped heteroatomic mesoporous Zeolite Socony Mobil-5 exhibited

TABLE 12.1 Summary of NMO catalysts for VOC oxidation.

Photocatalyst	Support	VOC pollutants	Temperature(°C)	Degradation efficiency (%)/Time	
MnAlO hydrotalcite-derived oxide	–	Acetone, acetaldehyde	150, 170	100	Sun et al. (2019)
Ni-based spinel oxide	–	Ethanol	250–550	100 (60 min)	Muroyama et al. (2010)
Pt–CeO_2	–	Toluene	–	100	Du et al. (2022)
Pt	CeO_2-loaded ZSM-5	Acetaldehyde	<200	100	Fuku et al. (2013)
Cu–Ce	BDC	Acetone	189	Ce_4Cu_1-40/qqqqqBDC 90	Liu et al. (2021)
Ceria and Pd–Pt oxides	γ-Al_2O_3	Benzene	200	0.2%Pd–Pt (1:1)/10%Ce/γ-Al_2O_3 100	Zhu et al. (2019)
Transition metal doped mesoporous ZSM-5	–	Hexane	420	Cu-HZ >90	Li et al. (2022)
$PtMn_{0.2}$@ZSM5	–	Acetone	165	95	Yang et al. (2022)
Pt/CeO_2–TiO_2	–	Benzene (BEN) ethyl acetate (EAC) acetonitrile (ACN) dichloroethane (DCE)	BEN 150 EAC 210 ACN 242 DCE 278	90	Shi et al. (2022)
3D@2D Pd/$CoMn_2O_4$	Al_2O_3	Toluene (MB) benzene (BEN), ethyl acetate (EAC)	MB 165, BEN 160, EAC 155	1wt%Pd/h-Al@4CMO 100	He et al. (2022)
Au/FeO_x	CeO_2	Methanol, toluene	180–400	100	Bonelli et al. (2011)
Ag	Mn_2O_3	Toluene	220	0.13Ag/Mn_2O_3-ms 100	Deng et al. (2015)
Nano-RuO_2	TiO_2	Propane	150–450	100	Eric et al. (2014)
Mesoporous SiO_2 doped TiO_2 nanofibers	–	Toluene	400	10%SiO_2 90.6	Sihui et al. (2014)
$NiMnO_4$	–	Toluene (MB), 2-propanol (IPA)	MB 350, IPA 250	100	Hosseini et al. (2012)

(*Continued*)

TABLE 12.1 Summary of NMO catalysts for VOC oxidation.—cont'd

Photocatalyst	Support	VOC pollutants	Temperature(°C)	Degradation efficiency (%)/Time	
Mn_6Al_2HT	–	Toluene	249	100	Genty et al. (2012)
$ZnO/NiMoO_4/C_3N_4$	–	Toluene	26	95 (90min)	Anwer et al. (2021)
$LaCoO_x$	ZrO_2	Methanol	<400	$LaCoO_3/ZrO_2$ 100	Kustov et al. (2011)
Pt	γ-Al_2O_3	Benzene	100–150	Pt with (220) facet 100	Pulleri et al. (2018)
α-MnO_2 nanotube	–	Formaldehyde	80	100 (30 min)	Zhao et al. (2016)
Pt/CeO_2	–	Toluene	225	0.5wt% Pt 99	Abbasi et al. (2011)

the highest catalytic activity (Yang et al., 2022). Pt nanosheets supported on alumina (gamma-Al_2O_3) showed the best catalytic oxidation of benzene (Pulleri et al., 2018). Bimetallic Pt–Mn nanoparticles were encapsulated in ZSM-5 zeolite with superior performance for acetone catalytic oxidation (Yang et al., 2022). Pt doped CeO_2–TiO_2 showed strong catalytic activity for benzene oxidation (Shi et al., 2022). 3D@2D Pd/$CoMn_2O_4$ nanosheets were supported on hollow Al_2O_3 microspheres for catalytic oxidation of toluene (Shi et al., 2022). Nano-sized FeO_x and Au/FeO_x catalysts were supported on CeO_2 for total oxidation of methanol and toluene (Bonelli et al., 2011). Nano-size platinum was added to iron oxide based catalyst to improve the toluene conversion with enhanced surface oxygen mobility (Jung et al., 2014). Fig. 12.2 shows the examples of noble metal doped NMOs with excellent catalytic activity.

NMOs have attracted great attention because of their favorable economic factors and their positive low-temperature oxidation and adsorption properties. However, the efficiency of single-metal-oxide catalysts in removing VOCs is lower (Kamal et al., 2016). The synergistic effect of two or more metal oxides can have a higher efficiency compared to a single-metal-oxide catalyst. For example, nano-RuO_2 deposited onto TiO_2 was tested in the total oxidation of propane and this composite catalyst was very active and highly selective (Eric et al., 2014). SiO_2 was doped in TiO_2 mesoporous nanofibers for rapid degradation of toxic toluene (Sihui et al., 2014). Nano nickel-manganite oxides were promising catalysts for remediation of VOCs. In the composite, crystal phase and morphologies of the MnO_2 predominated the catalytic oxidation of resorcinol and formaldehyde (Zhao et al., 2016). Nanocatalysts derived from hydrotalcite like compounds Mn_6Al_2HT hydrotalcites showed the best activity for toluene oxidation (Genty et al., 2012). ZnO nanorods grown on a stainless steel mesh with a porous $NiMoO_4/C_3N_4$ achieved a high degree of toluene adsorption (Anwer et al., 2021). $LaCoO_{(x)}$ nanoparticles were prepared inside the mesopores of $ZrO_{(2)}$, exhibiting a very high catalytic activity in the abatement of methanol and other light hydrocarbons (Kustov et al., 2011).

The advantages of NMO-based catalytic oxidation of VOCs include low temperature (200–500°C), controllable product selectivity; while the disadvantages are high energy

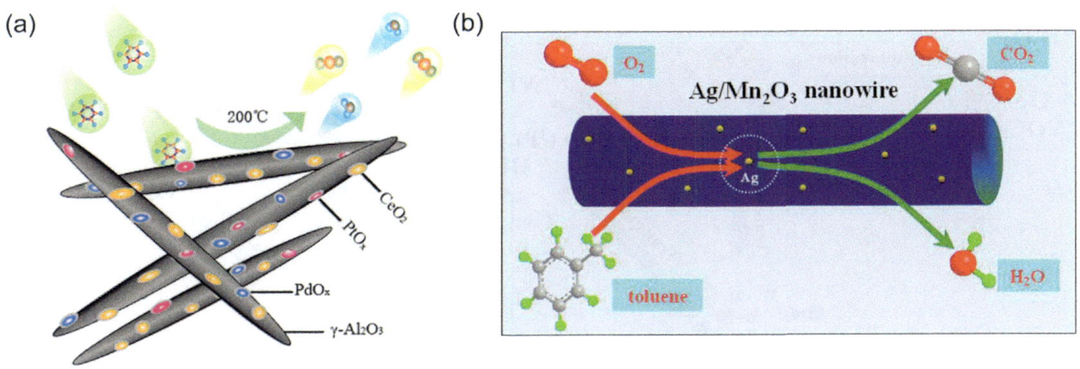

FIGURE 12.2 (a) Mesoporous γ-Al_2O_3 supported Pd–Pt nanocrystalline catalysts demonstrated the best activity for the complete oxidation of benzene at 200°C (Zhu et al., 2019), and (b) dispersion of silver nanoparticles on the surface of Mn_2O_3 nanowires improved catalytic activity (Deng et al., 2015).

consumption, incomplete oxidation of VOCs, and deactivating, poisoning, coking, and sintering of catalysts.

12.3.2 NMOs for plasma-catalytic oxidation of VOCs

Plasma catalytic oxidation is an attractive way to efficiently decompose dilute VOCs by injecting small amounts of energy. Fig. 12.3 shows reaction pathways of VOCs in plasma catalysis system. The plasmacatalysis which combines plasma and NMO catalysis is considered to be a cost- and energy-efficient process to remove diluted VOCs due to its merits of high degradation efficiency, good selectivity, low generation of unwanted by-products, and high energy efficiency. Numerous efforts have been made to develop efficient catalysts to enhance VOC abatement by utilizing the synergetic effect of plasma and catalyst (Veerapandian et al., 2017). The NMOs used for plasma oxidation include oxides of Co, Zn, Mg, Ba, Cu, and Mn (Chang et al., 2022). For example, the combination of Cu-doped MnO_2 catalysts and plasma significantly enhanced the removal of acetone and methyl ethyl ketone (Zeng et al., 2020). Ternary perovskite-type Co, Zn, Mg, Ba-x MnO_3 nanocatalyst in plasma-catalysis effectively degraded benzene, toluene, and xylene (Parvizi et al., 2020). CuO nanowires/Cu foam was prepared as a monolithic catalyst for plasma-catalytic oxidation of toluene (Zheng et al., 2017).

NMO-based catalysts in plasma catalytic removal of VOCs are attractive due to their obvious merits such as relatively low cost, environmentally friendly, strong poisoning resistance, the availability of a range of metal oxides in different sizes and shapes, the capability of regeneration, and highly collaborative interactions with plasma (Kamal et al., 2016). Normally, mixed metal oxides or supported catalysts are preferred over pure metal oxide due to their superior activity resulting from the interactions between different metal oxides or metal oxides and supports. It is well known that mixed metal oxides can accelerate the electron transfer through the lattice for surface oxygen mobility to facilitate plasma catalysis of

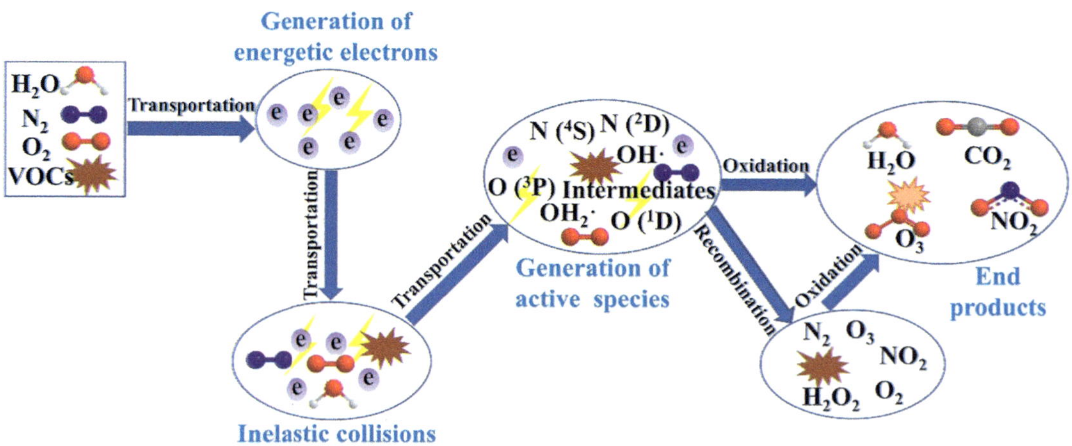

FIGURE 12.3 Schematic representation of the reaction pathways of VOCs in plasma catalysis system (Chang et al., 2022).

VOCs (Wang et al., 2019; Veerapandian et al., 2017). However, the practical challenges of nonnoble NMO-based catalysts for plasmas catalysis of VOCs include the poisoning and deactivation of the catalysts.

One of the most commonly used NMOs in plasma catalysis of VOCs is MnO_x. MnOx has demonstrated its high efficiency in the transition of O_3 to active oxygen to accelerate the plasma-catalytic oxidation of VOCs. The excellent performance of MnO_x catalysts is associated with the lattice oxygen and coexistence of mixed-valence states of Mn^{3+}/Mn^{4+} and/or Mn^{2+}/Mn^{3+} (Yang et al., 2020). Mixed CeO_2-MnOx catalysts showed better activity than that of pure oxides in the plasma removal of toluene, due to the interactions between Ce and Mn species, larger specific surface area, decreased grain size, more oxygen vacancies, and improved oxygen mobility (Wang et al., 2017). Yao et al. prepared Co-Mn catalyst to decompose hexanal, and the enhancement of redox properties caused by the high content of Co^{3+} and Mn^{3+} cations (Yao et al., 2019). Xiong et al. reported that the Fe–Mn catalysts showed excellent performance in the plasma-catalytic oxidation of ethylene oxide due to its large surface area, high reducibility, and a large amount of surface adsorbed oxygen (Xiong et al., 2021). Veerapandian et al. reported that the combination of Cu–Mn mixed oxides and nonthermal plasma enhanced the TCE conversion efficiency up to 80% at a low catalyst temperature of 150°C (Veerapandian et al., 2019).

Other NMOs, especially NMO-based composites, have also showed excellent performance in plasma catalysis of VOCs. Lu et al. reported that ~50% toluene removal efficiency was promoted by 3% FeOx/SBA-15 in comparison with plasma alone system (Lu et al., 2015). Xu et al. proposed that the benzene removal efficiency of plasma treatment was improved by the CuO/active carbon catalyst (Xu et al., 2014). Among various alumina-supported catalysts, the performance of Cu is highest followed by Mn, Co, Ni, and Ce in plasma removal of acetone (Zhu et al., 2015). Xu et al. employed Ni-SBA catalyst with different Ni loadings to decompose toluene using plasma (Xu et al., 2020). Xu et al. also found that Ca–Ni/ZSM-5 catalyst showed good plasma oxidation of toluene due to its strong and weak acidic sites and superior surface properties (Xu et al., 2020).

The most common support materials for plasma-catalysts include zeolites, carbon-based materials, Al_2O_3, SBA-15, MnO_2, and CeO_2, etc. because they have a large surface area and good thermal stability (Veerapandian et al., 2017; Xu et al., 2014; Liu et al., 2019). Noble metals are commonly loaded on supports to improve the dispersion and utilization of metal active sites (Veerapandian et al., 2017). Zhu et al. (2020) found that the combination of Au/CeO_2/Al_2O_3 nanocatalysts with plasma for toluene oxidation produced various reactive species to remove organic intermediates from catalyst surface at low temperatures, inhibiting the formation of secondary pollutants and the poisoning of catalysts. Zhu et al. (2021) obtained 100% toluene conversion through plasmas catalysis over Au/γ-Al_2O_3 (0.1 wt% Au loading) catalyst at a temperature of 257°C. Yuan et al. (2014) reported that the addition of Pt/Al_2O_3 catalyst was beneficial for the decomposition of naphthalene and n-butanol in comparison with the plasma alone system. Qin et al. reported that Ag–Mn/γ-Al_2O_3 catalyst showed good performance in plasma catalysis of toluene, with a carbon balance of 91%, and it also suppressed O_3 and N_2O formation (Qin et al., 2016).

The main advantage of plasma catalysis for VOC abatement lies in its ability to induce various chemical reactions at room temperature and atmospheric pressure, enabling rapid start-up and shutdown. Plasma utilizes oxygen and water vapor in the carrier gas to form oxidative species. In the plasma process, VOC molecules are dissociated by the high-energy electrons and oxidized by active species to form CO_2, H_2O and other by-products (Veerapandian et al., 2017). However, the disadvantage of this technology is the formation of unwanted by-products such as O_3, NOx, and other VOCs, which may even be more toxic than the treated VOCs (Huang et al., 2011). Thus, it is critical to develop more efficient, selective and low-cost NMO-based catalysts to further promote plasma-catalytic oxidation of VOCs.

12.3.3 Other NMO-based catalytic oxidation technologies for VOCs

Catalytic combustion is a method based on the mechanism by which VOC gas contacts with the catalyst during combustion. NMO catalysts in the VOC oxidation reaction can reduce the operating temperature of the direct combustion process. For example, NMOs of Ce, Mo, Ti, Mn, Co, Ni etc. and their derivative catalysts have shown excellent catalytic activities for VOC combustion (Zang et al., 2019). CeO_2 and MoO_3 loading on V−W/Ti catalyst represented the best simultaneous removal effect of benzene and toluene at 500°C (Chen et al., 2020). The addition of CeO_2 improved the oxygen supply performance of the Pd−Pt-based catalyst for the catalytic combustion of storage tank VOCs (Yang et al., 2020). Ru nanoparticles was deposited on CeO_2 to improve activity and stability for combustion of mixture of light hydrocarbon VOCs (Okal et al., 2018). Pt/TiO_2 cordierite monolithic catalyst with 0.1 wt‱ Pt content exhibited excellent catalytic combustion activity, stability, and firmness during catalytic combustion of VOCs (Xiao et al., 2018). Similarly, structured Co_3O_4 nano-flowers on silicalite-1/SiC foam catalysts exhibited excellent performance in complete thermocatalytic combustion of isopropanol (Guan et al., 2021).

NMOs have also been applied as catalysts in high temperature thermal incineration of VOCs (Tomatis et al., 2016). Thermal incineration is another technology that utilizes high temperatures over 800°C to reach the combustion of highly concentrated VOCs. Despite the high cost of this technique, the heat released by incineration may be recovered, and also various reaction intermediates (such as NO_x) can be generated during the combustion (Tomatis et al., 2016).

12.4 NMOs for photocatalytic oxidation of VOCs

The photocatalytic oxidation process is another advanced oxidation technology that is highly effective and economical for VOC removal (Fig. 12.4). Commonly used NMOs for photo oxidation of VOCs include TiO_2, ZnO, MnO_2, $BaWO_4$, $BiFeO_3$, and so on (Almaie et al., 2022).

In photocatalytic oxidation, effective separation and migration of excited electron−hole pairs and mass transfer of VOCs from the gas phase to the surface of a photocatalyst are key processes. Among the most investigated NMOs, the modification of TiO_2 as a high photocatalytic response under UV light to remove gas-phase mixed VOCs has recently gained much attention (Shayegan et al., 2018). In the past few decades, TiO_2 has been thoroughly considered owing to excellent properties such as its suitable bandgap energy, nontoxicity, abundance, relatively low price, high chemical stability, strong oxidizing power, and good photocatalytic activity (Vatanpour et al., 2017). It should be noted that oxide-based catalysts

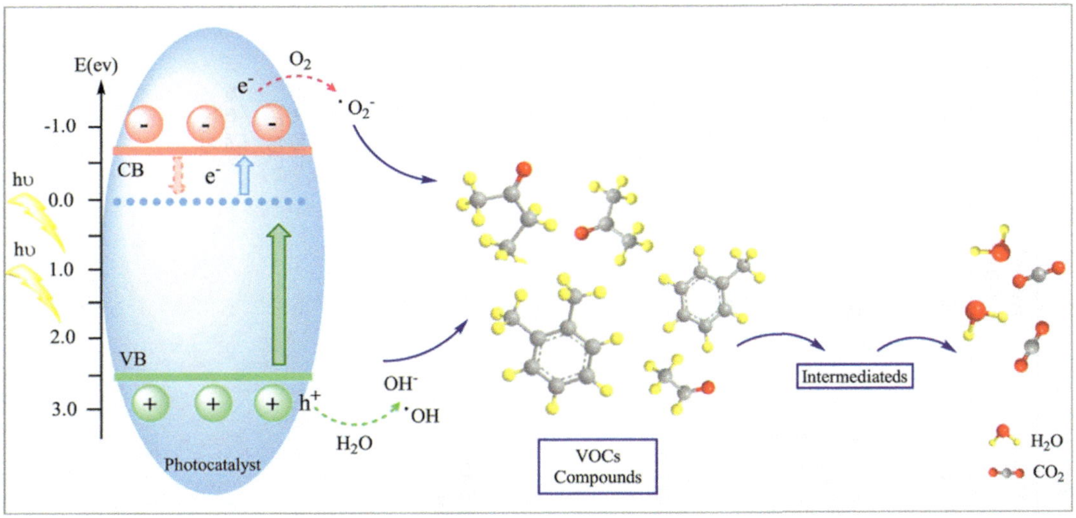

FIGURE 12.4 General mechanism of photocatalytic oxidation of VOCs (Almaie et al., 2022).

(especially TiO_2) for VOC oxidation rely on the effective properties, morphologies of supports, and the formation of active sites for VOC oxidation. For example, single metal oxide can degrade VOCs and UV light promotes titanium dioxide to degrade 17 molecules in mixture (Stucchi et al., 2018). Different $(TiO_2)_{(n)}$ clusters were applied for photocatalytic degradation of styrene (Wang et al., 2015).

ZnO nanoparticles are one of the most promising photocatalysts for removing VOCs due to their wide bandgap (3.37 eV), low cost, and good physical and chemical properties (Mahmodi et al., 2013). On the basis of that, it is proposed that the photocatalytic activities of ZnO semiconductors in the field of the PCO possess are controlled by various factors such as surface area, pore structure, supports, and relative humidity. Thus, more research is required to optimize the photocatalytic characteristics ZnO-supported nanocomposites for VOC removal for large scale industry applications. Examples of TiO-based photocatalysts for VOC removal under UV and visible lights are presented in Tables 12.2 and 12.3, respectively. Table 12.4 lists examples of ZnO-based photocatalysts for VOC removal.

One of the effective methods used to enhance the efficiency of photocatalytic activity is to combine metals with NMOs. For example, adding Co to TiO_2 significantly improved its photo oxidation of toluene (Almomani et al., 2020). Pt was deposited in microporous TiO_2 to improve its photocatalytic degradation of toluene (Li et al., 2018). Ag–TiO_2/$Sr_4Al_{14}O_2$:Eu^{2+}, Dy^{3+} photocatalyst was prepared to remove toluene (Eun et al., 2021). Hot electron-induced photoactivation promoted UV-vis-IR light-driven on a Pt/CeO_2 nanocomposite for benzene catalytic oxidation (Mao et al., 2017). Mg doping considerably promotes the photocatalytic activity of MnOx for the oxidation of benzene with full solar spectrum irradiation (Fang et al., 2017).

Combining NMOs with other materials can also obtain a higher efficiency. Nitrogen doping and pure TiO_2 nano-catalysts coated on glass beads significantly increased the removal efficiency of ethylbenzene under visible light irradiation (Kamaei et al., 2018).

TABLE 12.2 Summary of UV active photocatalytic degradation of VOCs by TiO_2-based nanocomposites.

Photocatalyst	Support	VOC pollutants	Light source	Degradation efficiency (%)/Time	Humidity relative (%)	References
TiO_2	Activated carbon fiber nickel foam filter	MEK	Low-pressure mercury UV lamp	TiO_2/ACF: 62.2 TiO_2/NFF: 36.3 (15–19 h)	20	Mamaghani et al. (2021)
F Doped TiO_2	Reduced GO, glass, carbon cloth, Ni foam	Toluene, ozone	Vacuum ultraviolet (VUV)	F–TiO_2/rGO 90.8, 99.7 F–TiO_2/G 77.3, 97.1 F–TiO_2/CC 76.6, 94.3 F–TiO_2/NiF 83, 92.8 (150 min)	50	Zhang et al. (2021)
Pd-loaded TiO_2	–	Benzene	UV light	65 (120 min)	15	Selishchev et al. (2021)
Four commercial TiO_2 (P25, PC500, UV100, and S5-300A)	Ni foam	MEK, toluene	UV light	Over 28 (0.02–0.1 s)	0–50	Mamaghani et al. (2018)
Fluorinated Ce-doped TiO_2	–	MEK	UV light (3.6 mW/cm^2)	80 (3 h)	60	Shayegan et al. (2020a)
$TiO_2/H_2Ti_3O_7$ composite	–	Toluene, acetone	UV light	For toluene 91.5 and for acetone 92 (3 h)	–	Lyu et al. (2020)
TiO_2/Pd	–	n-octane isooctane n-hexane cyclohexane	UV–Vis radiation range (200–800 nm)	90 (27 s)	30–80	Fujimoto et al. (2017)
$Cu_2O/CuO/TiO_2$	Polyurethane foam	Acetone	UV–Vis	86.24 130 min	–	Fujimoto et al. (2017)
Noble metals (Ag, Pd, Pt) doped on TiO_2 bentonite/M – TiO_2 nanocomposites	Clay	Chlorobenzene, benzaldehyde	UV lamp	80–90 140 min	–	Mishra et al. (2018)
Au doped mixed TiO_2–CeO_2 oxide	–	2-propanol	Hg lamp	60 80 min	–	Fiorenza et al. (2016)

C-doped TiO_2	Nickel foam	MEK	UV lamp	94 (0.025–0.076 s)	0–60	Shayegan et al. (2020b)
N and S Co-doped TiO_2 nanotube	–	VOCs degradation	UV lamp (9 W)	94 (3 h)	55–70	Yen et al. (2020)
TiO_2 immobilized	Nonwoven fabric	2-propanol	UV lamp (8 W)	75 (20 h)	–	Yasui et al. (2021)
Fluorinated Fe –TiO_2	–	MEK	UV lamp	80 (0.076 s)	0–60	Shayegan et al. (2021)
Nano- TiO_2	–	Methyl isobutyl ketone, MEK, acetone	UVC lamp	Conversion and mineralization of ketones	25–75	Assadi et al. (2021)
TiO_2	–	Ethylbenzene	UV lamp (100 W)	82.2–98 50 h	26–80	Saldanha et al. (2021)
Pd modified TiO_2	–	Formaldehyde –toluene/xylene mixtures	UV lamp (365 nm)	Generated CO_2 3 h	50 and 100	Wu et al. (2021)
TiO_2	Nanodiamond	Toluene	UV lamp (4 W, 365 nm)	10 35 h	–	Kim et al. (2011)
Mn doped TiO_2	–	Benzene	Vacuum ultraviolet (254 nm)	82 (300 min)	50	Shu et al. (2018)
TiO_2-WO_3	ZSM-5 zeolite	Xylene	UVA (356 nm)	47.80 (400 min)	35	Rangkooy et al. (2019)
Mn^{2+} grafted TiO_2	Carbon cloth	Toluene	Vacuum ultraviolet (4 W, 185 nm)	98 (20 min)	50	Zhao et al. (2022)
Ag_2CO_3– TiO_2 –GO	Blended in PVDF to fabricate electrospun fiber mat	Methanol acetaldehyde	UV lamp	Methanol (100 in 40 min) acetaldehyde (100 in 15 min)	–	Yang et al. (2020)
Rare-earth single atoms decorated 2D-TiO_2 nanosheets	Glass plate	O-xylene	Xenon lamp (400 W)	90 (1 h)	–	Chen et al. (2022)

TABLE 12.3 Summary of visible-light active photocatalytic degradation of VOCs by TiO_2-based nanocomposites.

Photocatalyst	Preparation method	VOC pollutants	Degradation efficiency (%)	References
CQDs/TiO_2	Hydrothermal	Benzene, p-xylene, and toluene	64 (total)	Mahmood et al. (2021)
$BiVO_4$/TiO_2	Sol-gel and hydrothermal	Benzene (260 mg/L)	41	Hu et al. (2018)
TiO_2/SiO_2/Bi_2O_3	Co-precipitation	Benzene (25 mg/L)	33	Ren et al. (2015)
Fe-doped TiO_2	Sol-gel	Toluene (370 mg/L)	96.5	Sun et al. (2012)
N-doped TiO_2	Sol-gel	Toluene (NA)	82.39	Kannangara et al. (2018)
Bi–Zn co-doped TiO_2 nanotubes	Solvothermal	Toluene (250 mg/L)	93	Li et al. (2017)
Cu_2O/TiO_2	Solvothermal	Toluene (45.5 mg/L)	62.94	Cheng et al. (2018)
Ni–TiO_2/SnO_2	Impregnation	Toluene (177 mg/L)	>50	Khan and Kim (2009)
F–TiO_2/SiO_2	Sol-gel	Toluene (60 mg/L)	91.2	Qiu et al. (2018)
Mesoporous TiO_2	Evaporation induced self-assembly	Formaldehyde (NA)	95.8	Li et al. (2014)
Ag-doped TiO_2	Hydrothermal	Formaldehyde (1.22–2.04 mg/L)	94.9	Gao et al. (2015)
rGO/TiO_2	Hydrothermal	Formaldehyde (0.5 mg/L)	88.3	Yu et al. (2018)
N-doped TiO_2	Hydrothermal	Formaldehyde (0.8 mg/L)	82.9	Qu et al. (2009)
Pt-doped TiO_2	Sol-gel	NOx (39–41 mg/L)	50	Hu et al. (2015)
TiO_2/hydroxyapatite	Hydrothermal	NOx (420 μg/L)	44.6	Yao et al. (2017)
UO_2^{2+}/TiO_2	Impregnation	Ethanol, propan-2-ol, acetone, cyclohexane (500 mg/L)	99	Selishchev et al. (2019)
g-C_3N_4/Ag–TiO_2	pH-adjusted chemical reduction	Acetaldehyde (25 mg/L)	69.5	Wang et al. (2021)
Ag and Ag_2O co-doped TiO_2nanosheets	Hydrothermal	Toluene (300 mg/L)	50	Xue et al. (2021)
Nd-TiO_2 coupled with supported ionic liquid membrane	Sol-gel	Toluene (1000 mg/L)	60–80	Li et al. (2021)

TABLE 12.4 Comparison of ZnO-based photocatalysts for VOC degradation.

Photocatalyst	Support	VOC pollutants	Light source	Degradation efficiency (%)/Time	Humidity relative (%)	References
$ZnO-SnO_2$ nanoparticles	AC	Toluene	UV-A (8 W) lamp	$ZnO-SnO_2$/ACF 40 (3 h)	–	Rangkooy et al. (2017)
TiO_2-ZnO	Expanded graphite (EG)	Chlorobenzene	UV lamp (253.7 nm, 8 W)	EG- TiO_2/ZnO 92 (42 s)	–	Rangkooy et al. (2017)
Branched Ag–ZnO nanocomposite	–	Acetaldehyde dye (RhB)	UV light (125 W)	CO_2 generation concentration (1100 mg/L) 99 photocatalytic activity (16 h)	–	Huang et al. (2015)
ZnO	rGO	Acetaldehyde	UV lamp (200 W, $\lambda = 300-450$ nm)	ZnO-rGO composite 50 (4 h)	–	Chen et al. (2015)
ZnO	Silica-GO/ fiberglass	Benzene	UV lamp (8 W) (365 nm and 1.52 mW/cm^2)	ZnO/Silica-GO/ Fiberglass 87.2 (210 min)	35	Jafari et al. (2018)
Nanodiamond-decorated ZnO	–	Toluene	UV-365 light	ND-decorated ZnO around 100 (2 h)	–	Liu et al. (2019)
ZnO	Zeolite	Benzene toluene ethylbenzene xylene	–	ZnO/Zeolite over 90 (0.1–0.4 s)	25–35	Shojaei et al. (2021)
$ZnO-TiO_2$ nanoparticles	AC	Xylene	UV-C (8 W) lamp	95.1 (120 min)	35	Rangkooy et al. (2020)
ZnO/Ag/cellulose composite	–	Phenol benzene	UV light (245 nm, 6 W)	ZnO/Ag/cellulose composite phenol and benzene degraded around 100, 19.8 (2.5 h)	–	Zou et al. (2019)
ZnO	Perlite granules	n-hexane	UV-254 light	95 (14 h)	–	Saucedo-Lucero and Arriaga (2015)
Nano-ZnO	Granular activated carbon	Benzene	UV light (365 nm, 4 W)	70.24 (5 h)	30–35	Jafari et al. (2018)

(*Continued*)

TABLE 12.4 Comparison of ZnO-based photocatalysts for VOC degradation.—cont'd

Photocatalyst	Support	VOC pollutants	Light source	Degradation efficiency (%)/Time	Humidity relative (%)	References
Nano-ZnO	–	Acetone	UV lamp (320 –400 nm)	58.7 (140 s)	–	Xie et al. (2012)
Anisotropic Au–ZnO	–	n-hexane	Visible light	>99 (405 nm) 36 (460 nm)	–	Bueno-Alejo et al. (2021)
ZnS/ZnO/$CuFeS_2$ films	–	Formaldehyde	Visible light	55 (1.5 h)	–	Luévano-Hipólito and Torres-Martínez (2021)
ZnO coated on glass plates	–	Benzene	Visible light	57 (4 h)	–	Jafari et al. (2018)
Ag/ZnO, Cd/ZnO Pb/ZnO	–	Chlorobenzene	Visible light	100 (2 h)	–	Nasrollahi et al. (2021)

$CeMnxOy/TiO_2$ nanocomposite showed Highly efficient UV-Vis-infrared catalytic purification of benzene in the full solar spectrum region (Liu et al., 2016). Zinc oxide (ZnO) nano-spines were sprouted on activated-carbon nanofibers (I@ZnO/ACNFs) for photocatalytic VOC oxidation (Kang et al., 2022). Silica-functionalized graphene oxide/ZnO coated on fiberglass was synthesis for application in photocatalytic removal of gaseous benzene (Jafari et al., 2018).

In summary, NMO-based photocatalytic oxidation technology is an effective VOC removal method due to its low cost, mild conditions, nontoxicity, and photo corrosion resistance. Despite these advantages, the deactivation of photocatalysts, insufficient active sites, and the generation of unwanted by-products are the major limitations on the application of NMOs in VOC abatement (Gholami et al., 2020; Khataee et al., 2019; Rad et al., 2020; Thanh Truc et al., 2019).

12.5 NMO-base adsorbents for VOC removal

In addition to catalysis, NMOs have also been used in VOC adsorption. Adsorption is an economical method that has received much attention owing to its potential to recover VOCs and reuse adsorbents. In addition, adsorption of VOCs can also facilitate their oxidation by NMOs (Fig. 12.5). Selecting an appropriate NMO-based adsorbent is crucial for efficient adsorption. An ideal adsorbent for VOC removal is expected to possess (a) high adsorption

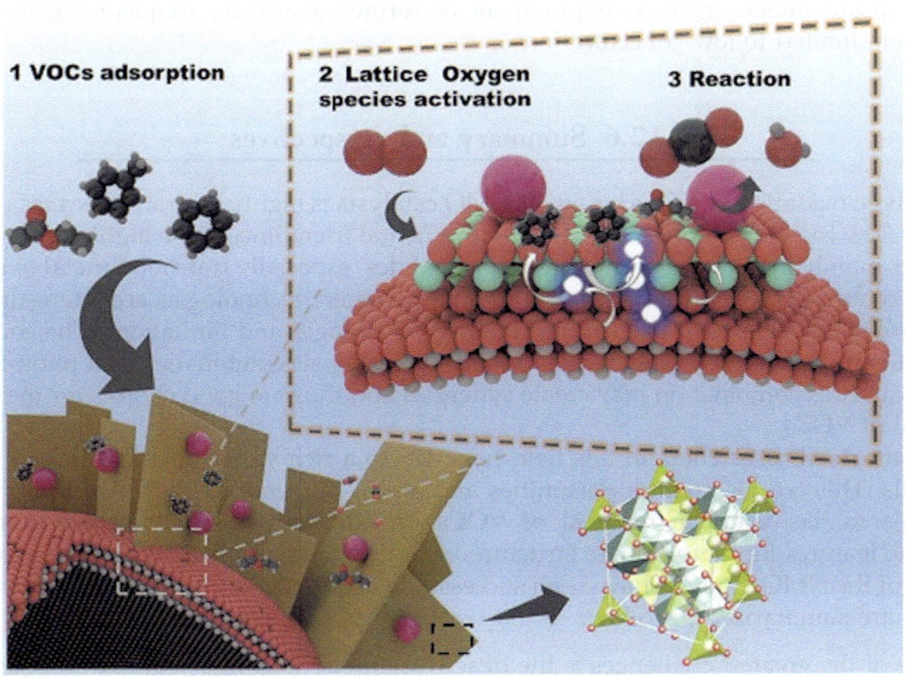

FIGURE 12.5 Adsorption facilitates catalytic oxidation of VOCs by NOMs (He et al., 2022).

capacity and (b) fast adsorption rate, which can be evaluated by adsorption isotherms and kinetics; (c) high hydrophobicity, to overcome the competitive adsorption from water vapor commonly present in the flue gas; (d) easy regeneration and (e) high thermal and hydrothermal stability, to ensure the cyclic adsorption-regeneration usage (Yang et al., 2019). To develop effective adsorbents for certain VOCs, modification or functionalization is commonly applied to improve the specific properties of the adsorbents.

Several NMOs such as Si, Al, Fe, Ce, and Ti oxides have been identified as effective VOC adsorbents (Almaie et al., 2022). For example, the presence of ultrahigh silica zeolites in the cellulose acetate materials increases their adsorption capacity for VOCs (Ojstršek et al., 2020). The adsorption ability of functionalized ordered mesoporous silica to VOCs follows the sequence of acetone $>$ ethyl acetate $>$ toluene (Chen et al., 2022). Beta zeolite nanostructures with a high SiO_2/Al_2O_3 ratio maintain excellent adsorption performance for toluene (Wang et al., 2021). The addition of Al_2O_3 as the support material strengthens the adsorption of toluene and oxygen on $CoMn_2O_4$, which promotes their activation (Shi et al., 2022). Packed-bed of alpha-Al_2O_3 can effectively adsorb acetaldehyde under atmospheric nonthermal plasma (Klett et al., 2014). Fe_3O_4 magnetic nanoparticles decorated with perylene bisimides can adsorb aromatic VOCs (de Las Nieves Pina et al., 2018). Styrene can be readily adsorbed onto (TiO_2) (n) clusters through its vinyl group with strong chemisorption (Wang et al., 2015). The adsorption of VOCs by NMO-based adsorbents can be further enhanced by increasing specific surface area, pore-volume, and surface functional groups, and reducing the pore size. The advantages of VOC removal by adsorption on NMOs are low energy consumption and running cost, facile operation, and no byproducts. However, there are also several disadvantages such as requirement of further treatment, frequent regeneration of adsorbent, limited to low concentrations.

12.6 Summary and perspectives

Catalytic oxidation of VOCs based on NMO catalysts is highly desirable because of energy savings, low cost, operation safety and environmental friendliness. The highly efficient NMO catalysts mainly include nanostructured metal oxides, especially transition metal oxides, and their derivatives. Hybrid processes that combine different technologies are interesting topics for future study since each of them has its own advantages and limitations. The coupling of multiple removal technologies, such as adsorption-catalysis combination and photocatalysis-thermocatalysis combination may create synergies and thus are good choices for the effective removal of VOCs.

The efforts of researchers in this field have led to a rich variety of NMO-based catalytic materials. This creates great opportunities and also a tremendous challenge to apply these materials in the complete removal of VOCs. In addition to structural advantages, the welcome features for a multiphase structure including the synergistic effect among different species of the NMO catalysts have been successfully acquired. Some of the challenges/opportunities are summarized as follows.

(1) One of the greatest challenges is the deactivation of NMO catalysts under realistic process conditions. In some industrial applications such as the catalytic combustion widely used in the tail gas processing and the oil refining process, NMO catalysts can maintain their activities in the reaction at only 300–400°C for most of the time. However, the reaction temperature sometimes may reach as high as nearly 1000°C at one moment owing to the inevitably excessive combustion or other reasons. Although this situation only lasts for several seconds, the momentary process brings serious irreversible degradation of NMO catalysts. So in the preparation process of the practical catalysts, NMOs must be aged under harsh pretreatment conditions before being used in order to make sure that they are durable and stable enough. Further investigations thus are needed to meet this challenge.

(2) The mechanistic understanding of the underlying chemistry of the NMO catalytic systems. For example, the synergetic effect and interface boundary sites may be valuable for the development of the highly efficient and stable NMO catalysts with interesting architectures and tailored functionalities. Currently, the mechanisms for catalytic oxidation of small VOC molecules, such as formaldehyde, have been well documented, and most of the intermediate products can be determined. However, the mechanisms for catalytic oxidation of large VOC molecules are yet to be fully investigated because the reaction pathways are too complicated. The investigation of reaction mechanisms can accelerate the fast development of characterization techniques, especially in situ analysis instruments for NMO catalysts.

(3) Only few studies have been carried out in terms of NMO-based catalytic oxidation of mixed VOCs from industrial processes and highly diluted VOCs from indoor environments. The performance and mechanism of catalytic oxidation of these two types of VOCs by NMOs are quite different from those of single VOC with high concentrations. Therefore, more academic and industrial efforts should be devoted before the extensive application of NMOs as catalysts for VOC removal.

References

Abbasi, Z., Haghighi, M., Fatehifar, E., Saedy, S., 2011. Synthesis and physicochemical characterization of nanostructured Pt/CeO_2 catalyst used for total oxidation of toluene. International Journal of Chemical Reactor Engineering 9.

Almaie, S., Vatanpour, V., Rasoulifard, M.H., Koyuncu, I., 2022. Volatile organic compounds (VOCs) removal by photocatalysts: a review. Chemosphere 306, 135655.

Almomani, F., Bhosale, R., Shawaqfah, M., 2020. Solar oxidation of toluene over Co doped nano-catalyst. Chemosphere 255, 126878.

Anwer, H., Ali, M., Lee, S., Jeong, S.H., Park, J.W., 2021. Simulating alveoli-inspired air pockets in a $ZnO/NiMoO_4/C_3N_4$ catalyst filter for toluene entrapment and photodecomposition. Journal of Hazardous Materials 409, 124497.

Assadi, H., Armaghan, F., Taheri, R.A., 2021. Photocatalytic oxidation of ketone group volatile organic compounds in an intensified fluidized bed reactor using nano-TiO_2/UV process: an experimental and modeling study. Chemical Engineering and Processing-Process Intensification 161, 108312.

Bonelli, R., Albonetti, S., Morandi, V., Ortolani, L., Riccobene, P.M., Scire, S., Zacchini, S., 2011. Design of nano-sized FeO_x and Au/FeO_x catalysts supported on CeO_2 for total oxidation of VOC. Appl Catal a-Gen 395, 10–18.

Bueno-Alejo, C.J., Graus, J., Arenal, R., Lafuente, M., Bottega-Pergher, B., Hueso, J.L., 2021. Anisotropic Au-ZnO photocatalyst for the visible-light expanded oxidation of n-hexane. Catalysis Today 362, 97–103.

Chang, T., Shen, Z.X., Ma, C.L., Lu, J.Q., Huang, Y., Veerapandian, S.K.P., De Geyter, N., Morent, R., 2021. Process optimization of plasma-catalytic formaldehyde removal using MnO_x-Fe_2O_3 catalysts by response surface methodology. Journal of Environmental Chemical Engineering 9.

Chang, T., Wang, Y., Wang, Y., Zhao, Z., Shen, Z., Huang, Y., Veerapandian, S.K.P., De Geyter, N., Wang, C., Chen, Q., Morent, R., 2022. A critical review on plasma-catalytic removal of VOCs: catalyst development, process parameters and synergetic reaction mechanism. The Science of the Total Environment 828, 154290.

Chang, T., Ma, C., Shen, Z., Veerapandian, S.K.P., Huang, Y., De Geyter, N., Morent, R., 2021. Mn-based catalysts for post non-thermal plasma catalytic abatement of VOCs: a review on experiments, simulations and modeling. Plasma Chemistry and Plasma Processing 41, 1239–1278.

Chen, H., Mu, X., Ma, J., Yao, M., Liu, F., 2022. Performance evaluation of functionalized ordered mesoporous silica for VOCs adsorption by molecular dynamics simulation. Arabian Journal of Chemistry 15, 104192.

Chen, L., Liao, Y., Xin, S., Song, X., Liu, G., Ma, X., 2020. Simultaneous removal of NO and volatile organic compounds (VOCs) by Ce/Mo doping-modified selective catalytic reduction (SCR) catalysts in denitrification zone of coal-fired flue gas. Fuel 262, 116485.

Chen, Y.-C., Katsumata, K.-ichi, Chiu, Y.-H., Okada, K., Matsushita, N., Hsu, Y.-J., 2015. ZnO–graphene composites as practical photocatalysts for gaseous acetaldehyde degradation and electrolytic water oxidation. Applied Catalysis A: General 490, 1–9.

Cheng, Y.Q., Gao, X., Zhang, X.X., Su, J.X., Wang, G.Q., Wang, L.Y., 2018. Synthesis of a TiO_2-Cu_2O composite catalyst with enhanced visible light photocatalytic activity for gas-phase toluene. New Journal of Chemistry 42, 9252–9259.

de Las Nieves Pina, M., Rodriguez, P., Gutierrez, M.S., Quinonero, D., Morey, J., Frontera, A., 2018. Adsorption and quantification of volatile organic compounds (VOCs) by using hybrid magnetic nanoparticles. Chemistry 24, 12820–12826.

Deng, J.G., He, S.N., Xie, S.H., Yang, H.G., Liu, Y.X., Guo, G.S., Dai, H.X., 2015. Ultralow loading of silver nanoparticles on Mn_2O_3 nanowires derived with molten salts: a high-efficiency catalyst for the oxidative removal of toluene. Environmental Science & Technology 49, 11089–11095.

Du, P., Wang, R.Y., Deng, B.H., He, X., Long, Y.Z., Yang, C., Wang, Z.W., Ge, B.H., Huang, K., Zhang, R., Lei, M., Wu, H., 2022. In-situ Joule-heating drives rapid and on-demand catalytic VOCs removal with ultralow energy consumption. Nano Energy 102.

Enesca, A., Cazan, C., 2020. Volatile organic compounds (VOCs) removal from indoor air by heterostructures/composites/doped photocatalysts: a mini-review. Nanomaterials 10.

Eric, M.,G., Farin, B., Damien, P., Debecker, 2014. Total oxidation of propane with a nano-RuO_2/TiO_2 catalyst. Applied Catalysis, A. General: An International Journal Devoted to Catalytic Science and Its Applications 481, 11–18.

Eun, S.R., Mavengere, S., Kim, J.S., 2021. Preparation of Ag-$TiO_2/Sr_4Al_{14}O_{25}$:Eu^{2+},Dy^{3+} photocatalyst on phosphor beads and its photoreaction characteristics. Catalysts 11.

Fang, S.M., Li, Y.Z., Yang, Y., Chen, J., Liu, H.H., Zhao, X.J., 2017. Mg-doped OMS-2 nanorods: a highly efficient catalyst for purification of volatile organic compounds with full solar spectrum irradiation. Environmental Science: Nano 4, 1798–1807.

Fiorenza, R., Bellardita, M., D'Urso, L., Compagnini, G., Palmisano, L., Scire, S., 2016. Au/TiO_2-CeO_2 catalysts for photocatalytic water splitting and VOCs oxidation reactions. Catalysts 6.

Fujimoto, T.M., Ponczek, M., Rochetto, U.L., Landers, R., Tomaz, E., 2017. Photocatalytic oxidation of selected gas-phase VOCs using UV light, TiO_2, and TiO_2/Pd. Environmental Science and Pollution Research 24, 6390–6396.

Fuku, K., Goto, M., Sakano, T., Kamegawa, T., Mori, K., Yamashita, H., 2013. Efficient degradation of CO and acetaldehyde using nano-sized Pt catalysts supported on CeO_2 and CeO_2/ZSM-5 composite. Catalysis Today 201, 57–61.

Gao, L.K., Gan, W.T., Xiao, S.L., Zhan, X.X., Li, J., 2015. Enhancement of photo-catalytic degradation of formaldehyde through loading anatase TiO_2 and silver nanoparticle films on wood substrates. RSC Advances 5, 52985–52992.

Genty, E., Cousin, R., Capelle, S., Gennequin, C., Siffert, S., 2012. Catalytic oxidation of toluene and CO over nanocatalysts derived from hydrotalcite-like compounds ($X_6^{2+}Al_2^{3+}$): effect of the bivalent cation. European Journal of Inorganic Chemistry 2012, 2802–2811.

Gholami, P., Khataee, A., Soltani, R.D.C., Dinpazhoh, L., Bhatnagar, A., 2020. Photocatalytic degradation of gemifloxacin antibiotic using Zn-Co-LDH@biochar nanocomposite. Journal of Hazardous Materials 382, 121070.

Guan, Y.A., Zhou, Y.T., Jiang, C.H., Xu, X.X., Yang, Z.M., Zhang, J.S., Fan, X.L., Jiao, Y.L., 2021. Catalytic combustion of volatile organic compounds (VOCs) over structured Co_3O_4 nano-flowers on silicalite-1/SiC foam catalysts. Microporous and Mesoporous Materials 323, 111173.

He, C., Cheng, J., Zhang, X., Douthwaite, M., Pattisson, S., Hao, Z., 2019. Recent advances in the catalytic oxidation of volatile organic compounds: a review based on pollutant sorts and sources. Chemical Reviews 119, 4471–4568.

He, J., Zheng, F., Zhou, Y., Li, X., Wang, Y., Xiao, J., Li, Y., Chen, D., Lu, J., 2022. Catalytic oxidation of VOCs over 3D@2D Pd/$CoMn_2O_4$ nanosheets supported on hollow Al_2O_3 microspheres. Journal of Colloid and Interface Science 613, 155–167.

Hosseini, S.A., Niaei, A., Salari, D., Nabavi, S.R., 2012. Nanocrystalline AMn_2O_4 (A=Co, Ni, Cu) spinels for remediation of volatile organic compounds—synthesis, characterization and catalytic performance. Ceramics International 38, 1655–1661.

Hu, Y., Chen, W., Fu, J., Ba, M., Sun, F., Zhang, P., Zou, J., 2018. Hydrothermal synthesis of $BiVO_4/TiO_2$ composites and their application for degradation of gaseous benzene under visible light irradiation. Applied Surface Science 436, 319–326.

Hu, Y., Song, X., Jiang, S., Wei, C., 2015. Enhanced photocatalytic activity of Pt-doped TiO_2 for NOx oxidation both under UV and visible light irradiation: a synergistic effect of lattice Pt^{4+} and surface PtO. Chemical Engineering Journal 274, 102–112.

Huang, H., Ye, D., Leung, D.Y.C., Feng, F., Guan, X., 2011. Byproducts and pathways of toluene destruction via plasma-catalysis. Journal of Molecular Catalysis A: Chemical 336, 87–93.

Huang, Q.L., Zhang, Q.T., Yuan, S.S., Zhang, Y.C., Zhang, M., 2015. One-pot facile synthesis of branched Ag-ZnO heterojunction nanostructure as highly efficient photocatalytic catalyst. Applied Surface Science 353, 949–957.

Jafari, A.J., Kalantari, R.R., Kermani, M., Firooz, M.H., 2018. Photocatalytic degradation data of benzene and toluene by ZnO coated on glass plates under simulated sunlight. Data in Brief 20, 490–495.

Jafari, A.J., Arfaeinia, H., Badi, M.Y., Kalantary, R.R., Kermani, M., 2018. Ozone-Assisted photocatalytic degradation of benzene using nano-zinc oxide impregnated granular activated carbon (ZnO–GAC) in a continuous fluidized bed reactor. Environmental Progress & Sustainable Energy 38, 13082.

Jafari, A.J., Kalantary, R.R., Esrafili, A., Arfaeinia, H., 2018. Synthesis of silica-functionalized graphene oxide/ZnO coated on fiberglass and its application in photocatalytic removal of gaseous benzene. Process Safety and Environmental Protection 116, 377–387.

Jiang, N., Qiu, C., Guo, L., Shang, K., Lu, N., Li, J., Zhang, Y., Wu, Y., 2019. Plasma-catalytic destruction of xylene over Ag-Mn mixed oxides in a pulsed sliding discharge reactor. Journal of Hazardous Materials 369, 611–620.

Jung, S.C., Nahm, S.W., Park, Y.K., Kim, S.C., 2014. Additive effect of nano-size platinum to pretreated iron based catalyst on complete oxidation of toluene. Journal of Nanoscience and Nanotechnology 14, 6390–6393.

Kamaei, M., Rashedi, H., Dastgheib, S.M.M., Tasharrofi, S., 2018. Comparing photocatalytic degradation of gaseous ethylbenzene using N-doped and pure TiO_2 nano-catalysts coated on glass beads under both UV and visible light irradiation. Catalysts 8, 466.

Kamal, M.S., Razzak, S.A., Hossain, M.M., 2016. Catalytic oxidation of volatile organic compounds (VOCs) – a review. Atmospheric Environment 140, 117–134.

Kang, S., Park, D.H., Hwang, J., 2022. Hierarchical ZnO nano-spines grown on a carbon fiber seed layer for efficient VOC removal and airborne virus and bacteria inactivation. Journal of Hazardous Materials 424, 127262.

Kannangara, Y.Y., Wijesena, R., Rajapakse, R.M.G., de Silva, K.M.N., 2018. Heterogeneous photocatalytic degradation of toluene in static environment employing thin films of nitrogen-doped nano-titanium dioxide. International Nano Letters 8, 31–39.

Khan, R., Kim, T.-J., 2009. Preparation and application of visible-light-responsive Ni-doped and SnO_2-coupled TiO_2 nanocomposite photocatalysts. Journal of Hazardous Materials 163, 1179–1184.

Khataee, A., Rad, T.S., Nikzat, S., Hassani, A., Aslan, M.H., Kobya, M., Demirbas, E., 2019. Fabrication of NiFe layered double hydroxide/reduced graphene oxide (NiFe-LDH/rGO) nanocomposite with enhanced sonophotocatalytic activity for the degradation of moxifloxacin. Chemical Engineering Journal 375, 122102.

Kim, K.D., Dey, N.K., Seo, H.O., Kim, Y.D., Lim, D.C., Lee, M., 2011. Photocatalytic decomposition of toluene by nanodiamond-supported TiO_2 prepared using atomic layer deposition. Applied Catalysis A: General 408, 148–155.

Klett, C., Duten, X., Tieng, S., Touchard, S., Jestin, P., Hassouni, K., Vega-Gonzalez, A., 2014. Acetaldehyde removal using an atmospheric non-thermal plasma combined with a packed bed: role of the adsorption process. Journal of Hazardous Materials 279, 356–364.

Kustov, A.L., Tkachenko, O.P., Kustov, L.M., Romanovsky, B.V., 2011. Lanthanum cobaltite perovskite supported onto mesoporous zirconium dioxide: nature of active sites of VOC oxidation. Environment International 37, 1053–1056.

Lee, H., Song, M.Y., Ryu, S., Park, Y.K., 2020. Acetaldehyde oxidation under high humidity using a catalytic non-thermal plasma system over Mn-loaded Y zeolites. Materials Letters 262.

Li, C.-y, Jia, Y.-r, Zhang, X.-c, Zhang, S.-y, Tang, A.-d, 2014. Photocatalytic degradation of formaldehyde using mesoporous TiO_2 prepared by evaporation-induced self-assembly. Journal of Central South University 21, 4066–4070.

Li, J.L., Li, B.X., Sui, G.Z., Du, L.J., Zhuang, Y., Zhang, Y.L., Zou, Y.F., 2021. Removal of volatile organic compounds from air using supported ionic liquid membrane containing ultraviolet-visible light-driven Nd-TiO_2 nanoparticles. Journal of Molecular Structure 1231.

Li, J., Wang, Y.H., Tian, Y.Y., He, X., Yang, P.P., Yuan, M.H., Cao, Y.Q., Lyu, J.Z., 2018. Crystallization of microporous TiO_2 through photochemical deposition of Pt for photocatalytic degradation of volatile organic compounds. Environmental Science and Pollution Research 25, 15662–15670.

Li, J., Zhao, J.J., Hu, J.N., Shi, Y.J., Wang, F., Deng, S., Cui, Y.T., 2022. Effects of the metal ionic radius at lattice site on surface active oxygen and reducibility over doped mesoporous ZSM-5 for catalytic oxidation VOC. Microporous and Mesoporous Materials 339.

Li, J.-J., Cai, S.-C., Xu, Z., Chen, X., Chen, J., Jia, H.-P., Chen, J., 2017. Solvothermal syntheses of Bi and Zn co-doped TiO_2 with enhanced electron-hole separation and efficient photodegradation of gaseous toluene under visible-light. Journal of Hazardous Materials 325, 261–270.

Li, Y.Z., Fan, Z.Y., Shi, J.W., Liu, Z.Y., Shangguan, W.F., 2014. Post plasma-catalysis for VOCs degradation over different phase structure MnO_2 catalysts. Chemical Engineering Journal 241, 251–258.

Liu, B., Ji, J., Zhang, B., Huang, W., Gan, Y., Leung, D.Y.C., Huang, H., 2022. Catalytic ozonation of VOCs at low temperature: a comprehensive review. Journal of Hazardous Materials 422, 126847.

Liu, H.H., Li, Y.Z., Yang, Y., Mao, M.Y., Zeng, M., Lan, L., Yun, L., Zhao, X.J., 2016. Highly efficient UV-Vis-infrared catalytic purification of benzene on $CeMn_xO_y/TiO_2$ nanocomposite, caused by its high thermocatalytic activity and strong absorption in the full solar spectrum region. Journal of Materials Chemistry A 4, 9890–9899.

Liu, J., Wang, P.L., Qu, W.Q., Li, H.R., Shi, L.Y., Zhang, D.S., 2019. Nanodiamond-decorated ZnO catalysts with enhanced photocorrosion-resistance for photocatalytic degradation of gaseous toluene. Applied Catalysis B: Environmental 257, 117880.

Liu, L., Zhang, Z., Das, S., Kawi, S., 2019. Reforming of tar from biomass gasification in a hybrid catalysis-plasma system: a review. Applied Catalysis B: Environmental 250, 250–272.

Liu, Q., Wu, X., Han, R., Pang, C., Ma, D., 2021. Confined growth of Cu-doped CeO_2 nanostructures in boron-rich carbon frameworks for acetone oxidation. ACS Applied Nano Materials 4, 10275–10286.

Liu, S.Y., Mei, D.H., Nahil, M.A., Gadkari, S., Gu, S., Williams, P.T., Tu, X., 2017. Hybrid plasma-catalytic steam reforming of toluene as a biomass tar model compound over Ni/Al_2O_3 catalysts. Fuel Processing Technology 166, 269–275.

Liu, Y., Deng, J., Xie, S., Wang, Z., Dai, H., 2016. Catalytic removal of volatile organic compounds using ordered porous transition metal oxide and supported noble metal catalysts. Chinese Journal of Catalysis 37, 1193–1205.

Lu, M.J., Huang, R., Wu, J.L., Fu, M.L., Chen, L.M., Ye, D.Q., 2015. On the performance and mechanisms of toluene removal by FeO_x/SBA-15-assisted non-thermal plasma at atmospheric pressure and room temperature. Catalysis Today 242, 274–286.

Luévano-Hipólito, E., Torres-Martínez, L.M., 2021. Earth-abundant $ZnS/ZnO/CuFeS_2$ films for air purification and solar fuels production. Materials Science in Semiconductor Processing 134, 106029.

Lyu, J.Z., Zhou, L.L., Shao, J.W., Zhou, Z., Gao, J.X., Li, J., Dong, Y.M., Wang, Z.Y., 2020. Synthesis of $TiO_2/H_2Ti_3O_7$ composite with nanoscale spiny hollow hierarchical structure for photocatalytic mineralization of VOCs. Chemical Engineering Journal 400, 125927.

Mahmodi, G., Sharifnia, S., Madani, M., Vatanpour, V., 2013. Photoreduction of carbon dioxide in the presence of H_2, H_2O and CH_4 over TiO_2 and ZnO photocatalysts. Solar Energy 97, 186–194.

Mahmood, A., Shi, G., Wang, Z., Rao, Z., Xiao, W., Xie, X., Sun, J., 2021. Carbon quantum dots-TiO_2 nanocomposite as an efficient photocatalyst for the photodegradation of aromatic ring-containing mixed VOCs: an experimental and DFT studies of adsorption and electronic structure of the interface. Journal of Hazardous Materials 401, 123402.

Mamaghani, A.H., Haghighat, F., Lee, C.S., 2021. Effect of titanium dioxide properties and support material on photocatalytic oxidation of indoor air pollutants. Building and Environment 189, 107518.

Mamaghani, A.H., Haghighat, F., Lee, C.-S., 2018. Photocatalytic degradation of VOCs on various commercial titanium dioxides: impact of operating parameters on removal efficiency and by-products generation. Building and Environment 138, 275–282.

Mao, M., Li, Y., Lv, H., Hou, J., Zeng, M., Ren, L., Huang, H., Zhao, X., 2017. Efficient UV-vis-IR light-driven thermocatalytic purification of benzene on a Pt/CeO_2 nanocomposite significantly promoted by hot electron-induced photoactivation. Environmental Science: Nano 4, 373–384.

Mishra, A., Mehta, A., Kainth, S., Basu, S., 2018. Effect of different plasmonic metals on photocatalytic degradation of volatile organic compounds (VOCs) by bentonite/M-TiO_2 nanocomposites under UV/visible light. Applied Clay Science 153, 144–153.

Muroyama, H., Nakase, R., Matsui, T., Eguchi, K., 2010. Ethanol steam reforming over Ni-based spinel oxide. International Journal of Hydrogen Energy 35, 1575–1581.

Nasrollahi, N., Ghalamchi, L., Vatanpour, V., Khataee, A., 2021. Photocatalytic-membrane technology: a critical review for membrane fouling mitigation. Journal of Industrial and Engineering Chemistry 93, 101–116.

Ojstršek, A., Fakin, D., Hribernik, S., Fakin, T., Kurečič, M., 2020. Electrospun nanofibrous composites from cellulose acetate/ultra-high silica zeolites and their potential for VOC adsorption from air. Carbohydrate Polymers 236, 116071.

Okal, J., Zawadzki, M., Kraszkiewicz, P., Adamska, K., 2018. Ru/CeO_2 catalysts for combustion of mixture of light hydrocarbons: effect of preparation method and metal salt precursors. Applied Catalysis A: General 549, 161–169.

Parvizi, N., Rahemi, N., Allahyari, S., Tasbihi, M., 2020. Plasma-catalytic degradation of BTX over ternary perovskite-type $La_{1-x}(Co, Zn, Mg, Ba)_xMnO_3$ nanocatalysts. Journal of Industrial and Engineering Chemistry 84.

Pulleri, J.K., Yearwar, D., Saravanan, G., Rayalu, S., Labhasetwar, N., 2018. Effect of morphology of platinum nanoparticles on benzene oxidation activity. Journal of Nanoscience and Nanotechnology 18, 419–425.

Qin, C., Huang, X., Dang, X., Huang, J., Teng, J., Kang, Z., 2016. Toluene removal by sequential adsorption-plasma catalytic process: effects of Ag and Mn impregnation sequence on Ag-Mn/gamma-Al_2O_3. Chemosphere 162, 125–130.

Qiu, L., Wang, Y., Li, H., Cao, G., Ouyang, F., Zhu, R., 2018. Photocatalytic oxidation of toluene on fluorine doped TiO_2/SiO_2 catalyst under simulant sunlight in a flat reactor. Catalysts 8 (12), 596.

Qu, X.G., Liu, W.X., Ma, J., Cao, W.B., 2009. Research on photodegradation of formaldehyde by nanocrystalline N-TiO_2 powders under visible light irradiation. Research on Chemical Intermediates 35, 313–320.

Rad, T.S., Ansarian, Z., Soltani, R.D.C., Khataee, A., Orooji, Y., Vafaei, F., 2020. Sonophotocatalytic activities of FeCuMg and CrCuMg LDHs: influencing factors, antibacterial effects, and intermediate determination. Journal of Hazardous Materials 399, 123062.

Rangkooy, H.A., Ghaedi, H., Jahani, F., 2019. Removal of xylene vapor pollutant from the air using new hybrid substrates of TiO_2-WO_3 nanoparticles immobilized on the ZSM-5 zeolite under UV radiation at ambient temperature: experimental towards modeling. Journal of Environmental Chemical Engineering 7.

Rangkooy, H.A., Jahani, F., Ahangar, A.S., 2020. Photocatalytic removal of xylene as a pollutant in the air using ZnO-activated carbon, TiO_2-activated carbon, and TiO_2/ZnOactivated carbon nanocomposites. Environmental Health Engineering and Management Journal 7, 41–47.

Rangkooy, H., Tanha, F., Jaafarzadeh, N., Mohammadbeigi, A., 2017. The influence of ZnO-SnO_2 nanoparticles and activated carbon on the photocatalytic degradation of toluene using continuous flow mode. Medical Gas Research 7, 260–264.

Ren, C., Qiu, W., Zhang, H., He, Z., Chen, Y., 2015. Degradation of benzene on $TiO_2/SiO_2/Bi_2O_3$ photocatalysts under UV and visible light. Journal of Molecular Catalysis A: Chemical 398, 215–222.

Saldanha, L.A.S., das Graças Santos, N.T., Tomaz, E., 2021. Photocatalytic ethylbenzene degradation associated with ozone ($TiO_2/UV/O_3$) under different percentages of catalytic coating area: evaluation of process parameters. Separation and Purification Technology 263, 118344.

Saucedo-Lucero, J.O., Arriaga, S., 2015. Study of ZnO-photocatalyst deactivation during continuous degradation of n-hexane vapors. Journal of Photochemistry and Photobiology A: Chemistry 312, 28–33.

Selishchev, D., Svintsitskiy, D., Kovtunova, L., Gerasimov, E., Gladky, A., Kozlov, D., 2021. Surface modification of TiO_2 with Pd nanoparticles for enhanced photocatalytic oxidation of benzene micropollutants. Colloids and Surfaces A: Physicochemical and Engineering Aspects 612.

Selishchev, D.S., Filippov, T.N., Lyulyukin, M.N., Kozlov, D.V., 2019. Uranyl-modified TiO_2 for complete photocatalytic oxidation of volatile organic compounds under UV and visible light. Chemical Engineering Journal 370, 1440–1449.

Shah, K.W., Li, W., 2019. A review on catalytic nanomaterials for volatile organic compounds VOC removal and their applications for healthy buildings. Nanomaterials 9.

Shang, K., Ren, J., Zhang, Q., Lu, N., Jiang, N., Li, J., 2021. Successive treatment of benzene and derived byproducts by a novel plasma catalysis-adsorption process. Journal of Environmental Chemical Engineering 9, 105767.

Shayegan, Z., Haghighat, F., Lee, C.S., 2020a. Surface fluorinated Ce-doped TiO_2 nanostructure photocatalyst: a trap and remove strategy to enhance the VOC removal from indoor air environment. Chemical Engineering Journal 401, 125932.

Shayegan, Z., Haghighat, F., Lee, C.-S., 2020b. Carbon-doped TiO_2 film to enhance visible and UV light photocatalytic degradation of indoor environment volatile organic compounds. Journal of Environmental Chemical Engineering 8, 104162.

Shayegan, Z., Haghighat, F., Lee, C.-S., 2021. Anatase/brookite biphasic surface fluorinated Fe−TiO_2 photocatalysts to enhance photocatalytic removal of VOCs under visible and UV light. Journal of Cleaner Production 287, 125462.

Shayegan, Z., Lee, C.-S., Haghighat, F., 2018. TiO_2 photocatalyst for removal of volatile organic compounds in gas phase − a review. Chemical Engineering Journal 334, 2408−2439.

Shi, Y.J., Wan, J., Kong, F.Z., Wang, Y.Y., Zhou, R.X., 2022. Influence of Pt dispersibility and chemical states on catalytic performance of Pt/CeO_2-TiO_2 catalysts for VOCs low-temperature removal. Colloids and Surfaces A: Physicochemical and Engineering Aspects 652.

Shojaei, A., Ghafourian, H., Yadegarian, L., Lari, K., Sadatipour, M.T., 2021. Removal of volatile organic compounds (VOCs) from waste air stream using ozone assisted zinc oxide (ZnO) nanoparticles coated on zeolite. Journal of Environmental Health Science and Engineering 19, 771−780.

Shu, Y.J., Ji, J., Xu, Y., Deng, J.G., Huang, H.B., He, M., Leung, D.Y.C., Wu, M.Y., Liu, S.W., Liu, S.L., Liu, G.Y., Xie, R.J., Feng, Q.Y., Zhan, Y.J., Fang, R.M., Ye, X.G., 2018. Promotional role of Mn doping on catalytic oxidation of VOCs over mesoporous TiO_2 under vacuum ultraviolet (VUV) irradiation. Applied Catalysis B: Environmental 220, 78−87.

Sihui, Z., Yang, Y., Gao, X., Yu, H., Yang, S., Zhu, D., Li, Y., 2014. Rapid degradation of toxic toluene using novel mesoporous SiO_2 doped TiO_2 nanofibers. Catalysis Today 225, 10−17.

Stucchi, M., Galli, F., Bianchi, C.L., Pirola, C., Boffito, D.C., Biasioli, F., Capucci, V., 2018. Simultaneous photodegradation of VOC mixture by TiO_2 powders. Chemosphere 193, 198−206.

Sultana, S., Ye, Z.P., Veerapandian, S.K.P., Lofberg, A., De Geyter, N., Morent, R., Giraudon, J.M., Lamonier, J.F., 2018. Synthesis and catalytic performances of K-OMS-2, Fe/K-OMS-2 and Fe-K-OMS-2 in post plasma-catalysis for dilute TCE abatement. Catalysis Today 307, 20−28.

Sun, S., Ding, J., Bao, J., Gao, C., Qi, Z., Yang, X., He, B., Li, C., 2012. Photocatalytic degradation of gaseous toluene on Fe-TiO_2 under visible light irradiation: a study on the structure, activity and deactivation mechanism. Applied Surface Science 258, 5031−5037.

Sun, Y., Li, N., Xing, X., Zhang, X., Zhang, Z., Wang, G., Cheng, J., Hao, Z., 2019. Catalytic oxidation performances of typical oxygenated volatile organic compounds (acetone and acetaldehyde) over MAlO (M = Mn, Co, Ni, Fe) hydrotalcite-derived oxides. Catalysis Today 327, 389−397.

Thanh Truc, N.T., Pham, T.-D., Van Thuan, D., Son, L.T., Tran, D.T., Nguyen, M.V., Nguyen, V.N., Dang, N.M., Hoang, T.T., 2019. Superior activity of Cu-$NiWO_4$/g-C_3N_4 Z direct system for photocatalytic decomposition of VOCs in aerosol under visible light. Journal of Alloys and Compounds 798, 12−18.

Tomatis, M., Xu, H.-H., He, J., Zhang, X.-D., 2016. Recent development of catalysts for removal of volatile organic compounds in flue gas by combustion: a review. Journal of Chemistry 2016, 1−15.

Vatanpour, V., Karami, A., Sheydaei, M., 2017. Central composite design optimization of Rhodamine B degradation using TiO_2 nanoparticles/UV/PVDF process in continuous submerged membrane photoreactor. Chemical Engineering and Processing: Process Intensification 116, 68−75.

Veerapandian, S.K.P., Ye, Z., Giraudon, J.M., De Geyter, N., Morent, R., Lamonier, J.F., 2019. Plasma assisted Cu-Mn mixed oxide catalysts for trichloroethylene abatement in moist air. Journal of Hazardous Materials 379, 120781.

Veerapandian, S.K.P., Leys, C., De Geyter, N., Morent, R., 2017. Abatement of VOCs using packed bed non-thermal plasma reactors: a review. Catalysts 7 (4), 113.

Wang, B.W., Chi, C.M., Xu, M., Wang, C., Meng, D.J., 2017. Plasma-catalytic removal of toluene over CeO_2-MnO_x catalysts in an atmosphere dielectric barrier discharge. Chemical Engineering Journal 322, 679−692.

Wang, C., Rao, Z., Mahmood, A., Wang, X., Wang, Y., Xie, X., Sun, J., 2021. Improved photocatalytic oxidation performance of gaseous acetaldehyde by ternary g-C_3N_4/Ag-TiO_2 composites under visible light. Journal of Colloid and Interface Science 602, 699−711.

Wang, H., Ji, Y., Chen, J., Li, G., An, T., 2015. Theoretical investigation on the adsorption configuration and (*)OH-initiated photocatalytic degradation mechanism of typical atmospheric VOCs styrene onto $(TiO_2)n$ clusters. Scientific Reports 5, 15059.

Wang, J., Cao, S., Sun, Y., Meng, X., Wei, J., Ge, Y., Liu, B., Gong, Y., Li, Z., Mo, G., 2021. β zeolite nanostructures with a high SiO_2/Al_2O_3 ratio for the adsorption of volatile organic compounds. ACS Applied Nano Materials 4, 13257–13266.

Wang, X.Q., Wu, J.L., Wang, J.L., Xiao, H.L., Chen, B.X., Peng, R.S., Fu, M.L., Chen, L.M., Ye, D.Q., Wen, W., 2019. Methanol plasma-catalytic oxidation over CeO_2 catalysts: effect of ceria morphology and reaction mechanism. Chemical Engineering Journal 369, 233–244.

Wu, Q., Ye, J., Qiao, W., Li, Y., Niemantsverdriet, J.W., Richards, E., Pan, F., Su, R., 2021. Inhibit the formation of toxic methylphenolic by-products in photo-decomposition of formaldehyde–toluene/xylene mixtures by Pd cocatalyst on TiO_2. Applied Catalysis B: Environmental 291, 120118.

Xiao, C., Zhao, Z., Ying, Z., Zhu, Q., Pan, Z., Lu, H., 2018. A facile route for spraying preparation of Pt/TiO_2 monolithic catalysts toward VOCs combustion. Applied Catalysis A: General 566.

Xie, W., Li, Y., Shi, W., Zhao, L., Zhao, X., Fang, P., Zheng, F., Wang, S., 2012. Novel effect of significant enhancement of gas-phase photocatalytic efficiency for nano ZnO. Chemical Engineering Journal 213, 218–224.

Xiong, H., Zhu, X., Lu, S., Zhou, C., Xu, W., Zhou, Z., 2021. Enhancement of plasma-catalytic oxidation of ethylene oxide (EO) over FeMn catalysts in a dielectric barrier discharge reactor. Science of The Total Environment 788, 147675.

Xu, N., Fu, W., He, C., Cao, L., Liu, X., Zhao, J., Pan, H., 2014. Benzene removal using non-thermal plasma with CuO/AC catalyst: reaction condition optimization and decomposition mechanism. Plasma Chemistry and Plasma Processing 34, 1387–1402.

Xu, W., Chen, B., Jiang, X., Xu, F., Chen, X., Chen, L., Wu, J., Fu, M., Ye, D., 2020. Effect of calcium addition in plasma catalysis for toluene removal by Ni/ZSM-5 : acidity/basicity, catalytic activity and reaction mechanism. Journal of Hazardous Materials 387, 122004.

Xu, W., Jiang, X., Chen, H., Chen, X., Chen, L., Wu, J., Fu, M., Ye, D., 2020. Adsorption-discharge plasma system for toluene decomposition over Ni-SBA catalyst: in situ observation and humidity influence study. Chemical Engineering Journal 382, 122950.

Xue, X.L., Chen, X.Y., Gong, X.W., 2021. Fast electron transfer and enhanced visible light photocatalytic activity of silver and Ag_2O co-doped titanium dioxide with the doping of electron mediator for removing gaseous toluene. Materials Science in Semiconductor Processing 132, 105901.

Yang, C.T., Miao, G., Pi, Y.H., Xia, Q.B., Wu, J.L., Li, Z., Xiao, J., 2019. Abatement of various types of VOCs by adsorption/catalytic oxidation: a review. Chemical Engineering Journal 370, 1128–1153.

Yang, H., Li, W., Liu, J., Sun, Y., Liu, W., 2019. Polyethylenimine-impregnated resins: the effect of support structures on selective adsorption for CO_2 from simulated biogas. Chemical Engineering Journal 355, 822–829.

Yang, L.Z., Liu, Q.L., Han, R., Fu, K.X., Su, Y., Zheng, Y.F., Wu, X.Q., Song, C.F., Ji, N., Lu, X.B., Ma, D.G., 2022. Confinement and synergy effect of bimetallic Pt-Mn nanoparticles encapsulated in ZSM-5 zeolite with superior performance for acetone catalytic oxidation. Applied Catalysis B: Environmental 309, 121224.

Yang, S., Yang, H., Yang, J., Qi, H., Kong, J., Bo, Z., Li, X., Yan, J., Cen, K., Tu, X., 2020. Three-dimensional hollow urchin α-MnO_2 for enhanced catalytic activity towards toluene decomposition in post-plasma catalysis. Chemical Engineering Journal 402, 126154.

Yang, Y., Wang, G., Fang, D., Han, J., Yang, M., 2020. Study of the use of a Pd–Pt-based catalyst for the catalytic combustion of storage tank VOCs. International Journal of Hydrogen Energy 45.

Yao, J., Zhang, Y., Wang, Y., Chen, M., Huang, Y., Cao, J., Ho, W., Lee, S.C., 2017. Enhanced photocatalytic removal of NO over titania/hydroxyapatite (TiO_2/HAp) composites with improved adsorption and charge mobility ability. RSC Advances 7, 24683–24689.

Yao, X., Gao, M., Wei, Z., Chen, M., Shangguan, W., 2019. Removal of hexanal in cooking fume by combination of storage and plasma-catalytic oxidation on alkali-modified Co-Mn solid solution. Chemosphere 220, 738–747.

Yasui, F., Sekiguchi, K., Yoshida, T., Kim, K.H., Tamura, H., 2021. Complete decomposition of 2-propanol using TiO_2 immobilized on a nonwoven fabric under UV light irradiation by adding H_2O_2 and O_3 microbubbles. Water, Air, & Soil Pollution 232, 175.

Yen, L.T.H., Thuan, D.V., Hanh, N.T., Vy, N.H.T., Hang, T.T.M., Ha, H.V., Pham, T.D., Sharma, A.K., Nguyen, M.V., Dang, N.M., Truc, N.T.T., 2020. Synthesis of N and S Co-doped $TiO_{(2)}$Nanotubes for advanced photocatalytic degradation of volatile organic compounds (VOCs) in gas phase. Topics in Catalysis 63, 1077–1085.

Yu, L., Wang, L., Sun, X., Ye, D., 2018. Enhanced photocatalytic activity of rGO/$TiO_{(2)}$ for the decomposition of formaldehyde under visible light irradiation. Journal of Environmental Sciences 73, 138–146.

Yuan, M.H., Chang, C.C., Chang, C.Y., Lin, Y.Y., Shie, J.L., Wu, C.H., Tseng, J.Y., Ji, D.R., 2014. Radio-frequency-powered atmospheric pressure plasma jet for the destruction of binary mixture of naphthalene and n-butanol with Pt/Al_2O_3 catalyst. Journal of the Taiwan Institute of Chemical Engineers 45, 468–474.

Zang, M., Zhao, C.C., Wang, Y.Q., Chen, S.Q., 2019. A review of recent advances in catalytic combustion of VOCs on perovskite-type catalysts. Journal of Saudi Chemical Society 23, 645–654.

Zeng, X., Li, B., Liu, R., Li, X., Zhu, T., 2020. Investigation of promotion effect of Cu doped MnO_2 catalysts on ketone-type VOCs degradation in a one-stage plasma-catalysis system. Chemical Engineering Journal 384, 123362.

Zhang, Y., Wu, M., Wang, Y., Kwok, Y.H., Pan, W., Szeto, W., Huang, H., Leung, D.Y.C., 2021. Fluorinated TiO_2 coupling with α-MnO_2 nanowires supported on different substrates for photocatalytic VOCs abatement under vacuum ultraviolet irradiation. Applied Catalysis B: Environmental 280, 119388.

Zhao, T., Duan, F., Jiang, X., Xu, H., Gao, X., Wang, J., 2016. Hydrothermal synthesis of three structured MnO_2 nanomaterials and their performance for catalytic oxidation of resorcinol and formaldehyde. Chemistry Letters 45, 457–459.

Zhao, X.L., Wu, M.Y., Zhang, Y.G., Szeto, W., Wang, Y.F., Pan, W.D., Li, J.T., Leung, D.Y.C., 2022. Bifunctional Mn^{2+} grafted Ultra-small TiO_2 nanoparticles on carbon cloth with efficient toluene degradation in a continuous flow reactor. Chemical Engineering Science 250, 117389.

Zheng, M.F., Yu, D.Q., Duan, L.J., Yu, W.H., Huang, L., 2017. In-situ fabricated CuO nanowires/Cu foam as a monolithic catalyst for plasma-catalytic oxidation of toluene. Catalysis Communications 100, 187–190.

Zhu, B., Zhang, L.-Y., Li, M., Yan, Y., Zhang, X.-M., Zhu, Y.-M., 2020. High-performance of plasma-catalysis hybrid system for toluene removal in air using supported Au nanocatalysts. Chemical Engineering Journal 381, 122599.

Zhu, D., Chen, Z., Li, J., Wu, Z., Gao, E., Wang, W., Yao, S., 2021. Evaluation of Au/gamma-Al(2)O(3) nanocatalyst for plasma-catalytic decomposition of toluene. Chemosphere 285, 131474.

Zhu, L., Jing, C., Yang, P., Cheng, Z., Li, J., Zuo, S., 2019. Ce-modified mesoporous γ-Al_2O_3 supported Pd-Pt nanoparticle catalysts and their structure-function relationship in complete benzene oxidation. Chemical Engineering Journal 356, 255–261.

Zhu, X.B., Gao, X., Yu, X.N., Zheng, C.H., Tu, X., 2015. Catalyst screening for acetone removal in a single-stage plasma-catalysis system. Catalysis Today 256, 108–114.

Zhu, X., Gao, X., Qin, R., Zeng, Y., Qu, R., Zheng, C., Tu, X., 2015. Plasma-catalytic removal of formaldehyde over Cu–Ce catalysts in a dielectric barrier discharge reactor. Applied Catalysis B: Environmental 170–171, 293–300.

Zhu, X., Zhang, S., Yang, Y., Zheng, C., Zhou, J., Gao, X., Tu, X., 2017. Enhanced performance for plasma-catalytic oxidation of ethyl acetate over $La_{1-x}Ce_xCoO_{3+\delta}$ catalysts. Applied Catalysis B: Environmental 213, 97–105.

Zou, W., Gao, B., Ok, Y.S., Dong, L., 2019. Integrated adsorption and photocatalytic degradation of volatile organic compounds (VOCs) using carbon-based nanocomposites: a critical review. Chemosphere 218, 845–859.

Index

Note: "Page numbers followed by *f* indicate figures and *t* indicate tables."

D

E

F

9780443156748